Applied Statistics Using SPSS, STATISTICA, MATLAB and R

T0181167

Joaquim P. Marques de Sá

Applied Statistics
Using SPSS, STATISTICA, MATLAB and R

With 195 Figures and a CD

 Springer

Editors

Prof. Dr. Joaquim P. Marques de Sá
Universidade do Porto
Fac. Engenharia
Rua Dr. Roberto Frias s/n
4200-465 Porto
Portugal
e-mail: jmsa@fe.up.pt

ISBN 978-3-642-43744-1   Springer Berlin Heidelberg New York

Additional material to this book can be downloaded from http://extras.springger.com.

Springer is a part of Springer Science+Business Media
springer.com
© Springer-Verlag Berlin Heidelberg 2007
Softcover re-print of the Hardcover 2nd edition 2007

Production: Integra Software Services Pvt. Ltd., India
Cover design: WMX design, Heidelberg

Printed on acid-free paper   SPIN: 11908944   42/3100/Integra   5 4 3 2 1 0

To
Wiesje
and Carlos.

Contents

Preface to the Second Edition   xv

Preface to the First Edition   xvii

Symbols and Abbreviations   xix

1 Introduction   1

1.1   Deterministic Data and Random Data 1
1.2   Population, Sample and Statistics 5
1.3   Random Variables 8
1.4   Probabilities and Distributions 10
     1.4.1   Discrete Variables 10
     1.4.2   Continuous Variables 12
1.5   Beyond a Reasonable Doubt 13
1.6   Statistical Significance and Other Significances 17
1.7   Datasets 19
1.8   Software Tools 19
     1.8.1   SPSS and STATISTICA 20
     1.8.2   MATLAB and R 22

2 Presenting and Summarising the Data   29

2.1   Preliminaries 29
     2.1.1   Reading in the Data 29
     2.1.2   Operating with the Data 34
2.2   Presenting the Data 39
     2.2.1   Counts and Bar Graphs 40
     2.2.2   Frequencies and Histograms 47
     2.2.3   Multivariate Tables, Scatter Plots and 3D Plots 52
     2.2.4   Categorised Plots 56
2.3   Summarising the Data 58
     2.3.1   Measures of Location 58
     2.3.2   Measures of Spread 62
     2.3.3   Measures of Shape 64

2.3.4   Measures of Association for Continuous Variables....................66
2.3.5   Measures of Association for Ordinal Variables........................69
2.3.6   Measures of Association for Nominal Variables.......................73
Exercises...77

3   Estimating Data Parameters   81

3.1   Point Estimation and Interval Estimation.................................81
3.2   Estimating a Mean 85
3.3   Estimating a Proportion 92
3.4   Estimating a Variance...95
3.5   Estimating a Variance Ratio..97
3.6   Bootstrap Estimation..99
Exercises...107

4   Parametric Tests of Hypotheses   111

4.1   Hypothesis Test Procedure...111
4.2   Test Errors and Test Power 115
4.3   Inference on One Population...121
         4.3.1   Testing a Mean 121
         4.3.2   Testing a Variance 125
4.4   Inference on Two Populations 126
         4.4.1   Testing a Correlation 126
         4.4.2   Comparing Two Variances...129
         4.4.3   Comparing Two Means 132
4.5   Inference on More than Two Populations....................................141
         4.5.1   Introduction to the Analysis of Variance............................141
         4.5.2   One-Way ANOVA 143
         4.5.3   Two-Way ANOVA 156
Exercises...166

5   Non-Parametric Tests of Hypotheses   171

5.1   Inference on One Population...172
         5.1.1   The Runs Test..172
         5.1.2   The Binomial Test 174
         5.1.3   The Chi-Square Goodness of Fit Test 179
         5.1.4   The Kolmogorov-Smirnov Goodness of Fit Test 183
         5.1.5   The Lilliefors Test for Normality 187
         5.1.6   The Shapiro-Wilk Test for Normality 187
5.2   Contingency Tables..189
         5.2.1   The 2×2 Contingency Table 189
         5.2.2   The rxc Contingency Table 193

	5.2.3	The Chi-Square Test of Independence	195
	5.2.4	Measures of Association Revisited	197
5.3	Inference on Two Populations		200
	5.3.1	Tests for Two Independent Samples	201
	5.3.2	Tests for Two Paired Samples	205
5.4	Inference on More Than Two Populations		212
	5.4.1	The Kruskal-Wallis Test for Independent Samples	212
	5.4.2	The Friedmann Test for Paired Samples	215
	5.4.3	The Cochran Q test	217
Exercises			218

6 Statistical Classification   **223**

6.1	Decision Regions and Functions		223
6.2	Linear Discriminants		225
	6.2.1	Minimum Euclidian Distance Discriminant	225
	6.2.2	Minimum Mahalanobis Distance Discriminant	228
6.3	Bayesian Classification		234
	6.3.1	Bayes Rule for Minimum Risk	234
	6.3.2	Normal Bayesian Classification	240
	6.3.3	Dimensionality Ratio and Error Estimation	243
6.4	The ROC Curve		246
6.5	Feature Selection		253
6.6	Classifier Evaluation		256
6.7	Tree Classifiers		259
Exercises			268

7 Data Regression   **271**

7.1	Simple Linear Regression		272
	7.1.1	Simple Linear Regression Model	272
	7.1.2	Estimating the Regression Function	273
	7.1.3	Inferences in Regression Analysis	279
	7.1.4	ANOVA Tests	285
7.2	Multiple Regression		289
	7.2.1	General Linear Regression Model	289
	7.2.2	General Linear Regression in Matrix Terms	289
	7.2.3	Multiple Correlation	292
	7.2.4	Inferences on Regression Parameters	294
	7.2.5	ANOVA and Extra Sums of Squares	296
	7.2.6	Polynomial Regression and Other Models	300
7.3	Building and Evaluating the Regression Model		303
	7.3.1	Building the Model	303
	7.3.2	Evaluating the Model	306
	7.3.3	Case Study	308
7.4	Regression Through the Origin		314

7.5   Ridge Regression 316
7.6   Logit and Probit Models 322
Exercises...327

8 Data Structure Analysis   **329**

8.1   Principal Components...329
8.2   Dimensional Reduction...337
8.3   Principal Components of Correlation Matrices...................................339
8.4   Factor Analysis 347
Exercises...350

9 Survival Analysis   **353**

9.1   Survivor Function and Hazard Function 353
9.2   Non-Parametric Analysis of Survival Data..354
         9.2.1   The Life Table Analysis 354
         9.2.2   The Kaplan-Meier Analysis..359
         9.2.3   Statistics for Non-Parametric Analysis................................362
9.3   Comparing Two Groups of Survival Data 364
9.4   Models for Survival Data...367
         9.4.1   The Exponential Model 367
         9.4.2   The Weibull Model..369
         9.4.3   The Cox Regression Model 371
Exercises...373

10 Directional Data   **375**

10.1   Representing Directional Data 375
10.2   Descriptive Statistics..380
10.3   The von Mises Distributions 383
10.4   Assessing the Distribution of Directional Data..................................387
         10.4.1   Graphical Assessment of Uniformity 387
         10.4.2   The Rayleigh Test of Uniformity 389
         10.4.3   The Watson Goodness of Fit Test 392
         10.4.4   Assessing the von Misesness of Spherical Distributions...........393
10.5   Tests on von Mises Distributions...395
         10.5.1   One-Sample Mean Test 395
         10.5.2   Mean Test for Two Independent Samples 396
10.6   Non-Parametric Tests..397
         10.6.1   The Uniform Scores Test for Circular Data...........................397
         10.6.2   The Watson Test for Spherical Data.....................................398
         10.6.3   Testing Two Paired Samples 399
Exercises...400

Appendix A - Short Survey on Probability Theory   **403**

A.1   Basic Notions 403
        A.1.1   Events and Frequencies 403
        A.1.2   Probability Axioms..404
A.2   Conditional Probability and Independence 406
        A.2.1   Conditional Probability and Intersection Rule..........................406
        A.2.2   Independent Events 406
A.3   Compound Experiments..408
A.4   Bayes' Theorem 409
A.5   Random Variables and Distributions 410
        A.5.1   Definition of Random Variable 410
        A.5.2   Distribution and Density Functions..411
        A.5.3   Transformation of a Random Variable 413
A.6   Expectation, Variance and Moments 414
        A.6.1   Definitions and Properties 414
        A.6.2   Moment-Generating Function 417
        A.6.3   Chebyshev Theorem 418
A.7   The Binomial and Normal Distributions...418
        A.7.1   The Binomial Distribution..418
        A.7.2   The Laws of Large Numbers 419
        A.7.3   The Normal Distribution 420
A.8   Multivariate Distributions 422
        A.8.1   Definitions...422
        A.8.2   Moments..425
        A.8.3   Conditional Densities and Independence..................................425
        A.8.4   Sums of Random Variables 427
        A.8.5   Central Limit Theorem 428

Appendix B - Distributions   **431**

B.1   Discrete Distributions 431
        B.1.1   Bernoulli Distribution..431
        B.1.2   Uniform Distribution 432
        B.1.3   Geometric Distribution 433
        B.1.4   Hypergeometric Distribution...434
        B.1.5   Binomial Distribution..435
        B.1.6   Multinomial Distribution...436
        B.1.7   Poisson Distribution 438
B.2   Continuous Distributions 439
        B.2.1   Uniform Distribution 439
        B.2.2   Normal Distribution..441
        B.2.3   Exponential Distribution..442
        B.2.4   Weibull Distribution 444
        B.2.5   Gamma Distribution 445
        B.2.6   Beta Distribution 446
        B.2.7   Chi-Square Distribution...448

B.2.8   Student's *t* Distribution...449
B.2.9   F Distribution 451
B.2.10 Von Mises Distributions...452

Appendix C - Point Estimation   **455**

C.1   Definitions...455
C.2   Estimation of Mean and Variance.......................................457

Appendix D - Tables   **459**

D.1   Binomial Distribution 459
D.2   Normal Distribution 465
D.3   Student's *t* Distribution 466
D.4   Chi-Square Distribution 467
D.5   Critical Values for the *F* Distribution 468

Appendix E - Datasets   **469**

E.1   Breast Tissue..469
E.2   Car Sale..469
E.3   Cells 470
E.4   Clays 470
E.5   Cork Stoppers...471
E.6   CTG 472
E.7   Culture 473
E.8   Fatigue 473
E.9   FHR..474
E.10   FHR-Apgar 474
E.11   Firms 475
E.12   Flow Rate...475
E.13   Foetal Weight...475
E.14   Forest Fires...476
E.15   Freshmen..476
E.16   Heart Valve...477
E.17   Infarct..478
E.18   Joints 478
E.19   Metal Firms...479
E.20   Meteo 479
E.21   Moulds 479
E.22   Neonatal 480
E.23   Programming..480
E.24   Rocks 481
E.25   Signal & Noise...481

E.26   Soil Pollution 482
E.27   Stars 482
E.28   Stock Exchange..483
E.29   VCG 484
E.30   Wave 484
E.31   Weather..484
E.32   Wines 485

Appendix F - Tools   487

F.1   MATLAB Functions..487
F.2   R Functions 488
F.3   Tools EXCEL File 489
F.4   SCSize Program 489

References   491

Index   499

Preface to the Second Edition

Four years have passed since the first edition of this book. During this time I have had the opportunity to apply it in classes obtaining feedback from students and inspiration for improvements. I have also benefited from many comments by users of the book. For the present second edition large parts of the book have undergone major revision, although the basic concept – concise but sufficiently rigorous mathematical treatment with emphasis on computer applications to real datasets –, has been retained.

The second edition improvements are as follows:

- Inclusion of R as an application tool. As a matter of fact, R is a free software product which has nowadays reached a high level of maturity and is being increasingly used by many people as a statistical analysis tool.

- Chapter 3 has an added section on bootstrap estimation methods, which have gained a large popularity in practical applications.

- A revised explanation and treatment of tree classifiers in Chapter 6 with the inclusion of the QUEST approach.

- Several improvements of Chapter 7 (regression), namely: details concerning the meaning and computation of multiple and partial correlation coefficients, with examples; a more thorough treatment and exemplification of the ridge regression topic; more attention dedicated to model evaluation.

- Inclusion in the book CD of additional MATLAB functions as well as a set of R functions.

- Extra examples and exercises have been added in several chapters.

- The bibliography has been revised and new references added.

I have also tried to improve the quality and clarity of the text as well as notation. Regarding notation I follow in this second edition the more widespread use of denoting random variables with italicised capital letters, instead of using small cursive font as in the first edition. Finally, I have also paid much attention to correcting errors, misprints and obscurities of the first edition.

J.P. Marques de Sá
Porto, 2007

Preface to the First Edition

This book is intended as a reference book for students, professionals and research workers who need to apply statistical analysis to a large variety of practical problems using STATISTICA, SPSS and MATLAB. The book chapters provide a comprehensive coverage of the main statistical analysis topics (data description, statistical inference, classification and regression, factor analysis, survival data, directional statistics) that one faces in practical problems, discussing their solutions with the mentioned software packages.

The only prerequisite to use the book is an undergraduate knowledge level of mathematics. While it is expected that most readers employing the book will have already some knowledge of elementary statistics, no previous course in probability or statistics is needed in order to study and use the book. The first two chapters introduce the basic needed notions on probability and statistics. In addition, the first two Appendices provide a short survey on Probability Theory and Distributions for the reader needing further clarification on the theoretical foundations of the statistical methods described.

The book is partly based on tutorial notes and materials used in data analysis disciplines taught at the Faculty of Engineering, Porto University. One of these disciplines is attended by students of a Master's Degree course on information management. The students in this course have a variety of educational backgrounds and professional interests, which generated and brought about datasets and analysis objectives which are quite challenging concerning the methods to be applied and the interpretation of the results. The datasets used in the book examples and exercises were collected from these courses as well as from research. They are included in the book CD and cover a broad spectrum of areas: engineering, medicine, biology, psychology, economy, geology, and astronomy.

Every chapter explains the relevant notions and methods concisely, and is illustrated with practical examples using real data, presented with the distinct intention of clarifying sensible practical issues. The solutions presented in the examples are obtained with one of the software packages STATISTICA, SPSS or MATLAB; therefore, the reader has the opportunity to closely follow what is being done. The book is not intended as a substitute for the STATISTICA, SPSS and MATLAB user manuals. It does, however, provide the necessary guidance for applying the methods taught without having to delve into the manuals. This includes, for each topic explained in the book, a clear indication of which STATISTICA, SPSS or MATLAB tools to be applied. These indications appear in specific "Commands" frames together with a complementary description on how to use the tools, whenever necessary. In this way, a comparative perspective of the capabilities of those software packages is also provided, which can be quite useful for practical purposes.

STATISTICA, SPSS or MATLAB do not provide specific tools for some of the statistical topics described in the book. These range from such basic issues as the choice of the optimal number of histogram bins to more advanced topics such as directional statistics. The book CD provides these tools, including a set of MATLAB functions for directional statistics.

I am grateful to many people who helped me during the preparation of the book. Professor Luís Alexandre provided help in reviewing the book contents. Professor Willem van Meurs provided constructive comments on several topics. Professor Joaquim Góis contributed with many interesting discussions and suggestions, namely on the topic of data structure analysis. Dr. Carlos Felgueiras and Paulo Sousa gave valuable assistance in several software issues and in the development of some software tools included in the book CD. My gratitude also to Professor Pimenta Monteiro for his support in elucidating some software tricks during the preparation of the text files. A lot of people contributed with datasets. Their names are mentioned in Appendix E. I express my deepest thanks to all of them. Finally, I would also like to thank Alan Weed for his thorough revision of the texts and the clarification of many editing issues.

J.P. Marques de Sá
Porto, 2003

Symbols and Abbreviations

Sample Sets

A	event
$\mathcal{A}$	set (of events)
$\{A_1, A_2,...\}$	set constituted of events $A_1, A_2,...$
$\overline{A}$	complement of $\{A\}$
$A \cup B$	union of $\{A\}$ with $\{B\}$
$A \cap B$	intersection of $\{A\}$ with $\{B\}$
$\mathcal{E}$	set of all events (universe)
ϕ	empty set

Functional Analysis

$\exists$	there is
$\vee$	for every
$\in$	belongs to
$\notin$	doesn't belong to
$\equiv$	equivalent to
$\| \ \|$	Euclidian norm (vector length)
$\Rightarrow$	implies
$\rightarrow$	converges to
$\mathfrak{R}$	real number set
$\mathfrak{R}^+$	$[0, +\infty\ [$
$[a, b]$	closed interval between and including a and b
$]a, b]$	interval between a and b, excluding a
$[a, b[$	interval between a and b, excluding b

$]a, b[$	open interval between a and b (excluding a and b)
$\sum_{i=1}^{n}$	sum for index $i = 1,\dots, n$
$\prod_{i=1}^{n}$	product for index $i = 1,\dots, n$
$\int_{a}^{b}$	integral from a to b
$k!$	factorial of k, $k! = k(k{-}1)(k{-}2)\dots 2.1$
$\binom{n}{k}$	combinations of n elements taken k at a time
$\lvert x \rvert$	absolute value of x
$\lfloor x \rfloor$	largest integer smaller or equal to x
$g_X(a)$	function g of variable X evaluated at a
$\dfrac{dg}{dX}$	derivative of function g with respect to X
$\left.\dfrac{d^n g}{dX^n}\right\rvert_a$	derivative of order n of g evaluated at a
$\ln(x)$	natural logarithm of x
$\log(x)$	logarithm of x in base 10
$\mathrm{sgn}(x)$	sign of x
$\mathrm{mod}(x,y)$	remainder of the integer division of x by y

Vectors and Matrices

$\mathbf{x}$	vector (column vector), multidimensional random vector
$\mathbf{x}'$	transpose vector (row vector)
$[x_1\ x_2\dots x_n]$	row vector whose components are $x_1, x_2,\dots,x_n$
x_i	i-th component of vector $\mathbf{x}$
$x_{k,i}$	i-th component of vector $\mathbf{x}_k$
$\Delta\mathbf{x}$	vector $\mathbf{x}$ increment
$\mathbf{x}'\mathbf{y}$	inner (dot) product of $\mathbf{x}$ and $\mathbf{y}$
$\mathbf{A}$	matrix
a_{ij}	i-th row, j-th column element of matrix $\mathbf{A}$
$\mathbf{A}'$	transpose of matrix $\mathbf{A}$
$\mathbf{A}^{-1}$	inverse of matrix $\mathbf{A}$

$|\mathbf{A}|$	determinant of matrix $\mathbf{A}$
$\text{tr}(\mathbf{A})$	trace of $\mathbf{A}$ (sum of the diagonal elements)
$\mathbf{I}$	unit matrix
λ_i	eigenvalue i

Probabilities and Distributions

X	random variable (with value denoted by the same lower case letter, x)
$P(A)$	probability of event A
$P(A|B)$	probability of event A conditioned on B having occurred
$P(\mathbf{x})$	discrete probability of random vector $\mathbf{x}$
$P(\omega_i|\mathbf{x})$	discrete conditional probability of ω_i given $\mathbf{x}$
$f(\mathbf{x})$	probability density function f evaluated at $\mathbf{x}$
$f(\mathbf{x}\,|\omega_i)$	conditional probability density function f evaluated at $\mathbf{x}$ given ω_i
$X \sim f$	X has probability density function f
$X \sim F$	X has probability distribution function (is distributed as) F
Pe	probability of misclassification (error)
Pc	probability of correct classification
df	degrees of freedom
$x_{df,\alpha}$	α-percentile of X distributed with df degrees of freedom
$b_{n,p}$	binomial probability for n trials and probability p of success
$B_{n,p}$	binomial distribution for n trials and probability p of success
u	uniform probability or density function
U	uniform distribution
g_p	geometric probability (Bernoulli trial with probability p)
G_p	geometric distribution (Bernoulli trial with probability p)
$h_{N,D,n}$	hypergeometric probability (sample of n out of N with D items)
$H_{N,D,n}$	hypergeometric distribution (sample of n out of N with D items)
p_λ	Poisson probability with event rate λ
P_λ	Poisson distribution with event rate λ
$n_{\mu,\sigma}$	normal density with mean μ and standard deviation σ

$N_{\mu,\sigma}$	normal distribution with mean μ and standard deviation σ
ε_λ	exponential density with spread factor λ
E_λ	exponential distribution with spread factor λ
$w_{\alpha,\beta}$	Weibull density with parameters α, β
$W_{\alpha,\beta}$	Weibull distribution with parameters α, β
$\gamma_{a,p}$	Gamma density with parameters a, p
$\Gamma_{a,p}$	Gamma distribution with parameters a, p
$\beta_{p,q}$	Beta density with parameters p, q
$B_{p,q}$	Beta distribution with parameters p, q
χ^2_{df}	Chi-square density with df degrees of freedom
X^2_{df}	Chi-square distribution with df degrees of freedom
t_{df}	Student's t density with df degrees of freedom
T_{df}	Student's t distribution with df degrees of freedom
f_{df_1,df_2}	F density with df_1, df_2 degrees of freedom
F_{df_1,df_2}	F distribution with df_1, df_2 degrees of freedom

Statistics

$\hat{x}$	estimate of x
$E[X]$	expected value (average, mean) of X
$V[X]$	variance of X
$E[\mathbf{x} \mid \mathbf{y}]$	expected value of $\mathbf{x}$ given $\mathbf{y}$ (conditional expectation)
m_k	central moment of order k
μ	mean value
σ	standard deviation
σ_{XY}	covariance of X and Y
ρ	correlation coefficient
$\boldsymbol{\mu}$	mean vector

Σ	covariance matrix
$\bar{x}$	arithmetic mean
v	sample variance
s	sample standard deviation
x_α	α-quantile of X ($F_X(x_\alpha) = \alpha$)
med(X)	median of X (same as $x_{0.5}$)
S	sample covariance matrix
α	significance level ($1-\alpha$ is the confidence level)
x_α	α-percentile of X
ε	tolerance

Abbreviations

FNR	False Negative Ratio
FPR	False Positive Ratio
iff	if an only if
i.i.d.	independent and identically distributed
IRQ	inter-quartile range
pdf	probability density function
LSE	Least Square Error
ML	Maximum Likelihood
MSE	Mean Square Error
PDF	probability distribution function
RMS	Root Mean Square Error
r.v.	Random variable
ROC	Receiver Operating Characteristic
SSB	Between-group Sum of Squares
SSE	Error Sum of Squares
SSLF	Lack of Fit Sum of Squares
SSPE	Pure Error Sum of Squares
SSR	Regression Sum of Squares

SST   Total Sum of Squares

SSW   Within-group Sum of Squares

TNR   True Negative Ratio

TPR   True Positive Ratio

VIF   Variance Inflation Factor

Tradenames

EXCEL   Microsoft Corporation

MATLAB   The MathWorks, Inc.

SPSS   SPSS, Inc.

STATISTICA   Statsoft, Inc.

WINDOWS   Microsoft Corporation

1 Introduction

1.1 Deterministic Data and Random Data

Our daily experience teaches us that some data are generated in accordance to known and precise laws, while other data seem to occur in a purely haphazard way. Data generated in accordance to known and precise laws are called *deterministic data*. An example of such type of data is the fall of a body subject to the Earth's gravity. When the body is released at a height h, we can calculate precisely where the body stands at each time t. The physical law, assuming that the fall takes place in an empty space, is expressed as:

$$h = h_0 - \tfrac{1}{2}gt^2,$$

where h_0 is the initial height and g is the Earth's gravity acceleration at the point where the body falls.

Figure 1.1 shows the behaviour of h with t, assuming an initial height of 15 meters.

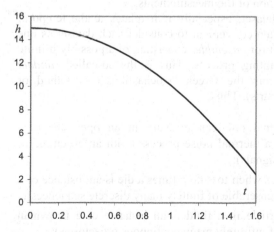

t	h
0.00	15.00
0.20	14.80
0.40	14.22
0.60	13.24
0.80	11.86
1.00	10.10
1.20	7.94
1.40	5.40
1.60	2.46

Figure 1.1. Body in free-fall, with height in meters and time in seconds, assuming $g = 9.8$ m/s^2. The h column is an example of deterministic data.

In the case of the body fall there is a law that allows the exact computation of one of the variables h or t (for given h_0 and g) as a function of the other one. Moreover, if we repeat the body-fall experiment under identical conditions, we consistently obtain the same results, within the precision of the measurements. These are the attributes of deterministic data: *the same data will be obtained, within the precision of the measurements, under repeated experiments in well-defined conditions.*

Imagine now that we were dealing with Stock Exchange data, such as, for instance, the daily share value throughout one year of a given company. For such data there is no known law to describe how the share value evolves along the year. Furthermore, the possibility of experiment repetition with identical results does not apply here. We are, thus, in presence of what is called *random data.*

Classical examples of random data are:

- Thermal noise generated in electrical resistances, antennae, etc.;
- Brownian motion of tiny particles in a fluid;
- Weather variables;
- Financial variables such as Stock Exchange share values;
- Gambling game outcomes (dice, cards, roulette, etc.);
- Conscript height at military inspection.

In none of these examples can a precise mathematical law describe the data. Also, there is no possibility of obtaining the same data in repeated experiments, performed under similar conditions. This is mainly due to the fact that several unforeseeable or immeasurable causes play a role in the generation of such data. For instance, in the case of the Brownian motion, we find that, after a certain time, the trajectories followed by several particles that have departed from exactly the same point, are completely different among them. Moreover it is found that such differences largely exceed the precision of the measurements.

When dealing with a random dataset, especially if it relates to the temporal evolution of some variable, it is often convenient to consider such dataset as one realization (or one *instance*) of a set (or *ensemble*) consisting of a possibly infinite number of realizations of a generating process. This is the so-called *random process* (or *stochastic process*, from the Greek "stochastikos" = method or phenomenon composed of random parts). Thus:

- The wandering voltage signal one can measure in an open electrical resistance is an instance of a thermal noise process (with an ensemble of infinitely many continuous signals);
- The succession of face values when tossing n times a die is an instance of a die tossing process (with an ensemble of finitely many discrete sequences).
- The trajectory of a tiny particle in a fluid is an instance of a Brownian process (with an ensemble of infinitely many continuous trajectories);

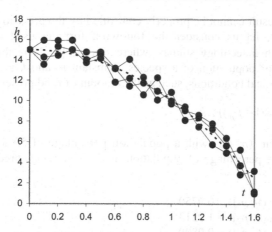

Figure 1.2. Three "body fall" experiments, under identical conditions as in Figure 1.1, with measurement errors (*random data* components). The dotted line represents the theoretical curve (*deterministic data* component). The solid circles correspond to the measurements made.

We might argue that if we knew all the causal variables of the "random data" we could probably find a deterministic description of the data. Furthermore, if we didn't know the mathematical law underlying a deterministic experiment, we might conclude that a random dataset were present. For example, imagine that we did not know the "body fall" law and attempted to describe it by running several experiments in the same conditions as before, performing the respective measurement of the height h for several values of the time t, obtaining the results shown in Figure 1.2. The measurements of each single experiment display a random variability due to measurement errors. These are always present in any dataset that we collect, and we can only hope that by averaging out such errors we get the "underlying law" of the data. This is a central idea in *statistics*: that certain quantities give the "big picture" of the data, averaging out random errors. As a matter of fact, statistics were first used as a means of summarising data, namely social and state data (the word "statistics" coming from the "science of state").

Scientists' attitude towards the "deterministic vs. random" dichotomy has undergone drastic historical changes, triggered by major scientific discoveries. Paramount of these changes in recent years has been the development of the quantum description of physical phenomena, which yields a granular-all-connectedness picture of the universe. The well-known "uncertainty principle" of Heisenberg, which states a limit to our capability of ever decreasing the measurement errors of experiment related variables (e.g. position and velocity), also supports a critical attitude towards determinism.

Even now the "deterministic vs. random" phenomenal characterization is subject to controversies and often statistical methods are applied to deterministic data. A good example of this is the so-called *chaotic phenomena*, which are described by a precise mathematical law, i.e., such phenomena are deterministic. However, the sensitivity of these phenomena on changes of causal variables is so large that the precision of the result cannot be properly controlled by the precision of the causes. To illustrate this, let us consider the following formula used as a model of population growth in ecology studies, where $p(n) \in [0, 1]$ is the fraction of a limiting number of population of a species at instant n, and k is a constant that depends on ecological conditions, such as the amount of food present:

$$p_{n+1} = p_n(1 + k(1 - p_n)), \quad k > 0.$$

Imagine we start ($n = 1$) with a population percentage of 50% ($p_1 = 0.5$) and wish to know the percentage of population at the following three time instants, with $k = 1.9$:

$$p_2 = p_1(1 + 1.9 \times (1 - p_1)) = 0.9750$$
$$p_3 = p_2(1 + 1.9 \times (1 - p_2)) = 1.0213$$
$$p_4 = p_3(1 + 1.9 \times (1 - p_3)) = 0.9800$$

It seems that after an initial growth the population dwindles back. As a matter of fact, the evolution of p_n shows some oscillation until stabilising at the value 1, the limiting number of population. However, things get drastically more complicated when $k = 3$, as shown in Figure 1.3. A mere deviation in the value of p_1 of only 10^{-6} has a drastic influence on p_n. For practical purposes, for k around 3 we are unable to predict the value of the p_n after some time, since it is so sensitive to very small changes of the initial condition p_1. In other words, the deterministic p_n process can be dealt with as a random process for some values of k.

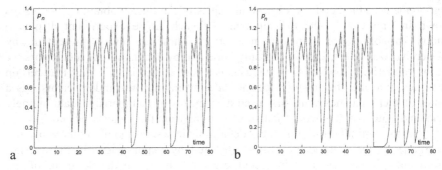

Figure 1.3. Two instances of the population growth process for $k = 3$: a) $p_1 = 0.1$; b) $p_1 = 0.100001$.

The random-like behaviour exhibited by some iterative series is also present in the so-called "random number generator routine" used in many computer programs. One such routine iteratively generates x_n as follows:

$$x_{n+1} = \alpha x_n \bmod m.$$

Therefore, the next number in the "random number" sequence is obtained by computing the remainder of the integer division of α times the previous number by a suitable constant, m. In order to obtain a convenient "random-like" behaviour of this purely deterministic sequence, when using numbers represented with p binary digits, one must use $m = 2^p$ and $\alpha = 2^{\lfloor p/2 \rfloor} + 3$, where $\lfloor p/2 \rfloor$ is the nearest integer smaller than $p/2$. The periodicity of the sequence is then 2^{p-2}. Figure 1.4 illustrates one such sequence.

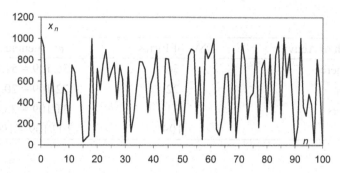

Figure 1.4. "Random number" sequence using $p = 10$ binary digits with $m = 2^p = 1024$, $\alpha = 35$ and initial value $x(0) = 2^p - 3 = 1021$.

1.2 Population, Sample and Statistics

When studying a collection of data as a random dataset, the basic assumption being that no law explains any *individual* value of the dataset, we attempt to study the data by means of some *global* measures, known as *statistics*, such as frequencies (of data occurrence in specified intervals), means, standard deviations, etc.

Clearly, these same measures can be applied to a deterministic dataset, but, after all, the mean height value in a set of height measurements of a falling body, among other things, is irrelevant.

Statistics had its beginnings and key developments during the last century, especially the last seventy years. The need to compare datasets and to infer from a dataset the process that generated it, were and still are important issues addressed by statisticians, who have made a definite contribution to forwarding scientific knowledge in many disciplines (see e.g. Salsburg D, 2001). In an inferential study, from a dataset to the process that generated it, the statistician considers the dataset as a *sample* from a vast, possibly infinite, collection of data called *population*. Each individual item of a sample is a *case* (or *object*). The sample itself is a list of values of one or more *random variables*.

The population data is usually not available for study, since most often it is either infinite or finite but very costly to collect. The data sample, obtained from the population, should be *randomly drawn*, i.e., any individual in the population is supposed to have an equal chance of being part of the sample. Only by studying randomly drawn samples can one expect to arrive at legitimate conclusions, about the whole population, from the data analyses.

Let us now consider the following three examples of datasets:

Example 1.1

The following Table 1.1 lists the number of firms that were established in town X during the year 2000, in each of three branches of activity.

☐

Table 1.1

Branch of Activity	No. of Firms	Frequencies
Commerce	56	56/109 = 51.4 %
Industry	22	22/109 = 20.2 %
Services	31	31/109 = 28.4 %
Total	109	109/109 = 100 %

Example 1.2

The following Table 1.2 lists the classifications of a random sample of 50 students in the examination of a certain course, evaluated on a scale of 1 to 5.

☐

Table 1.2

Classification	No. of Occurrences	Accumulated Frequencies
1	3	3/50 = 6.0%
2	10	13/50 = 26.0%
3	12	25/50 = 50.0%
4	15	40/50 = 80.0%
5	10	50/50 = 100.0%
Total	50	100.0%
Median[a] = 3		

[a] Value below which 50% of the cases are included.

Example 1.3

The following Table 1.3 lists the measurements performed in a random sample of 10 electrical resistances, of nominal value 100 Ω (ohm), produced by a machine.

☐

Table 1.3

Case #	Value (in Ω)
1	101.2
2	100.3
3	99.8
4	99.8
5	99.9
6	100.1
7	99.9
8	100.3
9	99.9
10	100.1
Mean	(101.2+100.3+99.8+...)/10 = 100.13

In Example 1.1 the random variable is the "number of firms that were established in town X during the year 2000, in each of three branches of activity". Population and sample are the same. In such a case, besides the summarization of the data by means of the frequencies of occurrence, not much more can be done. It is clearly a situation of limited interest. In the other two examples, on the other hand, we are dealing with samples of a larger population (potentially infinite in the case of Example 1.3). It's these kinds of situations that really interest the statistician – those in which the whole population is characterised based on statistical values computed from samples, the so-called *sample statistics*, or just *statistics* for short. For instance, how much information is obtainable about the *population mean* in Example 1.3, knowing that the *sample mean* is 100.13 Ω?

A statistic is a function, t_n, of the n sample values, x_i:

$$t_n(x_1, x_2, \ldots, x_n).$$

The sample mean computed in Table 1.3 is precisely one such function, expressed as:

$$\bar{x} \equiv m_n(x_1, x_2, \ldots, x_n) = \sum_{i=1}^{n} x_i / n.$$

We usually intend to draw some conclusion about the population based on the statistics computed in the sample. For instance, we may want to infer about the population mean based on the sample mean. In order to achieve this goal the x_i must be considered values of *independent random variables having the same probabilistic distribution as the population*, i.e., they constitute what is called a *random sample*. We sometimes encounter in the literature the expression "representative sample of the population". This is an incorrect term, since it conveys the idea that the composition of the sample must somehow mimic the composition of the population. This is not true. What must be achieved, in order to obtain a random sample, is to simply select elements of the population at random.

This can be done, for instance, with the help of a random number generator. In practice this "simple" task might not be so simple after all (as when we conduct statistical studies in a human population). The sampling topic is discussed in several books, e.g. (Blom G, 1989) and (Anderson TW, Finn JD, 1996). Examples of statistical malpractice, namely by poor sampling, can be found in (Jaffe AJ, Spirer HF, 1987). The sampling issue is part of the planning phase of the statistical investigation. The reader can find a good explanation of this topic in (Montgomery DC, 1984) and (Blom G, 1989).

In the case of temporal data a subtler point has to be addressed. Imagine that we are presented with a list (sequence) of voltage values originated by thermal noise in an electrical resistance. This sequence should be considered as an instance of a random process capable of producing an infinite number of such sequences. Statistics can then be computed either for the ensemble of instances or for the time sequence of the voltage values. For instance, one could compute a mean voltage value in two different ways: first, assuming one has available a sample of voltage sequences randomly drawn from the ensemble, one could compute the mean voltage value at, say, $t = 3$ seconds, for all sequences; and, secondly, assuming one such sequence lasting 10 seconds is available, one could compute the mean voltage value for the duration of the sequence. In the first case, the sample mean is an estimate of an *ensemble mean* (at $t = 3$ s); in the second case, the sample mean is an estimate of a *temporal mean*. Fortunately, in a vast number of situations, corresponding to what are called *ergodic* random processes, one can derive ensemble statistics from temporal statistics, i.e., one can limit the statistical study to the study of only one time sequence. This applies to the first two examples of random processes previously mentioned (as a matter of fact, thermal noise and dice tossing are ergodic processes; Brownian motion is not).

1.3 Random Variables

A random dataset presents the values of *random variables*. These establish a mapping between an event domain and some conveniently chosen value domain (often a subset of $\mathfrak{R}$). A good understanding of what the random variables are and which mappings they represent is a preliminary essential condition in any statistical analysis. A rigorous definition of a random variable (sometimes abbreviated to r.v.) can be found in Appendix A.

Usually the value domain of a random variable has a direct correspondence to the outcomes of a random experiment, but this is not compulsory. Table 1.4 lists random variables corresponding to the examples of the previous section. Italicised capital letters are used to represent random variables, sometimes with an identifying subscript. The Table 1.4 mappings between the event and the value domain are:

X_F: {commerce, industry, services} $\rightarrow$ {1, 2, 3}.
X_E: {bad, mediocre, fair, good, excellent} $\rightarrow$ {1, 2, 3, 4, 5}.
X_R: [90 Ω, 110 Ω] $\rightarrow$ [90, 110].

Table 1.4

Dataset	Variable	Value Domain	Type
Firms in town X, year 2000	X_F	$\{1, 2, 3\}$[a]	Discrete, Nominal
Classification of exams	X_E	$\{1, 2, 3, 4, 5\}$	Discrete, Ordinal
Electrical resistances (100 Ω)	X_R	$[90, 110]$	Continuous

[a] $1 \equiv$ Commerce, $2 \equiv$ Industry, $3 \equiv$ Services.

One could also have, for instance:

X_F: {commerce, industry, services} $\rightarrow$ $\{-1, 0, 1\}$.
X_E: {bad, mediocre, fair, good, excellent} $\rightarrow$ $\{0, 1, 2, 3, 4\}$.
X_R: $[90\,\Omega, 110\,\Omega] \rightarrow [-10, 10]$.

The value domains (or domains for short) of the variables X_F and X_E are discrete. These variables are *discrete random variables*. On the other hand, variable X_R is a *continuous random variable*.

The values of a *nominal* (or categorial) discrete variable are mere symbols (even if we use numbers) whose only purpose is to distinguish different categories (or classes). Their value domain is unique up to a biunivocal (one-to-one) transformation. For instance, the domain of X_F could also be codified as {A, B, C} or {I, II, III}.

Examples of nominal data are:

- Class of animal: bird, mammal, reptile, etc.;
- Automobile registration plates;
- Taxpayer registration numbers.

The only statistics that make sense to compute for nominal data are the ones that are invariable under a biunivocal transformation, namely: category counts; frequencies (of occurrence); mode (of the frequencies).

The domain of *ordinal* discrete variables, as suggested by the name, supports a total order relation ("larger than" or "smaller than"). It is unique up to a strict monotonic transformation (i.e., preserving the total order relation). That is why the domain of X_E could be {0, 1, 2, 3, 4} or {0, 25, 50, 75, 100} as well.

Examples of ordinal data are abundant, since the assignment of ranking scores to items is such a widespread practice. A few examples are:

- Consumer preference ranks: "like", "accept", "dislike", "reject", etc.;
- Military ranks: private, corporal, sergeant, lieutenant, captain, etc.;
- Certainty degrees: "unsure", "possible", "probable", "sure", etc.

Several statistics, whose only assumption is the existence of a total order relation, can be applied to ordinal data. One such statistic is the median, as shown in Example 1.2.

Continuous variables have a real number interval (or a reunion of intervals) as domain, which is unique up to a linear transformation. One can further distinguish between *ratio* type variables, supporting linear transformations of the $y = ax$ type, and *interval* type variables supporting linear transformations of the $y = ax + b$ type. The domain of ratio type variables has a fixed zero. This is the most frequent type of continuous variables encountered, as in Example 1.3 (a zero ohm resistance is a zero resistance in whatever measurement scale we choose to elect). The whole panoply of statistics is supported by continuous ratio type variables. The less common interval type variables do not have a fixed zero. An example of interval type data is temperature data, which can either be measured in degrees Celsius (X_C) or in degrees Fahrenheit (X_F), satisfying the relation $X_F = 1.8X_C + 32$. There are only a few, less frequent statistics, requiring a fixed zero, not supported by this type of variables.

Notice that, strictly speaking, there is no such thing as continuous data, since all data can only be measured with finite precision. If, for example, one is dealing with data representing people's height in meters, "real-flavour" numbers such as 1.82 m may be used. Of course, if the highest measurement precision is the millimetre, one is in fact dealing with integer numbers such as 182 mm, i.e., the height data is, in fact, ordinal data. In practice, however, one often assumes that there is a continuous domain underlying the ordinal data. For instance, one often assumes that the height data can be measured with arbitrarily high precision. Even for rank data such as the examination scores of Example 1.2, one often computes an average score, obtaining a value in the continuous interval [0, 5], i.e., one is implicitly assuming that the examination scores can be measured with a higher precision.

1.4 Probabilities and Distributions

The process of statistically analysing a dataset involves operating with an appropriate measure expressing the randomness exhibited by the dataset. This measure is the *probability measure*. In this section, we will introduce a few topics of Probability Theory that are needed for the understanding of the following material. The reader familiar with Probability Theory can skip this section. A more detailed survey (but still a brief one) on Probability Theory can be found in Appendix A.

1.4.1 Discrete Variables

The beginnings of Probability Theory can be traced far back in time to studies on chance games. The work of the Swiss mathematician Jacob Bernoulli (1654-1705), *Ars Conjectandi*, represented a keystone in the development of a Theory of

Probability, since for the first time, mathematical grounds were established and the application of probability to statistics was presented. The notion of probability is originally associated with the notion of frequency of occurrence of one out of k events in a sequence of trials, in which each of the events can occur by pure chance.

Let us assume a sample dataset, of size n, described by a discrete variable, X. Assume further that there are k distinct values x_i of X each one occurring n_i times. We define:

– *Absolute frequency* of x_i: n_i ;

– *Relative frequency* (or simply *frequency* of x_i): $f_i = \dfrac{n_i}{n}$ with $n = \sum_{i=1}^{k} n_i$.

In the classic frequency interpretation, *probability* is considered a limit, for large n, of the relative frequency of an event: $P_i \equiv P(X = x_i) = \lim_{n \to \infty} f_i \in [0, 1]$. In Appendix A, a more rigorous definition of probability is presented, as well as properties of the convergence of such a limit to the probability of the event (Law of Large Numbers), and the justification for computing $P(X = x_i)$ as the "ratio of the number of favourable events over the number of possible events" when the event composition of the random experiment is known beforehand. For instance, the probability of obtaining two heads when tossing two coins is ¼ since only one out of the four possible events (head-head, head-tail, tail-head, tail-tail) is favourable. As exemplified in Appendix A, one often computes probabilities of events in this way, using enumerative and combinatorial techniques.

The values of P_i constitute the *probability function* values of the random variable X, denoted $P(X)$. In the case the discrete random variable is an ordinal variable the accumulated sum of P_i is called the *distribution function*, denoted $F(X)$. Bar graphs are often used to display the values of probability and distribution functions of discrete variables.

Let us again consider the classification data of Example 1.2, and assume that the frequencies of the classifications are correct estimates of the respective probabilities. We will then have the probability and distribution functions represented in Table 1.5 and Figure 1.5. Note that the probabilities add up to 1 (total certainty) which is the largest value of the monotonic increasing function $F(X)$.

Table 1.5. Probability and distribution functions for Example 1.2, assuming that the frequencies are correct estimates of the probabilities.

x_i	Probability Function $P(X)$	Distribution Function $F(X)$
1	0.06	0.06
2	0.20	0.26
3	0.24	0.50
4	0.30	0.80
5	0.20	1.00

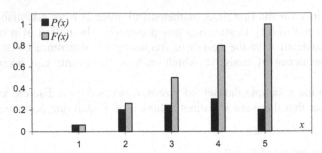

Figure 1.5. Probability and distribution functions for Example 1.2, assuming that the frequencies are correct estimates of the probabilities.

Several discrete distributions are described in Appendix B. An important one, since it occurs frequently in statistical studies, is the *binomial distribution*. It describes the probability of occurrence of a "success" event k times, in n independent trials, performed in the same conditions. The complementary "failure" event occurs, therefore, $n - k$ times. The probability of the "success" in a single trial is denoted p. The complementary probability of the failure is $1 - p$, also denoted q. Details on this distribution can be found in Appendix B. The respective probability function is:

$$P(X = k) = \binom{n}{k} p^k (1-p)^{n-k} = \binom{n}{k} p^k q^{n-k}. \qquad 1.1$$

1.4.2 Continuous Variables

We now consider a dataset involving a continuous random variable. Since the variable can assume an infinite number of possible values, the probability associated to each particular value is zero. Only probabilities associated to intervals of the variable domain can be non-zero. For instance, the probability that a gunshot hits a particular point in a target is zero (the variable domain is here two-dimensional). However, the probability that it hits the "bull's-eye" area is non-zero.

For a continuous variable, X (with value denoted by the same lower case letter, x), one can assign infinitesimal probabilities $\Delta p(x)$ to infinitesimal intervals Δx:

$$\Delta p(x) = f(x)\Delta x, \qquad 1.2$$

where $f(x)$ is the *probability density function*, computed at point x.

For a finite interval $[a, b]$ we determine the corresponding probability by adding up the infinitesimal contributions, i.e., using:

$$P(a < X \le b) = \int_a^b f(x)dx. \qquad 1.3$$

Therefore, the probability density function, $f(x)$, must be such that:
$\int_D f(x)dx = 1$, where D is the domain of the random variable.

Similarly to the discrete case, the *distribution function*, $F(x)$, is now defined as:

$$F(u) = P(X \le u) = \int_{-\infty}^{u} f(x)dx .$$ 1.4

Sometimes the notations $f_X(x)$ and $F_X(x)$ are used, explicitly indicating the random variable to which respect the density and distribution functions.

The reader may wish to consult Appendix A in order to learn more about continuous density and distribution functions. Appendix B presents several important continuous distributions, including the most popular, the *Gauss* (or *normal*) *distribution*, with density function defined as:

$$n_{\mu,\sigma}(x) = \frac{1}{\sqrt{2\pi}\sigma} e^{-\frac{(x-\mu)^2}{2\sigma^2}} .$$ 1.5

This function uses two parameters, μ and σ, corresponding to the mean and standard deviation, respectively. In Appendices A and B the reader finds a description of the most important aspects of the normal distribution, including the reason of its broad applicability.

1.5 Beyond a Reasonable Doubt...

We often see movies where the jury of a Court has to reach a verdict as to whether the accused is found "guilty" or "not guilty". The verdict must be consensual and established beyond any reasonable doubt. And like the trial jury, the statistician has also to reach objectively based conclusions, "beyond any reasonable doubt"...

Consider, for instance, the dataset of Example 1.3 and the statement "the 100 Ω electrical resistances, manufactured by the machine, have a (true) mean value in the interval [95, 105]". If one could measure all the resistances manufactured by the machine during its whole lifetime, one could compute the *population mean* (true mean) and assign a True or False value to that statement, i.e., a conclusion with entire certainty would then be established. However, one usually has only available a *sample* of the population; therefore, the best one can produce is a conclusion of the type "... have a mean value in the interval [95, 105] with probability δ"; i.e., one has to deal not with total certainty but with a degree of certainty:

$P(\text{mean} \in [95, 105]) = \delta = 1 - \alpha .$

We call δ (or $1-\alpha$) the *confidence level* (α is the *error* or *significance level*) and will often present it in percentage (e.g. $\delta = 95\%$). We will learn how to establish confidence intervals based on *sample statistics* (*sample mean* in the above example) and on appropriate *models* and/or *conditions* that the datasets must satisfy.

Let us now look in more detail what a confidence level really means. Imagine that in Example 1.2 we were dealing with a random sample extracted from a population of a very large number of students, attending the course and subject to an examination under the same conditions. Thus, only one random variable plays a role here: the student variability in the apprehension of knowledge. Consider, further, that we wanted to statistically assess the statement "the student performance is 3 or above". Denoting by *p* the probability of the event "the student performance is 3 or above" we derive from the dataset an estimate of *p*, known as *point estimate* and denoted $\hat{p}$, as follows:

$$\hat{p} = \frac{12+15+10}{50} = 0.74.$$

The question is how reliable this estimate is. Since the random variable representing such an estimate (with random samples of 50 students) takes value in a continuum of values, we know that the probability that the true mean is exactly that particular value (0.74) is zero. We then loose a bit of our innate and candid faith in exact numbers, relax our exigency, and move forward to thinking in terms of intervals around $\hat{p}$ (*interval estimate*). We now ask with which degree of certainty (confidence level) we can say that the true proportion *p* of students with "performance 3 or above" is, for instance, between 0.72 and 0.76, i.e., with a deviation – or *tolerance* – of $\varepsilon = \pm 0.02$ from that estimated proportion?

In order to answer this question one needs to know the so-called *sampling distribution* of the following random variable:

$$P_n = (\sum_{i=1}^{n} X_i)/n,$$

where the X_i are *n* independent random variables whose values are 1 in case of "success" (student performance ≥ 3 in this example) and 0 in case of "failure".

When the *np* and $n(1-p)$ quantities are "reasonably large" P_n has a distribution well approximated by the normal distribution with mean equal to *p* and standard deviation equal to $\sqrt{p(1-p)/n}$. This topic is discussed in detail in Appendices A and B, where what is meant by "reasonably large" is also presented. For the moment, it will suffice to say that using the normal distribution approximation (*model*), one is able to compute confidence levels for several values of the tolerance, ε, and *sample size*, *n*, as shown in Table 1.6 and displayed in Figure 1.6.

Two important aspects are illustrated in Table 1.6 and Figure 1.6: first, the confidence level always converges to 1 (absolute certainty) with increasing *n*; second, when we want to be more precise in our interval estimates by decreasing the tolerance, then, for fixed *n*, we have to lower the confidence levels, i.e., simultaneous and arbitrarily good precision and certainty are impossible (some trade-off is always necessary). In the "jury verdict" analogy it is the same as if one said the degree of certainty increases with the number of evidential facts (tending to absolute certainty if this number tends to infinite), and that if the jury wanted to increase the precision (details) of the verdict, it would then lose in degree of certainty.

Table 1.6. Confidence levels (δ) for the interval estimation of a proportion, when $\hat{p} = 0.74$, for two different values of the tolerance (ε).

n	δ for $\varepsilon = 0.02$	δ for $\varepsilon = 0.01$
50	0.25	0.13
100	0.35	0.18
1000	0.85	0.53
10000	≈ 1.00	0.98

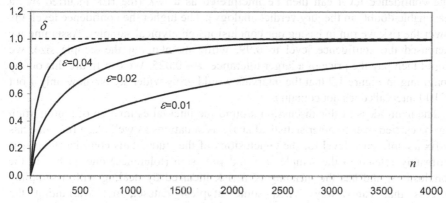

Figure 1.6. Confidence levels for the interval estimation of a proportion, when $\hat{p} = 0.74$, for three different values of the tolerance.

There is also another important and subtler point concerning confidence levels. Consider the value of $\delta = 0.25$ for a $\varepsilon = \pm 0.02$ tolerance in the $n = 50$ sample size situation (Table 1.6). When we say that the proportion of students with performance ≥ 3 lies somewhere in the interval $\hat{p} \pm 0.02$, with the confidence level 0.25, it really means that if we were able to infinitely repeat the experiment of randomly drawing $n = 50$ sized samples from the population, we would then find that 25% of the times (in 25% of the samples) the true proportion p lies in the interval $\hat{p}_k \pm 0.02$, where the $\hat{p}_k$ ($k = 1, 2, ...$) are the several sample estimates (from the ensemble of all possible samples). Of course, the "25%" figure looks too low to be reassuring. We would prefer a much higher degree of certainty; say 95% – a very popular value for the confidence level. We would then have the situation where 95% of the intervals $\hat{p}_k \pm 0.02$ would "intersect" the true value p, as shown in Figure 1.7.

Imagine then that we were dealing with random samples from a random experiment in which we knew beforehand that a "success" event had a $p = 0.75$ probability of occurring. It could be, for instance, randomly drawing balls with replacement from an urn containing 3 black balls and 1 white "failure" ball. Using the normal approximation of P_n, one can compute the needed sample size in order to obtain the 95% confidence level, for an $\varepsilon = \pm 0.02$ tolerance. It turns out to be $n \approx 1800$. We now have a sample of 1800 drawings of a ball from the urn, with an estimated proportion, say $\hat{p}_0$, of the success event. Does this mean that when dealing with a large number of samples of size $n = 1800$ with estimates $\hat{p}_k$ ($k = 1$, 2,...), 95% of the $\hat{p}_k$ will lie somewhere in the interval $\hat{p}_0 \pm 0.02$? No. It means, as previously stated and illustrated in Figure 1.7, that 95% of the intervals $\hat{p}_k \pm 0.02$ will contain p. As we are (usually) dealing with a single sample, we could be unfortunate and be dealing with an "atypical" sample, say as sample #3 in Figure 1.7. Now, it is clear that 95% of the time p does not fall in the $\hat{p}_3 \pm 0.02$ interval. The confidence level can then be interpreted as a *risk* (the risk incurred by "a reasonable doubt" in the jury verdict analogy). The higher the confidence level, the lower the risk we run in basing our conclusions on atypical samples. Assuming we increased the confidence level to 0.99, while maintaining the sample size, we would then pay the price of a larger tolerance, $\varepsilon = 0.025$. We can figure this out by imagining in Figure 1.7 that the intervals would grow wider so that now only 1 out of 100 intervals does not contain p.

The main ideas of this discussion around the interval estimation of a proportion can be carried over to other statistical analysis situations as well. As a rule, one has to fix a confidence level for the conclusions of the study. This confidence level is intimately related to the sample size and precision (tolerance) one wishes in the conclusions, and has the meaning of a risk incurred by dealing with a sampling process that can always yield some atypical dataset, not warranting the conclusions. After losing our innate and candid faith in exact numbers we now lose a bit of our certainty about intervals...

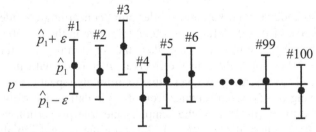

Figure 1.7. Interval estimation of a proportion. For a 95% confidence level only roughly 5 out of 100 samples, such as sample #3, are atypical, in the sense that the respective $\hat{p} \pm \varepsilon$ interval does not contain p.

The choice of an appropriate confidence level depends on the problem. The 95% value became a popular figure, and will be largely used throughout the book, because it usually achieves a "reasonable" tolerance in our conclusions (say, $\varepsilon < 0.05$) for a not too large sample size (say, $n > 200$), and it works well in many applications. For some problem types, where a high risk can have serious consequences, one would then choose a higher confidence level, 99% for example. Notice that arbitrarily small risks (arbitrarily small "reasonable doubt") are often impractical. As a matter of fact, a zero risk – no "doubt" at all – means, usually, either an infinitely large, useless, tolerance, or an infinitely large, prohibitive, sample. A compromise value achieving a useful tolerance with an affordable sample size has to be found.

1.6 Statistical Significance and Other Significances

Statistics is surely a recognised and powerful data analysis tool. Because of its recognised power and its pervasive influence in science and human affairs people tend to look to statistics as some sort of recipe book, from where one can pick up a recipe for the problem at hand. Things get worse when using statistical software and particularly in inferential data analysis. A lot of papers and publications are plagued with the "computer *dixit*" syndrome when reporting statistical results. People tend to lose any critical sense even in such a risky endeavour as trying to reach a general conclusion (law) based on a data sample: the inferential or inductive reasoning.

In the book of A. J. Jaffe and Herbert F. Spirer (Jaffe AJ, Spirer HF 1987) many misuses of statistics are presented and discussed in detail. These authors identify four common sources of misuse: incorrect or flawed data; lack of knowledge of the subject matter; faulty, misleading, or imprecise interpretation of the data and results; incorrect or inadequate analytical methodology. In the present book we concentrate on how to choose adequate analytical methodologies and give precise interpretation of the results. Besides theoretical explanations and words of caution the book includes a large number of examples that in our opinion help to solidify the notions of adequacy and of precise interpretation of the data and the results. The other two sources of misuse – flawed data and lack of knowledge of the subject matter – are the responsibility of the practitioner.

In what concerns statistical inference the reader must exert extra care of not applying statistical methods in a mechanical and mindless way, taking or using the software results uncritically. Let us consider as an example the comparison of foetal heart rate baseline measurements proposed in Exercise 4.11. The heart rate "baseline" is roughly the most stable heart rate value (expressed in beats per minute, bpm), after discarding rhythm acceleration or deceleration episodes. The comparison proposed in Exercise 4.11 respects to measurements obtained in 1996 against those obtained in other years (CTG dataset samples). Now, the popular two-sample t-test presented in chapter 4 does not detect a statiscally significant diference between the means of the measurements performed in 1996 and those performed in other years. If a statistically significant diference was detected did it mean that the 1996 foetal population was different, in that respect, from the population of other years? Common sense (and other senses as well) rejects such a claim. If a statistically significant difference was detected one should look carefully to the conditions presiding the data collection: can the samples be considered as being random?; maybe the 1996 sample was collected in at-risk foetuses with lower baseline measurements; and so on. As a matter of fact, when dealing with large samples even a small compositional difference may sometimes produce statistically significant results. For instance, for the sample sizes of the CTG dataset even a difference as small as 1 bpm produces a result usually considered as statistically significant ($p = 0.02$). However, obstetricians only attach practical meaning to rhythm differences above 5 bpm; i.e., the statistically significant difference of 1 bpm has no practical significance.

Inferring causality from data is even a riskier endeavour than simple comparisons. An often encountered example is the inference of causality from a statistically significant but spurious correlation. We give more details on this issue in section 4.4.1.

One must also be very careful when performing goodness of fit tests. A common example of this is the normality assessment of a data distribution. A vast quantity of papers can be found where the authors conclude the normality of data distributions based on very small samples. (We have found a paper presented in a congress where the authors claimed the normality of a data distribution based on a sample of four cases!) As explained in detail in section 5.1.6, even with 25-sized samples one would often be wrong when admitting that a data distribution is normal because a statistical test didn't reject that possibility at a 95% confidence level. More: one would often be accepting the normality of data generated with asymmetrical and even bimodal distributions! Data distribution modelling is a difficult problem that usually requires large samples and even so one must bear in mind that most of the times and beyond a reasonable doubt one only has evidence of a *model*; the true distribution remains unknown.

Another misuse of inferential statistics arrives in the assessment of classification or regression models. Many people when designing a classification or regression model that performs very well in a training set (the set used in the design) suffer from a kind of love-at-first-sight syndrome that leads to neglecting or relaxing the evaluation of their models in test sets (independent of the training sets). Research literature is full with examples of improperly validated models that are later on dropped out when more data becomes available and the initial optimism plunges down. The love-at-first-sight is even stronger when using computer software that automatically searches for the best set of variables describing the model. The book of Chamont Wang (Wang C, 1993), where many illustrations and words of caution on the topic of inferential statistics can be found, mentions an experiment where 51 data samples were generated with 100 random numbers each and a regression model was searched for "explaining" one of the data samples (playing the role of dependent variable) as a function of the other ones (playing the role of independent variables). The search finished by finding a regression model with a significant R-square and six significant coefficients at 95% confidence level. In other words, a functional model was found explaining a relationship between noise and noise! Such a model would collapse had proper validation been applied. In the present book we will pay attention to the topic of model validation both in classification and regression.

1.7 Datasets

A statistical data analysis project starts, of course, by the data collection task. The quality with which this task is performed is a major determinant of the quality of the overall project. Issues such as reducing the number of missing data, recording the pertinent documentation on what the problem is and how the data was collected and inserting the appropriate description of the meaning of the variables involved must be adequately addressed.

Missing data – failure to obtain for certain objects/cases the values of one or more variables – will always undermine the degree of certainty of the statistical conclusions. Many software products provide means to cope with missing data. These can be simply coding missing data by symbolic numbers or tags, such as "na" ("not available") which are neglected when performing statistical analysis operations. Another possibility is the substitution of missing data by average values of the respective variables. Yet another solution is to simply remove objects with missing data. Whatever method is used the quality of the project is always impaired.

The collected data should be stored in a tabular form ("data matrix"), usually with the rows corresponding to objects and the columns corresponding to the variables. A spreadsheet such as the one provided by EXCEL (a popular application of the WINDOWS systems) constitutes an adequate data storing solution. An example is shown in Figure 2.1. It allows to easily performing simple calculations on the data and to store an accompanying data description sheet. It also simplifies data entry operations for many statistical software products.

All the statistical methods explained in this book are illustrated with real-life problems. The real datasets used in the book examples and exercises are stored in EXCEL files. They are described in Appendix E and included in the book CD. Dataset names correspond to the respective EXCEL file names. Variable identifiers correspond to the column identifiers of the EXCEL files.

There are also many datasets available through the Internet which the reader may find useful for practising the taught matters. We particularly recommend the datasets of the UCI Machine Learning Repository (http://www.ics.uci.edu/~mlearn/MLRepository.html). In these (and other) datasets data is presented in text file format. Conversion to EXCEL format is usually straightforward since EXCEL provides means to read in text files with several types of column delimitation.

1.8 Software Tools

There are many software tools for statistical analysis, covering a broad spectrum of possibilities. At one end we find "closed" products where the user can only perform menu operations. SPSS and STATISTICA are examples of "closed" products. At the other end we find "open" products allowing the user to program any arbitrarily complex sequence of statistical analysis operations. MATLAB and R are examples of "open" products providing both a programming language and an environment for statistical and graphic operations.

This book explains how to apply SPSS, STATISTICA, MATLAB or R to solving statistical problems. The explanation is guided by solved examples where we usually use one of the software products and provide indications (in specific "Commands" frames) on how to use the other ones. We use the releases SPSS STATISTICA 7.0, MATLAB 7.1 with the *Statistics Toolbox* and R 2.2.1 for the Windows operating system; there is, usually, no significant difference when using another release of these products (especially if it is a more advanced one), or running these products in other non-Windows based platforms. All book figures obtained with these software products are presented in greyscale, therefore sacrificing some of the original display quality.

The reader must bear in mind that the present book is not intended as a substitute of the user manuals or on-line helps of SPSS, STATISTICA, MATLAB and R. However, we do provide the really needed information and guidance on how to use these software products, so that the reader will be able to run the examples and follow the taught matters with a minimum effort. As a matter of fact, our experience using this book as a teaching aid is that usually those explanations are sufficient for solving most practical problems. Anyway, besides user manuals and on-line helps, the reader interested in deepening his/her knowledge of particular topics may also find it profitable to consult the specific bibliography on these software products mentioned in the References. In this section we limit ourselves to describing a few basic aspects that are essential as a first hands-on.

1.8.1 SPSS and STATISTICA

SPSS from *SPSS Inc.* and STATISTICA from *StatSoft Inc.* are important and popularised software products of the menu-driven type on window environments with user-friendly facilities of data edition, representation and graphical support in an interactive way. Both products require minimal time for familiarization and allow the user to easily perform statistical analyses using a spreadsheet-based philosophy for operating with the data.

Both products reveal a lot of similarities, starting with the menu bars shown in Figures 1.8 and 1.9, namely the individual options to manage files, to edit the data spreadsheets, to manage graphs, to perform data operations and to apply statistical analysis procedures.

Concerning flexibility, both SPSS and STATISTICA provide command language and macro construction facilities. As a matter of fact STATISTICA is close to an "open" product type, since it provides advanced programming facilities such as the use of external code (DLLs) and application programming interfaces (API), as well as the possibility of developing specific routines in a Basic-like programming language.

In the following we use courier type font for denoting SPSS and STATISTICA commands.

1.8.1.1 SPSS

The menu bar of the SPSS user interface is shown in Figure 1.8 (with the data file Meteo.sav in current operation). The contents of the menu options (besides the obvious Window and Help), are as follows:

File: Operations with data files (*.sav), syntax files (*.sps), output files (*.spo), print operations, etc.
Edit: Spreadsheet edition.
View: View configuration of spreadsheets, namely of value labels and gridlines.
Data: Insertion and deletion of variables and cases, and operations with the data, namely sorting and transposition.
Transform: More operations with data, such as recoding and computation of new variables.
Analyze: Statistical analysis tools.
Graphs: Operations with graphs.
Utilities: Variable definition reports, running scripts, etc.

Besides the menu options there are alternative ways to perform some operations using icons.

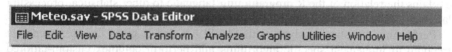

Figure 1.8. Menu bar of SPSS user interface (the dataset being currently operated is Meteo.sav).

1.8.1.2 STATISTICA

The menu bar of STATISTICA user interface is shown in Figure 1.9 (with the data file Meteo.sta in current operation). The contents of the menu options (besides the obvious Window and Help) are as follows:

File: Operations with data files (*.sta), scrollsheet files (*.scr), graphic files (*.stg), print operations, etc.
Edit: Spreadsheet edition, screen catching.
View: View configuration of spreadsheets, namely of headers, text labels and case names.
Insert: Insertion and copy of variables and cases.
Format: Format specifications of spreadsheet cells, variables and cases.
Statistics: Statistical analysis tools and STATISTICA Visual Basic.

Graphs: Operations with graphs.
Tools: Selection conditions, macros, user options, etc.
Data: Several operations with the data, namely sorting, recalculation
               and recoding of data.

Besides the menu options there are alternative ways to perform a given
operation using icons and key combinations (using underlined characters).

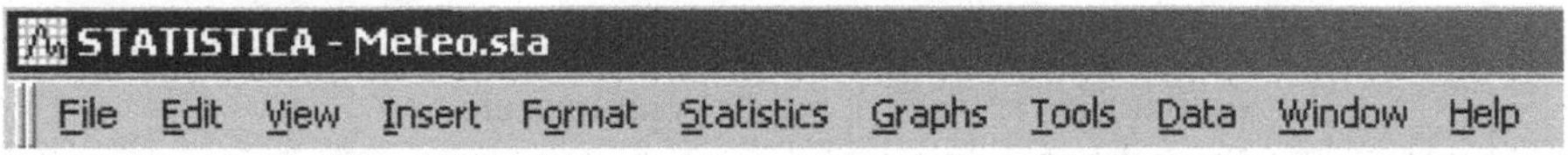

STATISTICA - Meteo.sta

| File | Edit | View | Insert | Format | Statistics | Graphs | Tools | Data | Window | Help |

Figure 1.9. Menu bar of STATISTICA user interface (the dataset being currently
operated is Meteo.sta).

1.8.2 MATLAB and R

MATLAB, a mathematical software product from *The MathWorks, Inc.*, and R (R:
A Language and Environment for Statistical Computing) from the *R Development
Core Team* (R Foundation for Statistical Computing, Vienna, Austria, ISBN 3-
900051-07-0), a free software product for statistical computing, are popular
examples of "open" products. R can be downloaded from the Internet URL
http://www.r-project.org/. This site explains the R history and indicates a set of
URLs (the so-called CRAN mirrors) that can be used for downloading R. It also
explains the relation of the R programming language to other statistical processing
languages such as S and S-Plus.

Performing statistical analysis with MATLAB and R gives the user complete
freedom to implement specific algorithms and perform complex custom-tailored
operations. MATLAB and R are also especially useful when the statistical
operations are part of a larger project. For instance, when developing a signal or
image classification project one may have to first compute signal or image features
using specific MATLAB or R toolboxes, followed by the application of
appropriate statistical classification procedures. The penalty to be paid for this
flexibility is that the user must learn how to program with the MATLAB or R
language. In this book we restrict ourselves to present the essentials of MATLAB
and R command-driven operations and will not enter into programming topics.

We use courier type font for denoting MATLAB and R commands. When
needed, we will clarify the correspondence between the mathematical and the
software symbols. For instance MATLAB or R matrix x will often correspond to
the mathematical matrix **X**.

1.8.2.1 MATAB

MATLAB command lines are written with appropriate arguments following the
prompt, », in a MATLAB console as shown in Figure 1.10. This same Figure

illustrates that after writing down the command `help stats` (ending with the "Return" or the "Enter" key), one obtains a list of all available commands (functions) of the MATLAB Statistical toolbox. One could go on and write, for instance, `help betafit`, getting help about the `betafit` function.

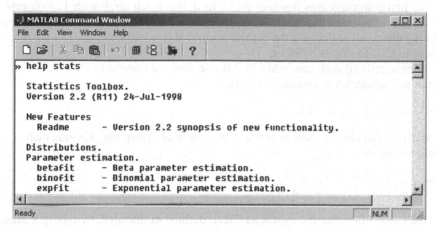

Figure 1.10. The command window of MATLAB showing the list of available statistical functions (obtained with the help command).

Note that MATLAB is case-sensitive. For instance, `Betafit` is not the same as `betafit`.

The basic data type in MATLAB and the one that will use more often are matrices. Matrix values can be directly typed in the MATLAB console. For instance, the following command defines a 2×2 matrix x with the typed in values:

```
» x=[1 2
     3 4];
```

The "=" symbol is an assignment operator. The symbol "x" is the matrix identifier. Object identifiers in MATLAB can be arbitrary strings not starting by a digit; exception is made to reserved MATLAB words.

Indexing in MATLB is straightforward using the parentheses as index qualifier. Thus, for example x(2,1) is the element of the second row and first column of x with value 3.

A vector is just a special matrix that can be thought of as a 1×n (row vector) or as an n×1 (column vector) matrix.

MATLAB allows the definition of character vectors (e.g. `c=['abc']`) and also of vectors of strings. In this last case one must use the so-called "cell array" which is simply an object recipient array. Consider the following sequence of commands:

```
>> c=cell(1,3);
>> c(1,1)={'Pmax'};
```

```
>> c(1,2)={'T80'};
>> c(1,3)={'T82'};
>> c
c =
    'Pmax'      'T80'      'T82'
```

The first command uses function cell to define a cell array with 1×3 objects. These are afterwards assigned some string values (delimited with '). When printing the c values one gets the confirmation that c is a row vector with the three strings (e.g., c(1,2) is 'T80').

When specifying matrices in MATLAB one may use comma to separate column values and semicolon to separate row values as in:

```
» x=[1, 2 ; 3, 4];
```

Matrices can also be used to define other matrices. Thus, the previous matrix x could also be defined as:

```
» x=[[1 2] ; [3 4]];
» x=[[1; 3], [2; 4]];
```

One can confirm that the matrix has been defined as intended, by typing x after the prompt, and obtaining:

```
x =
    1       2
    3       4
```

The same result could be obtained by removing the semicolon terminating the previous command. In MATLAB a semicolon inhibits the production of screen output. Also MATLAB commands can either be used in a procedure-like manner, producing output (as "answers", denoted ans), or in a function-like manner producing a value assigned to a variable (considered to be a matrix). This is illustrated next, with the command that computes the mean of a sequence of values structured as a row vector:

```
» v=[1 2 3 4 5 6];
» mean(v)
ans =
    3.5000
» y=mean(v)
y =
    3.5000
```

Whenever needed one may know which objects (e.g. matrices) are currently in the console environment by issuing who. Object removal is performed by writing clear followed by the name of the object. For instance, clear x removes matrix x from the environment; it will no longer be available. The use of clear without arguments removes all objects from the environment.

On-line help about general or specific topics of MATLAB can be obtained from the Help menu option. On-line help about a specific function can be obtained by just typing it after the `help` command, as seen above.

1.8.2.2   R

R command lines are written with appropriate arguments following the R prompt, >, in the R Gui interface (R console) as shown in Figure 1.11. As in MATLAB command lines must be terminated with the "Return" or the "Enter" key.

Data is represented in R by means of vectors, matrices and data frames. The basic data representation in R is a column vector but for statistical analyses one mostly uses data frames. Let us start with vectors. The command

```
> x <- c(1,2,3,4,5,6)
```

defines a column vector named x containing the list of values between parentheses. The "<-" symbol is the assignment operator. The "c" function fills the vector with the list of values. The symbol "x" is the vector identifier. Object identifiers in R can be arbitrary strings not starting by a digit; exception is made to reserved R words.

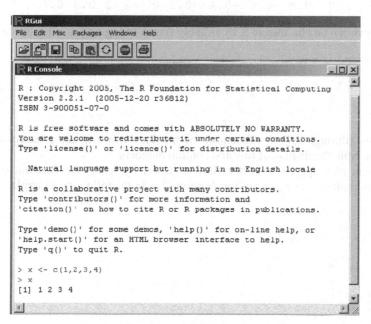

Figure 1.11. The R Gui showing the definition of a vector.

We may list the contents of x just by issuing it as a command:

```
> x
[1] 1 2 3 4 5 6
```

The [1] means the first element of x. For instance,

```
> y <- rnorm(12)
> y
 [1] -0.1354 -0.2519  0.5716  0.6845 -1.5148 -0.1190
 [7]  0.7328 -1.0274  0.3319 -0.3468 -1.2619  0.7146
```

generates and lists a vector with 12 normally distributed random numbers. The 1st and 7th elements are indicated. (The numbers are represented here with four digits after the decimal point because of page width constraints. In R the representation is with seven digits.) One could also obtain the previous list by just issuing: > rnorm(12). Most R functions also behave as procedures in that way, displaying lists of values in the R console.

A vector can be filled with strings (delimited with "), as in v <- c("Pmax","T80","T82"). Now v is a vector containing three strings. The second vector element, v[2], is "T80"

R also provides a function, named seq, to define evenly spaced number sequences, as in the following example:

```
> seq(-1,1,0.2)
 [1] -1.0 -0.8 -0.6 -0.4 -0.2 0.0 0.2 0.4 0.6 0.8 1.0
```

A matrix can be obtained in R by suitably transforming a vector. For instance,

```
> dim(x) <- c(2,3)
> x
     [,1] [,2] [,3]
[1,]   1    3    5
[2,]   2    4    6
```

transforms (through the dim function) the previous vector x into a matrix of 2×3 elements. Note the display of row and column numbers.

One can also aggregate vectors into a matrix by using the function cbind ("column binding") or rbind ("row binding") as in the following example:

```
> u <- c(1,2,3)
> v <- c(-1,-2,-3)
> m <- cbind(u,v)
> m
     u  v
[1,] 1 -1
[2,] 2 -2
[3,] 3 -3
```

Matrix indexing in R uses square brackets as index qualifier. As an example, m[2,2] has the value -2.

Note that R is case-sensitive. For instance, Cbind cannot be used as a replacement for cbind.

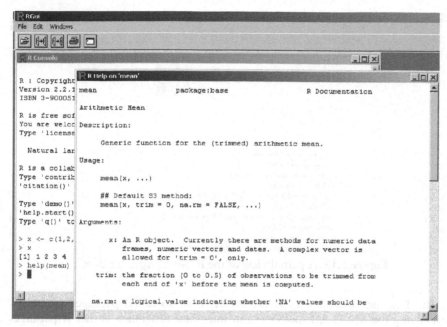

Figure 1.12. An illustration of R on-line help of function mean. The "Help on 'mean'" is displayed in a specific window.

An R data frame is a recipient for a list of objects. We mostly use data frames that are simply data matrices with appropriate column names, as in the above matrix m.

Operations on data are obtained by using suitable R functions. For instance,

```
> mean(x)
[1] 3.5
```

displays the mean value of the x vector on the console. Of course one could also assign this mean value to a new variable, say mu, by issuing the command mu <- mean(x).

Whenever needed one may obtain the information on which objects are currently in the console environment by using ls() ("list"). (Be sure to include the parentheses; otherwise R will interpret it as you wishing to obtain the ls function code.) Object removal is performed by applying the function rm ("remove") to a list of object identifiers. For instance, rm(x) removes matrix x from the environment; it will no longer be available.

On-line help about general topics of R, namely command constructs and available functions, can be obtained from the Help menu option of the R Gui. On-line help about a specific function can be obtained using the R help function as illustrated in Figure 1.12.

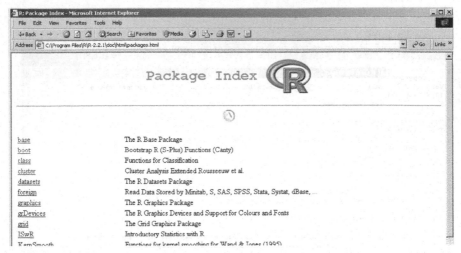

Figure 1.13. A partial view of the R "`Package Index`".

The functions available in R are collected in so-called packages (somehow resembling the MATLAB toolboxes; an important difference is that R packages may also include datasets). One can inspect which packages are currently loaded by issuing the `search()` command (with no arguments). Consider that you have done that and obtained:

```
> search()
[1]".GlobalEnv"         "package:methods"    "package:stats"
[4]"package:graphics"    "package:grDevices"  "package:utils"
[7]"package:datasets"    "Autoloads"          "package:base"
```

We will often use functions of the `stats` package. In order to get the information of which functions are available in the `stats` package one may issue the `help.start()` command. An Internet window pops up from where one clicks on "Packages" and obtains the "Package Index" window partially shown in Figure 1.13.

By clicking on `stats` of the "Package Index" one obtains a complete list of the available `stats` functions. The same procedure can be followed to obtain function (and dataset) lists of other packages.

The command `library()` issues a list of the packages installed at one's site. One of the listed packages is the `boot` package. In order to have it currently loaded one should issue `library(boot)`. A following `search()` would display:

```
> search()
[1] ".GlobalEnv"         "package:boot"       "package:methods"
[4] "package:stats"      "package:graphics"   "package:grDevices"
[7] "package:utils"      "package:datasets"   "Autoloads"
[10]"package:base"
```

2 Presenting and Summarising the Data

Presenting and summarising the data is certainly the introductory task in any statistical analysis project and comprehends a set of topics and techniques, collectively known as *descriptive statistics*.

2.1 Preliminaries

2.1.1 Reading in the Data

Data is usually gathered and arranged in tables. The spreadsheet approach followed by numerous software products is a convenient tabular approach to deal with the data. Consider the meteorological dataset Meteo (see Appendix E for a description). It is provided in the book CD as an EXCEL file (Meteo.xls) with the *cases* (meteorological stations) along the rows and the *random variables* (weather variables) along the columns, as shown in Figure 2.1. The first column is the cases column, containing numerical codes or, as in Figure 2.1, names of cases. The first row is usually a header row containing names of variables. This is a convenient way to store the data.

Notice also the indispensable *Description* datasheet, where all the necessary information concerning the meaning of the data, the definitions of the variables and of the cases, as well as the source and possible authorship of the data should be supplied.

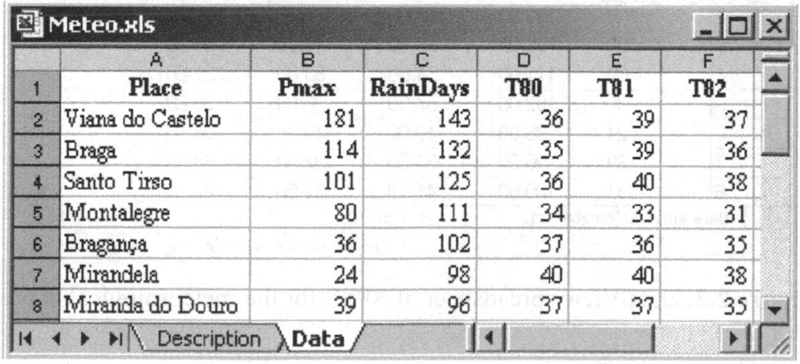

	A	B	C	D	E	F
1	**Place**	**Pmax**	**RainDays**	**T80**	**T81**	**T82**
2	Viana do Castelo	181	143	36	39	37
3	Braga	114	132	35	39	36
4	Santo Tirso	101	125	36	40	38
5	Montalegre	80	111	34	33	31
6	Bragança	36	102	37	36	35
7	Mirandela	24	98	40	40	38
8	Miranda do Douro	39	96	37	37	35

Meteo.xls — Description \ Data

Figure 2.1. The meteorological dataset presented as an EXCEL file.

Carrying out this dataset into SPSS, STATISTICA or MATLAB is an easy task. The basic thing to do is to select the data in the usual way (mouse dragging between two corners of the data spreadsheet), copy the data (e.g., using the CTRL+C keys) and paste it (e.g., using the CTRL+V keys). In R data has to be read from a text file. One can also, of course, type in the data directly into the SPSS or STATISTICA spreadsheets or into the MATLAB command window or the R console. This is usually restricted to small datasets. In the following subsections we present the basics of data entry in SPSS, STATISTICA, MATLAB and R.

2.1.1.1 SPSS Data Entry

When first starting SPSS a file specification box may be displayed and the user asked whether a (last operated) data file should be opened. One can cancel this file specification box and proceed to define a new data file (File, New), where the data can be pasted (from EXCEL) or typed in. The SPSS data spreadsheet starts with a comfortably large number of variables and cases. Further variables and cases may be added when needed (use the Insert Variable or Insert Case options of the Data menu). One can then proceed to add specifications to the variables, either by double clicking with the mouse left button over the column heading or by clicking on the Variable View tab underneath (this is a toggle tab, toggling between the Variable View and the Data View). The Variable View and Data View spreadsheets for the meteorological data example are shown in Figure 2.2 and 2.3, respectively. Note that the variable identifiers in SPSS use only lower case letters.

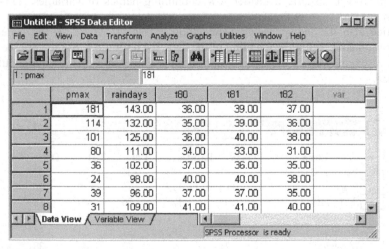

Figure 2.2. Data View spreadsheet of SPSS for the meteorological data.

The data can then be saved with Save As (File menu), specifying the data file name (Meteo.sav) which will appear in the title heading of the data spreadsheet. This file can then be comfortably opened in a following session with the Open option of the File menu.

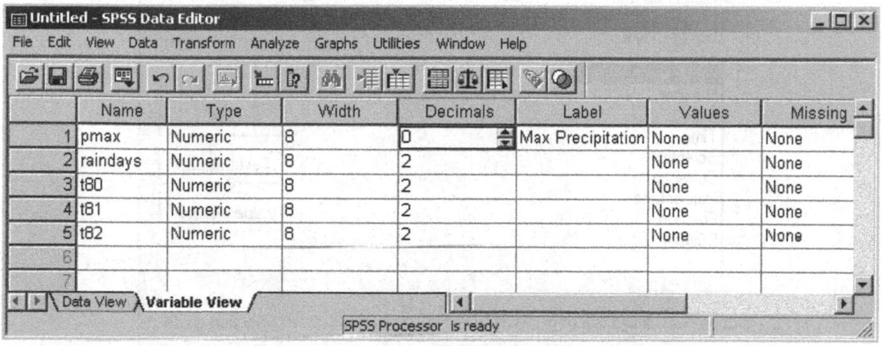

Figure 2.3. Variable View spreadsheet of SPSS for the meteorological data. Notice the fields for filling in variable labels and missing data codes.

2.1.1.2   STATISTICA Data Entry

With STATISTICA one starts by creating a new data file (File, New) with the desired number of variables and cases, before pasting or typing in the data. There is also the possibility of using any previous template data file and adjusting the number of variables and cases (click the right button of the mouse over the variable column(s) or case row(s) or, alternatively, use Insert). One may proceed to define the variables, by assigning them a specific name and declaring their type. This can be done by double clicking the mouse left button over the respective column heading. The specification box shown in Figure 2.4 is then displayed. Note the possibility of specifying a variable label (describing the variable meaning) or a formula (this last possibility will be used later). Missing data (MD) codes and text labels assigned to variable values can also be specified. Figure 2.5 shows the data spreadsheet corresponding to the Meteo.xls dataset. The similarity with Figure 2.1 is evident.

After building the data spreadsheet, it is advisable to save it using the Save As of the File menu. In this case we specify the filename Meteo, creating thus a Meteo.sta STATISTICA file that can be easily opened at another session with the Open option of File. Once the data filename is specified, it will appear in the title heading of the data spreadsheet and in this case, instead of "Data: Spreadsheet2*", "Data: Meteo.sta" will appear. The notation 5v by 25c indicates that the file is composed of 5 variables with 25 cases.

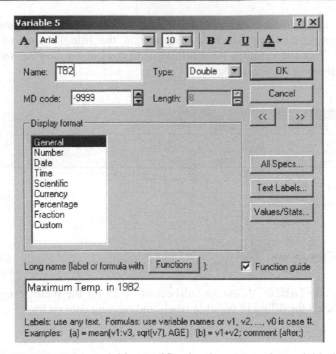

Figure 2.4. STATISTICA variable specification box. Note the variable label at the bottom, describing the meaning of the variable T82.

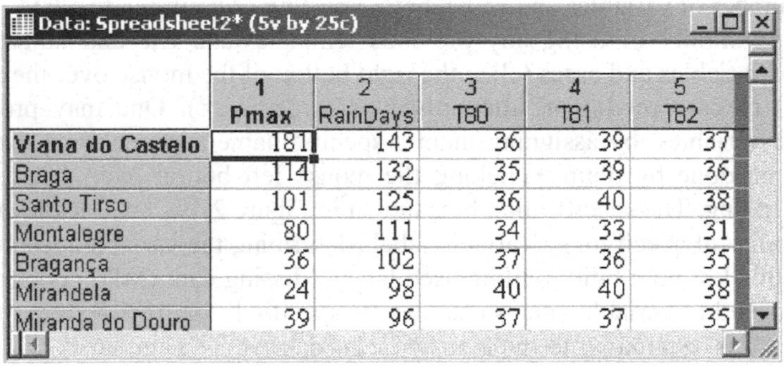

	1 Pmax	2 RainDays	3 T80	4 T81	5 T82
Viana do Castelo	181	143	36	39	37
Braga	114	132	35	39	36
Santo Tirso	101	125	36	40	38
Montalegre	80	111	34	33	31
Bragança	36	102	37	36	35
Mirandela	24	98	40	40	38
Miranda do Douro	39	96	37	37	35

Figure 2.5. STATISTICA spreadsheet corresponding to the meteorological data.

2.1.1.3 *MATLAB Data Entry*

In MATLAB, one can also directly paste data from an EXCEL file, inside a matrix definition typed in the MATLAB command window. For the meteorological data one would have (the "…" denotes part of the listing that is not shown; the % symbol denotes a MATLAB user comment):

```
» meteo=[
181 143 36 39 37    % Pasting starts here
114 132 35 39 36
101 125 36 40 38
...
14 70 35 37 39      % and ends here.
];                  % Typed after the pasting.
```

One would then proceed to save the meteo matrix with the save command. In order to save the data file (as well as other files) in a specific directory, it is advisable to change the directory with the cd command. For instance, imagine one wanted to save the data in a file named Meteodata, residing in the c:\experiments directory. One would then specify:

```
» cd('c:\experiments');
» save Meteodata meteo;
```

The MATLAB dir command would then list the presence of the MATLAB file Meteodata.mat in that directory.

In a later session the user can retrieve the matrix variable meteo by simply using the load command: » load Meteodata.

This will load the meteo matrix from the Meteodata.mat file as can be confirmed by displaying its contents with: » meteo.

2.1.1.4   R Data Entry

The tabular form of data in R is called *data frame*. A data frame is an aggregate of column vectors, corresponding to the variables related across the same objects (cases). In addition it has a unique set of row names. One can create an R data frame from a text file (direct data entry from an EXCEL file is not available). Let us illustrate the whole procedure using the meteo.xls file shown in Figure 2.1 as an example. The first thing to do is to convert the numeric data area of meteo.xls to a tab-delimited text file, e:meteo.txt, say, from within EXCEL (with Save As). We now issue the following command in the R console:

```
> meteo <- read.table(file("e:meteo.txt"))
```

The argument of file is the path to the file we want to read in. As a result of read.table a data frame is created with the same numeric information as the meteo.xls file. We can see this with:

```
> meteo
    V1  V2 V3 V4 V5
1  181 143 36 39 37
2  114 132 35 39 36
3  101 125 36 40 38
...
```

For future use we may now proceed to save this data frame in e:meteo, say, with save(meteo,file="e:meteo"). At a later session we can immediately load in the data frame with load("e:meteo").

It is often convenient to have appropriate column names for the data, instead of the default V1, V2, etc. One way to do this is to first create a string vector and pass it to the read.table function as a col.names parameter value. For the meteo data we could have:

```
> l <- c("PMax","RainDays","T80","T81","T82")
> meteo<-read.table(file("e:meteo.txt"),col.names=l)
> meteo
    PMax RainDays T80 T81 T82
1    181      143  36  39  37
2    114      132  35  39  36
3    101      125  36  40  38
...
```

Column names and row names[1] can also be set or retrieved with the functions colnames and rownames, respectively. For instance, the following sequence of commands assigns row names to meteo corresponding to the names of the places where the meteorological data was collected (see Figure 2.1):

```
>   r   <-   c("V.   Castelo",   "Braga",   "S.   Tirso",
"Montalegre",   "Bragança",   "Mirandela",   "M.   Douro",
"Régua",   "Viseu",   "Guarda",   "Coimbra",   "C.   Branco",
"Pombal",   "Santarém",   "Dois   Portos",   "Setúbal",
"Portalegre",   "Elvas",   "Évora",   "A.   Sal",   "Beja",
"Amareleja",   "Alportel",   "Monchique",   "Tavira");
> rownames(meteo) <- r
> meteo
             PMax RainDays T80 T81 T82
V. Castelo    181      143  36  39  37
Braga         114      132  35  39  36
S. Tirso      101      125  36  40  38
Montalegre     80      111  34  33  31
Bragança       36      102  37  36  35
Mirandela      24       98  40  40  38
M. Douro       39       96  37  37  35
Régua          31      109  41  41  40
...
```

2.1.2 Operating with the Data

After having read in a data set, one is often confronted with the need of defining new variables, according to a certain formula. Sometimes one also needs to manage the data in specific ways; for instance, sorting cases according to the values of one or more variables, or transposing the data, i.e., exchanging the roles of columns and rows. In this section, we will present only the fundamentals of such operations, illustrated for the meteorological dataset. We further assume that we

---

[1] Column or row names should preferably not use reserved R words.

are interested in defining a new variable, PClass, that categorises the maximum rain precipitation (variable PMax) into three categories:

1. PMax ≤ 20 (low);
2. 20 < PMax ≤ 80 (moderate);
3. PMax > 80 (high).

Variable PClass can be expressed as

PClass = 1 + (PMax > 20) + (PMax > 80),

whenever logical values associated to relational expressions such as "PMax > 20" are represented by the arithmetical values 0 and 1, coding False and True, respectively. That is precisely how SPSS, STATISTICA, MATLAB and R handle such expressions. The reader can easily check that PClass values are 1, 2 and 3 in correspondence with the low, moderate and high categories.

In the following subsections we will learn the essentials of data operation with SPSS, STATISTICA, MATLAB and R.

2.1.2.1   SPSS

The addition of a new variable is made in SPSS by using the `Insert Variable` option of the `Data` menu. In the case of the previous categorisation variable, one would then proceed to compute its values by using the `Compute` option of the `Transform` menu. The `Compute Variable` window shown in Figure 2.6 will then be displayed, where one would fill in the above formula using the respective variable identifiers; in this case: `1+(pmax>20)+(pmax>80)`.

Looking to Figure 2.6 one may rightly suspect that a large number of functions are available in SPSS for building arbitrarily complex formulas.

Other data management operations such as sorting and transposing can be performed using specific options of the SPSS `Data` menu.

2.1.2.2   STATISTICA

The addition of a new variable in STATISTICA is made with the `Add Variable` option of the `Insert` menu. The variable specification window shown in Figure 2.7 will then be displayed, where one would fill in, namely, the number of variables to be added, their names and the formulas used to compute them. In this case, the formula is:

`1+(v1>20)+(v1>80)`.

In STATISTICA variables are symbolically denoted by v followed by a number representing the position of the variable column in the spreadsheet. Since `Pmax` happens to be the first column, it is then denoted `v1`. The cases column is `v0`. It is also possible to use variable identifiers in formulas instead of v-notations.

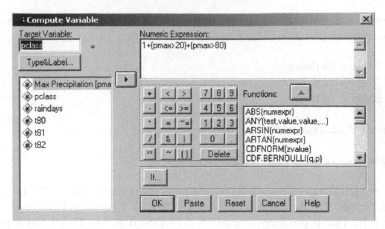

Figure 2.6. Computing, in SPSS, the new variable PClass in terms of the variable pmax.

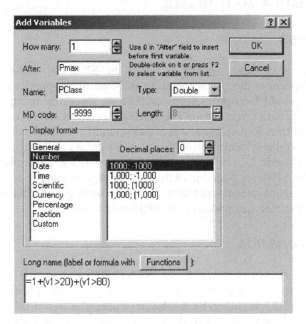

Figure 2.7. Specification of a new (categorising) variable, PClass, inserted after PMax in STATISTICA.

The presence of the equal sign, preceding the expression, indicates that one wants to compute a formula and not merely assign a text label to a variable. One can also build arbitrarily complex formulas in STATISTICA, using a large number of predefined functions (see button Functions in Figure 2.7).

Besides the insertion of new variables, one can also perform other operations such as sorting the entire spreadsheet based on column values, or transposing columns and cases, using the appropriate STATISTICA Data menu options.

2.1.2.3 MATLAB

In order to operate with the matrix data in MATLAB we need to first learn some basic ingredients. We have already mentioned that a matrix element is accessed through its indices, separated by comma, between parentheses. For instance, for the previous meteo matrix, one can find out the value of the maximum precipitation (1st column) for the 3rd case, by typing:

```
» meteo(3,1)

ans =
   101
```

If one wishes a list of the PMax values from the 3rd to the 5th cases, one would write:

```
» meteo(3:5,1)

ans =
   101
    80
    36
```

Therefore, a range in cases (or columns) is obtained using the range values separated by a colon. The use of the colon alone, without any range values, means the complete range, i.e., the complete column (or row). Thus, in order to extract the PMax column vector from the meteo matrix we need only specify:

```
» pmax = meteo(:,1);
```

We may now proceed to compute the new column vector, PClass:

```
» pclass = 1+(pmax>20)+(pmax>80);
```

and join it to the meteo matrix, with:

```
» meteo = [meteo pclass]
```

Transposition of a matrix in MATLAB is straightforward, using the apostrophe as the transposition operation. For the meteo matrix one would write:

```
» meteotransp = meteo';
```

Sorting the rows of a matrix, as a group and in ascending order, is performed with the sortrows command:

```
» meteo = sortrows(meteo);
```

2.1.2.4 R

Let us consider the `meteo` data frame created in 2.1.1.4. Every data column can be extracted from this data frame using its name followed by the column name with the "$" symbol in between. Thus:

```
> meteo$PMax
```

lists the values of the `PMax` column. We may then proceed as follows:

```
PClass <- 1 + (meteo$PMax>20) + (meteo$PMax>80)
```

creating a vector for the needed new variable. The only thing remaining to be done is to bind this new vector to the data frame, as follows:

```
> meteo <- cbind(meteo,PClass)
> meteo
    PMax RainDays T80 T81 T82 PClass
1    181      143  36  39  37      3
2    114      132  35  39  36      3
...
```

One can get rid of the clumsy $-notation to qualify data frame variables by using the `attach` command:

```
> attach(meteo)
```

In this way variable names always respect to the attached data frame. From now on we will always assume that an attach operation has been performed. (Whenever needed one may undo it with `detach.`)

Indexing data frames is straightforward. One just needs to specify the indices between square brackets. Some examples: `meteo[2,5]` and `T82[2]` mean the same thing: the value of T82, 36, for the second row (case); `meteo[2,]` is the whole second row; `meteo[3:5,2]` is the sub-vector containing the RainDays values for the cases 3 through 5, i.e., 125, 111 and 102.

Sometimes one may need to transpose a data frame. R provides the `t` ("transpose") function to do that:

```
> meteo <- t(meteo)
> meteo
                1   2   3   4   5   6   7   8    9  10 11 12
13 14 15 16 17 18 19 20 21 22 23 24 25
    PMax      181 114 101  80  36  24  39  31   49  57 72 60
36 45 36 28 41 13 14 16  8 18 24 37 14
    RainDays 143 132 125 111 102  98  96 109  102 104 95 85
92 90 83 81 79 77 75 80 72 72 71 71 70
    T80        36  35  36  34  37  40  37  41   38  32 36 39
36 40 37 37 38 40 37 39 39 41 38 38 35
    ...
```

Sorting a vector can be performed with the function `sort`. One often needs to sort data frame variables according to a certain ordering of one or more of its variables. Imagine that one wanted to get the sorted list of the maximum precipitation variable, `PMax`, of the `meteo` data frame. The procedure to follow for this purpose is to first use the `order` function:

```
> order(PMax)
 [1] 21 18 19 25 20 22  6 23 16  8  5 13 15 24  7 17
14  9 10 12 11  4  3  2  1
```

The `order` function supplies a permutation list of the indices corresponding to an increasing order of its argument(s). In the above example the 21st element of the `PMax` variable is the smallest one, followed by the 18th element and so on up to the 1st element which is the largest. One may obtain a decreasing order sort and store the permutation list as follows:

```
> o <- order(PMax, decreasing=TRUE)
```

The permutation list can now be used to perform the sorting of `PMax` or any other variable of `meteo`:

```
> PMax[o]
  [1] 181 114 101  80  72  60  57  49  45  41  39  37
 36  36  36  31  28  24  24  18  16  14  14
 [24]  13   8
```

2.2 Presenting the Data

A general overview of the data in terms of the frequencies with which a certain interval of values occurs, both in tabular and in graphical form, is usually advisable as a preliminary step before proceeding to the computation of specific statistics and performing statistical analysis. As a matter of fact, one usually obtains some insight on what to compute and what to do with the data by first looking to frequency tables and graphs. For instance, if from the inspection of such a table and/or graph one gets a clear idea that an asymmetrical distribution is present, one may drop the intent of performing a normal distribution goodness-of-fit test.

After the initial familiarisation with the software products provided by the previous sections, the present and following sections will no longer split explanations by software product but instead they will include specific frames, headed by a "Commands" caption and ending with "■", where we present which commands (or functions in the MATLAB and R cases) to use in order to perform the explained statistical operations. The MATLAB functions listed in "Commands" are, except otherwise stated, from the MATLAB Base or Statistics Toolbox. The R functions are, except otherwise stated, from the R Base, Graphics or Stats packages. We also provide in the book CD many MATLAB and R *implemented* functions for specific tasks. They are listed in Appendix F and appear in italic in the "Commands" frames. SPSS and STATISTICA commands are described in terms of menu options separated by ";" in the "Commands" frames. In this case one may read "," as "followed by". For MATLAB and R functions ";" is simply a separator. Alternative menu options or functions are separated by "|".

In the following we also provide many examples illustrating the statistical analysis procedures. We assume that the datasets used throughout the examples are available as conveniently formatted data files (*.sav for SPSS, *.sta for STATISTICA, *.mat for MATLAB, files containing data frames for R). "Example" frames end with □.

2.2.1 Counts and Bar Graphs

Tables of counts and bar graphs are used to present discrete data. Denoting by X the discrete random variable associated to the data, the table of counts – also know as *tally sheet* – gives us:

- The *absolute frequencies* (*counts*), n_k;
- The *relative frequencies* (or simply, *frequencies*) of occurrence $f_k = n_k/n$,

for each *discrete value* (*category*), x_k, of the random variable X (n is the total number of cases).

Example 2.1

Q: Consider the Meteo dataset (see Appendix E). We assume that this data has been already read in by SPSS, STATISTICA, MATLAB or R. Obtain a tally sheet showing the counts of maximum precipitation categories (discrete variable PClass). What is the category with higher frequency?

A: The tally sheet can be obtained with the commands listed in Commands 2.1. Table 2.1 shows the results obtained with SPSS. The category with higher rate of occurrence is category 2 (64%). The Valid Percent column will differ from the Percent column, only in the case of missing data, with the Valid Percent removing the *missing data* from the computations.

Table 2.1. Frequency table for the discrete variable PClass, obtained with SPSS.

		Frequency	Percent	Valid Percent	Cumulative Percent
Valid	1.00	6	24.0	24.0	24.0
	2.00	16	64.0	64.0	88.0
	3.00	3	12.0	12.0	100.0
	Total	25	100.0	100.0	

In Table 2.1 the counts are shown in the column headed by `Frequency`, and the frequencies, given in percentage, are in the column headed by `Percent`. These last ones are unbiased and consistent point estimates of the corresponding probability values p_k. For more details see A.1 and the Appendix C.

Commands 2.1. SPSS, STATISTICA, MATLAB and R commands used to obtain frequency tables. For SPSS and STATISTICA the semicolon separates menu options that must be used in sequence.

SPSS	`Analyze; Descriptive Statistics; Frequencies`
STATISTICA	`Statistics; Basic Statistics and Tables; Descriptive Statistics; Frequency Tables`
MATLAB	`tabulate(x)`
R	`table(x); prop.table(x)`

When using SPSS or STATISTICA, one has to specify, in appropriate windows, the variables used in the statistical analysis. Figure 2.8 shows the windows used for that purpose in the present "Descriptive Statistics" case.

With SPSS the variable specification window pops up immediately after choosing `Frequencies` in the menu `Descriptive Statistics`. Using a select button that toggles between select (▶) and remove (◀), one can specify which variables to use in the analysis. The frequency table is outputted into the *output sheet*, which constitutes a session logbook, that can be saved (*.spo file) and opened at a later session. From the output sheet the frequency table can be copied into the clipboard in the usual way (e.g., using the CTRL+C keys) by first selecting it with the mouse (just click the mouse left button over the table).

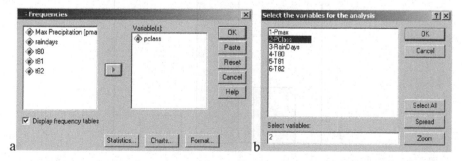

Figure 2.8. Variable specification windows for descriptive statistics: a) SPSS; b) STATISTICA.

With STATISTICA, the variable specification window pops up when clicking the `Variables` tab in the `Descriptive Statistics` window. One can select variables with the mouse or edit their identification numbers in a text box. For instance, editing "2-4", means that one wishes the analysis to be performed starting from variable `v2` up to variable `v4`. There is also a `Select All` variables button. The frequency table is outputted into a specific scroll-sheet that is part of a session *workbook* file, which constitutes a session logbook that can be saved (`*.stw` file) and opened at a later session. The entire scroll-sheet (or any part of the screen) can be copied to the clipboard (from where it can be pasted into a document in the normal way), using the `Screen Catcher` tool of the `Edit` menu. As an alternative, one can also copy the contents of the table alone in the normal way.

The MATLAB `tabulate` function computes a 3-column matrix, such that the first column contains the different values of the argument, the second column values are absolute frequencies (counts), and the third column are these frequencies in percentage. For the PClass example we have:

```
» t=tabulate(PClass)
t =
        1       6      24
        2      16      64
        3       3      12
```

Text output of MATLAB can be copied and pasted in the usual way.

The R table function – `table(PClass)` for the example – computes the counts. The function `prop.table(x)` computes proportions of each vector x element. In order to obtain the information of the above last column one should use `prop.table(table(PClass))`. Text output of the R console can be copied and pasted in the usual way. ∎

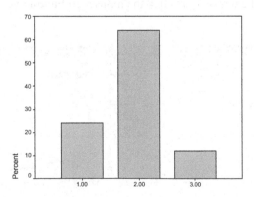

PCLASS

Figure 2.9. Bar graph, obtained with SPSS, representing the frequencies (in percentage values) of PClass.

With SPSS, STATISTICA, MATLAB and R one can also obtain a graphic representation of a tally sheet, which constitutes for the example at hand an estimate of the probability function of the associated random variable X_{PClass}, in the form of a *bar graph* (see Commands 2.2). Figure 2.9 shows the bar graph obtained with SPSS for Example 2.1. The heights of the bars represent estimates of the discrete probabilities (see Appendix B for examples of bar graph representations of discrete probability functions).

Commands 2.2. SPSS, STATISTICA, MATLAB and R commands used to obtain bar graphs. The "|" symbol separates alternative options or functions.

SPSS	`Graphs; Bar Charts`
STATISTICA	`Graphs; Histograms`
MATLAB	`bar(f)	hist(y,x)`
R	`barplot(x)	hist(x)`

With SPSS, after selecting the `Simple` option of `Bar Charts` one proceeds to choose the variable (or variables) to be represented graphically in the `Define Simple Bar` window by selecting it for the `Category Axis`, as shown in Figure 2.10. For the frequency bar graph one must check the "`% of cases`" option in this window. The graph output appears in the SPSS output sheet in the form of a resizable object, which can be copied (select it first with the mouse) and pasted in the usual way. By double clicking over this object, the `SPSS Chart Editor` pops up (see Figure 2.11), with many options for the user to tailor the graph to his/her personal preferences.

With STATISTICA one can obtain a bar graph using the `Histograms` option of the `Graphs` menu. A `2D Histograms` window pops up, where the user must specify the variable (or variables) to be represented graphically (using the `Variables` button), and, in this case, the `Regular` type for the bar graph. The user must also select the `Codes` option, and specify the codes for the variable categories (clicking in the respective button), as shown in Figure 2.12. In this case, the `Normal fit` box is left unchecked. Figure 2.13 shows the bar graph obtained with STATISTICA for the PClass variable.

Any graph in STATISTICA is a resizable object that can be copied (and pasted) in the usual way. One can also completely customise the graph by clicking over it and modifying the required specifications in the `All Options` window, shown in Figure 2.14. For instance, the bar graph of Figure 2.13 was obtained by: choosing the white background in the `Graph Window` sub-window; selecting black hatched fill in the `Plot Bars` sub-window; leaving the `Gridlines` box unchecked in the `Axis Major Units` sub-window (shown in Figure 2.14).

MATLAB has a routine for drawing histograms (to be described in the following section) that can also be used for obtaining bar graphs. The routine, hist(y,x), plots a bar graph of the y frequencies, using a vector x with the categories. For the PClass variable one would have to write down the following commands:

```
» cat=[1 2 3];          %vector with categories
» hist(pclass,cat)
```

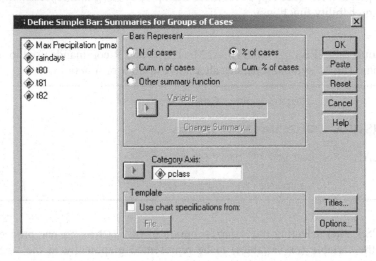

Figure 2.10. SPSS Define Simple Bar window, for specifying bar charts.

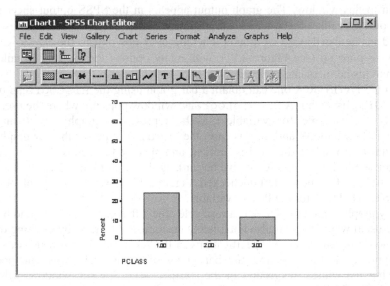

Figure 2.11. The SPSS Chart Editor, with which the user can configure the graphic output (in the present case, Figure 2.9). For instance, by using Color from the Format menu one can modify the bar colour.

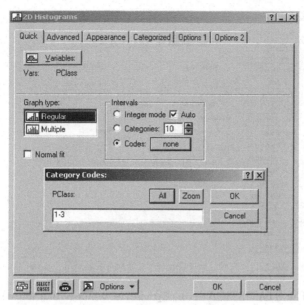

Figure 2.12. Specification of a bar chart for variable PClass (Example 2.1) using STATISTICA. The category codes can be filled in directly or by clicking the `All` button.

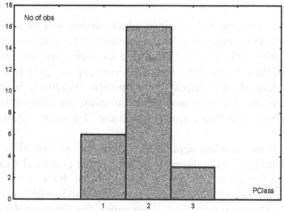

Figure 2.13. Bar graph, obtained with STATISTICA, representing the frequencies (counts) of variable PClass (Example 2.1).

If one has available the vector with the counts, it is then also possible to use the `bar` command. In the present case, after obtaining the previously mentioned `t` vector (see Commands 2.1), one would proceed to obtain the bar graph corresponding to column 3 of `t`, with:

```
» colormap([.5 .5 .5]); bar(t(:,3))
```

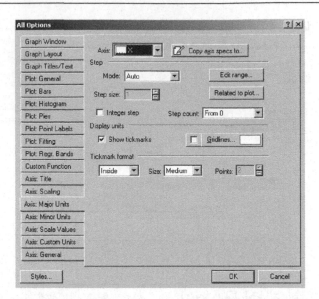

Figure 2.14. The STATISTICA All Options window that allows the user to completely customise the graphic output. This window has several sub-windows that can be opened with the left tabs. The sub-window corresponding to the axis units is shown.

The colormap command determines which colour will be used for the bars. Its argument is a vector containing the composition rates (between 0 and 1) of the red, green and blue colours. In the above example, as we are using equal composition of all the colours, the graph, therefore, appears grey in colour.

Figures in MATLAB are displayed in specific windows, as exemplified in Figure 2.15. They can be customised using the available options in Tools. The user can copy a resizable figure using the Copy Figure option of the Edit menu.

The R hist function when applied to a discrete variable plots its bar graph. Instead of providing graphical editing operations in the graphical window, as in the previous software products, R graphical functions have a whole series of configuration arguments. Figure 2.16a was obtained with hist(PClass, col="gray"). The argument col determines the filling colour of the bars. There are arguments for specifying shading lines, the border colour of the bars, the labels, and so on. For instance, Figure 2.16b was obtained with hist(PClass, density = 10, angle = 30, border = "black", col = "gray", labels = TRUE). From now on we assume that the reader will browse through the on-line help of the graphical functions in order to obtain the proper guidance on how to set argument values. Graphical plots in R can be copied as bitmaps or metafiles using menu options popped up with the mouse right button.

∎

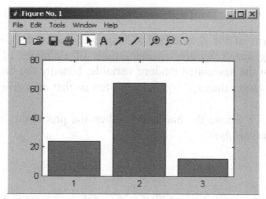

Figure 2.15. MATLAB figure window, containing the bar graph of PClass. The graph itself can be copied to the clipboard using the `Copy Figure` option of the `Edit` menu.

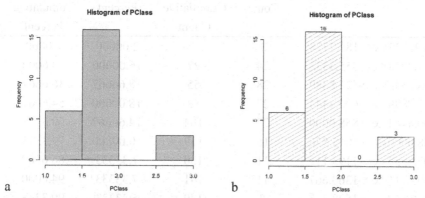

Figure 2.16. Bar graphs of PClass obtained with R: a) Using grey bars; b) Using dashed gray lines and count labels.

2.2.2 Frequencies and Histograms

Consider now a continuous variable. Instead of a tally sheet/bar graph, representing an estimate of a discrete probability function, we now want a tabular and graphical representation of an estimate of a probability density function. For this purpose, we establish a certain number of equal[2] length intervals of the random variable and compute the frequency of occurrence in each of these intervals (also known as *bins*). In practice, one determines the lowest, x_l, and highest, x_h, sample values and divides the range, $x_h - x_l$, into r equal length bins, h_k, $k = 1, 2, \ldots, r$. The computed frequencies are now:

---

[2] Unequal length intervals are seldom used.

$f_k = n_k/n$, where n_k is the number of sample values (observations) in bin h_k.

The tabular form of the f_k is called a *frequency table*; the graphical form is known as a *histogram*. They are representations of estimates of the probability density function of the associated random variable. Usually the histogram range is chosen somewhat larger than $x_h - x_l$, and adjusted so that convenient limits for the bins are obtained.

Let $d = (x_h - x_l)/r$ denote the bin length. Then the probability density estimate for each of the intervals h_k is:

$$\hat{p}_k = \frac{f_k}{d}$$

The areas of the h_k intervals are therefore f_k and they sum up to 1 as they should.

Table 2.2. Frequency table of the cork stopper PRT variable using 10 bins (table obtained with STATISTICA).

	Count	Cumulative Count	Percent	Cumulative Percent
20.22222<x<=187.7778	3	3	2.00000	2.0000
187.7778<x<=355.3333	24	27	16.00000	18.0000
355.3333<x<=522.8889	28	55	18.66667	36.6667
522.8889<x<=690.4444	27	82	18.00000	54.6667
690.4444<x<=858.0000	22	104	14.66667	69.3333
858.0000<x<=1025.556	15	119	10.00000	79.3333
1025.556<x<=1193.111	11	130	7.33333	86.6667
1193.111<x<=1360.667	11	141	7.33333	94.0000
1360.667<x<=1528.222	8	149	5.33333	99.3333
1528.222<x<=1695.778	1	150	0.66667	100.0000
Missing	0	150	0.00000	100.0000

Example 2.2

Q: Consider the variable PRT of the Cork Stoppers' dataset (see Appendix E). This variable measures the total perimeter of cork defects, and can be considered a continuous (ratio type) variable. Determine the frequency table and the histogram of this variable, using 10 and 6 bins, respectively.

A: The frequency table and histogram can be obtained with the commands listed in Commands 2.1 and Commands 2.3, respectively.

Table 2.2 shows the frequency table of PRT using 10 bins. Figure 2.17 shows the histogram of PRT, using 6 bins.

□

Let X denote the random variable associated to PRT. Then, the histogram of the frequency values represents an estimate, $\hat{f}_X(x)$, of the unknown probability density function $f_X(x)$.

The number of bins to use in a histogram (or in a frequency table) depends on its goodness of fit to the true density function $f_X(x)$, in terms of bias and variance. In order to clarify this issue, let us consider the histograms of PRT using $r = 3$ and $r = 50$ bins as shown in Figure 2.18. Consider in both cases the $\hat{f}_X(x)$ estimate represented by a polygonal line passing through the mid-point values of the histogram bars. Notice that in the first case ($r = 3$) the $\hat{f}_X(x)$ estimate is quite smooth and lacks detail, corresponding to a large bias of the expected value of $\hat{f}_X(x) - f_X(x)$; i.e., in average terms (for an ensemble of similar histograms associated to X) the histogram will give a point estimate of the density that can be quite far from the true density. In the second case ($r = 50$) the $\hat{f}_X(x)$ estimate is too rough; our polygonal line may pass quite near the true density values, but the $\hat{f}_X(x)$ values vary widely (large variance) around the $f_X(x)$ curve (corresponding to an average of a large number of such histograms).

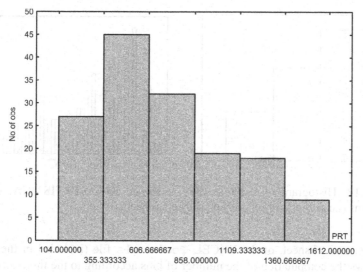

Figure 2.17. Histogram of variable PRT (cork stopper dataset) obtained with STATISTICA using $r = 6$ bins.

Some formulas for selecting a "reasonable" number of bins, r, achieving a trade-off between large bias and large variance, have been divulged in the literature, namely:

$r = 1 + 3.3 \log(n)$ (Sturges, 1926);   2.1

$r = 1 + 2.2 \log(n)$ (Larson, 1975).   2.2

The choice of an optimal value for r was studied by Scott (Scott DW, 1979), using as optimality criterion the minimisation of the global mean square error:

$$MSE = \int_D E[(\hat{f}_X(x) - f_X(x))^2] dx,$$

where D is the domain of the random variable.

The MSE minimisation leads to a formula for the optimal choice of a bin width, $h(n)$, which for the Gaussian density case is:

$$h(n) = 3.49 s n^{-1/3},$$ 2.3

where s is the sample standard deviation of the data.

Although the $h(n)$ formula was derived for the Gaussian density case, it was experimentally verified to work well for other densities too. With this $h(n)$ one can compute the optimal number of bins using the data range:

$$r = (x_h - x_l)/h(n).$$ 2.4

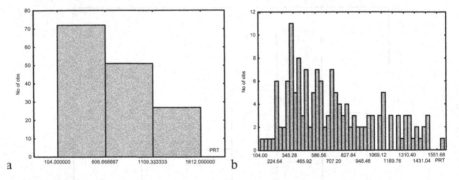

a   b

Figure 2.18. Histogram of variable PRT, obtained with STATISTICA, using: a) $r = 3$ bins (large bias); b) $r = 50$ bins (large variance).

The `Bins` worksheet, of the EXCEL `Tools.xls` file (included in the book CD), allows the computation of the number of bins according to the three formulas 2.1, 2.2 and 2.4. In the case of the PRT variable, we obtain the results of Table 2.3, legitimising the use of 6 bins as in Figure 2.17.

Table 2.3. Recommended number of bins for the PRT data ($n = 150$ cases, $s = 361$, range = 1508).

Formula	Number of Bins
Sturges	8
Larson	6
Scott	6

Commands 2.3. SPSS, STATISTICA, MATLAB and R commands used to obtain histograms.

SPSS	`Graphs; Histogram |Interactive; Histogram`
STATISTICA	`Graphs; Histograms`
MATLAB	`hist(y,x)`
R	`hist(x)`

The commands used to obtain histograms of continuous type data, are similar to the ones already described in Commands 2.2.

In order to obtain a histogram with SPSS, one can use the `Histogram` option of `Graphs`, or preferably, use the sequence of commands `Graphs; Interactive; Histogram`. One can then select the appropriate number of bins, or alternatively, set the bin width. It is also possible to choose the starting point of the bins.

With STATISTICA, one simply defines the bins in appropriate windows as previously mentioned. Besides setting the desired number of bins, there is instead also the possibility of defining the bin width (`Step size`) and the starting point of the bins.

With MATLAB one obtains both the frequencies and the histogram with the `hist` command. Consider the following commands applied to the cork stopper data stored in the MATLAB `cork` matrix:

```
» prt = cork(:,4)
» [f,x] = hist(prt,6);
```

In this case the `hist` command generates an `f` vector containing the frequencies counted in 6 bins and an `x` vector containing the bin locations. Listing the values of `f` one gets:

```
» f
f =
    27    45    32    19    18    9    ,
```

which are precisely the values shown in Figure 2.17. One can also use the `hist` command with specifications of bins stored in a vector b, as `hist(prt, b)`.

With R one can use the `hist` function either for obtaining a histogram or for obtaining a frequency list. The frequency list is obtained by assigning the outcome of the function to a variable identifier, which then becomes a "histogram" object. Assuming that a data frame has been created (and attached) for cork stoppers we get a "histogram" object for PRT issuing the following command:

```
> h <- hist(PRT)
```

By listing the contents of h one gets among other things the information of the break points of the histogram bins, the counts and the densities. The densities represent the probability density estimate for a given bin. We can list de densities of PRT as follows:

```
> h$density
[1] 1.333333e-04 1.033333e-03 1.166667e-03
[4] 9.666667e-04 5.666667e-04 4.666667e-04
[7] 4.333333e-04 2.000000e-04 3.333333e-05
```

Thus, using the formula previously mentioned for the probability density estimates, we compute the relative frequencies using the bin length (200 in our case) as follows:

```
> h$density*200
[1] 0.026666661 0.206666667 0.233333333 0.193333333
[5] 0.113333333 0.093333333 0.086666667 0.040000000
[9] 0.006666667
```

■

2.2.3 Multivariate Tables, Scatter Plots and 3D Plots

Multivariate tables display the frequencies of multivariate data. Figure 2.19 shows the format of a bivariate table displaying the counts n_{ij} corresponding to the several combinations of categories of two random variables. Such a bivariate table is called a *cross table* or *contingency table*.

When dealing with continuous variables, one can also build cross tables using categories in accordance to the bins that would be assigned to a histogram representation of the variables.

	x_1	x_2	$\cdots$	x_c	
y_1	n_{11}	n_{12}	$\cdots$	n_{1c}	r_1
y_2	n_{21}	n_{22}	$\cdots$	n_{2c}	r_2
$\vdots$	$\cdots$	$\cdots$	$\cdots$	$\cdots$	$\vdots$
y_r	n_{r1}	n_{r2}	$\cdots$	n_{rc}	r_r
	c_1	c_2	$\cdots$	c_c	

Figure 2.19. An $r \times c$ contingency table with the observed absolute frequencies (counts n_{ij}). The row and column totals are r_i and c_j, respectively.

Example 2.3

Q: Consider the variables SEX and Q4 (4th enquiry question) of the Freshmen dataset (see Appendix E). Determine the cross table for these two categorical variables.

A: The cross table can be obtained with the commands listed in Commands 2.4. Table 2.4 shows the counts and frequencies for each pair of values of the two categorical variables. Note that the variable Q4 can be considered an ordinal variable if we assign ordered scores, e.g. from 1 till 5, from "fully disagree" through "fully agree", respectively.

A cross table is an estimate of the respective bivariate probability or density function. Notice the total percentages across columns (last row in Table 2.4) and across rows (last column in Table 2.4), which are estimates of the respective marginal probability functions (see section A.8.1).

□

Table 2.4. Cross table (obtained with SPSS) of variables SEX and Q4 of the Freshmen dataset.

			Fully disagree	Disagree	Q4 No comment	Agree	Fully agree	Total
SEX	male	Count	3	8	18	37	31	97
		% of Total	2.3%	6.1%	13.6%	28.0%	23.5%	73.5%
	female	Count	1	2	4	13	15	35
		% of Total	.8%	1.5%	3.0%	9.8%	11.4%	26.5%
Total		Count	4	10	22	50	46	132
		% of Total	3.0%	7.6%	16.7%	37.9%	34.8%	100.0%

Table 2.5. Trivariate cross table (obtained with SPSS) of variables SEX, LIKE and DISPL of the Freshmen dataset.

DISPL				LIKE like	dislike	no comment	Total
yes	SEX	male	Count	25			25
			% of Total	67.6%			67.6%
		female	Count	10		2	12
			% of Total	27.0%		5.4%	32.4%
	Total		Count	35		2	37
			% of Total	94.6%		5.4%	100.0%
no	SEX	male	Count	64	1	6	71
			% of Total	68.1%	1.1%	6.4%	75.5%
		female	Count	21		2	23
			% of Total	22.3%		2.1%	24.5%
	Total		Count	85	1	8	94
			% of Total	90.4%	1.1%	8.5%	100.0%

Example 2.4

Q: Determine the trivariate table for the variables SEX, LIKE and DISPL of the Freshmen dataset.

A: In order to represent cross tables for more than two variables, one builds sub-tables for each value of one of the variables in excess of 2, as illustrated in Table 2.5.   □

Commands 2.4. SPSS, STATISTICA, MATLAB and R commands used to obtain cross tables.

SPSS	Analyze; Descriptive Statistics; Crosstabs
STATISTICA	Statistics; Basic Statistics and Tables; Descriptive Statistics; (Tables and banners | Multiple Response Tables)
MATLAB	crosstab(x,y)
R	table(x,y) | xtabs(~x+y)

The MATLAB function crosstab and the R functions table and xtabs generate cross-tabulations of the variables passed as arguments. Supposing that the dataset Freshmen has been read into the R data frame freshmen, one would obtain Table 2.4 as follows (the ## symbol denotes an R user comment):

```
> attach(freshmen)
> table(SEX,Q4)          ## or xtabs(~SEX+Q4)
     Q4
SEX   1  2  3  4  5
   1  3  8 18 37 31
   2  1  2  4 13 15
```
■

Commands 2.5. SPSS, STATISTICA, MATLAB and R commands used to obtain scatter plots and 3D plots.

SPSS	Graphs; Scatter; Simple Graphs; Scatter; 3-D
STATISTICA	Graphs; Scatterplots Graphs; 3D XYZ Graphs; Scatterplots
MATLAB	scatter(x,y,s,c) scatter3(x,y,z,s,c)
R	plot.default(x,y)

The s, c arguments of MATLAB scatter and scatter3 are the size and colour of the marks, respectively.

The plot.default function is the x-y scatter plot function of R and has several configuration parameters available (colours, type of marks, etc.). The R Graphics package has no 3D plot available.   ∎

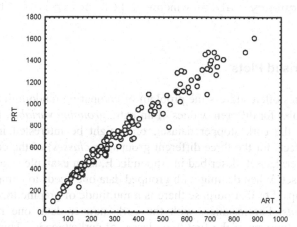

Figure 2.20. Scatter plot (obtained with STATISTICA) of the variables ART and PRT of the cork stopper dataset.

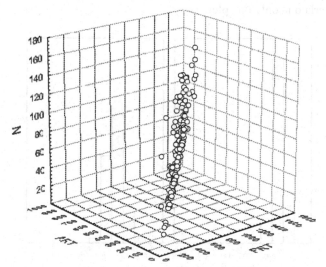

Figure 2.21. 3D plot (obtained with STATISTICA) of the variables ART, PRT and N of the cork stopper dataset.

The most popular graphical tools for multivariate data are the *scatter plots* for bivariate data and the *3D plots* for trivariate data. Examples of these plots, for the cork stopper data, are shown in Figures 2.20 and 2.21. As a matter of fact, the 3D plot is often not so easy to interpret (as in Figure 2.21); therefore, in normal practice, one often inspects multivariate data graphically through scatter plots of the variables grouped in pairs.

Besides scatter plots and 3D plots, it may be convenient to inspect bivariate histograms or bar plots (such as the one shown in Figure A.1, Appendix A). STATISTICA affords the possibility of obtaining such bivariate histograms from within the Frequency Tables window of the Descriptive Statistics menu.

2.2.4 Categorised Plots

Statistical studies often address the problem of comparing random distributions of the same variables for different values of an extra *grouping variable*. For instance, in the case of the cork stopper dataset, one might be interested in comparing numbers of defects for the three different groups (or *classes*) of the cork stoppers. The cork stopper dataset, described in Appendix E, is an example of a *grouped* (or *classified*) dataset. When dealing with grouped data one needs to compare the data across the groups. For that purpose there is a multitude of graphic tools, known as *categorised plots*. For instance, with the cork stopper data, one may wish to compare the histograms of the first two classes of cork stoppers. This comparison is shown as a categorised histogram plot in Figure 2.22, for the variable ART. Instead of displaying the individual histograms, it is also possible to display all histograms overlaid in only one plot.

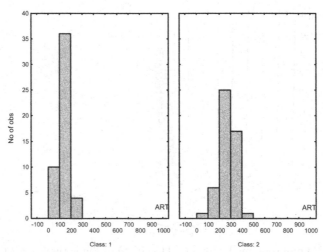

Figure 2.22. Categorised histogram plot obtained with STATISTICA for variable ART and the first two classes of cork stoppers.

When the number of groups is high, the visual comparison of the histograms may be rather difficult. The situation usually worsens if one uses overlaid histograms. A better alternative to comparing data distributions for several groups is to use the so-called *box plot* (or *box-and-whiskers plot*). As illustrated in Figure 2.23, a box plot uses a distinct rectangular box for each group, where each box corresponds to the central 50% of the cases, the so-called *inter-quartile range* (IQR). A central mark or line inside the box indicates the *median*, i.e., the value below which 50% of the cases are included. The boxes are prolonged with lines (whiskers) covering the range of the non-outlier cases, i.e., cases that do not exceed, by a certain factor of the IQR, the above or below box limits. A usual IQR factor for outliers is 1.5. Sometimes box plots also indicate, with an appropriate mark, the extreme cases, similarly defined as the outliers, but using a larger IQR factor, usually 3. As an alternative to using the central 50% range of the cases around the median, one can also use the mean ± standard deviation.

There is also the possibility of obtaining categorised scatter plots or categorised 3D plots. Their real usefulness is however questionable.

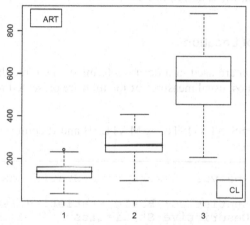

Figure 2.23. Box plot of variable ART, obtained with R, for the three classes of the cork stoppers data. The "o" sign for Class 1 indicates an outlier, i.e., a case exceeding the top of the box by more than 1.5×IQR.

Commands 2.6. SPSS, STATISTICA, MATLAB and R commands used to obtain box plots.

SPSS	`Graphs; Boxplot`
STATISTICA	`Graphs; 2D Graphs; Boxplots`
MATLAB	`boxplot(x)`
R	`boxplot(x~y); legend(x,y,label)`

The R boxplot function uses the so-called x~y "formula" to create a box plot of x grouped by y. The legend function places label as a legend at the (x,y) position of the plot. The graph of Figure 2.23 (CL is the Class variable) was obtained with:

```
> boxplot(ART~CL)
> legend(3.2,100,legend="CL")
> legend(0.5,900,legend="ART")
```

2.3   Summarising the Data

When analysing a dataset, one usually starts by determining some indices that give a global picture on where and how the data is concentrated and what is the shape of its distribution, i.e., indices that are useful for the purpose of summarising the data. These indices are known as *descriptive statistics*.

2.3.1 Measures of Location

Measures of *location* are used in order to determine where the data distribution is concentrated. The most usual measures of location are presented next.

Commands 2.7. SPSS, STATISTICA, MATLAB and R commands used to obtain measures of location.

SPSS	`Analyze; Descriptive Statistics`
STATISTICA	`Statistics; Basic Statistics/Tables; Descriptive Statistics`
MATLAB	`mean(x) ; trimmean(x,p) ; median(x); prctile(x,p)`
R	`mean(x, trim) ; median(x); summary(x); quantile(x,seq(...))`

2.3.1.1   Arithmetic Mean

Let $x_1, \ldots, x_n$ be the data. The arithmetic mean (or simply *mean*) is:

$$\bar{x} = \frac{1}{n}\sum_{i=1}^{n} x_i \ .$$

$$2.5$$

The arithmetic mean is the sample estimate of the mean of the associated random variable (see Appendices B and C). If one has a tally sheet of a discrete type data, one can also compute the mean using the absolute frequencies (counts), n_k, of each distinct value x_k:

$$\bar{x} = \frac{1}{n}\sum_{k=1}^{n} n_k x_k \quad \text{with} \quad n = \sum_{k=1}^{n} n_k \,. \tag{2.6}$$

If one has a frequency table of a continuous type data (also known in some literature as grouped data), with r bins, one can obtain an estimate of $\bar{x}$, using the frequencies f_j of the bins and the mid-bin values, $\dot{x}_j$, as follows:

$$\hat{\bar{x}} = \frac{1}{r}\sum_{j=1}^{n} f_j \dot{x}_j \,. \tag{2.7}$$

This mean estimate used to be presented as an expedite way of calculating the arithmetic mean for long tables of data. With the advent of statistical software the interest of such a method is at least questionable. We will proceed no further with such a "grouped data" approach.

Sometimes, when in presence of datasets exhibiting outliers and extreme cases (see 2.2.4) that can be suspected to be the result of rough measurement errors, one can use a *trimmed mean* by neglecting a certain percentage of the tail cases (e.g., 5%).

The arithmetic mean is a point estimate of the expected value (true mean) of the random variable associated to the data and has the same properties as the true mean (see A.6.1). Note that the expected value can be interpreted as the center of gravity of a weightless rod with probability mass-points, in the case of discrete variables, or of a rod whose mass-density corresponds to the probability density function, in the case of continuous variables.

2.3.1.2 Median

The median of a dataset is that value of the data below which lie 50% of the cases. It is an estimate of the median, med(X), of the random variable, X, associated to the data, defined as:

$$F_X(x) = \frac{1}{2} \quad \Rightarrow \quad \text{med}(X)\,, \tag{2.8}$$

where $F_X(x)$ is the distribution function of X.

Note that, using the previous rod analogy for the continuous variable case, the median divides the rod into equal mass halves corresponding to equal areas under the density curve:

$$\int_{-\infty}^{\text{med}(X)} f_X(x) = \int_{\text{med}(X)}^{\infty} f_X(x) = \frac{1}{2}\,.$$

The median satisfies the same linear property as the mean (see A.6.1), but not the other properties (e.g. additivity). Compared to the mean, the median has the advantage of being quite insensitive to outliers and extreme cases.

Notice that, if we sort the dataset, the sample median is the central value if the number of the data values is odd; if it is even, it is computed as the average of the two most central values.

2.3.1.3 Quantiles

The quantile of order α $(0 < \alpha < 1)$ of a random variable distribution $F_X(x)$ is defined as the root of the equation (see A.5.2):

$$F_X(x) = \alpha .\qquad\qquad 2.9$$

We denote the root as: x_α.

Likewise we compute the quantile of order α of a dataset as the value below which lies a percentage α of cases of the dataset. The median is therefore the 50% quantile, or $x_{0.5}$. Often used quantiles are:

–   *Quartiles*, corresponding to multiples of 25% of the cases. The box plot mentioned in 2.2.4 uses the quartiles and the inter-quartile range (IQR) in order to determine the outliers of the dataset distribution.

–   *Deciles*, corresponding to multiples of 10% of the cases.

–   *Percentiles*, corresponding to multiples of 1% of the cases. We will often use the percentile $p = 2.5\%$ and its complement $p = 97.5\%$.

2.3.1.4 Mode

The mode of a dataset is its maximum value. It is an estimate of the probability or density function maximum.

For continuous type data one should determine the midpoint of the modal bin of the data grouped into an appropriate number of bins.

When a data distribution exhibits several relative maxima of almost equal value, we say that it is a *multimodal* distribution.

Example 2.5

Q: Consider the Cork Stoppers' dataset. Determine the measures of location of the variable PRT. Comment the results. Imagine that we had a new variable, PRT1, obtained by the following linear transformation of PRT: PRT1 = 0.2 PRT + 5. Determine the mean and median of PRT1.

A: Table 2.6 shows some measures of location of the variable PRT. Notice that as a mode estimate we can use the midpoint of the bin [355.3 606.7] as shown in Figure 2.17, i.e., 481. Notice also the values of the lower and upper quartiles delimiting 50% of the cases. The large deviation of the 95% percentile from the upper quartile, when compared to the deviation of the 5% percentile from the lower quartile, is evidence of a right skewed asymmetrical distribution.

By the linear properties of the mean and the median, we have:

Mean(PRT1)   = 0.2 Mean(PRT) + 5   = 147;
Median(PRT1) = 0.2 Median(PRT) + 5 = 131.

                                                                          ▯

Table 2.6. Location measures (computed with STATISTICA) for variable PRT of the cork stopper dataset (150 cases).

Mean	Median	Lower Quartile	Upper Quartile	Percentile 5%	Percentile 95%
710.3867	629.0000	410.0000	974.0000	246.0000	1400.000

An important aspect to be considered, when using values computed with statistical software, is the precision of the results expressed by the number of *significant digits*. Almost every software product will produce results with a large number of digits, independent of whether or not they mean something. For instance, in the case of the PRT variable (Table 2.6) it would be foolish to publish that the mean of the total perimeter of the defects of the cork stoppers is 710.3867. First of all, the least significant digit is, in this case, the unit (no perimeter can be measured in fractions of the pixel unit; see Appendix E). Thus, one would have to publish a value rounded up to the units, in this case 710. Second, there are omnipresent measurement errors that must be accounted for. Assuming that the perimeter measurement error is of one unit, then the mean is 710 ± 1[3]. As a matter of fact, even this one unit precision for the mean is somewhat misleading, as we will see in the following chapter. From now on the published results will take this issue into consideration and may, therefore, appropriately round the results obtained with the software products.

The R functions also provide a large number of digits, as when calculating the mean of PRT:

```
> mean(PRT)
[1] 710.3867
```

However, the summary function provides a reasonable rounding:

```
> summary(PRT)
   Min. 1st Qu.  Median    Mean 3rd Qu.    Max.
  104.0   412.0   629.0   710.4   968.5  1612.0
```

---

[3] Denoting by Δx a single data measurement error, the mean of n measurements has an error of $\pm(n.\mathrm{abs}(\Delta x))/n = \pm\Delta x$ in the worst case.

2.3.2 Measures of Spread

The measures of *spread* (or *dispersion*) give an indication of how concentrated a data distribution is. The most usual measures of spread are presented next.

Commands 2.8. SPSS, STATISTICA, MATLAB and R commands used to obtain measures of spread and shape.

SPSS	`Analyze; Descriptive Statistics`
STATISTICA	`Statistics; Basic Statistics/Tables; Descriptive Statistics`
MATLAB	`iqr(x) ;| range(x) ; std(x) ; var(x) ; skewness(x) ; kurtosis(x)`
R	`IQR(x) ; range(x) | sd(x) | var(x)| skewness(x) ; kurtosis(x)`

■

2.3.2.1 Range

The range of a dataset is the difference between its maximum and its minimum, i.e.:

$$R = x_{max} - x_{min}.$$

(2.10)

The basic disadvantage of using the range as measure of spread is that it is dependent on the extreme cases of the dataset. It also tends to increase with the sample size, which is an additional disadvantage.

2.3.2.2 Inter-quartile range

The inter-quartile range is defined as (see also section 2.2.4):

$$IQR = x_{0.75} - x_{0.25}.$$

(2.11)

The IQR is less influenced than the range by outliers and extreme cases. It tends also to be less influenced by the sample size (and can either increase or decrease).

2.3.2.3 Variance

The variance of a dataset $x_1, \dots, x_n$ (sample variance) is defined as:

$$v = \sum_{i=1}^{n} (x_i - \bar{x})^2 /(n-1).$$

(2.12)

The sample variance is the point estimate of the associated random variable variance (see Appendices B and C). It can be interpreted as the mean square deviation (or *mean square error*, MSE) of the sample values from their mean. The use of the $n - 1$ factor, instead of n as in the usual computation of a mean, is explained in C.2. Notice also that given $\bar{x}$, only $n - 1$ cases can vary independently in order to achieve the same variance. We say that the variance has $df = n - 1$ *degrees of freedom*. The mean, on the other hand, has n degrees of freedom.

2.3.2.4 Standard Deviation

The standard deviation of a dataset is the root square of its variance. It is, therefore, a *root mean square error* (RMSE):

$$s = \sqrt{v} = [\sum\nolimits_{i=1}^{n}(x_i - \bar{x})^2 /(n-1)]^{1/2} .$$ (2.13)

The standard deviation is preferable than the variance as a measure of spread, since it is expressed in the same units as the original data. Furthermore, many interesting results about the spread of a distribution are expressed in terms of the standard deviation. For instance, for any random variable X, the Chebyshev Theorem tall us that (see A.6.3):

$$P(|X - \mu| > k\sigma) \le \frac{1}{k^2} .$$

Using s as point estimate of σ, we can then expect that for any dataset distribution at least 75 % of the cases lie within 2 standard deviations of the mean.

Example 2.6

Q: Consider the Cork Stoppers' dataset. Determine the measures of spread of the variable PRT. Imagine that we had a new variable, PRT1, obtained by the following linear transformation of PRT: PRT1 = 0.2 PRT + 5. Determine the variance of PRT1.

A: Table 2.7 shows measures of spread of the variable PRT. The sample variance enjoys the same linear transformation property as the true variance (see A.6.1). For the PRT1 variable we have:

variance(PRT1) = $(0.2)^2$ variance(PRT) = 5219.

Note that the addition of a constant to PRT (i.e., a scale translation) has no effect on the variance.

□

Table 2.7. Spread measures (computed with STATISTICA) for variable PRT of the cork stopper dataset (150 cases).

Range	Inter-quartile range	Variance	Standard Deviation
1508	564	130477	361

2.3.3 Measures of Shape

The most popular measures of shape, exemplified for the PRT variable of the Cork Stoppers' dataset (see Table 2.8), are presented next.

2.3.3.1 Skewness

A continuous symmetrical distribution around the mean, μ, is defined as a distribution satisfying:

$$f_X(\mu + x) = f_X(\mu - x).$$

This applies similarly for discrete distributions, substituting the density function by the probability function.

A useful asymmetry measure around the mean is the *coefficient of skewness*, defined as:

$$\gamma = \mathrm{E}[(X - \mu)^3]/\sigma^3.$$
                                                                                2.14

This measure uses the fact that any central moment of odd order is zero for symmetrical distributions around the mean. For asymmetrical distributions γ reflects the unbalance of the density or probability values around the mean. The formula uses a σ^3 standardization factor, ensuring that the same value is obtained for the same unbalance, independently of the spread. Distributions that are skewed to the right (*positively skewed distributions*) tend to produce a positive value of γ, since the longer rightward tail will positively dominate the third order central moment; distributions skewed to the left (*negatively skewed distributions*) tend to produce a negative value of γ, since the longer leftward tail will negatively dominate the third order central moment (see Figure 2.24). The coefficient γ, however, has to be interpreted with caution, since it may produce a false impression of symmetry (or asymmetry) for some distributions. For instance, the probability function $p_k = \{0.1, 0.15, 0.4, 0.35\}$, $k = \{1, 2, 3, 4\}$, has $\gamma = 0$, although it is an asymmetrical distribution.

The skewness of a dataset $x_1, \ldots, x_n$ is the point estimate of γ, defined as:

$$g = n\sum_{i=1}^{n}(x_i - \bar{x})^3 /[(n-1)(n-2)s^3].$$
                                                                                2.15

Note that:

- For symmetrical distributions, if the mean exists, it will coincide with the median. Based on this property, one can also measure the skewness using g = (mean – median)/(standard deviation). It can be proved that $-1 \leq g \leq 1$.

- For asymmetrical distributions, with only one maximum (which is then the mode), the median is between the mode and the mean as shown in Figure 2.24.

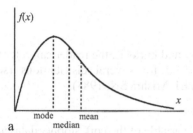
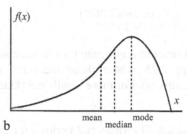

a   b

Figure 2.24. Two asymmetrical distributions: a) Skewed to the right (usually with $\gamma > 0$); b) Skewed to the left (usually with $\gamma < 0$).

2.3.3.2 Kurtosis

The degree of flatness of a probability or density function near its center, can be characterised by the so-called *kurtosis*, defined as:

$$\kappa = E[(X - \mu)^4]/\sigma^4 - 3.$$ 2.16

The factor 3 is introduced in order that $\kappa = 0$ for the normal distribution. As a matter of fact, the κ measure as it stands in formula 2.16, is often called *coefficient of excess* (excess compared to the normal distribution). Distributions flatter than the normal distribution have $\kappa < 0$; distributions more peaked than the normal distribution have $\kappa > 0$.

The sample estimate of the kurtosis is computed as:

$$k = [n(n+1)M_4 - 3(n-1)M_2^2]/[(n-1)(n-2)(n-3)s^4],$$ 2.17

with: $M_j = \sum_{i=1}^{n}(x_i - \bar{x})^j$.

Note that the kurtosis measure has the same shortcomings as the skewness measure. It does not always measure what it is supposed to.

The skewness and the kurtosis have been computed for the PRT variable of the Cork Stoppers' dataset as shown in Table 2.8. The PRT variable exhibits a positive skewness indicative of a rightward skewed distribution and a positive kurtosis indicative of a distribution more peaked than the normal one.

There are no functions in the R stats package to compute the skewness and kurtosis. We provide, however, as stated in Commands 2.8, R functions for that purpose in text file format in the book CD (see Appendix F). The only thing to be done is to copy the function text from the file and paste it in the R console, as in the following example:

```
> skewness <- function(x){
+ n <- length(x)
+ y <- (x-mean(x))^3
+ n*sum(y)/((n-1)*(n-2)*sd(x)^3)
+ }
> skewness(PRT)
[1] 0.592342
```

In order to appreciate the obtained skewness and kurtosis, the reader can refer to Figure 2.25 where these measures are plotted for several distributions (see Appendix B). For more details see (Dudewicz EJ, Mishra SN, 1988).

Table 2.8. Skewness and kurtosis for the PRT variable of the cork stopper dataset.

Skewness	Kurtosis
0.59	−0.63

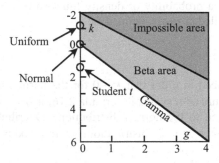

Figure 2.25. Skewness and kurtosis coefficients for several distributions.

2.3.4 Measures of Association for Continuous Variables

The *correlation coefficient* is the most popular measure of association for continuous type data. For a dataset with two variables, X and Y, the sample estimate of the correlation coefficient ρ_{XY} (see definition in A.8.2) is computed as:

$$r \equiv r_{XY} = \frac{s_{XY}}{s_X s_Y}, \qquad\qquad 2.18$$

where s_{XY}, the sample covariance of X and Y, is computed as:

$$s_{XY} = \sum_{i=1}^{n} (x_i - \bar{x})(y_i - \bar{y}) / (n-1).$$ 2.19

Note that the correlation coefficient (also known as *Pearson correlation*) is a dimensionless measure of the degree of linear association of two r.v., with value in the interval $[-1, 1]$, with:

0 :   No linear association (X and Y are linearly uncorrelated);
1 :   Total linear association, with X and Y varying in the same direction;
−1:   Total linear association, with X and Y varying in the opposite direction.

Figure 2.26 shows scatter plots exemplifying several situations of correlation. Figure 2.26f illustrates a situation where, although there is an evident association between X and Y, the correlation coefficient fails to measure it since X and Y are not *linearly* associated.

Note that, as described in Appendix A (section A.8.2), adding a constant or multiplying by a constant any or both variables does not change the magnitude of the correlation coefficient. Only a change of sign can occur if one of the multiplying constants is negative.

The correlation coefficients can be arranged, in general, into a symmetrical *correlation matrix*, where each element is the correlation coefficient of the respective column and row variables.

Table 2.9. Correlation matrix of five variables of the cork stopper dataset.

	N	ART	PRT	ARTG	PRTG
N	1.00	0.80	0.89	0.68	0.72
ART	0.80	1.00	0.98	0.96	0.97
PRT	0.89	0.98	1.00	0.91	0.93
ARTG	0.68	0.96	0.91	1.00	0.99
PRTG	0.72	0.97	0.93	0.99	1.00

Example 2.7

Q: Compute the correlation matrix of the following five variables of the Cork Stoppers' dataset: N, ART, PRT, ARTG, PRTG.

A: Table 2.9 shows the (symmetric) correlation matrix corresponding to the five variables of the cork stopper dataset (see Commands 2.9). Notice that the main diagonal elements (from the upper left corner to the right lower corner) are all equal to one. In a later chapter, we will learn how to correctly interpret the correlation values displayed.

□

In multivariate problems, concerning datasets described by n random variables, $X_1, X_2, \ldots, X_n$, one sometimes needs to assess what is the degree of association of two variables, say X_1 and X_2, under the hypothesis that they are linearly estimated by the remaining $n-2$ variables. For this purpose, the correlation $\rho_{X_1 X_2}$ is defined in terms of the marginal distributions of X_1 or X_2 given the other variables, and is then called the *partial correlation* of X_1 and X_2 given the other variables. Details on partial correlations will be postponed to Chapter 7.

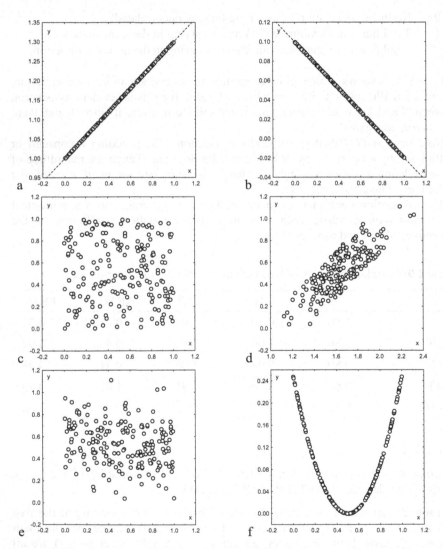

Figure 2.26. Sample correlation values for different datasets: a) $r = 1$; b) $r = -1$; c) $r = 0$; d) $r = 0.81$; e) $r = -0.21$; f) $r = 0.04$.

STATISTICA and SPSS afford the possibility of computing partial correlations as indicated in Commands 2.9. For the previous example, the partial correlation of PRTG and ARTG, given PRT and ART, is 0.79. We see, therefore, that PRT and ART can "explain" about 20% of the high correlation (0.99) of those two variables.

Another measure of association for continuous variables is the *multiple correlation* coefficient, which measures the degree of association of one variable Y in relation to a set of variables, $X_1, X_2, \ldots, X_n$, that linearly "predict" Y. Details on multiple correlation will be postponed to Chapter 7.

Commands 2.9. SPSS, STATISTICA, MATLAB and R commands used to obtain measures of association for continuous variables.

SPSS	`Analyze; Correlate; Bivariate	Partial`
STATISTICA	`Statistics; Basic Statistics/Tables; Correlation matrices (Quick |Advanced; Partial Correlations)`
MATLAB	`corrcoef(x) ; cov(x)`
R	`cor(x,y) ; cov(x,y)`

Partial correlations are computed in MATLAB and R as part of the regression functions (see Chapter 7). ∎

2.3.5 Measures of Association for Ordinal Variables

2.3.5.1 The Spearman Rank Correlation

When dealing with ordinal data the correlation coefficient, previously described, can be computed in a simplified way. Consider the ordinal variables X and Y with ranks between 1 and N. It seems natural to measure the lack of agreement between X and Y by means of the difference of the ranks $d_i = x_i - y_i$ for each data pair (x_i, y_i). Using these differences we can express 2.18 as:

$$r = \frac{\sum_{i=1}^{n} x_i^2 + \sum_{i=1}^{n} y_i^2 - \sum_{i=1}^{n} d_i^2}{2\sqrt{\sum_{i=1}^{n} x_i^2 \sum_{i=1}^{n} y_i^2}}.$$

2.20

Assuming the values of x_i and y_i are ranked from 1 through N and that there are no tied ranks in any variable, we have:

$$\sum_{i=1}^{n} x_i^2 = \sum_{i=1}^{n} y_i^2 = (N^3 - N)/12.$$

Applying this result to 2.20, the following *Spearman's rank correlation* (also known as *rank correlation coefficient*) is derived:

$$r_s = 1 - \frac{6\sum_{i=1}^{n} d_i^2}{N(N^2 - 1)}, \qquad\qquad 2.21$$

When tied ranks occur – i.e., two or more cases receive the same rank on the same variable –, each of those cases is assigned the average of the ranks that would have been assigned had no ties occurred. When the proportion of tied ranks is small, formula 2.21 can still be used. Otherwise, the following correction factor is computed:

$$T = \sum_{i=1}^{g} (t_i^3 - t_i),$$

where g is the number of groupings of different tied ranks and t_i is the number of tied ranks in the ith grouping. The Spearman's rank correlation with correction for tied ranks is now written as:

$$r_s = 1 - \frac{(N^3 - N) - 6\sum_{i=1}^{n} d_i^2 - (T_x + T_y)/2}{\sqrt{(N^3 - N)^2 - (T_x + T_y)(N^3 - N) + T_x T_y}}, \qquad\qquad 2.22$$

where T_x and T_y are the correction factors for the variables X and Y, respectively.

Table 2.10. Contingency table obtained with SPSS of the NC, PRTGC variables (cork stopper dataset).

				PRTGC			Total
			0	1	2	3	
NC	0	Count	25	9	4	1	39
		% of Total	16.7%	6.0%	2.7%	.7%	26.0%
	1	Count	12	13	10	1	36
		% of Total	8.0%	8.7%	6.7%	.7%	24.0%
	2	Count	1	13	15	9	38
		% of Total	.7%	8.7%	10.0%	6.0%	25.3%
	3	Count	1	1	9	26	37
		% of Total	.7%	.7%	6.0%	17.3%	24.7%
Total		Count	39	36	38	37	150
		% of Total	26.0%	24.0%	25.3%	24.7%	100.0%

Example 2.8

Q: Compute the rank correlation for the variables N and PRTG of the Cork Stopper' dataset, using two new variables, NC and PRTGC, which rank N and PRTG into 4 categories, according to their value falling into the 1st, 2nd, 3rd or 4th quartile intervals.

A: The new variables NC and PRTGC can be computed using formulas similar to the formula used in 2.1.6 for computing PClass. Specifically for NC, given the values of the three N quartiles, 59 (25%), 78.5 (50%) and 95 (75%), respectively, NC coded in {0, 1, 2, 3} is computed as:

NC = (N>59) + (N>78.5) + (N>95)

The corresponding contingency table is shown in Table 2.10. Note that NC and PRTGC are ordinal variables since their ranks do indeed satisfy an order relation.

The rank correlation coefficient computed for this table (see Commands 2.10) is 0.715 which agrees fairly well with the 0.72 correlation computed for the corresponding continuous variables, as shown in Table 2.9.

□

2.3.5.2 The Gamma Statistic

Another measure of association for ordinal variables is based on a comparison of the values of both variables, X and Y, for all possible pairs of cases (x, y). Pairs of cases can be:

–   Concordant (in rank order): The values of both variables for one case are higher (or are both lower) than the corresponding values for the other case. For instance, in Table 2.10 ($X = $ NC; $Y = $ PRTGC), the pair $\{(0, 0), (2, 1)\}$ is concordant.

–   Discordant (in rank order): The value of one variable for one case is higher than the corresponding value for the other case, and the direction is reversed for the other variable. For instance, in Table 2.10, the pair $\{(0, 2), (3, 1)\}$ is discordant.

–   Tied (in rank order): The two cases have the same value on one or on both variables. For instance, in Table 2.10, the pair $\{(1, 2), (3, 2)\}$ are tied.

The following γ measure of association (gamma coefficient) is defined:

$$\gamma = \frac{P(\text{Concordant}) - P(\text{Discordant})}{1 - P(\text{Tied})} = \frac{P(\text{Concordant}) - P(\text{Discordant})}{P(\text{Concordant}) + P(\text{Discordant})}. \qquad 2.23$$

Let P and Q represent the total counts for the concordant and discordant cases, respectively. A point estimate of γ is then:

$$G = \frac{P-Q}{P+Q}, \qquad 2.24$$

with P and Q computed from the counts n_{ij} (of table cell ij), of a contingency table with r rows and c columns, as follows:

$$P = \sum_{i=1}^{r-1}\sum_{j=1}^{c-1} n_{ij} N_{ij}^{+} \quad ; \quad Q = \sum_{i=1}^{r-1}\sum_{j=2}^{c} n_{ij} N_{ij}^{-}, \qquad 2.25$$

where the N_{ij}^+ is the sum of all counts below and to the right of the ijth cell, and the N_{ij}^- is the sum of all counts below and to the left of the ijth cell.

The gamma measure varies, as does the correlation coefficient, in the interval [−1, 1]. It will be 1 if all the frequencies lie in the main diagonal of the table (from the upper left corner to the lower right corner), as for all cases where there are no discordant contributions (see Figure 2.27a). It will be −1 if all the frequencies lie in the other diagonal of the table, and also for all cases where there are no concordant contributions (see Figure 2.27b). Finally, it will be zero when the concordant contributions balance the discordant ones.

The G value for the example of Table 2.10 is 0.785. We will see in Chapter 5 the significance of the G statistic.

There are other measures of association similar to the gamma coefficient that are applicable to ordinal data. For more details the reader can consult e.g. (Siegel S, Castellan Jr NJ, 1988).

Commands 2.10. SPSS, STATISTICA, MATLAB and R commands used to obtain measures of association for ordinal variables.

SPSS	Analyze; Descriptive Statistics; Crosstabs
STATISTICA	Statistics; Basic Statistics/Tables; Tables and Banners; Options
MATLAB	corrcoef(x) ; *gammacoef(t)*
R	cor(x) ; *gammacoef(t)*

Measures of association for ordinal variables are obtained in SPSS and STATISTICA as a result of applying contingency table analysis with the commands listed in Commands 5.7.

MATLAB Statistics toolbox and R stats package do not provide a function for computing the gamma statistic. We provide, however, MATLAB and R functions for that purpose in the book CD (see Appendix F). ∎

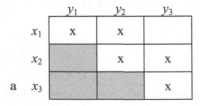

Figure 2.27. Examples of contingency table formats for: a) $G = 1$ (N_{ij}^- cells are shaded gray); b) $G = -1$ (N_{ij}^+ cells are shaded gray).

2.3.6 Measures of Association for Nominal Variables

Assume we have a multivariate dataset whose variables are of nominal type and we intend to measure their level of association. In this case, the correlation coefficient approach cannot be applied, since covariance and standard deviations are not applicable to nominal data. We need another approach that uses the contingency table information in a similar way as when we computed the gamma coefficient for the ordinal data.

Commands 2.11. SPSS, STATISTICA, MATLAB and R commands used to obtain measures of association for nominal variables.

SPSS	`Analyze; Descriptive Statistics; Crosstabs`
STATISTICA	`Statistics; Basic Statistics/Tables; Tables and Banners; Options`
MATLAB	`kappa(x,alpha)`
R	`kappa(x,alpha)`

Measures of association for nominal variables are obtained in SPSS and STATISTICA as a result of applying contingency table analysis (see Commands 5.7).

The kappa statistic can be computed with SPSS only when the values of the first variable match the values of the second variable. STATISTICA does not provide the kappa statistic.

MATLAB Statistics toolbox and R stats package do not provide a function for computing the kappa statistic. We provide, however, MATLAB and R functions for that purpose in the book CD (see Appendix F).

∎

2.3.6.1 The Phi Coefficient

Let us first consider a bivariate dataset with nominal variables that only have two values (dichotomous variables), as in the case of the 2×2 contingency table shown in Table 2.11.

In the case of a full association of both variables one would obtain a 100% frequency for the values along the main diagonal of the table, and 0% otherwise. Based on this observation, the following index of association, ϕ (*phi coefficient*), is defined:

$$\phi = \frac{ad - bc}{\sqrt{(a+b)(c+d)(a+c)(b+d)}}.$$

2.26

Note that the denominator of ϕ will ensure a value in the interval $[-1, 1]$ as with the correlation coefficient, with $+1$ representing a perfect positive association and -1 a perfect negative association. As a matter of fact the phi coefficient is a special case of the Pearson correlation.

Table 2.11. A general cross table for the bivariate dichotomous case.

	y_1	y_2	Total
x_1	a	b	$a + b$
x_2	c	d	$c + d$
Total	$a + c$	$b + d$	$a + b + c + d$

Example 2.9

Q: Consider the 2×2 contingency table for the variables SEX and INIT of the Freshmen dataset, shown in Table 2.12. Compute their phi coefficient.

A: The computed value of phi using 2.26 is 0.15, suggesting a very low degree of association. The significance of the phi values will be discussed in Chapter 5.

□

Table 2.12. Cross table (obtained with SPSS) of variables SEX and INIT of the freshmen dataset.

			INIT		Total
			yes	no	
SEX	male	Count	91	5	96
		% of Total	69.5%	3.8%	73.3%
	female	Count	30	5	35
		% of Total	22.9%	3.8%	26.7%
Total		Count	121	10	131
		% of Total	92.4%	7.6%	100.0%

2.3.6.2   The Lambda Statistic

Another useful measure of association, for multivariate nominal data, attempts to evaluate how well one of the variables predicts the outcome of the other variable. This measure is applicable to any nominal variables, either dichotomous or not. We will explain it using Table 2.4, by attempting to estimate the contribution of variable SEX in lowering the prediction error of Q4 ("liking to be initiated"). For that purpose, we first note that if nothing is known about the sex, the best prediction of the Q4 outcome is the "agree" category, the so-called *modal category*, with the highest frequency of occurrence (37.9%). In choosing this modal category, we expect to be in error 62.1% of the times. On the other hand, if we know the sex (i.e., we know the full table), we would choose as prediction outcome the "agree" category if it is a male (expecting then $73.5 - 28 = 45.5\%$ of errors), and the "fully agree" category if it is a female (expecting then $26.5 - 11.4 = 15.1\%$ of errors).

Let us denote:

i. $Pe_c \equiv$ Percentage of errors using only the columns $= 100 -$ percentage of modal column category.

ii. $Pe_{cr} \equiv$ Percentage of errors using also the rows $=$ sum along the rows of $(100 -$ percentage of modal column category in each row).

The λ measure (*Goodman and Kruskal lambda*) of *proportional reduction of error*, when using the columns depending from the rows, is defined as:

$$\lambda_{cr} = \frac{Pe_c - Pe_{cr}}{Pe_c}.$$ 2.27

Similarly, for the prediction of the rows depending from the columns, we have:

$$\lambda_{rc} = \frac{Pe_r - Pe_{rc}}{Pe_r}.$$ 2.28

The *coefficient of mutual association* (also called *symmetric lambda*) is a weighted average of both lambdas, defined as:

$$\lambda = \frac{\text{average reduction in errors}}{\text{average number of errors}} = \frac{(Pe_c - Pe_{cr}) + (Pe_r - Pe_{rc})}{Pe_c + Pe_r}.$$ 2.29

The lambda measure always ranges between 0 and 1, with 0 meaning that the independent variable is of no help in predicting the dependent variable and 1 meaning that the independent variable perfectly specifies the categories of the dependent variable.

Example 2.10

Q: Compute the lambda statistics for Table 2.4.

A: Using formula 2.27 we find $\lambda_{cr} = 0.024$, suggesting a non-helpful contribution of the sex in determining the outcome of Q4. We also find $\lambda_{rc} = 0$ and $\lambda = 0.017$. The significance of the lambda statistic will be discussed in Chapter 5.

□

2.3.6.3   The Kappa Statistic

The *kappa statistic* is used to measure the *degree of agreement* for categorical variables. Consider the cross table shown in Figure 2.19 where the r rows are objects to be assigned to one of c categories (columns). Furthermore, assume that k judges assigned the objects to the categories, with n_{ij} representing the number of judges that assigned object i to category j.

The sums of the counts along the rows totals k. Let c_j denote the sum of the counts along the column j. If all the judges were in perfect agreement one would find a column filed in with k and the others with zeros, i.e., one of the c_j would be rk and the others zero. The proportion of objects assigned to the jth category is:

$$p_j = c_j / (rk).$$

If the judges make their assignments at random, the expected proportion of agreement for each category is p_j^2 and the total expected agreement for all categories is:

$$P(E) = \sum_{j=1}^{c} p_j^2 . \qquad\qquad 2.30$$

The extent of agreement, s_i, concerning the ith object, is the proportion of the number of pairs for which there is agreement to the possible pairs of agreement:

$$s_i = \sum_{j=1}^{c} \binom{n_{ij}}{2} / \binom{k}{2} .$$

The total proportion of agreement is the average of these proportions across all objects:

$$P(A) = \frac{1}{r} \sum_{i=1}^{r} s_i . \qquad\qquad 2.31$$

The κ (kappa) statistic, based on the formulas 2.30 and 2.31, is defined as:

$$\kappa = \frac{P(A) - P(E)}{1 - P(E)} . \qquad\qquad 2.32$$

If there is complete agreement among the judges, then $\kappa = 1$ ($P(A) = 1$, $P(E) = 0$). If there is no agreement among the judges other than what would be expected by chance, then $\kappa = 0$ ($P(A) = P(E)$).

Example 2.11

Q: Consider the FHR dataset, which includes 51 foetal heart rate cases, classified by three human experts (E1C, E2C, E3C) and an automatic diagnostic system (SPC) into three categories: normal (0), suspect (1) and pathologic (2). Determine the degree of agreement among all 4 classifiers (experts and automatic system).

A: We use the N, S and P variables, which contain the data in the adequate contingency table format, shown in Table 2.13. For instance, object #1 was classified N by one of the classifiers (judges) and S by three of the classifiers.

Running the function kappa(x, 0.05) in MATLAB or R, where x is the data matrix corresponding to the N-S-P columns of Table 2.13, we obtain $\kappa = 0.213$, which suggests some agreement among all 4 classifiers. The significance of the kappa values will be discussed in Chapter 5.

▯

Table 2.13. Contingency table for the N, S and P categories of the FHR dataset.

Object #	N	S	P	Total
1	1	3	0	4
2	1	3	0	4
3	1	3	0	4
...	...	...	...	...
51	1	2	1	4

Exercises

2.1 Consider the "Team Work" evaluation scores of the Metal Firms' dataset:
    a) What type of data is it? Does it make sense to use the mean as location measure of this data?
    b) Compute the median value of "Evaluation of Competence" of the same dataset, with and without the lowest score value.

2.2 Does the median have the additive property of the mean (see A.6.1)? Explain why.

2.3 Variable EF of the Infarct dataset contains "ejection fraction" values (proportion of ejected blood between diastole and systole) of the heart left ventricle, measured in a random sample of 64 patients with some symptom of myocardial infarction.
    a) Determine the histogram of the data using an appropriate number of bins.
    b) Determine the corresponding frequency table and use it to estimate the proportion of patients that are expected to have an ejection fraction below 50%.
    c) Determine the mean, median and standard deviation of the data.

2.4 Consider the Freshmen dataset used in Example 2.3.
    a) What type of variables are Course and Exam 1?
    b) Determine the bar chart of Course. What category occurs most often?
    c) Determine the mean and median of Exam 1 and comment on the closeness of the values obtained.
    d) Based on the frequency table of Exam 1, estimate the number of flunking students.

2.5 Determine the histograms of variables LB, ASTV, MSTV, ALTV and MLTV of the
     CTG dataset using Sturges' rule for the number of bins. Compute the skewness and
     kurtosis of the variables and check the following statements:
     a)   The distribution of LB is well modelled by the normal distribution.
     b)   The distribution of ASTV is symmetric, bimodal and flatter than the normal
          distribution.
     c)   The distribution of ALTV is left skewed and more peaked than the normal
          distribution.

2.6 Taking into account the values of the skewness and kurtosis computed for variables
     ASTV and ALTV in the previous Exercise, which distributions should be selected as
     candidates for modelling these variables (see Figure 2.24)?

2.7 Consider the bacterial counts in three organs – the spleen, liver and lungs - included in
     the Cells dataset (datasheet CFU). Using box plots, compare the cell counts in the
     three organs 2 weeks and 2 months after infection. Also, determine which organs have
     the lowest and highest spread of bacterial counts.

2.8 The inter-quartile ranges of the bacterial counts in the spleen and in the liver after 2
     weeks have similar values. However, the range of the bacterial counts is much smaller
     in the spleen than in the liver. Explain what causes this discrepancy and comment on
     the value of the range as spread measure.

2.9 Determine the overlaid scatter plot of the three types of clays (Clays' dataset), using
     variables SiO_2 and Al_2O_3. Also, determine the correlation between both variables and
     comment on the results.

2.10 The Moulds' dataset contains measurements of bottle bottoms performed by three
     methods. Determine the correlation matrix for the three methods before and after
     subtracting the nominal value of 34 mm and explain why the same correlation results
     are obtained. Also, express your judgement on the measurement methods taking into
     account their low correlation.

2.11 The Culture dataset contains percentages of budget assigned to cultural activities in
     several Portuguese boroughs randomly sampled from three regions, coded 1, 2 and 3.
     Determine the correlations among the several cultural activities and consider them to be
     significant if they are higher than 0.4. Comment on the following statements:
     a)   The high negative correlation between "Halls" and "Sport" is due to chance alone.
     b)   Whenever there is a good investment in "Cine", there is also a good investment
          either in "Music" or in "Fine Arts".
     c)   In the northern boroughs, a high investment in "Heritage" causes a low investment
          in "Sport".

2.12 Consider the "Halls" variable of the Culture dataset:
     a)   Determine the overall frequency table and histogram, starting at zero and with bin
          width 0.02.
     b)   Determine the mean and median. Which of these statistics should be used as
          location measure and why?

2.13 Determine the box plots of the Breast Tissue variables I0 through PERIM, for the 6 classes of breast tissue. By visual inspection of the results, organise a table describing which class discriminations can be expected to be well accomplished by each variable.

2.14 Consider the two variables MH = "neonatal mortality rate at home" and MI = "neonatal mortality rate at Health Centre" of the Neonatal dataset. Determine the histograms and compare both variables according to the skewness and kurtosis.

2.15 Determine the scatter plot and correlation coefficient of the MH and MI variables of the previous exercise. Comment on the results.

2.16 Determine the histograms, skewness and kurtosis of the BPD, CP and AP variables of the Foetal Weight dataset. Which variable is better suited to normal modelling? Why?

2.17 Determine the correlation matrix of the BPD, CP and AP variables of the previous exercise. Comment on the results.

2.18 Determine the correlation between variables I0 and HFS of the Breast Tissue dataset. Check with the scatter plot that the very low correlation of those two variables does not mean that there is no relation between them. Compute the new variable I0S = $(I0 - 1235)^2$ and show that there is a significant correlation between this new variable and HFS.

2.19 Perform the following statistical analyses on the Rocks' dataset:
   a)   Determine the histograms, skewness and kurtosis of the variables and categorise them into the following categories: left asymmetric; right asymmetric; symmetric; symmetric and almost normal.
   b)   Compute the correlation matrix for the mechanical test variables and comment on the high correlations between RMCS and RCSG and between AAPN and PAOA.
   c)   Compute the correlation matrix for the chemical composition variables and determine which variables have higher positive and negative correlation with silica (SiO_2) and which variable has higher positive correlation with titanium oxide (TiO_2).

2.20 The student performance in a first-year university course on Programming can be partly explained by previous knowledge on such matter. In order to assess this statement, use the SCORE and PROG variables of the Programming dataset, where the first variable represents the final examination score on Programming (in [0, 20]) and the second variable categorises the previous knowledge. Using three SCORE categories – Poor, if SCORE<10, Fair if 10 ≤SCORE< 15, and Good if SCORE≥ 15 –, determine:
   a)   The Spearman correlation between the two variables.
   b)   The contingency table of the two variables.
   c)   The gamma statistic.

2.21 Show examples of 2×2 contingency tables for nominal data corresponding to ϕ = 1, −1, 0 and to λ, λ_{rc} and λ_{cr} = 1 and 0.

2.22 Consider the classifications of foetal heart rate performed by the human expert 3 (variable E3C) and by an automatic system (variable SPC) contained in the FHR dataset.
   a)   Determine two new variables, E3CB and SPCB, which dichotomise the classifications in {Normal} vs. {Suspect, Pathologic}.
   b)   Determine the 2×2 contingency table of E3CB and SPCB.
   c)   Determine appropriate association measures and assess whether knowing the automatic system classification helps predicting the human expert classification.

2.23 Redo Example 2.9 and 2.10 for the variables Q1 and Q4 and comment on the results obtained.

2.24 Consider the leadership evaluation of metallurgic firms, included in the Metal Firms' dataset, performed by means of seven variables, from TW = "Team Work" through DC = "Dialogue with Collaborators". Compute the coefficient of agreement of the seven variables, verifying that they do not agree in the assessment of leadership evaluation.

2.25 Determine the contingency tables and degrees of association between variable TW = "Team Work" and all the other leadership evaluation variables of the Metal Firms' dataset.

2.26 Determine the contingency table and degree of association between variable AB = "Previous knowledge of Boole's Algebra" and BA = "Previous knowledge of binary arithmetic" of the Programming dataset.

3 Estimating Data Parameters

Making inferences about a population based upon a random sample is a major task in statistical analysis. *Statistical inference* comprehends two inter-related problems: *parameter estimation* and *test of hypotheses*. In this chapter, we describe the estimation of several distribution parameters, using sample estimates that were presented as descriptive statistics in the preceding chapter. Because these descriptive statistics are single values, determined by appropriate formulas, they are called *point estimates*. Appendix C contains an introductory survey on how such point estimators may be derived and which desirable properties they should have. In this chapter, we also introduce the notion and methodology of *interval estimation*. In this and later chapters, we always assume that we are dealing with random samples. By definition, in a random sample x_1, ..., x_n from a population with probability density function $f_X(x)$, the random variables associated with the sample values, X_1, ..., X_n, are i.i.d., hence the random sample has a joint density given by:

$$f_{X_1,X_2,...,X_n}(x_1,x_2,...,x_n) = f_X(x_1)f_X(x_2)...f_X(x_n).$$

A similar result applies to the joint probability function when the variables are discrete. Therefore, we rule out sampling from a finite population without replacement since, then, the random variables X_1, ..., X_n are not independent.

Note, also, that in the applications one must often carefully distinguish between *target population* and *sampled population*. For instance, sometimes in the newspaper one finds estimation results concerning the proportion of votes on political parties. These results are usually presented as estimates for the whole population of a given country. However, careful reading discloses that the sample (hopefully a random one) was drawn using a telephone enquiry from the population residing in certain provinces. Although the target population is the population of the whole country, any inference made is only legitimate for the sampled population, i.e., the population residing in those provinces and that use telephones.

3.1 Point Estimation and Interval Estimation

Imagine that someone wanted to weigh a certain object using spring scales. The object has an unknown weight, ω. The weight measurement, performed with the scales, has usually two sources of error: a calibration error, because of the spring's loss of elasticity since the last calibration made at the factory, and exhibiting, therefore, a permanent deviation (bias) from the correct value; a random parallax error, corresponding to the evaluation of the gauge needle position, which can be considered normally distributed around the correct position (variance). The situation is depicted in Figure 3.1.

The weight measurement can be considered as a "bias + variance" situation. The bias, or *systematic error*, is a constant. The source of variance is a *random error*.

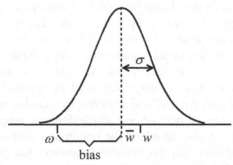

Figure 3.1. Measurement of an unknown quantity ω with a systematic error (bias) and a random error (variance σ^2). One measurement instance is w.

Figure 3.1 also shows one weight measurement instance, w. Imagine that we performed a large number of weight measurements and came out with the average value of $\overline{w}$. Then, the difference $\omega - \overline{w}$ measures the bias or *accuracy* of the weighing device. On the other hand, the standard deviation, σ, measures the *precision* of the weighing device. Accurate scales will, on average, yield a measured weight that is in close agreement with the true weight. High precision scales yield weight measurements with very small random errors.

Let us now turn to the problem of estimating a data parameter, i.e., a quantity θ characterising the distribution function of the random variable X, describing the data. For that purpose, we assume that there is available a random sample $\mathbf{x} = [x_1, x_2, \ldots, x_n]'$ – our dataset in vector format –, and determine a value $t_n(\mathbf{x})$, using an appropriate function t_n. This single value is a *point estimate* of θ.

The estimate $t_n(\mathbf{x})$ is a value of a random variable, that we denote T, called *point estimator* or *statistic*, $T \equiv t_n(X)$, where X denotes the n-dimensional random variable corresponding to the sampling process. The point estimator T is, therefore, a random variable function of X. Thus, $t_n(X)$ constitutes a sort of measurement device of θ. As with any measurement device, we want it to be simultaneously accurate and precise. In Appendix C, we introduce the topic of obtaining unbiased and consistent estimators. The unbiased property corresponds to the accuracy notion. The consistency corresponds to a growing precision for increasing sample sizes.

When estimating a data parameter the point estimate is usually insufficient. In fact, in all the cases that the point estimator is characterised by a probability density function the probability that the point estimate actually equals the true value of the parameter is zero. Using the spring scales analogy, we see that no matter how accurate and precise the scales are, the probability of obtaining the exact weight (with arbitrary large number of digits) is zero. We need, therefore, to attach some measure of the possible error of the estimate to the point estimate. For that purpose, we attempt to determine an interval, called *confidence interval*, containing the true parameter value θ with a given probability $1-\alpha$, the so-called *confidence level*:

$$P\big(t_{n,1}(\mathbf{x}) < \theta < t_{n,2}(\mathbf{x})\big) = 1-\alpha,\qquad\qquad 3.1$$

where α is a *confidence risk*.

The endpoints of the interval (also known as *confidence limits*), depend on the available sample and are determined taking into account the *sampling distribution*:

$$F_T(\mathbf{x}) \equiv F_{t_n(X)}(\mathbf{x}).$$

We have assumed that the interval endpoints are finite, the so-called *two-sided* (or *two-tail*) *interval estimation*. Sometimes we will also use *one-sided* (or *one-tail*) *interval estimation* by setting $t_{n,1}(\mathbf{x}) = -\infty$ or $t_{n,2}(\mathbf{x}) = +\infty$.

Let us now apply these ideas to the spring scales example. Imagine that, as happens with unbiased point estimators, there were no systematic error and furthermore the measured errors follow a known normal distribution; therefore, the measurement error is a one-dimensional random variable distributed as $N_{0,\sigma}$, with known σ. In other words, the distribution function of the random weight variable, W, is $F_W(w) \equiv F(w) = N_{\omega,\sigma}(w)$. We are now able to determine the two-sided 95% confidence interval of ω, given a measurement w, by first noticing, from the normal distribution tables, that the percentile 97.5% (i.e., $100-\alpha/2$, with α in percentage) corresponds to 1.96σ.

Thus:

$$F(w) = 0.975 \quad \Rightarrow \quad w_{0.975} = 1.96\sigma.\qquad\qquad 3.2$$

Given the symmetry of the normal distribution, we have:

$$P(w < \omega + 1.96\sigma) = 0.975 \quad \Rightarrow \quad P(\omega - 1.96\sigma < w < \omega + 1.96\sigma) = 0.95,$$

leading to the following 95% confidence interval:

$$\omega - 1.96\sigma < w < \omega + 1.96\sigma.\qquad\qquad 3.3$$

Hence, we expect that in a long run of measurements 95% of them will be inside the $\omega \pm 1.96\sigma$ interval, as shown in Figure 3.2a.

Note that the inequalities 3.3 can also be written as:

$$w - 1.96\sigma < \omega < w + 1.96\sigma ,$$ 3.4

allowing us to define the 95% confidence interval for the unknown weight (parameter) ω given a particular measurement w. (Comparing with expression 3.1 we see that in this case θ is the parameter ω, $t_{1,1} = w - 1.96\sigma$ and $t_{1,2} = w + 1.96\sigma$.) As shown in Figure 3.2b, the equivalent interpretation is that in a long run of measurements, 95% of the $w \pm 1.96\sigma$ intervals will cover the true and unknown weight ω and the remaining 5% will miss it.

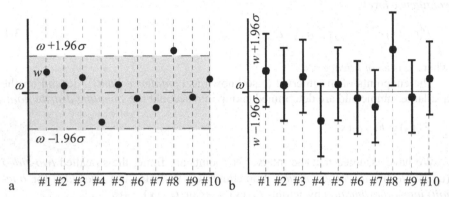

a   #1 #2 #3 #4 #5 #6 #7 #8 #9 #10   b   #1 #2 #3 #4 #5 #6 #7 #8 #9 #10

Figure 3.2. Two interpretations of the confidence interval: a) A certain percentage of the w measurements (#1,…, #10) is inside the $\omega \pm 1.96\sigma$ interval; b) A certain percentage of the $w \pm 1.96\sigma$ intervals contains the true value ω.

Note that when we say that the 95% confidence interval of ω is $w \pm 1.96\sigma$, it does not mean that "the probability that ω falls in the confidence interval is 95%". This is a misleading formulation since ω is not a random variable but an unknown parameter. In fact, it is the confidence interval endpoints that are random variables.

For an arbitrary risk, α, we compute from the standardised normal distribution the $1-\alpha/2$ percentile:

$$N_{0,1}(z) = 1 - \alpha/2 \quad \Rightarrow \quad z_{1-\alpha/2}.^{1}$$ 3.5

We now use this percentile in order to establish the confidence interval:

$$w - z_{1-\alpha/2}\sigma < \omega < w + z_{1-\alpha/2}\sigma .$$ 3.6

The factor $z_{1-\alpha/2}\sigma$ is designated as *tolerance*, ε, and is often expressed as a percentage of the measured value w, i.e., $\varepsilon = 100\, z_{1-\alpha/2}\sigma / w\,\%$.

---

[1] It is customary to denote the values obtained with the standardised normal distribution by the letter z, the so called *z-scores*.

In Chapter 1, section 1.5, we introduced the notions of confidence level and interval estimates, in order to illustrate the special nature of statistical statements and to advise taking precautions when interpreting them. We will now proceed to apply these concepts to several descriptive statistics that were presented in the previous chapter.

3.2 Estimating a Mean

We now estimate the mean of a random variable X using a confidence interval around the sample mean, instead of a single measurement as in the previous section. Let $\mathbf{x} = [x_1, x_2, \ldots, x_n]'$ be a random sample from a population, described by the random variable X with mean μ and standard deviation σ. Let $\bar{x}$ be the arithmetic mean:

$$\bar{x} = \sum_{i=1}^{n} x_i / n \ .$$
3.7

Therefore, $\bar{x}$ is a function $t_n(\mathbf{x})$ as in the general formulation of the previous section. The sampling distribution of $\bar{X}$ (whose values are $\bar{x}$), taking into account the properties of a sum of i.i.d. random variables (see section A.8.4), has the same mean as X and a standard deviation given by:

$$\sigma_{\bar{X}} = \sigma_X / \sqrt{n} \equiv \sigma / \sqrt{n} \ .$$
3.8

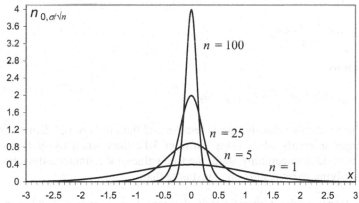

Figure 3.3. Normal distribution of the arithmetic mean for several values of n and with $\mu = 0$ ($\sigma = 1$ for $n = 1$).

Assuming that X is normally distributed, i.e., $X \sim N_{\mu,\sigma}$, then $\bar{X}$ is also normally distributed with mean μ and standard deviation $\sigma_{\bar{X}}$. The confidence interval, following the procedure explained in the previous section, is now computed as:

$$\bar{x} - z_{1-\alpha/2}\sigma/\sqrt{n} < \mu < \bar{x} + z_{1-\alpha/2}\sigma/\sqrt{n}.$$ 3.9

As shown in Figure 3.3, with increasing n, the distribution of $\bar{X}$ gets more peaked; therefore, the confidence intervals decrease with $\sqrt{n}$ (the precision of our estimates of the mean increase). This is precisely why computing averages is so popular!

In normal practice one does not know the exact value of σ, using the previously mentioned (2.3.2) point estimate s instead. In this case, the sampling distribution is not the normal distribution any more. However, taking into account Property 3 described in section B.2.8, the following random variable:

$$T_{n-1} = \frac{\bar{X} - \mu}{s/\sqrt{n}},$$

has a Student's t distribution with $df = n - 1$ degrees of freedom. The sample standard deviation of $\bar{X}$, $s/\sqrt{n}$, is known as the *standard error* of the statistic $\bar{x}$ and denoted SE.

We now compute the $1-\alpha/2$ percentile for the Student's t distribution with $df = n - 1$ degrees of freedom:

$$T_{n-1}(t) = 1 - \alpha/2 \quad \Rightarrow \quad t_{df,1-\alpha/2},$$ 3.10

and use this percentile in order to establish the two-sided confidence interval:

$$-t_{df,1-\alpha/2} < \frac{\bar{x} - \mu}{SE} < t_{df,1-\alpha/2},$$ 3.11

or, equivalently:

$$\bar{x} - t_{df,1-\alpha/2}SE < \mu < \bar{x} + t_{df,1-\alpha/2}SE.$$ 3.12

Since the Student's t distribution is less peaked than the normal distribution, one obtains larger intervals when using formula 3.12 than when using formula 3.9, reflecting the added uncertainty about the true value of the standard deviation.

When applying these results one must note that:

– For large n, the Central Limit theorem (see sections A.8.4 and A.8.5) legitimises the assumption of normal distribution of $\bar{X}$ even when X is not normally distributed (under very general conditions).

– For large n, the Student's t distribution does not deviate significantly from the normal distribution, and one can then use, for unknown σ, the same percentiles derived from the normal distribution, as one would use in the case of known σ.

There are several values of n in the literature that are considered "large", from 20 to 30. In what concerns the normality assumption of $\overline{X}$, the value $n = 20$ is usually enough. As to the deviation between $z_{1-\alpha/2}$ and $t_{1-\alpha/2}$ it is about 5% for $n = 25$ and $\alpha = 0.05$. In the sequel, we will use the threshold $n = 25$ to distinguish small samples from large samples. Therefore, when estimating a mean we adopt the following procedure:

1. Large sample ($n \geq 25$): Use formulas 3.9 (substituting σ by s) or 3.12 (if improved accuracy is needed). No normality assumption of X is needed.

2. Small sample ($n < 25$) and population distribution can be assumed to be normal: Use formula 3.12.

For simplicity most of the software products use formula 3.12 irrespective of the values of n (for small n the normality assumption has to be checked using the goodness of fit tests described in section 5.1).

Example 3.1

Q: Consider the data relative to the variable PRT for the first class (CLASS=1) of the Cork Stoppers' dataset. Compute the 95% confidence interval of its mean.

A: There are $n = 50$ cases. The sample mean and sample standard deviation are $\overline{x} = 365$ and $s = 110$, respectively. The standard error is $SE = s / \sqrt{n} = 15.6$. We apply formula 3.12, obtaining the confidence interval:

$$\overline{x} \pm t_{49,0.975} \times SE = \overline{x} \pm 2.01 \times 15.6 = 365 \pm 31.$$

Notice that this confidence interval corresponds to a tolerance of $31/365 \approx 8\%$. If we used in this large sample situation the normal approximation formula 3.9 we would obtain a very close result.

Given the interpretation of confidence interval (sections 3.1 and 1.5) we expect that in a large number of repetitions of 50 PRT measurements, in the same conditions used for the presented dataset, the respective confidence intervals such as the one we have derived will cover the true PRT mean 95% of the times. In other words, when presenting [334, 396] as a confidence interval for the PRT mean, we are incurring only on a 5% risk of being wrong by basing our estimate on an atypical dataset.

□

Example 3.2

Q: Consider the subset of the previous PRT data constituted by the first $n = 20$ cases. Compute the 95% confidence interval of its mean.

A: The sample mean and sample standard deviation are now $\overline{x} = 351$ and $s = 83$, respectively. The standard error is $SE = s / \sqrt{n} = 18.56$. Since $n = 20$, we apply the small sample estimate formula 3.12 assuming that the PRT distribution can be well approximated by the normal distribution. (This assumption should have to be checked with the methods described in section 5.1.) In these conditions the confidence interval is:

$$\bar{x} \pm t_{19,0.975} \times SE = \bar{x} \pm 2.09 \times SE \quad \Rightarrow \quad [312, 390].$$

If the 95% confidence interval were computed with the z percentile, one would wrongly obtain a narrower interval: [315, 387].   □

Example 3.3

Q: How many cases should one have of the PRT data in order to be able to establish a 95% confidence interval for its mean, with a tolerance of 3%?

A: Since the tolerance is smaller than the one previously obtained in Example 3.1, we are clearly in a large sample situation. We have:

$$\frac{z_{1-\alpha/2} s}{\bar{x}\sqrt{n}} \le \varepsilon \quad \Rightarrow \quad n \ge \left(\frac{z_{1-\alpha/2} s}{\varepsilon \bar{x}}\right)^2. \qquad\qquad 3.13$$

Using the previous sample mean and sample standard deviation and with $z_{0.975} = 1.96$, one obtains:

$$n \ge 558.$$

Note the growth of n with the square of $1/\varepsilon$.

                                                                                 □

The solutions of all the previous examples can be easily computed using Tools.xls (see Appendix F).

An often used tool in *Statistical Quality Control* is the *control chart* for the sample mean, the so-called *x-bar chart*. The x-bar chart displays means, e.g. of measurements performed on equal-sized samples of manufactured items, randomly drawn along the time. The chart also shows the *centre line* (CL), corresponding to the nominal value or the grand mean in a large sequence of samples, and lines of the *upper control limit* (UCL) and *lower control limit* (LCL), computed as a *ks* deviation from the mean, usually with $k = 3$ and s the sample standard deviation. Items above UCL or below LCL are said to be *out of control*. Sometimes, lines corresponding to a smaller deviation of the grand mean, e.g. with $k = 2$, are also drawn, corresponding to the so-called *upper warning line* (UWL) and *lower warning line* (LWL).

Example 3.4

Q: Consider the first 48 measurements of total area of defects, for the first class of the Cork Stoppers dataset, as constituting 16 samples of 3 cork stoppers randomly drawn at successive times. Draw the respective x-bar chart with 3-sigma control lines and 2-sigma warning lines.

A: Using MATLAB command xbarplot (see Commands 3.1) the x-bar chart shown in Figure 3.4 is obtained. We see that a warning should be issued for sample #1 and sample #12. No sample is out of control.

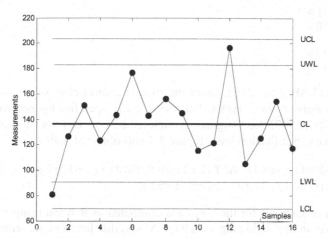

Figure 3.4. Control chart of the sample mean obtained with MATLAB for variable ART of the first cork stopper class.

Commands 3.1. SPSS, STATISTICA, MATLAB and R commands used to obtain confidence intervals of the mean.

SPSS	`Analyze; Descriptive Statistics; Explore; Statistics; Confidence interval for mean`
STATISTICA	`Statistics; Descriptive Statistics; Conf. limits for means`
MATLAB	`[m s mi si]=normfit(x,delta)` `xbarplot(data,conf,specs)`
R	`t.test(x) ; cimean(x,alpha)`

SPSS, STATISTICA, MATLAB and R compute confidence intervals for the mean using Student's *t* distribution, even in the case of large samples.

The MATLAB normfit command computes the mean, m, standard deviation, s, and respective confidence intervals, mi and si, of a data vector x, using confidence level delta (95%, by default). For instance, assuming that the PRT data was stored in vector prt, Example 3.2 would be solved as:

```
» prt20 = prt(1:20);
» [m s mi si] = normfit(prt20)
```

```
m =
   350.6000
s =
    82.7071
mi =
   311.8919
   389.3081
si =
    62.8979
   120.7996
```

The MATLAB xbarplot command plots a control chart of the sample mean for the successive rows of data. Parameter conf specifies the percentile for the control limits (0.9973 for 3-sigma); parameter specs is a vector containing the values of extra specification lines. Figure 3.4 was obtained with:

```
» y=[ART(1:3:48) ART(2:3:48) ART(3:3:48)];
» xbarplot(y,0.9973,[89 185])
```

Confidence intervals for the mean are computed in R when using t.test (to be described in the following chapter). A specific function for computing the confidence interval of the mean, cimean(x, alpha) is included in Tools (see Appendix F).

■

Commands 3.2. SPSS, STATISTICA, MATLAB and R commands for case selection.

SPSS	Data; Select cases
STATISTICA	Tools; Selection Conditions; Edit
MATLAB	x(x(:,i) == a,:)
R	x[col == a,]

In order to solve Examples 3.1 and 3.2 one needs to select the values of PRT for CLASS=1 and, inside this class, to select the first 20 cases. Selection of cases is an often-needed operation in statistical analysis. STATISTICA and SPSS make available specific windows where the user can fill in the needed conditions for case selection (see e.g. Figure 3.5a corresponding to Example 3.2). Selection can be accomplished by means of logical conditions applied to the variables and/or the cases, as well as through the use of especially defined *filter variables*.

There is also the possibility of selecting random subsets of cases, as shown in Figures 3.5a (Subset/Random Sampling tab) and 3.5b (Random sample of cases option).

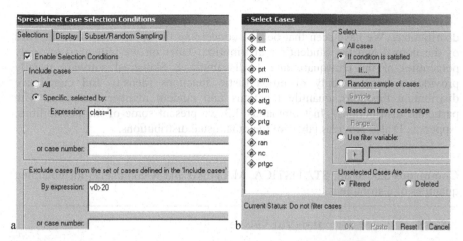

Figure 3.5. Selection of cases: a) Partial view of STATISTICA "Case Selection Conditions" window; b) Partial view of SPSS "Select Cases" window.

In MATLAB one may select a submatrix of matrix x based on a particular value, a, of a column i using the construction $x(x(:,i)==a,:)$. For instance, assuming the first column of `cork` contains the classifications of the cork stoppers, $c = cork(cork(:,1)==1,:)$ will retrieve the submatrix of `cork` corresponding to the first 50 cases of class 1. Other relational operators can be used instead of the equality operator "==". (Attention: "=" is an assignment operator, an equality operator.) For instance, $c = cork(cork(:,1)<2,:)$ will have the same effect.

The selection of cases in R is usually based on the construction $x[col == a,]$, which selects the submatrix whose column `col` is equal to a certain value a. For instance, `cork[CL == 1,]` selects the first 50 cases of class 1 of the data frame `cork`. As in MATLAB other relational operators can be used instead of the equality operator "==".

Selection of random subsets in MATLAB and R can be performed through the generation of filter variables using random number generators. An example is shown in Table 3.1. First, a filter variable with 150 random 0s and 1s is created by rounding random numbers with uniform distribution in [0,1]. Next, the filter variable is used to select a subset of the 150 cases of the cork data.

Table 3.1. Selecting a random subset of the `cork stoppers`' dataset.

MATLAB	`>> filter = round(unifrnd(0,1,150,1));` `>> fcork = cork(filter==1,:);`
R	`> filter <- round(runif(150,0,1))` `> fcork <- cork[filter==1,]`

In parameter estimation one often needs to use percentiles of random distributions. We have seen that before, concerning the application of percentiles of the normal and the Student's t distribution. Later on we will need to apply percentiles of the chi-square and F distributions. Statistical software usually provides a large panoply of probabilistic functions (density and cumulative distribution functions, quantile functions and random number generators with particular distributions). In Commands 3.3 we present some of the possibilities. Appendix D also provides tables of the most usual distributions.

Commands 3.3. SPSS, STATISTICA, MATLAB and R commands for obtaining quantiles of distributions.

SPSS	`Compute Variable`
STATISTICA	`Statistics; Probability Calculator`
MATLAB	`norminv(p,mu,sigma) ; tinv(p,df) ;` `chi2inv(p,df) ; finv(p,df1,df2)`
R	`qnorm(p,mean,sd) ; qt(p,df) ;` `qchisq(p,df) ; qf(p,df1,df2)`

The `Compute Variable` window of SPSS allows the use of functions to compute percentiles of distributions, namely the functions `Idf.IGauss`, `Idf.T`, `Idf.Chisq` and `Idf.F` for the normal, Student's t, chi-square and F distributions, respectively.

STATISTICA provides a versatile `Probability Calculator` allowing among other things the computation of percentiles of many common distributions.

The MATLAB and R functions allow the computation of quantiles of the normal, t, chi-square and F distributions, respectively.

■

3.3 Estimating a Proportion

Imagine that one wished to estimate the probability of occurrence, p, of a "success" event in a series of n Bernoulli trials. A Bernoulli trial is a dichotomous outcome experiment (see B.1.1). Let k be the number of occurrences of the success event. Then, the unbiased and consistent point estimate of p is (see Appendix C):

$$\hat{p} = \frac{k}{n}.$$

For instance, if there are $k = 5$ successes in $n = 15$ trials, the point estimate of p (estimation of a proportion) is $\hat{p} = 0.33$. Let us now construct an interval estimation for p. Remember that the sampling distribution of the number of "successes" is the binomial distribution (see B.1.5). Given the discreteness of the binomial distribution, it may be impossible to find an interval which has exactly the desired confidence level. It is possible, however, to choose an interval which covers p with probability at least $1- \alpha$.

Table 3.2. Cumulative binomial probabilities for $n = 15, p = 0.33$.

k	0	1	2	3	4	5	6	7	8	9	10
$B(k)$	0.002	0.021	0.083	0.217	0.415	0.629	0.805	0.916	0.971	0.992	0.998

Consider the cumulative binomial probabilities for $n = 15, p = 0.33$, as shown in Table 3.2. Using the values of this table, we can compute the following probabilities for intervals centred at $k = 5$:

$$P(4 \leq k \leq 6) = B(6) - B(3) = 0.59$$
$$P(3 \leq k \leq 7) = B(7) - B(2) = 0.83$$
$$P(2 \leq k \leq 8) = B(8) - B(1) = 0.95$$
$$P(1 \leq k \leq 9) = B(9) - B(0) = 0.99$$

Therefore, a 95% confidence interval corresponds to:

$$2 \leq k \leq 8 \quad \Rightarrow \quad \frac{2}{15} \leq p \leq \frac{8}{15} \quad \Rightarrow \quad 0.13 \leq p \leq 0.53.$$

This is too large an interval to be useful. This example shows the inherent high degree of uncertainty when performing an interval estimation of a proportion with small n. For large n (say $n > 50$), we use the normal approximation to the binomial distribution as described in section A.7.3. Therefore, the sampling distribution of $\hat{p}$ is modelled as $N_{\mu,\sigma}$ with:

$$\mu = p; \quad \sigma = \sqrt{\frac{pq}{n}} \quad (q = p - 1; \text{ see A.7.3}). \tag{3.14}$$

Thus, the large sample confidence interval of a proportion is:

$$\hat{p} - z_{1-\alpha/2}\sqrt{pq/n} < p < \hat{p} + z_{1-\alpha/2}\sqrt{pq/n}. \tag{3.15}$$

This is the formula already alluded to in Chapter 1, when describing the "uncertainties" about the estimation of a proportion. Note that when applying formula 3.15, one usually substitutes the true standard deviation by its point estimate, i.e., computing:

$$\hat{p} - z_{1-\alpha/2}\sqrt{\hat{p}\hat{q}/n} < p < \hat{p} + z_{1-\alpha/2}\sqrt{\hat{p}\hat{q}/n}. \tag{3.16}$$

The deviation of this formula from the exact formula is negligible for large n (see e.g. Spiegel MR, Schiller J, Srinivasan RA, 2000, for details).

One can also assume a worst case situation for σ, corresponding to $p = q = \frac{1}{2}$ $\Rightarrow \sigma = (2\sqrt{n})^{-1}$. The approximate 95% confidence level is now easy to remember:

$$\hat{p} \pm 1/\sqrt{n}.$$

Also, note that if we decrease the tolerance while maintaining n, the confidence level decreases as already mentioned in Chapter 1 and shown in Figure 1.6.

Example 3.5

Q: Consider, for the Freshmen dataset, the estimation of the proportion of freshmen that are displaced from their home (variable DISPL). Compute the 95% confidence interval of this proportion.

A: There are $n = 132$ cases, 37 of which are displaced, i.e., $\hat{p} = 0.28$. Applying formula 3.15, we have:

$$\hat{p} - 1.96\sqrt{\hat{p}\hat{q}/n} < p < \hat{p} + 1.96\sqrt{\hat{p}\hat{q}/n} \quad \Rightarrow \quad 0.20 < p < 0.36.$$

Note that this confidence interval is quite large. The following example will give some hint as to when we start obtaining reasonably useful confidence intervals.

□

Example 3.6

Q: Consider the interval estimation of a proportion in the same conditions as the previous example, i.e., with estimated proportion $\hat{p} = 0.28$ and $\alpha = 5\%$. How large should the sample size be for the confidence interval endpoints deviating less than $\varepsilon = 2\%$?

A: In general, we must apply the following condition:

$$\frac{z_{1-\alpha/2}\sqrt{\hat{p}\hat{q}}}{\sqrt{n}} \le \varepsilon \quad \Rightarrow \quad n \ge \left(\frac{z_{1-\alpha/2}\sqrt{\hat{p}\hat{q}}}{\varepsilon}\right)^2. \tag{3.17}$$

In the present case, we must have $n > 1628$. As with the estimation of a mean, n grows with the square of $1/\varepsilon$. As a matter of fact, assuming the worst case situation for σ, as we did above, the following approximate formula for 95% confidence level holds: $n \gtrsim (1/\varepsilon)^2$.

□

Confidence intervals for proportions, and lower bounds on n achieving a desired deviation in proportion estimation, can be computed with Tools.xls.

Interval estimation of a proportion can be carried out with SPSS, STATISTICA, MATLAB and R in the same way as we did with means. The only preliminary step is to convert the variable being analysed into a Bernoulli type variable, i.e., a binary variable with 1 coding the "success" event, and 0 the "failure" event. As a matter of fact, a dataset $x_1, ..., x_n$, with k successes, represented as a sequence of values of Bernoulli random variables (therefore, with k ones and $n - k$ zeros), has the following sample mean and sample variance:

$$\bar{x} = \sum_{i=1}^{n} x_i / n = k / n \equiv \hat{p}.$$

$$v = \frac{\sum_{i=1}^{n} (x_i - \hat{p})^2}{n-1} = \frac{n\hat{p}^2 - 2k\hat{p} + k}{n-1} = \frac{n}{n-1}(\hat{p} - \hat{p}^2) \approx \hat{p}\hat{q}.$$

In Example 3.5, variable DISPL with values 1 for "Yes" and 2 for "No" is converted into a Bernoulli type variable, DISPLB, e.g. by using the formula DISPLB = 2 − DISPL. Now, the "success" event ("Yes") is coded 1, and the complement is coded 0. In SPSS and STATISTICA we can also use "if" constructs to build the Bernoulli variables. This is especially useful if one wants to create Bernoulli variables from continuous type variables. SPSS and STATISTICA also have a Rank command that can be useful for the purpose of creating Bernoulli variables.

Commands 3.4. MATLAB and R commands for obtaining confidence intervals of proportions.

MATLAB	*ciprop(n0,n1,alpha)*
R	*ciprop(n0,n1,alpha)*

There are no specific functions to compute confidence intervals of proportions in MATLAB and R. However, we provide for MATLAB and R the function ciprop(n0,n1,alpha) for that purpose (see Appendix F). For Example 3.5 we obtain in R:

```
> ciprop(95,37,0.05)

          [,1]
[1,]  0.2803030
[2,]  0.2036817
[3,]  0.3569244                           ■
```

3.4 Estimating a Variance

The point estimate of a variance was presented in section 2.3.2. This estimate is also discussed in some detail in Appendix C. We will address the problem of establishing a confidence interval for the variance only in the case that the population distribution follows a normal law. Then, the sampling distribution of the variance follows a chi-square law, namely (see Property 4 of section B.2.7):

$$\frac{(n-1)v}{\sigma^2} \sim \chi^2_{n-1} \qquad\qquad 3.18$$

The chi-square distribution is asymmetrical; therefore, in order to establish a two-sided confidence interval, we have to use two different values for the lower and upper percentiles. For the 95% confidence interval and $df = n - 1$, we have:

$$\chi^2_{df,0.025} \le \frac{df \times v}{\sigma^2} \le \chi^2_{df,0.975} \,, \qquad\qquad 3.19$$

where $\chi^2_{df,\alpha}$ means the α percentile of the chi-square distribution with df degrees of freedom. Therefore:

$$\frac{df \times v}{\chi^2_{df,0.975}} \le \sigma^2 \le \frac{df \times v}{\chi^2_{df,0.025}}. \qquad\qquad 3.20$$

Example 3.7

Q: Consider the distribution of the average perimeter of defects, variable PRM, of class 2 in the Cork Stoppers' dataset. Compute the 95% confidence interval of its standard deviation.

A: The assumption of normality for the PRM variable is acceptable, as will be explained in Chapter 5. There are, in class 2, $n = 50$ cases with sample standard variance $v = 0.7168$. The chi-square percentiles are:

$$\chi^2_{49,0.025} = 31.56; \quad \chi^2_{49,0.975} = 70.22.$$

Therefore:

$$\frac{49 \times v}{70.22} \le \sigma^2 \le \frac{49 \times v}{31.56} \quad\Rightarrow\quad 0.50 \le \sigma^2 \le 1.11 \quad\Rightarrow\quad 0.71 \le \sigma \le 1.06.$$

□

Confidence intervals for the variance are computed by SPSS, STATISTICA, MATLAB and R as part of hypothesis tests presented in the following chapter. They can be computed, however, either using Tools.xls or, in the case of the variance alone, using the MATLAB command normfit mentioned in section 3.2. We also provide the MATLAB and R function civar(v,n,alpha) for computing confidence intervals of a variance (see Appendix F).

Commands 3.5. MATLAB and R commands for obtaining confidence intervals of a variance.

MATLAB	`civar(v,n,alpha)`
R	`civar(v,n,alpha)`

As an illustration we show the application of the R function `civar` to the Example 3.7:

```
> civar(0.7168,50,0.05)
          [,1]
[1,] 0.5001708
[2,] 1.1130817
```

■

3.5 Estimating a Variance Ratio

In statistical tests of hypotheses, concerning more than one distribution, one often needs to compare the respective distribution variances. We now present the topic of estimating a confidence interval for the ratio of two variances, σ_1^2 and σ_2^2, based on sample variances, v_1 and v_2, computed on datasets of size n_1 and n_2, respectively. We assume normal distributions for the two populations from where the data samples were obtained. We use the sampling distribution of the ratio:

$$\frac{v_1 / \sigma_1^2}{v_2 / \sigma_2^2},$$
3.21

which has the F_{n_1-1,n_2-1} distribution as mentioned in the section B.2.9 (Property 6).

Thus, the $1-\alpha$ two-sided confidence interval of the variance ratio can be computed as:

$$F_{\alpha/2} \leq \frac{v_1 / \sigma_1^2}{v_2 / \sigma_2^2} \leq F_{1-\alpha/2} \quad \Rightarrow \quad \frac{1}{F_{1-\alpha/2}} \frac{v_1}{v_2} \leq \frac{\sigma_1^2}{\sigma_2^2} \leq \frac{1}{F_{\alpha/2}} \frac{v_1}{v_2},$$
3.22

where we dropped the mention of the degrees of freedom from the F percentiles in order to simplify notation. Note that due to the asymmetry of the F distribution, one needs to compute two different percentiles in two-sided interval estimation.

The confidence intervals for the variance ratio are computed by SPSS, STATISTICA, MATLAB and R as part of hypothesis tests presented in the following chapter. We also provide the MATLAB and R function `civar2(v1,n1,v2,n2,alpha)` for computing confidence intervals of a variance ratio (see Appendix F).

Example 3.8

Q: Consider the distribution of variable ASTV (percentage of abnormal beat-to-beat variability), for the first two classes of the cardiotocographic data (CTG). The respective dataset histograms are shown in Figure 3.6. Class 1 corresponds to "calm sleep" and class 2 to "rapid-eye-movement sleep". The assumption of normality for both distributions of ASTV is acceptable (to be discussed in Chapter 5). Determine and interpret the 95% one-sided confidence interval, $[r, \infty[$, of the ASTV standard deviation ratio for the two classes.

A: There are $n_1 = 384$ cases of class 1, and $n_2 = 579$ cases of class 2, with sample standard deviations $s_1 = 15.14$ and $s_2 = 13.58$, respectively. The 95% F percentile, computed by any of the means explained in section 3.2, is:

$$F_{383,578,0.95} = 1.164.$$

Therefore:

$$\frac{1}{F_{n_1-1,n_2-1,1-\alpha}} \frac{v_1}{v_2} \le \frac{\sigma_1^2}{\sigma_2^2} \quad \Rightarrow \quad \frac{1}{\sqrt{F_{383,578,0.95}}} \frac{s_1}{s_2} \le \frac{\sigma_1}{\sigma_2} \quad \Rightarrow \quad \frac{\sigma_1}{\sigma_2} \ge 1.03.$$

Thus, with 95% confidence level the standard deviation of class 1 is higher than the standard deviation of class 2 by at least 3%.

□

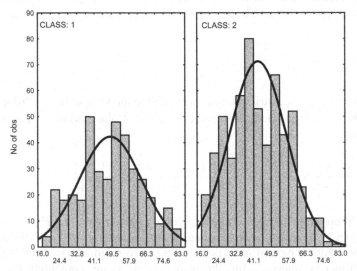

Figure 3.6. Histograms obtained with STATISTICA of the variable ASTV (percentage of abnormal beat-to-beat variability), for the first two classes of the cardiotocographic data, with superimposed normal fit.

When using F percentiles the following results can be useful:

i. $F_{df_2,df_1,1-\alpha} = 1/F_{df_1,df_2,\alpha}$. For instance, if in Example 3.8 we wished to compute a 95% one-sided confidence interval, $[0, r]$, for σ_2/σ_1, we would then have to compute $F_{578,383,0.05} = 1/F_{383,578,0.95} = 0.859$.

ii. $F_{df,\infty,\alpha} = \chi^2_{df,\alpha}/df$. Note that, in formula 3.21, with $n_2 \to \infty$ the sample variance v_2 converges to the true variance, s_2^2, yielding, therefore, the single-variance situation described by the chi-square distribution. In this sense the chi-square distribution can be viewed as a limiting case of the F distribution.

Commands 3.6. MATLAB and R commands for obtaining confidence intervals of a variance ratio.

MATLAB	`civar2(v1,n1,v2,n2,alpha)`
R	`civar2(v1,n1,v2,n2,alpha)`

The MATLAB and R function `civar2` returns a vector with three elements. The first element is the variance ratio, the other two are the confidence interval limits. As an illustration we show the application of the R function `civar2` to the Example 3.8:

```
> civar2(15.14^2,384,13.58^2,579,0.10)
         [,1]
[1,] 1.242946
[2,] 1.067629
[3,] 1.451063
```

Note that since we are computing a one-sided confidence interval we need to specify a double `alpha` value. The obtained lower limit, 1.068, is the square of 1.033, therefore in close agreement to the value we found in Example 3.8. ∎

3.6 Bootstrap Estimation

In the previous sections we made use of some assumptions regarding the sampling distributions of data parameters. For instance, we assumed the sample distribution of the variance to be a chi-square distribution in the case that the normal distribution assumption of the original data holds. Likewise for the F sampling distribution of the variance ratio. The exception is the distribution of the arithmetic mean which is always well approximated by the normal distribution, independently of the distribution law of the original data, whenever the data size is large enough. This is a result of the Central Limit theorem. However, no Central Limit theorem exists for parameters such as the variance, the median or the trimmed mean.

The bootstrap idea (Efron, 1979) is to mimic the sampling distribution of the statistic of interest through the use of many *resamples with replacement* of the original sample. In the present chapter we will restrict ourselves to illustrating the idea when applied to the computation of confidence intervals (bootstrap techniques cover a vaster area than merely confidence interval computation). Let us then illustrate the bootstrap computation of confidence intervals by referring it to the mean of the $n = 50$ PRT measurements for Class=1 of the `cork stoppers'` dataset (as in Example 3.1). The histogram of these data is shown in Figure 3.7a.

Denoting by X the associated random variable, we compute the sample mean of the data as $\bar{x} = 365.0$. The sample standard deviation of $\overline{X}$, the standard error, is $SE = s/\sqrt{n} = 15.6$. Since the dataset size, n, is not that large one may have some suspicion concerning the bias of this estimate and the accuracy of the confidence interval based on the normality assumption.

Let us now consider extracting at random and with replacement $m = 1000$ samples of size $n = 50$ from the original dataset. These resamples are called *bootstrap samples*. Let us further consider that for each bootstrap sample we compute its mean $\bar{x}$. Figure 3.7b shows the histogram[2] of the bootstrap distribution of the means. We see that this histogram looks similar to the normal distribution. As a matter of fact the bootstrap distribution of a statistic usually mimics the sample distribution of that statistic, which in this case happens to be normal.

Let us denote each bootstrap mean by $\bar{x}^*$. The mean and standard deviation of the 1000 bootstrap means are computed as:

$$\bar{x}_{\text{boot}} = \frac{1}{m}\sum \bar{x}^* = \frac{1}{1000}\sum \bar{x}^* = 365.1,$$

$$s_{\bar{x},\text{boot}} = \sqrt{\frac{1}{m-1}\sum\left(\bar{x}^* - \bar{x}_{\text{boot}}\right)^2} = 15.47,$$

where the summations extend to the $m = 1000$ bootstrap samples.

We see that the mean of the bootstrap distribution is quite close to the original sample mean. There is a bias of only $\bar{x}_{\text{boot}} - \bar{x} = 0.1$. It can be shown that this is usually the size of the bias that can be expected between $\bar{x}$ and the true population mean, μ. This property is not an exclusive of the bootstrap distribution of the mean. It applies to other statistics as well.

The sample standard deviation of the bootstrap distribution, called *bootstrap standard error* and denoted SE_{boot}, is also quite close to the theory-based estimate $SE = s/\sqrt{n}$. We could now use SE_{boot} to compute a confidence interval for the mean. In the case of the mean there is not much advantage in doing so (we should get practically the same result as in Example 3.1), since we have the Central Limit theorem in which to base our confidence interval computations. The good thing

---

[2] We should more rigorously say "one possible histogram", since different histograms are possible depending on the resampling process. For n and m sufficiently large they are, however, close to each other.

about the bootstrap technique is that it also often works for other statistics for which no theory on sampling distribution is available. As a matter of fact, the bootstrap distribution usually – for a not too small original sample size, say $n > 50$ – has the same shape and spread as the original sampling distribution, but is centred at the original statistic value rather than the true parameter value.

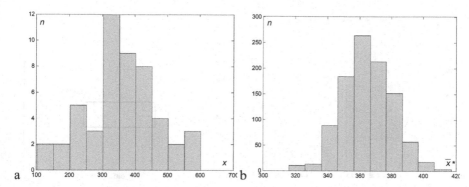

Figure 3.7. a) Histogram of the PRT data; b) Histogram of the bootstrap means.

Suppose that the bootstrap distribution of a statistic, w, is approximately normal and that the bootstrap estimate of bias is small. We then compute a two-sided bootstrap confidence interval at α risk, for the parameter that corresponds to the statistic, by the following formula:

$$w \pm t_{n-1,1-\alpha/2} SE_{boot}$$

We may use the percentiles of the normal distribution, instead of the Student's t distribution, whenever m is very large.

The question naturally arises on how large must the number of bootstrap samples be in order to obtain a reliable bootstrap distribution with reliable values of SE_{boot}? A good rule of thumb for m, based on theoretical and practical evidence, is to choose $m \geq 200$.

The following examples illustrate the computation of confidence intervals using the bootstrap technique.

Example 3.9

Q: Consider the percentage of lime, CaO, in the composition of clays, a sample of which constitutes the Clays' dataset. Compute the confidence interval at 95% level of the two-tail 5% trimmed mean and discuss the results. (The two-tail 5% trimmed mean disregards 10% of the cases, 5% at each of the tails.)

A: The histogram and box plot of the CaO data ($n = 94$ cases) are shown in Figure 3.8. Denoting the associated random variable by X we compute $\bar{x} = 0.28$.

We observe in the box plot a considerable number of "outliers" which leads us to mistrust the sample mean as a location measure and to use the two-tail 5% trimmed mean computed as (see Commands 2.7): $\bar{x}_{0.05} \equiv w = 0.2755$.

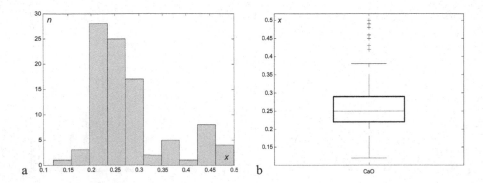

a   b

Figure 3.8. Histogram (a) and box plot (b) of the CaO data.

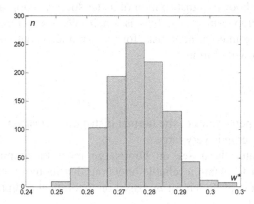

Figure 3.9. Histogram of the bootstrap distribution of the two-tail 5% trimmed mean of the CaO data (1000 resamples).

We now proceed to computing the bootstrap distribution with $m = 1000$ resamples. Figure 3.9 shows the histogram of the bootstrap distribution. It is clearly visible that it is well approximated by the normal distribution (methods not relying on visual inspection are described in section 5.1). From the bootstrap distribution we compute:

$w_{boot} = 0.2764$
$SE_{boot} = 0.0093$

The bias $w_{boot} - w = 0.2764 - 0.2755 = 0.0009$ is quite small (less than 10% of the standard deviation). We therefore compute the bootstrap confidence interval of the trimmed mean as:

$$w \pm t_{93,0.975} SE_{boot} = 0.2755 \pm 1.9858 \times 0.0093 = 0.276 \pm 0.018$$

$\square$

Example 3.10

Q: Compute the confidence interval at 95% level of the standard deviation for the data of the previous example.

A: The standard deviation of the original sample is $s \equiv w = 0.086$. The histogram of the bootstrap distribution of the standard deviation with $m = 1000$ resamples is shown in Figure 3.10. This empirical distribution is well approximated by the normal distribution. We compute:

$w_{boot} = 0.0854$
$SE_{boot} = 0.0070$

The bias $w_{boot} - w = 0.0854 - 0.086 = -0.0006$ is quite small (less than 10% of the standard deviation). We therefore compute the bootstrap confidence interval of the standard deviation as:

$$w \pm t_{93,0.975} SE_{boot} = 0.086 \pm 1.9858 \times 0.007 = 0.086 \pm 0.014$$

$\square$

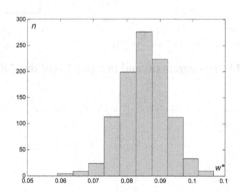

Figure 3.10. Histogram of the bootstrap distribution of the standard deviation of the CaO data (1000 resamples).

Example 3.11

Q: Consider the variable ART (total area of defects) of the cork stoppers' dataset. Using the bootstrap method compute the confidence interval at 95% level of its median.

A: The histogram and box plot of the ART data ($n = 150$ cases) are shown in Figure 3.11. The sample median and sample mean of ART are $med \equiv w = 263$ and $\bar{x} = 324$, respectively. The distribution of ART is clearly right skewed; hence, the mean is substantially larger than the median (almost one and half times the standard deviation). The histogram of the bootstrap distribution of the median with $m = 1000$ resamples is shown in Figure 3.12. We compute:

$w_{boot} = 266.1210$
$SE_{boot} = 20.4335$

The bias $w_{boot} - w = 266 - 263 = 3$ is quite small (less than 7% of the standard deviation). We therefore compute the bootstrap confidence interval of the median as:

$$w \pm t_{149,0.975} SE_{boot} = 263 \pm 1.976 \times 20.4335 = 263 \pm 40$$

☐

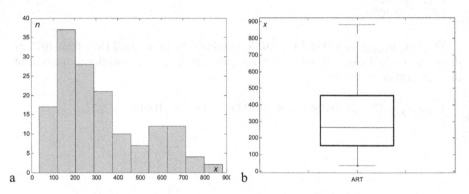

a   b

Figure 3.11. Histogram (a) and box plot (b) of the ART data.

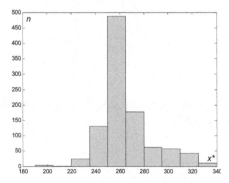

Figure 3.12. Histogram of the bootstrap distribution of the median of the ART data (1000 resamples).

In the above Example 3.11 we observe in Figure 3.12 a histogram that doesn't look to be well approximated by the normal distribution. As a matter of fact any goodness of fit test described in section 5.1 will reject the normality hypothesis. This is a common difficulty when estimating bootstrap confidence intervals for the median. An explanation of the causes of this difficulty can be found e.g. in (Hesterberg T *et al.*, 2003). This difficulty is even more severe when the data size n is small (see Exercise 3.20). Nevertheless, for data sizes larger then 100 cases, say, and for a large number of resamples, one can still rely on bootstrap estimates of the median as in Example 3.11.

Example 3.12

Q: Consider the variables Al2O3 and K2O of the Clays' dataset ($n = 94$ cases). Using the bootstrap method compute the confidence interval at 5% level of their Pearson correlation.

A: The sample Pearson correlation of Al2O3 and K2O is $r \equiv w = 0.6922$. The histogram of the bootstrap distribution of the Pearson correlation with $m = 1000$ resamples is shown in Figure 3.13. It is well approximated by the normal distribution. From the bootstrap distribution we compute:

$w_{boot} = 0.6950$
$SE_{boot} = 0.0719$

The bias $w_{boot} - w = 0.6950 - 0.6922 = 0.0028$ is quite small (about 0.4% of the correlation value). We therefore compute the bootstrap confidence interval of the Pearson correlation as:

$$w \pm t_{93,0.975} SE_{boot} = 0.6922 \pm 1.9858 \times 0.0719 = 0.69 \pm 0.14$$

☐

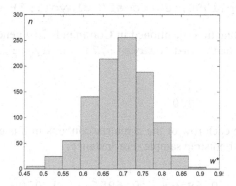

Figure 3.13. Histogram of the bootstrap distribution of the Pearson correlation between the variables Al2O3 and K2O of the Clays' dataset (1000 resamples).

We draw the reader's attention to the fact that when generating bootstrap samples of associated variables, as in the above Example 3.12, these have to be generated by drawing cases at random with replacement (and not the variables individually), therefore preserving the association of the variables involved.

Commands 3.7. MATLAB and R commands for obtaining bootstrap distributions.

MATLAB	`bootstrp(m,'statistic', arg1, arg2,...)`
R	`boot(x, statistic, m, stype="i",...)`

SPSS and STATISTICA don't have menu options for obtaining bootstrap distributions (although SPSS has a bootstrap macro to be used in its Output Management System and STATISTICA has a bootstrapping facility built into its Structural Equation Modelling module).

The bootstrap function of MATLAB can be used directly with one of MATLAB's statistical functions, followed by its arguments. For instance, the bootstrap distribution of Example 3.9 can be obtained with:

```
>> b = bootstrp(1000,'trimmean',cao,10);
```

Notice the name of the statistical function written as a string (the function `trimmean` is indicated in Commands 2.7). The function call returns the vector b with the 1000 bootstrap replicates of the trimmed mean from where one can obtain the histogram and other statistics.

Let us now consider Example 3.12. Assuming that columns 7 and 13 of the `clays'` matrix represent the variables Al2O3 and K2O, respectively, one obtains the bootstrap distribution with:

```
>> b=bootstrp(1000,'corrcoef',clays(:,7),clays(:,13))
```

The `corrcoef` function (mentioned in Commands 2.9) generates a correlation matrix. Specifically, `corrcoef(clays(:,7), clays(:,13))` produces:

```
ans =
    1.0000    0.6922
    0.6922    1.0000
```

As a consequence each row of the b matrix contains in this case the correlation matrix values of one bootstrap sample. For instance:

```
b =
    1.0000    0.6956    0.6956    1.0000
    1.0000    0.7019    0.7019    1.0000
    ...
```

Hence, one may obtain the histogram and the bootstrap statistics using `b(:,2)` or `b(:,3)`.

In order to obtain bootstrap distributions with R one must first install the boot package with library(boot). One can check if the package is installed with the search() function (see section 1.7.2.2).

The boot function of the boot package will generate m bootstrap replicates of a statistical function, denoted statistic, passed (its name) as argument. However, this function should have as second argument a vector of indices, frequencies or weights. In our applications we will use a vector of indices, which corresponds to setting the stype argument to its default value, stype="i". Since it is the default value we really don't need to mention it when calling boot. Anyway, the need to have the mentioned second argument obliges one to write the code of the statistical function. Let us consider Example 3.10. Supposing the clays data frame has been created and attached, it would be solved in R in the following way:

```
> sdboot <- function(x,i)sd(x[i])
> b <- boot(CaO,sdboot,1000)
```

The first line defines the function sdboot with two arguments. The first argument is the data. The second argument is the vector of indices which will be used to store the index information of the bootstrap samples. The function itself computes the standard deviation of those data elements whose indices are in the index vector i (see the last paragraph of section 2.1.2.4).

The boot function returns a so-called bootstrap object, denoted above as b. By listing b one may obtain:

```
Bootstrap Statistics :
       original        bias      std. error
t1* 0.08601075 -0.00082119 0.007099508
```

which agrees fairly well with the values computed with MATLAB in Example 3.10. One of the attributes of the bootstrap object is the vector with the bootstrap replicates, denoted t. The histogram of the bootstrap distribution can therefore be obtained with:

```
> hist(b$t)
```

∎

Exercises

3.1 Consider the $1-\alpha_1$ and $1-\alpha_2$ confidence intervals of a given statistic with $1-\alpha_1 > 1-\alpha_2$. Why is the confidence interval for $1-\alpha_1$ always larger than or equal to the interval for $1-\alpha_2$?

3.2 Consider the measurements of bottle bottoms of the Moulds dataset. Determine the 95% confidence interval of the mean and the x-charts of the three variables RC, CG and EG. Taking into account the x-chart, discuss whether the 95% confidence interval of the RC mean can be considered a reliable estimate.

3.3   Compute the 95% confidence interval of the mean and of the standard deviation of the RC variable of the previous exercise, for the samples constituted by the first 50 cases and by the last 50 cases. Comment on the results.

3.4   Consider the ASTV and ALTV variables of the CTG dataset. Assume that only a 15-case random sample is available for these variables. Can one expect to obtain reliable estimates of the 95% confidence interval of the mean of these variables using the Student's t distribution applied to those samples? Why? (Inspect the variable histograms.)

3.5   Obtain a 15-case random sample of the ALTV variable of the previous exercise (see Commands 3.2). Compute the respective 95% confidence interval assuming a normal and an exponential fit to the data and compare the results. The exponential fit can be performed in MATLAB with the function expfit.

3.6   Compute the 90% confidence interval of the ASTV and ALTV variables of the previous Exercise 3.4 for 10 random samples of 20 cases and determine how many times the confidence interval contains the mean value determined for the whole 2126 case set. In a long run of these 20-case experiments, which variable is expected to yield a higher percentage of intervals containing the whole-set mean?

3.7   Compute the mean with the 95% confidence interval of variable ART of the Cork Stoppers dataset. Perform the same calculations on variable LOGART = ln(ART). Apply the Gauss' approximation formula of A.6.1 in order to compare the results. Which point estimates and confidence intervals are more reliable? Why?

3.8   Consider the PERIM variable of the Breast Tissue dataset. What is the tolerance of the PERIM mean with 95% confidence for the carcinoma class? How many cases of the carcinoma class should one have available in order to reduce that tolerance to 2%?

3.9   Imagine that when analysing the TW="Team Work" variable of the Metal Firms dataset, someone stated that the team-work is at least good (score 4) for $3/8 = 37.5\%$ of the metallurgic firms. Does this statement deserve any credit? (Compute the 95% confidence interval of this estimate.)

3.10 Consider the Culture dataset. Determine the 95% confidence interval of the proportion of boroughs spending more than 20% of the budget for musical activities.

3.11 Using the CTG dataset, determine the percentage of foetal heart rate cases that have abnormal short term variability of the heart rate more than 50% of the time, during calm sleep (CLASS A). Also, determine the 95% confidence interval of that percentage and how many cases should be available in order to obtain an interval estimate with 1% tolerance.

3.12 A proportion $\hat{p}$ was estimated in 225 cases. What are the approximate worst-case 95% confidence interval limits of the proportion?

3.13 Redo Exercises 3.2 and 3.3 for the 99% confidence interval of the standard deviation.

3.14 Consider the CTG dataset. Compute the 95% and 99% confidence intervals of the standard deviation of the ASTV variable. Are the confidence interval limits equally away from the sample mean? Why?

3.15 Consider the computation of the confidence interval for the standard deviation performed in Example 3.6. How many cases should one have available in order to obtain confidence interval limits deviating less than 5% of the point estimate?

3.16 In order to represent the area values of the cork defects in a convenient measurement unit, the ART values of the Cork Stoppers dataset have been multiplied by 5 and stored into variable ART5. Using the point estimates and 95% confidence intervals of the mean and the standard deviation of ART, determine the respective statistics for ART5.

3.17 Consider the ART, ARM and N variables of the Cork Stoppers' dataset. Since ARM = ART/N, why isn't the point estimate of the ART mean equal to the ratio of the point estimates of the ART and N means? (See properties of the mean in A.6.1.)

3.18 Redo Example 3.8 for the classes C = "calm vigilance" and D = "active vigilance" of the CTG dataset.

3.19 Using the bootstrap technique compute confidence intervals at 95% level of the mean and standard deviation for the ART data of Example 3.11.

3.20 Determine histograms of the bootstrap distribution of the median of the river Cávado flow rate (see Flow Rate dataset). Explain why it is unreasonable to set confidence intervals based on these histograms.

3.21 Using the bootstrap technique compute confidence intervals at 95% level of the mean and the two-tail 5% trimmed mean for the BRISA data of the Stock Exchange dataset. Compare both results.

3.22 Using the bootstrap technique compute confidence intervals at 95% level of the Pearson correlation between variables CaO and MgO of the Clays' dataset.

4 Parametric Tests of Hypotheses

In statistical data analysis an important objective is the capability of making decisions about population distributions and statistics based on samples. In order to make such decisions a hypothesis is formulated, e.g. "is one manufacture method better than another?", and tested using an appropriate methodology. Tests of hypotheses are an essential item in many scientific studies. In the present chapter we describe the most fundamental tests of hypotheses, assuming that the random variable distributions are known – the so-called *parametric tests*. We will first, however, present a few important notions in section 4.1 that apply to parametric and to non-parametric tests alike.

4.1 Hypothesis Test Procedure

Any hypothesis test procedure starts with the formulation of an interesting hypothesis concerning the distribution of a certain random variable in the population. As a result of the test we obtain a decision rule, which allows us to either reject or accept the hypothesis with a certain probability of error, referred to as the *level of significance* of the test.

In order to illustrate the basic steps of the test procedure, let us consider the following example. Two methods of manufacturing a special type of drill, respectively A and B, are characterised by the following average lifetime (in continuous work without failure): μ_A = 1100 hours and μ_B = 1300 hours. Both methods have an equal standard deviation of the lifetime, σ = 270 hours. A new manufacturer of the same type of drills claims that his brand is of a quality identical to the best one, B, and with lower manufacture costs. In order to assess this claim, a sample of 12 drills of the new brand were tested and yielded an average lifetime of $\bar{x}$ = 1260 hours. The interesting hypothesis to be analysed is that there is *no difference* between the new brand and the old brand B. We call it the *null hypothesis* and represent it by H_0. Denoting by μ the average lifetime of the new brand, we then formalise the test as:

H_0: $\mu = \mu_B = 1300$.
H_1: $\mu = \mu_A = 1100$.

Hypothesis H_1 is a so-called *alternative hypothesis*. There can be many alternative hypotheses, corresponding to $\mu \neq \mu_B$. However, for the time being, we assume that $\mu = \mu_A$ is the only interesting alternative hypothesis. We also assume that the lifetime of the drills, X, for all the brands, follows a *normal distribution* with the same standard deviation[1]. We know, therefore, that the sampling distribution of $\overline{X}$ is also normal with the following standard error (see sections 3.2 and A.8.4):

$$\sigma_{\overline{X}} = \frac{\sigma}{\sqrt{12}} = 77.94 \,.$$

The sampling distributions (pdf's) corresponding to both hypotheses are shown in Figure 4.1. We seek a procedure to decide whether the 12-drill-sample provides statistically significant evidence leading to the acceptance of the null hypothesis H_0. Given the symmetry of the distributions, a "common sense" approach would lead us to establish a *decision threshold*, $\overline{x}_{\alpha}$, halfway between μ_A and μ_B, i.e. $\overline{x}_{\alpha}$ =1200 hours, and decide H_0 if $\overline{x}$ >1200, decide H_1 if $\overline{x}$ <1200, and arbitrarily if $\overline{x}$ =1200.

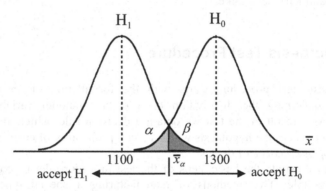

Figure 4.1. Sampling distribution (pdf) of $\overline{X}$ for the null and the alternative hypotheses.

Let us consider the four possible situations according to the truth of the null hypothesis and the conclusion drawn from the test, as shown in Figure 4.2. For the decision threshold $\overline{x}_{\alpha}$ =1200 shown in Figure 4.1, we then have:

$$\alpha = \beta = P(Z \le (1200 - 1300)/77.94) = N_{0,1}(-1.283) = 0.10 \,,$$

where Z is a random varable with standardised normal distribution.

---

[1] Strictly speaking the lifetime of the drills cannot follow a normal distribution, since $X > 0$. Also, as discussed in chapter 9, lifetime distributions are usually skewed. We assume, however, in this example, the distribution to be well approximated by the normal law.

Values of a normal random variable, standardised by subtracting the mean and dividing by the standard deviation, are called *z-scores*. In this case, the test errors α and β are evaluated using the *z*-score, -1.283.

In hypothesis tests, one is usually interested in that the probability of wrongly rejecting the null hypothesis is low; in other words, one wants to set a low value for the following *Type I Error*:

Type I Error: $\alpha = P(H_0$ is true and, based on the test, we reject $H_0)$.

This is the so-called *level of significance* of the test. The complement, $1-\alpha$, is the *confidence level*. A popular value for the level of significance that we will use throughout the book is $\alpha = 0.05$, often given in percentage, $\alpha = 5\%$. Knowing the α percentile of the standard normal distribution, one can easily determine the decision threshold for this level of significance:

$$P(Z \leq 0.05) = -1.64 \quad \Rightarrow \quad \bar{x}_\alpha = 1300 - 1.64 \times 77.94 = 1172.2 .$$

Decision

		Accept H_0	Accept H_1
Reality	H_0	Correct Decision	Type I Error α
	H_1	Type II Error β	Correct Decision

Figure 4.2. Types of error in hypothesis testing according to the reality and the decision drawn from the test.

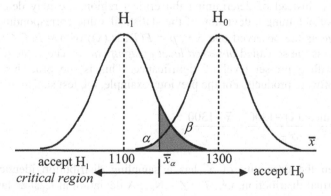

Figure 4.3. The critical region for a significance level of $\alpha = 5\%$.

Figure 4.3 shows the situation for this new decision threshold, which delimits the so-called *critical region* of the test, the region corresponding to a Type I Error. Since the computed sample mean for the new brand of drills, $\bar{x} = 1260$, falls in the non-critical region, we accept the null hypothesis at that level of significance (5%). In adopting this procedure, we expect that using it in a long run of sample-based tests, under identical conditions, we would be erroneously rejecting H_0 about 5% of the times.

In general, let us denote by C the critical region. If, as it happens in Figure 4.1 or 4.3, $\bar{x} \notin C$, we may say that "we accept the null hypothesis at that level of significance"; otherwise, we reject it.

Notice, however, that there is a non-null probability that a value as large as $\bar{x}$ could be obtained by type A drills, as expressed by the non-null β. Also, when we consider a wider range of alternative hypotheses, for instance $\mu < \mu_B$, there is always a possibility that a brand of drills with mean lifetime inferior to μ_B is, however, sufficiently close to yield with high probability sample means falling in the non-critical region. For these reasons, it is often advisable to adopt a conservative attitude stating that "there is no evidence to reject the null hypothesis at the α level of significance".

Any test procedure assessing whether or not H_0 should be rejected can be summarised as follows:

1. Choose a suitable *test statistic* $t_n(\mathbf{x})$, dependent on the n-dimensional sample $\mathbf{x} = [x_1, x_2, \ldots, x_n]'$, considered a value of a random variable, $T \equiv t_n(X)$, where X denotes the n-dimensional random variable associated to the sampling process.

2. Choose a level of significance α and use it together with the sampling distribution of T in order to determine the critical region C for H_0.

3. Test decision: If $t_n(\mathbf{x}) \in C$, then reject H_0, otherwise do not reject H_0. In the first case, the test is said to be *significant* (at level α); in the second case, the test is *non-significant*.

Frequently, instead of determining the critical region, we may determine the probability of obtaining a deviation of the statistical value corresponding to H_0 at least as large as the observed one, i.e., $p = P(T \geq t_n(\mathbf{x}))$ or $p = P(T \leq t_n(\mathbf{x}))$. The probability p is the so-called *observed level of significance*. The value of p is then compared with a pre-set level of significance. This is the procedure used by statistical software products. For the previous example, the test statistic is:

$$t_{12}(\mathbf{x}) = \frac{\text{mean}(\mathbf{x}) - 1300}{\sigma_{\bar{X}}} = \frac{\bar{x} - 1300}{\sigma_{\bar{X}}},$$

which, given the normality of X, has a sampling distribution identical to the standard normal distribution, i.e., $T = Z \sim N_{0,1}$. A deviation at least as large as the observed one in the left tail of the distribution has the observed significance:

$$p = P(Z \le (\overline{x} - \mu_B) / \sigma_{\overline{X}}) = P(Z \le (1260 - 1300) / 77.94) = 0.304 .$$

If we are basing our conclusions on a 5% level of significance, and since $p > 0.05$, we then have no evidence to reject the null hypothesis.

Note that until now we have assumed that we knew the true value of the standard deviation. This, however, is seldom the case. As already discussed in the previous chapter, when using the sample standard deviation – maintaining the assumption of normality of the random variable – one must use the Student's t distribution. This is the usual procedure, also followed by statistical software products, where these parametric tests of means are called t *tests*.

4.2   Test Errors and Test Power

As described in the previous section, any decision derived from hypothesis testing has, in general, a certain degree of uncertainty. For instance, in the drill example there is always a chance that the null hypothesis is incorrectly rejected. Suppose that a sample from the good quality of drills has $\overline{x} = 1190$ hours. Then, as can be seen in Figure 4.1, we would incorrectly reject the null hypothesis at a 10% significance level. However, we would not reject the null hypothesis at a 5% level, as shown in Figure 4.3. In general, by lowering the chosen level of significance, typically 0.1, 0.05 or 0.01, we decrease the Type I Error:

Type I Error: $\alpha = P(H_0$ is true and, based on the test, we reject $H_0)$.

The price to be paid for the decrease of the Type I Error is the increase of the *Type II Error*, defined as:

Type II Error: $\beta = P(H_0$ is false and, based on the test, we accept $H_0)$.

For instance, when in Figures 4.1 and 4.3 we decreased α from 0.10 to 0.05, the value of β increased from 0.10 to:

$$\beta = P(Z \ge (\overline{x}_\alpha - \mu_A) / \sigma_{\overline{X}}) = P(Z \ge (1172.8 - 1100) / 77.94) = 0.177 .$$

Note that a high value of β indicates that when the observed statistic does not fall in the critical region there is a good chance that this is due not to the verification of the null hypothesis itself but, instead, to the verification of a sufficiently close alternative hypothesis. Figure 4.4 shows that, for the same level of significance, α, as the alternative hypothesis approaches the null hypothesis, the value of β increases, reflecting a decreased protection against an alternative hypothesis.

The degree of protection against alternative hypotheses is usually measured by the so-called *power* of the test, $1 - \beta$, which measures the probability of rejecting the null hypothesis when it is false (and thus should be rejected). The values of the power for several alternative values of μ_A, using the computed values of β as shown above, are displayed in Table 4.1. The respective *power curve*, also called *operational characteristic* of the test, is shown with a solid line in Figure 4.5. Note that the power for the alternative hypothesis $\mu_A = 1100$ is somewhat higher than 80%. This is usually considered a lower limit of protection that one must have against alternative hypothesis.

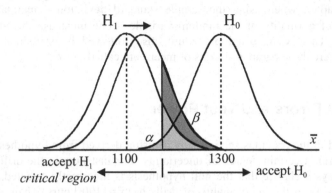

Figure 4.4. Increase of the Type II Error, β, for fixed α, when the alternative hypothesis approaches the null hypothesis.

Table 4.1. Type II Error and power for several alternative hypotheses of the drill example, with $n = 12$ and $\alpha = 0.05$.

μ_A	$z = (\mu_A - \bar{x}_{0.05})/\sigma_{\bar{X}}$	β	$1-\beta$
1100.0	0.93	0.18	0.82
1172.2	0.00	0.50	0.50
1200.0	−0.36	0.64	0.36
1250.0	−0.99	0.84	0.16
1300.0	−1.64	0.95	0.05

In general, for a given test and sample size, n, there is always a trade-off between either decreasing α or decreasing β. In order to increase the power of a test for a fixed level of significance, one is compelled to increase the sample size. For the drill example, let us assume that the sample size increased twofold, $n = 24$. We now have a reduction of $\sqrt{2}$ of the true standard deviation of the sample mean, i.e., $\sigma_{\bar{X}} = 55.11$. The distributions corresponding to the hypotheses are now more peaked; informally speaking, the hypotheses are better separated, allowing a smaller Type II Error for the same level of significance. Let us confirm this. The new decision threshold is now:

$$\bar{x}_\alpha = \mu_B - 1.64 \times \sigma_{\bar{X}} = 1300 - 1.64 \times 55.11 = 1209.6 ,$$

which, compared with the previous value, is less deviated from μ_B. The value of β for $\mu_A = 1100$ is now:

$$\beta = P(Z \geq (\bar{x}_\alpha - \mu_A)/\sigma_{\bar{X}}) = P(Z \geq (1209.6 - 1100)/55.11) = 0.023 .$$

Therefore, the power of the test improved substantially to 98%. Table 4.2 lists values of the power for several alternative hypotheses. The new power curve is shown with a dotted line in Figure 4.5. For increasing values of the sample size n, the power curve becomes steeper, allowing a higher degree of protection against alternative hypotheses for a small deviation from the null hypothesis.

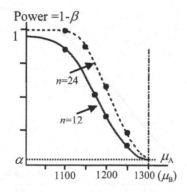

Figure 4.5. Power curve for the drill example, with $\alpha = 0.05$ and two values of the sample size n.

Table 4.2. Type II Error and power for several alternative hypotheses of the drill example, with $n = 24$ and $\alpha = 0.05$.

μ_A	$z = (\mu_A - \bar{x}_{0.05})/\sigma_{\bar{X}}$	β	$1-\beta$
1100	1.99	0.02	0.98
1150	1.08	0.14	0.86
1200	0.17	0.43	0.57
1250	−0.73	0.77	0.23
1300	−1.64	0.95	0.05

STATISTICA and SPSS have specific modules – Power Analysis and SamplePower, respectively – for performing power analysis for several types of tests. The R stats package also has a few functions for power calculations. Figure 4.6 illustrates the power curve obtained with STATISTICA for the last example. The power is displayed in terms of the *standardised effect*, E_s, which measures the deviation of the alternative hypothesis from the null hypothesis, normalised by the standard deviation, as follows:

$$E_s = \frac{\mu_B - \mu_A}{\sigma}.$$ 4.1

For instance, for $n = 24$ the protection against $\mu_A = 1100$ corresponds to a standardised effect of $(1300 - 1100)/260 = 0.74$ and the power graph of Figure 4.6 indicates a value of about 0.94 for $E_s = 0.74$. The difference from the previous value of 0.98 in Table 4.2 is due to the fact that, as already mentioned, STATISTICA uses the Student's t distribution.

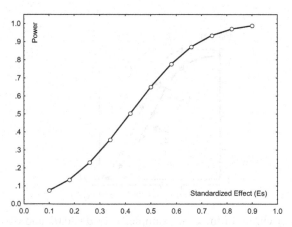

Figure 4.6. Power curve obtained with STATISTICA for the drill example with $\alpha = 0.05$ and $n = 24$.

In the work of Cohen (Cohen, 1983), some guidance is provided on how to qualify the standardised effect:

Small effect size: $E_s = 0.2$.
Medium effect size: $E_s = 0.5$.
Large effect size: $E_s = 0.8$.

In the example we have been discussing, we are in presence of a large effect size. As the effect size becomes smaller, one needs a larger sample size in order to obtain a reasonable power. For instance, imagine that the alternative hypothesis had precisely the same value as the sample mean, i.e., $\mu_A=1260$. In this case, the standardised effect is very small, $E_s = 0.148$. For this reason, we obtain very small values of the power for $n = 12$ and $n = 24$ (see the power for $\mu_A =1250$ in Tables 4.1 and 4.2). In order to "resolve" such close values (1260 and 1300) with low errors α and β, we need, of course, a much higher sample size. Figure 4.7 shows how the power evolves with the sample size in this example, for the fixed standardised effect $E_s = -0.148$ (the curve is independent of the sign of E_s). As can be appreciated, in order for the power to increase higher than 80%, we need $n > 350$.

Note that in the previous examples we have assumed alternative hypotheses that are always at one side of the null hypothesis: mean lifetime of the lower quality of drills. We then have a situation of *one-sided* or *one-tail* tests. We could as well contemplate alternative hypotheses of drills with better quality than the one corresponding to the null hypothesis. We would then have to deal with *two-sided* or *two-tail* tests. For the drill example a two-sided test is formalised as:

H_0: $\mu = \mu_B$.
H_1: $\mu \neq \mu_B$.

We will deal with two-sided tests in the following sections. For two-sided tests the power curve is symmetric. For instance, for the drill example, the two-sided power curve would include the reflection of the curves of Figure 4.5, around the point corresponding to the null hypothesis, μ_B.

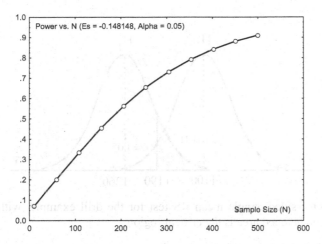

Figure 4.7. Evolution of the power with the sample size for the drill example, obtained with STATISTICA, with $\alpha = 0.05$ and $E_s = -0.148$.

A difficulty with tests of hypotheses is the selection of sensible values for α and β. In practice, there are two situations in which tests of hypotheses are applied:

1. The reject-support (RS) data analysis situation

This is by far the most common situation. The data analyst states H_1 as his belief, i.e., he seeks *to reject* H_0. In the drill example, the manufacturer of the new type of drills would formalise the test in a RS fashion if he wanted to claim that the new brand were better than brand A:

H_0: $\mu \leq \mu_A = 1100$.
H_1: $\mu > \mu_A$.

Figure 4.8 illustrates this one-sided, single mean test. The manufacturer is interested in a high power. In other words, he is interested that when H_1 is true (his belief) the probability of wrongly deciding H_0 (against his belief) is very low. In the case of the drills, for a sample size $n = 24$ and $\alpha = 0.05$, the power is 90% for the alternative $\mu = \bar{x}$, as illustrated in Figure 4.8. A power above 80% is often considered adequate to detect a reasonable departure from the null hypothesis.

On the other hand, society is interested in a low Type I Error, i.e., it is interested in a low probability of wrongly accepting the claim of the manufacturer when it is false. As we can see from Figure 4.8, there is again a trade-off between a low α and a low β. A very low α could have as consequence the inability to detect a new useful manufacturing method based on samples of reasonable size. There is a wide consensus that $\alpha = 0.05$ is an adequate value for most situations. When the sample sizes are very large (say, above 100 for most tests), trivial departures from H_0 may be detectable with high power. In such cases, one can consider lowering the value of α (say, $\alpha = 0.01$).

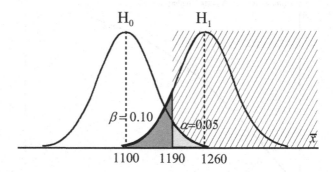

Figure 4.8. One-sided, single mean RS test for the drill example, with $\alpha = 0.05$ and $n = 24$. The hatched area is the critical region.

2. The accept-support (AS) data analysis situation

In this situation, the data analyst states H_0 as his belief, i.e., he seeks *to accept* H_0. In the drill example, the manufacturer of the new type of drills could formalise the test in an AS fashion if his claim is that the new brand is at least better than brand B:

H_0: $\mu \geq \mu_B = 1300$.
H_1: $\mu < \mu_B$.

Figure 4.9 illustrates this one-sided, single mean test. In the AS situation, lowering the Type I Error favours the manufacturer.

On the other hand, society is interested in a low Type II Error, i.e., it is interested in a low probability of wrongly accepting the claim of the manufacturer, H_0, when it is false. In the case of the drills, for a sample size $n = 24$ and $\alpha = 0.05$, the power is 17% for the alternative $\mu = \overline{x}$, as illustrated in Figure 4.9. This is an unacceptable low power. Even if we relax the Type I Error to $\alpha = 0.10$, the power is still unacceptably low (29%). Therefore, in this case, although there is no evidence supporting the rejection of the null hypothesis, there is also no evidence to accept it either.

In the AS situation, society should demand that the test be done with a sufficiently large sample size in order to obtain an adequate power. However, given the omnipresent trade-off between a low α and a low β, one should not impose a very high power because the corresponding α could then lead to the rejection of a hypothesis that explains the data almost perfectly. Again, a power value of at least 80% is generally adequate.

Note that the AS test situation is usually more difficult to interpret than the RS test situation. For this reason, it is also less commonly used.

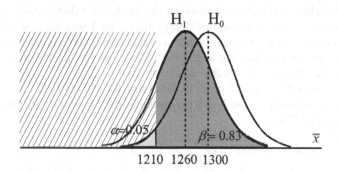

Figure 4.9. One-sided, single mean AS test for the drill example, with $\alpha = 0.05$ and $n = 24$. The hatched area is the critical region.

4.3   Inference on One Population

4.3.1 Testing a Mean

The purpose of the test is to assess whether or not the mean of a population, from which the sample was randomly collected, has a certain value. This single mean test was exemplified in the previous section 4.2. The hypotheses are:

$H_0: \mu = \mu_0$, $H_1: \mu \neq \mu_0$,   for a two-sided test;

$H_0: \mu \leq \mu_0$, $H_1: \mu > \mu_0$ or

$H_0: \mu \geq \mu_0$, $H_1: \mu < \mu_0$,   for a one-sided test.

We assume that the random variable being tested has a normal distribution. We then recall from section 3.2 that when the null hypothesis is verified, the following random variable:

$$T = \frac{\overline{X} - \mu_0}{s / \sqrt{n}} ,$$ 4.2

has a Student's t distribution with $n - 1$ degrees of freedom. We then use as the test statistic, $t_n(\mathbf{x})$, the following quantity:

$$t^* = \frac{\overline{x} - \mu_0}{s / \sqrt{n}}^2 .$$

When a statistic as t^* is standardised using the estimated standard deviation instead of the true standard deviation, it is called a *studentised statistic*.

For large samples, say $n > 25$, one could use the normal distribution instead, since it will yield a good approximation of the Student's t distribution. Even with small samples, we can use the normal distribution if we know the true value of the standard deviation. That's precisely what we have done in the preceding sections. However, in normal practice, the true value of the standard deviation is unknown and the test relies then on the Student's t distribution.

Assume a two-sided t test. In order to determine the critical region for a level of significance α, we compute the $1-\alpha/2$ percentile of the Student's t distribution with $df = n-1$ degrees of freedom:

$$T_{df}(t) = 1 - \alpha/2 \quad \Rightarrow \quad t_{df, 1-\alpha/2} ,$$ 4.3

and use this percentile in order to establish the non-critical region $\overline{C}$ of the test:

$$\overline{C} = \left[-t_{df, 1-\alpha/2}, +t_{df, 1-\alpha/2} \right].$$ 4.4

Thus, the two-sided probability of C is $2(\alpha/2) = \alpha$. The non-critical region can also be expressed in terms of $\overline{X}$, instead of T (formula 4.2):

$$\overline{C} = \left[\mu_0 - t_{df, 1-\alpha/2} \, s / \sqrt{n}, \, \mu_0 + t_{df, 1-\alpha/2} \, s / \sqrt{n} \right].$$ 4.4a

Notice how the test of a mean is similar to establishing a confidence interval for a mean.

---

[2] We use an asterisk to denote a test statistic.

Example 4.1

Q: Consider the Meteo (meteorological) dataset (see Appendix E). Perform the single mean test on the variable T81, representing the maximum temperature registered during 1981 at several weather stations in Portugal. Assume that, based on a large number of yearly records, a "typical" year has an average maximum temperature of 37.5°, which will be used as the test value. Also, assume that the Meteo dataset represents a random spatial sample and that the variable T81, for the population of an arbitrarily large number of measurements performed in the Portuguese territory, can be described by a normal distribution.

A: The purpose of the test is to assess whether or not 1981 was a "typical" year in regard to average maximum temperature. We then formalise the single mean test as:

H_0: $\mu_{T81} = 37.5$.

H_1: $\mu_{T81} \neq 37.5$.

Table 4.3 lists the results that can be obtained either with SPSS or with STATISTICA. The probability of obtaining a deviation from the test value, at least as large as $39.8 - 37.5$, is $p \approx 0$. Therefore, the test is significant, i.e., the sample does provide enough evidence to reject the null hypothesis at a very low α.

Notice that Table 4.3 also displays the values of t, the degrees of freedom, $df = n - 1$, and the *standard error* $s / \sqrt{n} = 0.548$.

□

Table 4.3. Results of the single mean t test for the T81 variable, obtained with SPSS or STATISTICA, with test value $\mu_0 = 37.5$.

Mean	Std. Dev.	n	Std. Err.	Test Value	t	df	p
39.8	2.739	25	0.548	37.5	4.199	24	0.0003

Example 4.2

Q: Redo previous Example 4.1, performing the test in its "canonical way", i.e., determining the limits of the critical region.

A: First we determine the t percentile for the set level of significance. In the present case, using $\alpha = 0.05$, we determine:

$t_{24,0.975} = 2.06$.

This determination can be done by either using the t distribution Tables (see Appendix D), or the probability calculator of the STATISTICA and SPSS, or the appropriate MATLAB or R functions (see Commands 3.3).

Using the t percentile value and the standard error, the non-critical region is the interval $[37.5 - 2.06 \times 0.548, 37.5 + 2.06 \times 0.548] = [36.4, 38.6]$. As the sample mean $\bar{x} = 39.8$ falls outside this interval, we also decide the rejection of the null hypothesis at that level of significance.

☐

Example 4.3

Q: Redo previous Example 4.2 in order to assess whether 1981 was a year with an atypically large average maximum temperature.

A: We now perform a one-sided test, using the alternative hypothesis:

H_1: $\mu_{T81} > 37.5$.

The critical region for this one-sided test, expressed in terms of $\bar{X}$, is:

$$C = [\mu_0 + t_{df, 1-\alpha} \ s / \sqrt{n}, \infty[.$$

Since $t_{24, 0.95} = 1.71$, we have $C = [37.5 + 1.71 \times 0.548, \infty [= [38.4, \infty [.$ Once again, the sample mean falls into the critical region leading to the rejection of the null hypothesis. Note that the alternative hypothesis $\mu_{T81} = 39.8$ in this Example 4.3 corresponds to a large effect size, $E_s = 0.84$, to which also corresponds a high power (larger than 95%; see Exercise 4.2).

☐

Commands 4.1. SPSS, STATISTICA, MATLAB and R commands used to perform the single mean t test.

SPSS	`Analyze; Compare Means; One-Sample T Test`
STATISTICA	`Statistics; Basic Statistics and Tables; t-test, single sample`
MATLAB	`[h,sig,ci]=ttest(x,m,alpha,tail)`
R	`t.test(x, alternative = c("two.sided", "less", "greater"), mu, conf.level)`

When using a statistical software product one obtains the probability of observing a value at least as large as the computed test statistic $t_n(\mathbf{x}) \equiv t^*$, assuming the null hypothesis. This probability is the so-called *observed significance*. The test decision is made comparing this observed significance with the chosen level of significance. Note that the published value of p corresponds to the two-sided observed significance. For instance, in the case of Table 4.3, the observed level of significance for the one-sided test is half of the published value, i.e., $p = 0.00015$.

When performing tests of hypotheses with MATLAB or R adequate percentiles for the critical region, the so-called *critical values*, are also computed.

MATLAB has a specific function for the single mean t test, which is shown in its general form in Commands 4.1. The best way to understand the meaning of the arguments is to run the previous Example 4.3 for T81. We assume that the sample is saved in the array t81 and perform the test as follows:

```
» [h,sig,ci]=ttest(t81,37.5,0.05,1)

h =
     1
sig =
   1.5907e-004
ci =
   38.8629    40.7371
```

The parameter tail can have the values 0, 1, −1, corresponding respectively to the alternative hypotheses $\mu \neq \mu_0$, $\mu > \mu_0$ and $\mu < \mu_0$. The value h = 1 informs us that the null hypothesis should be rejected (0 for not rejected). The variable sig is the observed significance; its value is practically the same as the above mentioned p. Finally, the vector ci is the 1 − alpha confidence interval for the true mean.

The same example is solved in R with:

```
> t.test(T81,alternative=("greater"),mu=37.5)

          One Sample t-test

data:  T81
t = 4.1992, df = 24, p-value = 0.0001591
alternative hypothesis: true mean is greater than
37.5
95 percent confidence interval:
   38.86291         Inf
sample estimates:
mean of x
     39.8
```

The conf.level of t.test is 0.95 by default.   ∎

4.3.2 Testing a Variance

The assessment of whether a random variable of a certain population has dispersion smaller or higher than a given "typical" value is an often-encountered task. Assuming that the random variable follows a normal distribution, this assessment can be performed by a test of a hypothesis involving a single variance, σ_0^2, as test value.

Let the sample variance, computed in the n-sized sample, be s^2. The test of a single variance is based on Property 5 of B.2.7, which states a chi-square sampling distribution for the ratio of the sample variance, $s_X^2 \equiv s^2(X)$, and the hypothesised variance:

$$s_X^2 / \sigma^2 ~\sim~ \chi_{n-1}^2 /(n-1) . \qquad\qquad 4.5$$

Example 4.4

Q: Consider the meteorological dataset and assume that a typical standard deviation for the yearly maximum temperature in the Portuguese territory is $\sigma = 2.2°$. This standard deviation reflects the spatial dispersion of maximum temperature in that territory. Also, consider the variable T81, representing the 1981 sample of 25 measurements of maximum temperature. Is there enough evidence, supported by the 1981 sample, leading to the conclusion that the standard deviation in 1981 was atypically high?

A: The test is formalised as:

$H_0: \sigma_{T81}^2 \le 4.84$.

$H_1: \sigma_{T81}^2 > 4.84$.

The sample variance in 1981 is $s^2 = 7.5$. Since the sample size of the example is $n = 25$, for a 5% level of significance we determine the percentile:

$$\chi_{24,0.95}^2 = 36.42 .$$

Thus, $\chi_{24,0.95}^2 / 24 = 1.52$.

This determination can be done in a variety of ways, as previously mentioned (in Commands 3.3): using the probability calculators of SPSS and STATISTICA, using MATLAB `chi2inv` function or R `qchisq` function, consulting tables (see D.4 for $P(\chi^2 > x) = 0.05$), etc.

Since $s^2 / \sigma^2 = 7.5 / 4.84 = 1.55$ lies in the critical region $[1.52, +\infty[$, we conclude that the test is significant, i.e., there is evidence supporting the rejection of the null hypothesis at the 5% level of significance.

[]

4.4   Inference on Two Populations

4.4.1 Testing a Correlation

When analysing two associated sample variables, one is often interested in knowing whether the sample provides enough evidence that the respective random variables are correlated. For instance, in data classification, when two variables are correlated and their correlation is high, one may contemplate the possibility of discarding one of the variables, since a highly correlated variable only conveys redundant information.

Let ρ represent the true value of the Pearson correlation mentioned in section 2.3.4. The correlation test is formalised as:

H_0: $\rho = 0$, H_1: $\rho \neq 0$, for a two-sided test.

For a one-sided test the alternative hypothesis is:

H_1: $\rho > 0$ or $\rho < 0$.

Let r represent the sample Pearson correlation when the null hypothesis is verified and the sample size is n. Furthermore, assume that the random variables are normally distributed. Then, the (r.v. corresponding to the) following test statistic:

$$t^* = r\sqrt{\frac{n-2}{1-r^2}} \, , \qquad\qquad\qquad 4.6$$

has a Student's t distribution with $n - 2$ degrees of freedom.

The Pearson correlation test can be performed as part of the computation of correlations with SPSS and STATISTICA. It can also be performed using the Correlation Test sheet of Tools.xls (see Appendix F) or the Probability Calculator; Correlations of STATISTICA (see also Commands 4.2).

Example 4.5

Q: Consider the variables PMax and T80 of the meteorological dataset (Meteo) for the "moderate" category of precipitation (PClass = 2) as defined in 2.1.2. We then have $n = 16$ measurements of the maximum precipitation and the maximum temperature during 1980, respectively. Is there evidence, at $\alpha = 0.05$, of a negative correlation between these two variables?

A: The distributions of PMax and T80 for "moderate" precipitation are reasonably well approximated by the normal distribution (see section 5.1). The sample correlation is $r = -0.53$. Thus, the test statistic is:

$r = -0.53, n = 16$ $\Rightarrow$ $t^* = -2.33$.

Since $t_{14,0.05} = -1.76$, the value of t^* falls in the critical region $] -\infty, -1.76]$; therefore, the null hypothesis is rejected, i.e., there is evidence of a negative correlation between PMax and T80 at that level of significance. Note that the observed significance of t^* is 0.0176, below α.

□

Commands 4.2. SPSS, STATISTICA, MATLAB and R commands used to perform the correlation test.

SPSS	Analyze; Correlate; Bivariate
STATISTICA	Statistics; Basic Statistics and Tables; Correlation Matrices
	Probability Calculator; Correlations
MATLAB	*[r,t,tcrit] = corrtest(x,y,alpha)*
R	cor.test(x, y, conf.level = 0.95, ...)

As mentioned above the Pearson correlation test can be performed as part of the computation of correlations with SPSS and STATISTICA. Also with the Correlations option of STATISTICA Probability Calculator.

MATLAB does not have a correlation test function. We do provide, however, a function for that purpose, corrtest (see Appendix F). Assuming that we have available the vector columns pmax, t80 and pclass as described in 2.1.2.3, Example 4.5 would be solved as:

```
>>[r,t,tcrit]=corrtest(pmax(pclass==2),t80(pclass==2)
,0.05)
  r =
    -0.5281
  t =
    -2.3268
  tcrit =
    -1.7613
```

The correlation test can be performed in R with the function cor.test. In Commands 4.2 we only show the main arguments of this function. As usual, by default conf.level=0.95. Example 4.5 would be solved as:

```
> cor.test(T80[Pclass==2],Pmax[Pclass==2])
        Pearson's product-moment correlation
data:  T80[Pclass == 2] and Pmax[Pclass == 2]
t = -2.3268, df = 14, p-value = 0.0355
alternative hypothesis: true correlation is not equal
to 0
95 percent confidence interval:
 -0.81138702 -0.04385491
sample estimates:
        cor
-0.5280802
```
■

As a final comment, we draw the reader's attention to the fact that correlation is by no means synonymous with causality. As a matter of fact, when two variables X and Y are correlated, one of the following situations can happen:

- One of the variables is the cause and the other is the effect. For instance, if X = "*nr of forest fires per year*" and Y = "*area of burnt forest per year*", then one usually finds that X is correlated with Y, since Y is the effect of X

- Both variables have an indirect cause. For instance, if X = "*% of persons daily arriving at a Hospital with yellow-tainted fingers*" and Y = "*% of persons daily arriving at the same Hospital with pulmonary carcinoma*", one finds that X is correlated with Y, but neither is cause or effect. Instead, there is another variable that is the cause of both – volume of inhaled tobacco smoke.

- The correlation is fortuitous and there is no causal link. For instance, one may eventually find a correlation between X = "*% of persons with blue eyes per household*" and Y = "*% of persons preferring radio to TV per household*". It would, however, be meaningless to infer causality between the two variables.

4.4.2 Comparing Two Variances

4.4.2.1   The F Test

In some comparison problems to be described later, one needs to decide whether or not two independent data samples A and B, with sample variances s_A^2 and s_B^2 and sample sizes n_A and n_B, were obtained from normally distributed populations with the same variance.

Using Property 6 of B.2.9, we know that:

$$\frac{s_A^2 / \sigma_A^2}{s_B^2 / \sigma_B^2} \sim F_{n_A-1, n_B-1}. \qquad 4.7$$

Under the null hypothesis "H_0: $\sigma_A^2 = \sigma_B^2$", we then use the test statistic:

$$F^* = s_A^2 / s_B^2 \sim F_{n_A-1, n_B-1}. \qquad 4.8$$

Note that given the asymmetry of the F distribution, one needs to compute the two $(1-\alpha/2)$-percentiles of F for a two-tailed test, and reject the null hypothesis if the observed F value is unusually large or unusually small. Note also that for applying the F test it is not necessary to assume that the populations have equal means.

Example 4.6

Q: Consider the two independent samples shown in Table 4.4 of normally distributed random variables. Test whether or not one should reject at a 5%

significance level the hypothesis that the respective population variances are unequal.

A: The sample variances are $v_1 = 1.680$ and $v_2 = 0.482$; therefore, $F^* = 3.49$, with an observed one-sided significance of $p = 0.027$. The 0.025 and 0.975 percentiles of $F_{9,11}$ are 0.26 and 3.59, respectively. Therefore, since the non-critical region [0.26, 3.59] contains p, we do not reject the null hypothesis at the 5% significance level.   □

Table 4.4. Two independent and normally distributed samples.

Case #	1	2	3	4	5	6	7	8	9	10	11	12
Group 1	4.7	3.7	5.2	6.3	6.2	6.7	2.8	4.8	6.1	3.9		
Group 2	10.1	8.6	10.9	9.7	9.7	10	9.4	10.1	9.9	10	10.8	8.7

Example 4.7

Q: Consider the meteorological data and test the validity of the following null hypothesis at a 5% level of significance:

H_0: $\sigma_{T81} = \sigma_{T80}$.

A: We assume, as in previous examples, that both variables are normally distributed. We then have to determine the percentiles of $F_{24,24}$ and the non-critical region:

$$\overline{C} = \left[F_{0.025},\ F_{0.975}\right] = \left[0.44, 2.27\right] .$$

Since $F^* = s_{T81}^2 / s_{T80}^2 = 7.5/4.84 = 1.55$ falls inside the non-critical region, the null hypothesis is not rejected at the 5% level of significance.

□

SPSS, STATISTICA and MATLAB do not include the test of variances as an individual option. Rather, they include this test as part of other tests, as will be seen in later sections. R has a function, `var.test`, which performs the F test of two variances. Running `var.test(T81,T80)` for the Example 4.7 one obtains:

```
F=1.5496, num df=24, denom df=24, p-value=0.2902
```

confirming the above results.

4.4.2.2 Levene's Test

A problem with the previous F test is that it is rather sensitive to the assumption of normality. A less sensitive test to the normality assumption (a more *robust* test) is

Levene's test, which uses deviations from the sample means. The test is carried out as follows:

1. Compute the means in the two samples: $\bar{x}_A$ and $\bar{x}_B$.

2. Let $d_{iA} = |x_{iA} - \bar{x}_A|$ and $d_{iB} = |x_{iB} - \bar{x}_B|$ represent the absolute deviations of the sample values around the respective mean.

3. Compute the sample means, $\bar{d}_A$ and $\bar{d}_B$, and sample variances, v_A and v_B of the previous absolute deviations.

4. Compute the *pooled variance*, v_p, for the two samples, with n_A and n_B cases, as the following weighted average of the individual variances:

$$s_p^2 \equiv v_p = \frac{(n_A - 1)v_A + (n_B - 1)v_B}{n_A + n_B - 2}.$$
4.9

5. Finally, perform a *t* test with the test statistic:

$$t^* = \frac{\bar{d}_A - \bar{d}_B}{s_p \sqrt{\frac{1}{n_A} + \frac{1}{n_B}}} \sim t_{n-2}.$$
4.10

There is a modification of the Levene's test that uses the deviations from the median instead of the mean (see section 7.3.3.2).

Example 4.8

Q: Redo the test of Example 4.7 using Levene's test.

A: The sample means are $\bar{x}_1 = 5.04$ and $\bar{x}_2 = 9.825$. Using these sample means, we compute the absolute deviations for the two groups shown in Table 4.5.
   The sample means and variances of these absolute deviations are: $\bar{d}_1 = 1.06$, $\bar{d}_2 = 0.492$; $v_1 = 0.432$, $v_2 = 0.235$. Applying formula 4.9 we obtain a pooled variance $v_p = 0.324$. Therefore, using formula 4.10, the observed test statistic is $t^* = 2.33$ with a two-sided observed significance of 0.03.
   Thus, we reject the null hypothesis of equal variances at a 5% significance level. Notice that this conclusion is the opposite of the one reached in Example 4.7.

□

Table 4.5. Absolute deviations from the sample means, computed for the two samples of Table 4.4.

Case #	1	2	3	4	5	6	7	8	9	10	11	12
Group 1	0.34	1.34	0.16	1.26	1.16	1.66	2.24	0.24	1.06	1.14		
Group 2	0.15	1.35	0.95	0.25	0.25	0.05	0.55	0.15	0.05	0.05	0.85	1.25

4.4.3 Comparing Two Means

4.4.3.1   Independent Samples and Paired Samples

Deciding whether two samples came from normally distributed populations with the same or with different means, is an often-met requirement in many data analysis tasks. The test is formalised as:

H_0: $\mu_A = \mu_B$ (or $\mu_A - \mu_B = 0$, whence the name "null hypothesis"),
H_1: $\mu_A \neq \mu_B$,   for a two-sided test;

H_0: $\mu_A \leq \mu_B$, H_1: $\mu_A > \mu_B$,   or
H_0: $\mu_A \geq \mu_B$, H_1: $\mu_A < \mu_B$,   for a one-sided test.

In tests of hypotheses involving two or more samples one must first clarify if the samples are *independent* or *paired*, since this will radically influence the methods used.

Imagine that two measurement devices, A and B, performed repeated and normally distributed measurements on the same object:

$x_1, x_2, \ldots, x_n$ with device A;
$y_1, y_2, \ldots, y_n$, with device B.

The sets $\mathbf{x} = [x_1\, x_2 \ldots x_n]$' and $\mathbf{y} = [\, y_1\, y_2 \ldots y_n]$', constitute *independent samples* generated according to N_{μ_A, σ_A} and N_{μ_B, σ_B}, respectively. Assuming that device B introduces a systematic deviation Δ, i.e., $\mu_B = \mu_A + \Delta$, our statistical model has 4 parameters: μ_A, Δ, σ_A and σ_B.

Now imagine that the n measurements were performed by A and B on a set of n different objects. We have a radically different situation, since now we must take into account the differences among the objects together with the systematic deviation Δ. For instance, the measurement of the object x_i is described in probabilistic terms by N_{μ_{Ai}, σ_A} when measured by A and by $N_{\mu_{Ai} + \Delta, \sigma_B}$ when measured by B. The statistical model now has $n + 3$ parameters: $\mu_{A1}, \mu_{A2}, \ldots, \mu_{An}$, Δ, σ_A and σ_B. The first n parameters reflect, of course, the differences among the n objects. Since our interest is the systematic deviation Δ, we apply the following trick. We compute the *paired* differences: $d_1 = y_1 - x_1$, $d_2 = y_2 - x_2$, $\ldots$, $d_n = y_n - x_n$. In this *paired samples* approach, we now may consider the measurements d_i as values of a random variable, D, described in probabilistic terms by N_{Δ, σ_D}. Therefore, the statistical model has now only two parameters.

The measurement device example we have been describing is a simple one, since the objects are assumed to be characterised by only one variable. Often the situation is more complex because several variables – known as *factors*, *effects* or *grouping variables* – influence the objects. The central idea in the "independent samples" study is that the cases are randomly drawn such that all the factors, except the one we are interested in, average out. For the "paired samples" study (also called *dependent* or *matched samples* study), the main precaution is that we pair truly comparable cases with respect to every important factor. Since this is an important topic, not only for the comparison of two means but for other tests as well, we present a few examples below.

Independent Samples:

i.   We wish to compare the sugar content of two sugar-beet breeds, A and B. For that purpose we collect random samples in a field of sugar-beet A and in another field of sugar-beet B. Imagine that the fields were prepared in the same way (e.g. same fertilizer, etc.) and the sugar content can only be influenced by exposition to the sun. Then, in order for the samples to be independent, we must make sure that the beets are drawn in a completely random way in what concerns the sun exposition. We then perform an "independent samples" test of variable "sugar content", dependent on factor "sugar-beet breed" with two categories, A and B.

ii. We are assessing the possible health benefit of a drug against a placebo. Imagine that the possible benefit of the drug depends on sex and age. Then, in an "independent samples" study, we must make sure that the samples for the drug and for the placebo (the so-called *control group*) are indeed random in what concerns sex and age. We then perform an "independent samples" test of variable "health benefit", dependent on factor "group" with two categories, "drug" and "placebo".

iii. We want to study whether men and women rate a TV program differently. Firstly, in an "independent samples" study, we must make sure that the samples are really random in what concerns other influential factors such as degree of education, environment, family income, reading habits, etc. We then perform an "independent samples" test of variable "TV program rate", dependent on factor "sex" with two categories, "man" and "woman".

Paired Samples:

i.   The comparison of sugar content of two breeds of sugar-beet, A and B, could also be studied in a "paired samples" approach. For that purpose, we would collect samples of beets A and B lying on nearby rows in the field, and would pair the neighbour beets.

ii. The study of the possible health benefit of a drug against a placebo could also be performed in a "paired samples" approach. For that purpose, the same group of patients is evaluated after taking the placebo and after taking the drug. Therefore, each patient is his/her own control. Of course, in clinical studies, ethical considerations often determine which kind of study must be performed.

iii. Studies of preference of a product, depending on sex, are sometimes performed in a "paired samples" approach, e.g. by pairing the enquiry results of the husband with those of the wife. The rationale being that husband and wife have similar ratings in what concerns influential factors such as degree of education, environment, age, reading habits, etc. Naturally, this assumption could be controversial.

Note that when performing tests with SPSS or STATISTICA for independent samples, one must have a datasheet column for the grouping variable that distinguishes the independent samples (groups). The grouping variable uses nominal codes (e.g. natural numbers) for that distinction. For paired samples, such a column does not exist because the variables to be tested are paired for each case.

4.4.3.2   Testing Means on Independent Samples

When two independent random variables X_A and X_B are normally distributed, as N_{μ_A,σ_A} and N_{μ_B,σ_B} respectively, then the variable $\overline{X}_A - \overline{X}_B$ has a normal distribution with mean $\mu_A - \mu_B$ and variance given by:

$$\sigma^2 = \frac{\sigma_A^2}{n_A} + \frac{\sigma_B^2}{n_B}. \qquad\qquad 4.11$$

where n_A and n_B are the sizes of the samples with means $\overline{x}_A$ and $\overline{x}_B$, respectively. Thus, when the variances are known, one can perform a comparison of two means much in the same way as in sections 4.1 and 4.2.

Usually the true values of the variances are unknown; therefore, one must apply a Student's t distribution. This is exactly what is assumed by SPSS, STATISTICA, MATLAB and R.

Two situations must now be considered:

1 – The variances σ_A and σ_B can be assumed to be equal.

Then, the following test statistic:

$$t^* = \frac{\overline{x}_A - \overline{x}_B}{\sqrt{\dfrac{v_p}{n_A} + \dfrac{v_p}{n_B}}}, \qquad\qquad 4.12$$

where v_p is the pooled variance computed as in formula 4.9, has a Student's t distribution with the following degrees of freedom:

$$df = n_A + n_B - 2. \qquad\qquad 4.13$$

2 – The variances σ_A and σ_B are unequal.

Then, the following test statistic:

$$t^* = \frac{\bar{x}_A - \bar{x}_B}{\sqrt{\frac{s_A^2}{n_A} + \frac{s_B^2}{n_B}}},$$

4.14

has a Student's t distribution with the following degrees of freedom:

$$df = \frac{(s_A^2 / n_A + s_B^2 / n_B)^2}{(s_A^2 / n_A)^2 / n_A + (s_B^2 / n_B)^2 / n_B}.$$

4.15

In order to decide which case to consider – equal or unequal variances – the F test or Levene's test, described in section 4.4.2, are performed. SPSS and STATISTICA do precisely this.

Example 4.9

Q: Consider the Wines' dataset (see description in Appendix E). Test at a 5% level of significance whether the variables ASP (aspartame content) and PHE (phenylalanine content) can distinguish white wines from red wines. The collected samples are assumed to be random. The distributions of ASP and PHE are well approximated by the normal distribution in both populations (white and red wines). The samples are described by the grouping variable TYPE (1 = white; 2 = red) and their sizes are $n_1 = 30$ and $n_2 = 37$, respectively.

A: Table 4.6 shows the results obtained with SPSS. In the interpretation of these results we start by looking to Levene's test results, which will decide if the variances can be assumed to be equal or unequal.

Table 4.6. Partial table of results obtained with SPSS for the independent samples t test of the wine dataset.

		Levene's Test		t-test				
		F	p	t	df	p (2-tailed)	Mean Difference	Std. Error Difference
ASP	Equal variances assumed	0.017	0.896	2.345	65	0.022	6.2032	2.6452
	Equal variances not assumed			2.356	63.16	0.022	6.2032	2.6331
PHE	Equal variances assumed	11.243	0.001	3.567	65	0.001	20.5686	5.7660
	Equal variances not assumed			3.383	44.21	0.002	20.5686	6.0803

For the variable ASP, we accept the null hypothesis of equal variances, since the observed significance is very high ($p = 0.896$). We then look to the t test results in the top row, which are based on the formulas 4.12 and 4.13. Note, particularly, that the number of degrees of freedom is $df = 30 + 37 - 2 = 65$. According to the results in the top row, we reject the null hypothesis of equal means with the observed significance $p = 0.022$. As a matter of fact, we also reject the one-sided hypothesis that aspartame content in white wines (sample mean 27.1 mg/l) is smaller or equal to the content in red wines (sample mean 20.9 mg/l). Note that the means of the two groups are more than two times the standard error apart.

For the variable PHE, we reject the hypothesis of equal variances; therefore, we look to the t test results in the bottom row, which are based on formulas 4.14 and 4.15. The null hypothesis of equal means is also rejected, now with higher significance since $p = 0.002$. Note that the means of the two groups are more than three times the standard error apart.

□

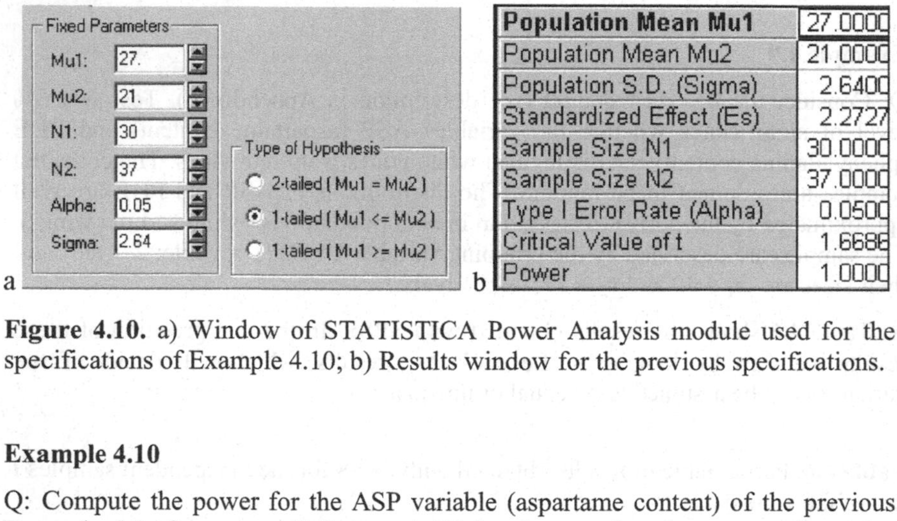

Figure 4.10. a) Window of STATISTICA Power Analysis module used for the specifications of Example 4.10; b) Results window for the previous specifications.

Example 4.10

Q: Compute the power for the ASP variable (aspartame content) of the previous Example 4.9, for a one-sided test at 5% level, assuming that as an alternative hypothesis white wines have more aspartame content than red wines. Determine what is the minimum distance between the population means that guarantees a power above 90% under the same conditions as the studied samples.

A: The one-sided test for this RS situation (see section 4.2) is formalised as:

H_0: $\mu_1 \leq \mu_2$;
H_1: $\mu_1 > \mu_2$. (White wines have more aspartame than red wines.)

The observed level of significance is half of the value shown in Table 4.6, i.e., $p = 0.011$; therefore, the null hypothesis is rejected at the 5% level. When the data analyst investigated the ASP variable, he wanted to draw conclusions with protection against a Type II Error, i.e., he wanted a low probability of wrongly not detecting the alternative hypothesis when true. Figure 4.10a shows the

STATISTICA specification window needed for the power computation. Note the specification of the one-sided hypothesis. Figure 4.10b shows that the power is very high when the alternative hypothesis is formalised with population means having the same values as the sample means; i.e., in this case the probability of erroneously deciding H_0 is negligible. Note the computed value of the standardised effect $(\mu_1 - \mu_2)/s = 2.27$, which is very large (see section 4.2).

Figure 4.11 shows the power curve depending on the standardised effect, from where we see that in order to have at least 90% power we need $E_s = 0.75$, i.e., we are guaranteed to detect aspartame differences of about 2 mg/l apart (precisely, $0.75 \times 2.64 = 1.98$).   □

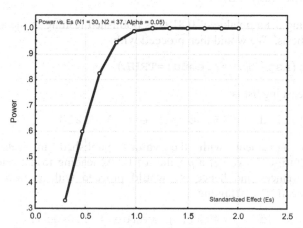

Figure 4.11. Power curve, obtained with STATISTICA, for the wine data Example 4.10.

Commands 4.3. SPSS, STATISTICA, MATLAB and R commands used to perform the two independent samples t test.

SPSS	`Analyze; Compare Means; Independent Samples T Test`
STATISTICA	`Statistics; Basic Statistics and Tables; t-test, independent, by groups`
MATLAB	`[h,sig,ci] = ttest2(x,y,alpha,tail]`
R	`t.test(formula, var.equal = FALSE)`

The MATLAB function `ttest2` works in the same way as the function `ttest` described in 4.3.1, with x and y representing two independent sample vectors. The function `ttest2` assumes that the variances of the samples are equal.

The R function t.test, already mentioned in Commands 4.1, can also be used to perform the two-sample *t* test. This function has several arguments the most important of which are mentioned above. Let us illustrate its use with Example 4.9. The first thing to do is to apply the two-variance *F* test with the var.test function mentioned in section 4.4.2.1. However, in this case we are analysing grouped data with a specific grouping (classification) variable: the wine type. For grouped data the function is applied as var.test(formula) where formula is written as var~group. In our Example 4.9, assuming variable CL represents the wine classification we would then test the equality of variances of variable Asp with:

```
> var.test(Asp~CL)
```

In the ensuing list a *p* value of 0.8194 is published leading to the acceptance of the null hypothesis. We would then proceed with:

```
> t.test(Asp~CL,var.equal=TRUE)
```

Part of the ensuing list is:

```
t = 2.3451, df = 65, p-value = 0.02208
```

which is in agreement with the values published in Table 4.6. For var.test(Phe~CL) we get a *p* value of 0.002 leading to the rejection of the equality of variances and hence we would proceed with t.test(Phe~CL, var.equal=FALSE) obtaining

```
t = 3.3828, df = 44.21, p-value = 0.001512
```

also in agreement with the values published in Table 4.6.

R stats package also has the following power.t.test function for performing power calculations of *t* tests:

```
power.t.test(n,  delta,  sd,  sig.level,  power,  type =
c("two.sample",  "one.sample",  "paired"),  alternative
= c("two.sided",  "one.sided"))
```

The arguments n, delta, sd are the number of cases, the difference of means and the standard deviation, respectively. The power calculation for the first part of Example 4.10 would then be performed with:

```
> power.t.test(30,  6,  2.64,  type=c("two.sample"),
  alternative=c("one.sided"))
```

A power of 1 is obtained. Note that the arguments of power.t.test have default values. For instance, in the above command we are assuming the default sig.level = 0.05. The power.t.test function also allows computing one parameter, passed as NULL, depending on the others. For instance, the second part of Example 4.10 would be solved with:

```
> power.t.test(30, delta=NULL, 2.64, power=0.9,
  type=c("two.sample"),alternative=c("one.sided"))
```

The result `delta = 2` would be obtained exactly as we found out in Figure 4.11.   ∎

4.4.3.3   Testing Means on Paired Samples

As explained in 4.4.3.1, given the sets $\mathbf{x} = [x_1 \, x_2 \ldots x_n]'$ and $\mathbf{y} = [y_1 \, y_2 \ldots y_n]'$, where the x_i, y_i refer to objects that can be paired, we then compute the *paired* differences: $d_1 = y_1 - x_1$, $d_2 = y_2 - x_2$, ..., $d_n = y_n - x_n$. Therefore, the null hypothesis:

H_0: $\mu_X = \mu_Y$,

is rewritten as:

H_0: $\mu_D = 0$ with $D = X - Y$.

The test is, therefore, converted into a single mean t test, using the studentised statistic:

$$t^* = \frac{\overline{d}}{s_d / \sqrt{n}} \quad \sim \quad t_{n-1},$$ 4.16

where s_d is the sample estimate of the variance of D, computed with the differences d_i. Note that since X and Y are not independent the additive property of the variances does not apply (see formula A.58c).

Example 4.11

Q: Consider the meteorological dataset. Use an appropriate test in order to compare the maximum temperatures of the year 1980 with those of the years 1981 and 1982.

A: Since the measurements are performed at the same weather stations, we are in adequate conditions for performing a paired samples t test. Based on the results shown in Table 4.7, we reject the null hypothesis for the pair T80-T81 and accept it for the pair T80-T82.

□

Table 4.7. Partial table of results, obtained with SPSS, in the paired samples t test for the meteorological dataset.

		Mean	Std. Deviation	Std. Error Mean	t	df	p (2-tailed)
Pair 1	T80 - T81	−2.360	2.0591	0.4118	−5.731	24	0.000
Pair 2	T80 - T82	0.000	1.6833	0.3367	0.000	24	1.000

Example 4.12

Q: Study the power of the tests performed in Example 4.11.

A: We use the STATISTICA Power Analysis module and the descriptive statistics shown in Table 4.8.

For the pair T80-T81, the standardised effect is $E_s = (39.8-37.44)/2.059 = 1.1$ (see Table 4.7 and 4.8). It is, therefore, a large effect – justifying a high power of the test.

Let us now turn our attention to the pair T80-T82, whose variables happen to have the same mean. Looking at Figure 4.12, note that in order to have a power $1-\beta = 0.8$, one must have a standardised effect of about $E_s = 0.58$. Since the standard deviation of the paired differences is 1.68, this corresponds to a deviation of the means computed as $E_s \times 1.68 = 0.97 \approx 1$. Thus, although the test does not reject the null hypothesis, we only have a reasonable protection against alternative hypotheses for a deviation in average maximum temperature of at least one degree centigrade.

□

Table 4.8. Descriptive statistics of the meteorological variables used in the paired samples t test.

	n	$\bar{x}$	s
T80	25	37.44	2.20
T81	25	39.80	2.74
T82	25	37.44	2.29

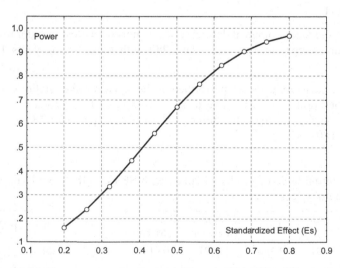

Figure 4.12. Power curve for the variable pair T80-T82 of Example 4.11.

Commands 4.4. SPSS, STATISTICA, MATLAB and R commands used to perform the paired samples *t* test.

SPSS	`Analyze; Compare Means; Paired-Samples T Test`
STATISTICA	`Statistics; Basic Statistics and Tables; t-test, dependent samples`
MATLAB	`[h,sig,ci]=ttest(x,m,alpha,tail]`
R	`t.test(x,y,paired = TRUE)`

With MATLAB the paired samples *t* test is performed using the single *t* test function `ttest`, previously described.

The R function `t.test`, already mentioned in Commands 4.1 and 4.3, is also used to perform the paired sample *t* test with the arguments mentioned above where x and y represent the paired data vectors. Thus, the comparison of T80 with T81 in Example 4.11 is solved with

```
> t.test(T80,T81,paired=TRUE)
```

obtaining the same values as in Table 4.7. The calculation of the difference of means for a power of 0.8 is performed with the `power.t.test` function (see Coomands 4.3) with:

```
> power.t.test(25,delta=NULL,1.68,power=0.8,
type=c("paired"),alternative=c("two.sided"))
```

yielding `delta = 0.98` in close agreement to the value found in Example 4.11 ∎

4.5   Inference on More than Two Populations

4.5.1 Introduction to the Analysis of Variance

In section 4.4.3, the two-means tests for independent samples and for paired samples were described. One could assume that, in order to infer whether more than two populations have the same mean, all that had to be done was to repeat the two-means test as many times as necessary. But in fact, this is not a commendable practice for the reason explained below.

Let us consider that we have c independent samples and we want to test whether the following null hypothesis is true:

$$H_0: \mu_1 = \mu_2 = \ldots = \mu_c ; \qquad\qquad 4.17$$

the alternative hypothesis being that there is at least one pair with unequal means, $\mu_i \neq \mu_j$.

We now assume that H_0 is assessed using two-means tests for all $\binom{c}{2}$ pairs of the c means. Moreover, we assume that every two-means test is performed at a 95% confidence level, i.e., the probability of not rejecting the null hypothesis when true, for every two-means comparison, is 95%:

$$P(\mu_i = \mu_j \mid H_{0ij}) = 0.95, \qquad\qquad 4.18$$

where H_{0ij} is the null hypothesis for the two-means test referring to the i and j samples.

The probability of rejecting the null hypothesis 4.17 for the c means, when it is true, is expressed as follows in terms of the two-means tests:

$$\begin{aligned} \alpha &= P(\text{reject } H_0 \mid H_0) \\ &= P(\mu_1 \neq \mu_2 \mid H_0 \text{ or } \mu_1 \neq \mu_3 \mid H_0 \text{ or} \ldots \text{or } \mu_{c-1} \neq \mu_c \mid H_0) \end{aligned} \qquad 4.19$$

Assuming the two-means tests are independent, we rewrite 4.19 as:

$$\alpha = 1 - P(\mu_1 = \mu_2 \mid H_0)P(\mu_1 = \mu_3 \mid H_0)\ldots P(\mu_{c-1} = \mu_c \mid H_0). \qquad 4.20$$

Since H_0 is more restrictive than any H_{0ij}, as it implies conditions on more than two means, we have $P(\mu_i \neq \mu_j \mid H_{0ij}) \geq P(\mu_i \neq \mu_j \mid H_0)$, or, equivalently, $P(\mu_i = \mu_j \mid H_{0ij}) \leq P(\mu_i = \mu_j \mid H_0)$.
Thus:

$$\alpha \geq 1 - P(\mu_1 = \mu_2 \mid H_{012})P(\mu_1 = \mu_3 \mid H_{013})\ldots P(\mu_{c-1} = \mu_c \mid H_{0c-1,c}). \qquad 4.21$$

For instance, for $c = 3$, using 4.18 and 4.21, we obtain a Type I Error $\alpha \geq 1 - 0.95^3 = 0.14$. For higher values of c the Type I Error degrades rapidly. Therefore, we need an approach that assesses the null hypothesis 4.17 in a "global" way, instead of assessing it using individual two-means tests.

In the following sections we describe the *analysis of variance* (ANOVA) approach, which provides a suitable methodology to test the "global" null hypothesis 4.17. We only describe the ANOVA approach for one or two grouping variables (*effects* or *factors*). Moreover, we only consider the so-called "fixed factors" model, i.e., we only consider making inferences on several fixed categories of a factor, observed in the dataset, and do not approach the problem of having to infer to more categories than the observed ones (the so called "random factors" model).

4.5.2 One-Way ANOVA

4.5.2.1   Test Procedure

The one-way ANOVA test is applied when only one grouping variable is present in the dataset, i.e., one has available c independent samples, corresponding to c categories (or *levels*) of an effect and wants to assess whether or not the null hypothesis should be rejected. As an example, one may have three independent samples of scores obtained by students in a certain course, corresponding to three different teaching methods, and want to assess whether or not the hypothesis of equality of student performance should be rejected. In this case, we have an effect – teaching method – with three categories.

A basic assumption for the variable X being tested is that the c independent samples are obtained from populations where X is normally distributed and with equal variance. Thus, the only possible difference among the populations refers to the means, μ_i. The equality of variance tests were already described in section 4.4.2. As to the normality assumption, if there are no "a priori" reasons to accept it, one can resort to goodness of fit tests described in the following chapter.

In order to understand the ANOVA approach, we start by considering a single sample of size n, subdivided in c subsets of sizes n_1, n_2, ..., n_c, with averages $\bar{x}_1, \bar{x}_2, ..., \bar{x}_k$, and investigate how the total variance, v, can be expressed in terms of the subset variances, v_i. Let any sample value be denoted x_{ij}, the first index referring to the subset, $i = 1, 2, ..., c$, and the second index to the case number inside the subset, $j = 1, 2, ..., n_i$. The total variance is related to the *total sum of squares*, SST, of the deviations from the *global sample mean*, $\bar{x}$:

$$\text{SST} = \sum_{i=1}^{c}\sum_{j=1}^{n_i}(x_{ij} - \bar{x})^2 .$$

$\qquad$ 4.22

Adding and subtracting $\bar{x}_i$ to the deviations, $x_{ij} - \bar{x}$, we derive:

$$\text{SST} = \sum_{i=1}^{c}\sum_{j=1}^{n_i}(x_{ij} - \bar{x}_i)^2 + \sum_{i=1}^{c}\sum_{j=1}^{n_i}(\bar{x}_i - \bar{x})^2 - 2\sum_{i=1}^{c}\sum_{j=1}^{n_i}(x_{ij} - \bar{x}_i)(\bar{x}_i - \bar{x}) .$$

$\qquad$ 4.23

The last term can be proven to be zero. Let us now analyse the other two terms. The first term is called the *within-group* (or *within-class*) *sum of squares*, SSW, and represents the contribution to the total variance of the errors due to the random scattering of the cases around their group means. This also represents an error term due to the scattering of the cases, the so-called *experimental error* or *error sum of squares*, SSE.

The second term is called the *between-group* (or *between-class*) *sum of squares*, SSB, and represents the contribution to the total variance of the deviations of the group means from the global mean.

Thus:

$$\text{SST} = \text{SSW} + \text{SSB}.$$

$\qquad$ 4.24

Let us now express these sums of squares, related by 4.24, in terms of variances:

$$SST = (n-1)v .$$ 4.25a

$$SSW \equiv SSE = \sum_{i=1}^{c} (n_i - 1)v_i = \left[\sum_{i=1}^{c} (n_i - 1) \right] v_W = (n-c)v_W .$$ 4.25b

$$SSB = (c-1)v_B .$$ 4.25c

Note that:

1. The *within-group variance*, v_W, is the *pooled variance* and corresponds to the generalization of formula 4.9:

$$v_W \equiv v_p = \frac{\sum_{i=1}^{c} (n_i - 1)v_i}{n-c} .$$ 4.26

This variance represents the stochastic behaviour of the cases around their group means. It is the point estimate of σ^2, the true variance of the population, and has $n - c$ degrees of freedom.

2. The within-group variance v_W represents a *mean square error*, MSE, of the observations:

$$MSE \equiv v_W = \frac{SSE}{n-c} .$$ 4.27

3. The *between-group variance*, v_B, represents the stochastic behaviour of the group means around the global mean. It is the point estimate of σ^2 when the null hypothesis is true, and has $c - 1$ degrees of freedom.
    When the number of cases per group is constant and equal to n, we get:

$$v_B = n \frac{\sum_{i=1}^{c} (\bar{x}_i - \bar{x})^2}{c-1} = nv_{\bar{X}} ,$$ 4.28

which is the sample expression of formula 3.8, allowing us to estimate the population variance, using the variance of the means.

4. The between-group variance, v_B, can be interpreted as a *mean between-group* or *classification sum of squares*, MSB:

$$MSB \equiv v_B = \frac{SSB}{c-1} .$$ 4.29

With the help of formula 4.24, we see that the *total sample variance*, v, can be broken down into two parts:

$$(n-1)v = (n-c)v_W + (c-1)v_B ,$$ 4.30

The ANOVA test uses precisely this "analysis of variance" property. Notice that the total number of degrees of freedom, $n - 1$, is also broken down into two parts: $n - c$ and $c - 1$.

Figure 4.13 illustrates examples for $c = 3$ of configurations for which the null hypothesis is true (a) and false (b). In the configuration of Figure 4.13a (null hypothesis is true) the three independent samples can be viewed as just one single sample, i.e., as if all cases were randomly extracted from a single population. The standard deviation of the population (shown in grey) can be estimated in two ways. One way of estimating the population variance is through the computation of the pooled variance, which assuming the samples are of equal size, n, is given by:

$$\hat{\sigma}^2 \equiv v \approx v_w = \frac{s_1^2 + s_2^2 + s_3^2}{3}. \qquad\qquad 4.31$$

The second way of estimating the population variance uses the variance of the means:

$$\hat{\sigma}^2 \equiv v \approx v_B = n v_{\overline{X}}. \qquad\qquad 4.32$$

When the null hypothesis is true, we expect both estimates to be near each other; therefore, their ratio should be close to 1. (If they are exactly equal 4.30 becomes an obvious equality.)

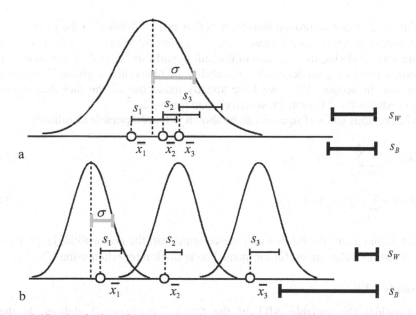

Figure 4.13. Analysis of variance, showing the means, $\overline{x}_i$, and the standard deviations, s_i, of three equal-sized samples in two configurations: a) H_0 is true; b) H_0 is false. On the right are shown the within-group and the between-group standard deviations (s_B is simply $s_{\overline{X}}$ multiplied by $\sqrt{n}$).

In the configuration of Figure 4.13b (null hypothesis is false), the between-group variance no longer represents an estimate of the population variance. In this case, we obtain a ratio v_B/v_W larger than 1. (In this case the contribution of v_B to the final value of v in 4.30 is smaller than the contribution of v_W.)

The *one-way ANOVA*, assuming the test conditions are satisfied, uses the following test statistic (see properties of the F distribution in section B.2.9):

$$F^* = \frac{v_B}{v_W} = \frac{MSB}{MSE} \sim F_{c-1,n-c} \text{ (under } H_0 \text{).} \tag{4.33}$$

If H_0 is not true, then F^* exceeds 1 in a statistically significant way.

The F distribution can be used even when there are mild deviations from the assumptions of normality and equality of variances. The equality of variances can be assessed using the ANOVA generalization of Levene's test described in the section 4.4.2.2.

Table 4.9. Critical F values at $\alpha = 0.05$ for $n = 25$ and several values of c.

c	2	3	4	5	6	7	8
$F_{c-1,n-c}$	4.26	3.42	3.05	2.84	2.71	2.63	2.58

For $c = 2$, it can be proved that the ANOVA test is identical to the t test for two independent samples. As c increases, the $1 - \alpha$ percentile of $F_{c-1,n-c}$ decreases (see Table 4.9), rendering the rejection of the null hypothesis "easier". Equivalently, for a certain level of confidence the probability of observing a given F^* under H_0 decreases. In section 4.5.1, we have already made use of the fact that the null hypothesis for $c > 2$ is more "restrictive" than for $c = 2$.

The previous sums of squares can be shown to be computable as follows:

$$SST = \sum_{i=1}^{c} \sum_{j=1}^{r_i} x_{ij}^2 - T^2 / n, \tag{4.34a}$$

$$SSB = \sum_{i=1}^{c} (T_i^2 / r_i) - T^2 / n, \tag{4.34b}$$

where T_i and T are the *totals* along the columns and the *grand total*, respectively. These last formulas are useful for manual computation (or when using EXCEL).

Example 4.13

Q: Consider the variable ART of the Cork Stoppers' dataset. Is there evidence, provided by the samples, that the three classes correspond to three different populations?

A: We use the one-way ANOVA test for the variable ART, with $c = 3$. Note that we can accept that the variable ART is normally distributed in the three classes using specific tests to be explained in the following chapter. For the moment, the reader has to rely on visual inspection of the normal fit curve to the histograms of ART.

Using MATLAB, one obtains the results shown in Figure 4.14. The box plot for the three classes, obtained with MATLAB, is shown in Figure 4.15. The MATLAB ANOVA results are obtained with the `anova1` command (see Commands 4.5) applied to vectors representing independent samples:

```
» x=[art(1:50),art(51:100),art(101:150)];
» p=anova1(x)
```

Note that the results table shown in Figure 4.14 has the classic configuration of the ANOVA tests, with columns for the total sums of squares (`SS`), degrees of freedom (`df`) and mean sums of squares (`MS`). The `source` of variance can be a between effect due to the `columns` (vectors) or a within effect due to the experimental `error`, adding up to a `total` contribution. Note particularly that MSB is much larger than MSE, yielding a significant (high F) test with the rejection of the null hypothesis of equality of means.

One can also compute the 95% percentile of $F_{2,147} = 3.06$. Since $F^* = 273.03$ falls within the critical region [3.06, +∞ [, we reject the null hypothesis at the 5% level.

Visual inspection of Figure 4.15 suggests that the variances of ART in the three classes may not be equal. In order to assess the assumption of equality of variances when applying ANOVA tests, it is customary to use the one-way ANOVA version of either of the tests described in section 4.4.2. For instance, Table 4.10 shows the results of the Levene test for *homogeneity of variances*, which is built using the breakdown of the total variance of the absolute deviations of the sample values around the means. The test rejects the null hypothesis of variance homogeneity. This casts a reasonable doubt on the applicability of the ANOVA test.

□

ANOVA Table					
Source	SS	df	MS	F	Prob>F
Columns	4.75959e+006	2	2379796.17	273.03	0
Error	1.2813e+006	147	8716.32		
Total	6.04089e+006	149			

Figure 4.14. One-way ANOVA test results, obtained with MATLAB, for the cork-stopper problem (variable ART).

Table 4.10. Levene's test results, obtained with SPSS, for the cork stopper problem (variable ART).

Levene Statistic	df1	df2	Sig.
27.388	2	147	0.000

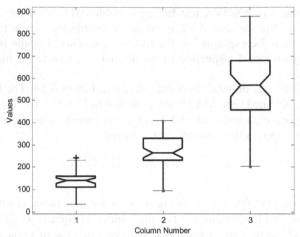

Figure 4.15. Box plot, obtained with MATLAB, for variable ART (Example 4.13).

As previously mentioned, a basic assumption of the ANOVA test is that the samples are independently collected. Another assumption, related to the use of the *F* distribution, is that the dependent variable being tested is normally distributed. When using large samples, say with the smallest sample size larger than 25, we can relax this assumption since the Central Limit Theorem will guarantee an approximately normal distribution of the sample means.

Finally, the assumption of equal variances is crucial, especially if the sample sizes are unequal. As a matter of fact, if the variances are unequal, we are violating the basic assumptions of what MSE and MSB are estimating. Sometimes when the variances are unequal, one can resort to a transformation, e.g. using the logarithm function of the dependent variable to obtain approximately equal variances. If this fails, one must resort to a non-parametric test, described in Chapter 5.

Table 4.11. Standard deviations of variables ART and ART1 = ln(ART) in the three classes of cork stoppers.

	Class 1	Class 2	Class3
ART	43.0	69.0	139.8
ART1	0.368	0.288	0.276

Example 4.14

Q: Redo the previous example in order to guarantee the assumption of equality of variances.

A: We use a new variable ART1 computed as: ART1 = ln(ART). The deviation of this new variable from the normality is moderate and the sample is large (50 cases per group), thereby allowing us to use the ANOVA test. As to the variances, Table 4.11 compares the standard deviation values before and after the logarithmic transformation. Notice how the transformation yielded approximate standard deviations, capitalising on the fact that the logarithm de-emphasises large values.

Table 4.12 shows the result of the Levene test, which authorises us to accept the hypothesis of equality of variances.

Applying the ANOVA test to ART1 the conclusions are identical to the ones reached in the previous example (see Table 4.13), namely we reject the equality of means hypothesis.

□

Table 4.12. Levene's test results, obtained with SPSS, for the cork-stopper problem (variable ART1 = ln(ART)).

Levene Statistic	df1	df2	Sig.
1.389	2	147	0.253

Table 4.13. One-way ANOVA test results, obtained with SPSS, for the cork-stopper problem (variable ART1 = ln(ART)).

	Sum of Squares	df	Mean Square	F	Sig.
Between Groups	51.732	2	25.866	263.151	0.000
Within Groups	14.449	147	9.829E-02		
Total	66.181	149			

Commands 4.5. SPSS, STATISTICA, MATLAB and R commands used to perform the one-way ANOVA test.

SPSS	`Analyze; Compare Means; Means|One-Way ANOVA` `Analyze; General Linear Model; Univariate`

STATISTICA	`Statistics; Basic Statistics and Tables; Breakdown & one-way ANOVA`
	`Statistics; ANOVA; One-way ANOVA`
	`Statistics; Advanced Linear/Nonlinear Models; General Linear Models; One-way ANOVA`

MATLAB	`[p,table,stats]=anova1(x,group,'dispopt')`
R	`anova(lm(X~f))`

The easiest commands to perform the one-way ANOVA test with SPSS and STATISTICA are with `Compare Means` and `ANOVA`, respectively.

"Post hoc" comparisons (e.g. Scheffé test), to be dealt with in the following section, are accessible using the `Post-hoc` tab in STATISTICA (click `More Results`) or clicking the `Post Hoc` button in SPSS. Contrasts can be performed using the `Planned comps` tab in STATISTICA (click `More Results`) or clicking the `Contrasts` button in SPSS.

Note that the ANOVA commands are also used in regression analysis, as explained in Chapter 7. When performing regression analysis, one often considers an "intercept" factor in the model. When comparing means, this factor is meaningless. Be sure, therefore, to check the `No intercept` box in STATISTICA (`Options` tab) and uncheck `Include intercept in the model` in SPSS (`General Linear Model`). In STATISTICA the `Sigma-restricted` box must also be unchecked.

The meanings of the arguments and return values of MATLAB `anova1` command are as follows:

p: *p* value of the null hypothesis;
table: matrix for storing the returned ANOVA table;
stats: test statistics, useful for performing multiple comparison of means with the `multcompare` function;
x: data matrix with each column corresponding to an independent sample;
group: optional character array with group names in each row;
dispopt: display option with two values, 'on' and 'off'. The default 'on' displays plots of the results (including the ANOVA table).

We now illustrate how to apply the one-way ANOVA test in R for the Example 4.14. The first thing to do is to create the ART1 variable with `ART1 <- log(ART)`. We then proceed to create a *factor* variable from the data frame classification variable denoted `CL`. The factor variable type in R is used to define a categorical variable with label values. The need of this step is that the ANOVA test can also be applied to continuous variables as we will see in Chapter 7. The creation of a factor variable from the numerical variable `CL` can be done with:

```
> CLf <- factor(CL,labels=c("I","II","III"))
```

Finally, we perform the one-way ANOVA with:

```
> anova(lm(ART1~CLf))
```

The `anova` call returns the following table similar to Table 4.13:

```
            Df Sum Sq Mean Sq F value     Pr(>F)
CLf          2 51.732  25.866  263.15 < 2.2e-16 ***
Residuals  147 14.449   0.098
---
Signif. codes:  0 '***' 0.001 '**' 0.01 '*' 0.05 '.'
        0.1 ' ' 1
```

■

4.5.2.2 *Post Hoc Comparisons*

Frequently, when performing one-way ANOVA tests resulting in the rejection of the null hypothesis, we are interested in knowing which groups or classes can then be considered as distinct. This knowledge can be obtained by a multitude of tests, known as post-hoc comparisons, which take into account pair-wise combinations of groups. These comparisons can be performed on individual pairs, the so-called *contrasts*, or considering all possible pair-wise combinations of means with the aim of detecting homogeneous groups of classes.

Software products such as SPSS and STATISTICA afford the possibility of analysing contrasts, using the *t* test. A contrast is specified by a *linear combination* of the population means:

$$H_0: a_1\mu_1 + a_2\mu_2 + \dots + a_k\mu_k = 0. \qquad 4.35$$

Imagine, for instance, that we wanted to compare the means of populations 1 and 2. The comparison is expressed as whether or not $\mu_1 = \mu_2$, or, equivalently, $\mu_1 - \mu_2 = 0$; therefore, we would use $a_1 = 1$ and $a_2 = -1$. We can also use groups of classes in contrasts. For instance, the comparison $\mu_1 = (\mu_3 + \mu_4)/2$ in a 5 class problem would use the contrast coefficients: $a_1 = 1$; $a_2 = 0$; $a_3 = -0.5$; $a_4 = -0.5$; $a_5 = 0$. We could also, equivalently, use the following integer coefficients: $a_1 = 2$; $a_2 = 0$; $a_3 = -1$; $a_4 = -1$; $a_5 = 0$.

Briefly, in order to specify a contrast (in SPSS or in STATISTICA), one assigns integer coefficients to the classes as follows:

i.   Classes omitted from the contrast have a coefficient of zero;
ii. Classes merged in one group have equal coefficients;
iii. Classes compared with each other are assigned positive or negative values, respectively;
iv. The total sum of the coefficients must be zero.

R has also the function `pairwise.t.test` that performs pair-wise comparisons of all levels of a factor with adjustment of the *p* significance for the multiple testing involved. For instance, `pairwise.t.test(ART1,CLf)` would perform all possible pair-wise contrasts for the example described in Commands 4.5.

It is possible to test a set of contrasts simultaneously based on the test statistic:

$$q = \frac{R_{\overline{x}}}{s_p / \sqrt{n}}, \qquad 4.36$$

where $R_{\overline{x}}$ is the observed range of the means. Tables of the sampling distribution of q, when the null hypothesis of equal means is true, can be found in the literature.

It can also be proven that the sampling distribution of q can be used to establish the following $1 - \alpha$ confidence intervals:

$$\frac{-q_{1-\alpha}s_p}{\sqrt{n}} < (a_1\bar{x}_1 + a_2\bar{x}_2 + \cdots + a_k\bar{x}_k)$$

$$-(a_1\mu_1 + a_2\mu_2 + \cdots + a_k\mu_k) < \frac{q_{1-\alpha}s_p}{\sqrt{n}} \,.$$

4.37

A popular test available in SPSS and STATISTICA, based on the result 4.37, is the Scheffé test. This test assesses simultaneously all possible pair-wise combinations of means with the aim of detecting homogeneous groups of classes.

Example 4.15

Q: Perform a one-way ANOVA on the Breast Tissue dataset, with post-hoc Scheffé test if applicable, using variable PA500. Discuss the results.

A: Using the goodness of fit tests to be described in the following chapter, it is possible to show that variable PA500 distribution can be well approximated by the normal distribution in the six classes of breast tissue. Levene's test and one-way ANOVA test results are displayed in Tables 4.14 and 4.15.

Table 4.14. Levene's test results obtained with SPSS for the breast tissue problem (variable PA500).

Levene Statistic	df1	df2	Sig.
1.747	5	100	0.131

Table 4.15. One-way ANOVA test results obtained with SPSS for the breast tissue problem (variable PA500).

	Sum of Squares	df	Mean Square	F	Sig.
Between Groups	0.301	5	6.018E-02	31.135	0.000
Within Groups	0.193	100	1.933E-03		
Total	0.494	105			

We see in Table 4.14 that the hypothesis of homogeneity of variances is not rejected at a 5% level. Therefore, the assumptions for applying the ANOVA test are fulfilled.

Table 4.15 justifies the rejection of the null hypothesis with high significance ($p < 0.01$). This result entitles us to proceed to a post-hoc comparison using the Scheffé test, whose results are displayed in Table 4.16. We see that the following groups of classes were found as distinct at a 5% significance level:

{CON, ADI, FAD, GLA}; {ADI, FAD, GLA, MAS}; {CAR}

These results show that variable PA500 can be helpful in the discrimination of carcinoma type tissues from other types.

□

Table 4.16. Scheffé test results obtained with SPSS, for the breast tissue problem (variable PA500). Values under columns "1", "2" and "3" are group means.

		Subset for alpha = 0.05		
CLASS	N	1	2	3
CON	14	7.029E-02		
ADI	22	7.355E-02	7.355E-02	
FAD	15	9.533E-02	9.533E-02	
GLA	16	0.1170	0.1170	
MAS	18		0.1231	
CAR	21			0.2199
Sig.		0.094	0.062	1.000

Example 4.16

Q: Taking into account the results of the previous Example 4.15, it may be asked whether or not class {CON} can be distinguished from the three-class group {ADI, FAD, GLA}, using variable PA500. Perform a contrast test in order to elucidate this issue.

A: We perform the contrast corresponding to the null hypothesis:

H_0: $\mu_{CON} = (\mu_{FAD} + \mu_{GLA} + \mu_{ADI})/3$,

i.e., we test whether or not the mean of class {CON} can be accepted equal to the mean of the joint class {FAD, GLA, ADI}. We therefore use the contrast coefficients shown in Table 4.17. Table 4.18 shows the t-test results for this contrast. The possibility of using variable PA500 for discrimination of class {CON} from the other three classes seems reasonable.

□

Table 4.17. Coefficients for the contrast {CON} vs. {FAD, GLA, ADI}.

CAR	FAD	MAS	GLA	CON	ADI
0	−1	0	−1	3	−1

Table 4.18. Results of the t test for the contrast specified in Table 4.17.

	Value of Contrast	Std. Error	t	df	Sig. (2-tailed)
Assume equal variances	−7.502E−02	3.975E−02	−1.887	100	0.062
Does not assume equal variances	−7.502E−02	2.801E−02	−2.678	31.79	0.012

4.5.2.3 Power of the One-Way ANOVA

In the one-way ANOVA, the null hypothesis states the equality of the means of c populations, $\mu_1 = \mu_2 = \ldots = \mu_c$, which are assumed to have a common value σ^2 for the variance. Alternative hypothesies correspond to specifying different values for the population means. In this case, the spread of the means can be measured as:

$$\sum_{i=1}^{c}(\mu_i - \overline{\mu})^2 /(c-1) .$$

4.38

It is convenient to standardise this quantity by dividing it by σ^2/n:

$$\phi^2 = \frac{\sum_{i=1}^{c}(\mu_i - \overline{\mu})^2 /(c-1)}{\sigma^2 / n},$$

4.39

where n is the number of observations from each population.

The square root of this quantity is known as the *root mean square standardised effect*, RMSSE $\equiv \phi$. The sampling distribution of RMSSE when the basic assumptions hold is available in tables and used by SPSS and STATISTICA power modules. R has the following `power.anova.test` function:

```
power.anova.test(g, n, between.var, within.var,
sig.level, power)
```

The parameters g and n are the number of groups and of cases per group, respectively. This functions works similarly to the `power.t.test` function described in Commands 4.4.

Example 4.17

Q: Determine the power of the one-way ANOVA test performed in Example 4.14 (variable ART1) assuming as an alternative hypothesis that the population means are the sample means.

A: Figure 4.16 shows the STATISTICA specification window for this power test. The RMSSE value can be specified using the `Calc. Effects` button and filling in the values of the sample means. The computed power is 1, therefore a good detection of the alternative hypothesis is expected. This same value is obtained in R issuing the command (see the between and within variance values in Table 4.13):

```
> power.anova.test(3, 50, between.var = 25.866,
  within.var = 0.098).
```

□

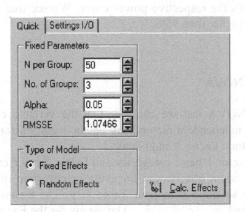

Figure 4.16. STATISTICA specification window for computing the power of the one-way ANOVA test of Example 4.17.

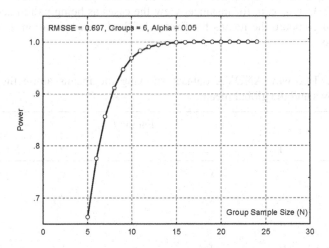

Figure 4.17. Power curve obtained with STATISTICA showing the dependence on n, for Example 4.18.

Example 4.18

Q: Consider the one-way ANOVA test performed in Example 4.15 (breast tissue). Compute its power assuming population means equal to the sample means and determine the minimum value of n that will guarantee a power of at least 95% in the conditions of the test.

A: We compute the power for the worst case of n: $n = 14$. Using the sample means as the means corresponding to the alternative hypothesis, and the estimate of the standard deviation s = 0.068, we obtain a standardised effect RMSSE = 0.6973. In these conditions, the power is 99.7%.

Figure 4.17 shows the respective power curve. We see that a value of $n \geq 10$ guarantees a power higher than 95%.

$\square$

4.5.3 Two-Way ANOVA

In the two-way ANOVA test we consider that the variable being tested, X, is categorised by two independent factors, say Factor 1 and Factor 2. We say that X depends on two factors: Factor 1 and Factor 2.

Assuming that Factor 1 has c categories and Factor 2 has r categories, and that there is only one random observation for every combination of categories of the factors, we get the situation shown in Table 4.19. The means for the Factor 1 categories are denoted $\bar{x}_{1.}$, $\bar{x}_{2.}$, ..., $\bar{x}_{c.}$. The means for the Factor 2 categories are denoted $\bar{x}_{.1}$, $\bar{x}_{.2}$, ..., $\bar{x}_{.r}$. The total mean for all observations is denoted $\bar{x}_{..}$.

Note that the situation shown in Table 4.19 constitutes a generalisation to multiple samples of the comparison of means for two paired samples described in section 4.4.3.3. One can, for instance, view the cases as being paired according to Factor 2 and compare the means for Factor 1. The inverse situation is, of course, also possible.

Table 4.19. Two-way ANOVA dataset showing the means along the columns, along the rows and the global mean.

	Factor 1				
Factor 2	1	2	...	c	Mean
1	x_{11}	x_{21}	...	x_{c1}	$\bar{x}_{.1}$
2	x_{12}	x_{22}	...	x_{c2}	$\bar{x}_{.2}$
...	...	...	...	...	...
r	x_{1r}	x_{2r}	...	x_{cr}	$\bar{x}_{.r}$
Mean	$\bar{x}_{1.}$	$\bar{x}_{2.}$	...	$\bar{x}_{c.}$	$\bar{x}_{..}$

Following the ANOVA approach of breaking down the total sum of squares (see formulas 4.22 through 4.30), we are now interested in reflecting the dispersion of the means along the rows and along the columns. This can be done as follows:

$$\mathrm{SST} = \sum_{i=1}^{c} \sum_{j=1}^{r} (x_{ij} - \bar{x}_{..})^2$$

$$= r \sum_{i=1}^{c} (\bar{x}_{i.} - \bar{x}_{..})^2 + c \sum_{j=1}^{r} (\bar{x}_{.j} - \bar{x}_{..})^2 + \sum_{i=1}^{c} \sum_{j=1}^{r} (x_{ij} - \bar{x}_{i.} - \bar{x}_{.j} + \bar{x}_{..})^2 \qquad 4.40$$

$$= \mathrm{SSC} + \mathrm{SSR} + \mathrm{SSE} \quad .$$

Besides the term SST described in the previous section, the sums of squares have the following interpretation:

1. SSC represents the sum of squares or *dispersion along the columns*, as the previous SSB. The variance along the columns is $v_c = \mathrm{SSC}/(c-1)$, has $c-1$ degrees of freedom and is the point estimate of $\sigma^2 + r\sigma_c^2$.

2. SSR represents the *dispersion along the rows*, i.e., is the row version of the previous SSB. The variance along the rows is $v_r = \mathrm{SSR}/(r-1)$, has $r-1$ degrees of freedom and is the point estimate of $\sigma^2 + c\sigma_r^2$.

3. SSE represents the *residual dispersion* or *experimental error*. The experimental variance associated to the randomness of the experiment is $v_e = \mathrm{SSE} / [(c-1)(r-1)]$, has $(c-1)(r-1)$ degrees of freedom and is the point estimate of σ^2.

Note that formula 4.40 can only be obtained when c and r are constant along the rows and along the columns, respectively. This corresponds to the so-called *orthogonal experiment*.

In the situation shown in Table 4.19, it is possible to consider every cell value as a random case from a population with mean μ_{ij}, such that:

$$\mu_{ij} = \mu + \mu_{i.} + \mu_{.j}, \text{ with } \sum_{i=1}^{c} \mu_{i.} = 0 \text{ and } \sum_{j=1}^{r} \mu_{.j} = 0, \qquad 4.41$$

i.e., the mean of the population corresponding to cell ij is obtained by adding to a global mean μ the means along the columns and along the rows. The sum of the means along the columns as well as the sum of the means along the rows, is zero. Therefore, when computing the mean of all cells we obtain the global mean μ. It is assumed that the variance for all cell populations is σ^2.

In this single observation, *additive effects* model, one can, therefore, treat the effects along the columns and along the rows independently, testing the following null hypotheses:

H_{01}: There are no column effects, $\mu_{i.} = 0$.
H_{02}: There are no row effects, $\mu_{.j} = 0$.

The null hypothesis H_{01} is tested using the ratio v_c/v_e, which, under the assumptions of independent sampling on normal distributions and with equal variances, follows the $F_{c-1,(c-1)(r-1)}$ distribution. Similarly, and under the same assumptions, the null hypothesis H_{02} is tested using the ratio v_r/v_e and the $F_{r-1,(c-1)(r-1)}$ distribution.

Let us now consider the more general situation where for each combination of column and row categories, we have several values available. This *repeated measurements experiment* allows us to analyse the data more fully. We assume that the number of repeated measurements per table cell (combination of column and row categories) is constant, *n*, corresponding to the so-called *factorial experiment*. An example of this sort of experiment is shown in Figure 4.18.

Now, the breakdown of the total sum of squares expressed by the equation 4.40, does not generally apply, and has to be rewritten as:

$$SST = SSC + SSR + SSI + SSE, \hspace{3cm} 4.42$$

with:

1. $$SST = \sum_{i=1}^{c}\sum_{j=1}^{r}\sum_{k=1}^{n}(x_{ijk} - \bar{x}_{...})^2 .$$

    Total sum of squares computed for all *n* cases in every combination of the *c×r* categories, characterising the dispersion of all cases around the global mean. The cases are denoted x_{ijk}, where *k* is the case index in each *ij* cell (one of the *c×r* categories with *n* cases).

2. $$SSC = rn\sum_{i=1}^{c}(\bar{x}_{i..} - \bar{x}_{...})^2 .$$

    Sum of the squares representing the dispersion along the columns. The variance along the columns is $v_c = SSC/(c-1)$, has $c-1$ degrees of freedom and is the point estimate of $\sigma^2 + rn\sigma_c^2$.

3. $$SSR = cn\sum_{j=1}^{r}(\bar{x}_{.j.} - \bar{x}_{...})^2 .$$

    Sum of the squares representing the dispersion along the rows. The variance along the rows is $v_r = SSR/(r-1)$, has $r-1$ degrees of freedom and is the point estimate of $\sigma^2 + cn\sigma_r^2$.

4. Besides the dispersion along the columns and along the rows, one must also consider the dispersion of the column-row combinations, i.e., one must consider the following sum of squares, known as *subtotal* or *model sum of squares* (similar to SSW in the one-way ANOVA):

    $$SSS = n\sum_{i=1}^{c}\sum_{j=1}^{r}(\bar{x}_{ij.} - \bar{x}_{...})^2 .$$

5. $SSE = SST - SSS.$

Sum of the squares representing the experimental error. The experimental variance is $v_e = \text{SSE}/[rc(n-1)]$, has $rc(n-1)$ degrees of freedom and is the point estimate of σ^2.

6. $\text{SSI} = \text{SSS} - (\text{SSC} + \text{SSR}) = \text{SST} - \text{SSC} - \text{SSR} - \text{SSE}$.

The SSI term represents the influence on the experiment of the *interaction* of the column and the row effects. The variance of the interaction, $v_i = \text{SSI}/[(c-1)(r-1)]$ has $(c-1)(r-1)$ degrees of freedom and is the point estimate of $\sigma^2 + n\sigma_I^2$.

Therefore, in the repeated measurements model, one can no longer treat independently the column and row factors; usually, a term due to the interaction of the columns with the rows has to be taken into account.

The ANOVA table for this experiment with additive and interaction effects is shown in Table 4.20. The "Subtotal" row corresponds to the explained variance due to both effects, Factor 1 and Factor 2, and their interaction. The "Residual" row is the experimental error.

Table 4.20. Canonical table for the two-way ANOVA test.

Variance Source	Sum of Squares	df	Mean Square	F
Columns	SSC	$c-1$	$v_c = \text{SSC}/(c-1)$	v_c / v_e
Rows	SSR	$r-1$	$v_r = \text{SSR}/(r-1)$	v_r / v_e
Interaction	SSI	$(c-1)(r-1)$	$v_i = \text{SSI}/[(c-1)(r-1)]$	v_i / v_e
Subtotal	SSS=SSC + SSR + SSI	$cr-1$	$v_m = \text{SSS}/(cr-1)$	v_m / v_e
Residual	SSE	$cr(n-1)$	$v_e = \text{SSE}/[cr(n-1)]$	
Total	SST	$crn-1$		

The previous sums of squares can be shown to be computable as follows:

$$\text{SST} = \sum_{i=1}^{c} \sum_{j=1}^{r} \sum_{k=1}^{n} x_{ijk}^2 - T_{...}^2 / (rcn), \qquad 4.43a$$

$$\text{SSS} = \sum_{i=1}^{c} \sum_{j=1}^{r} x_{ij.}^2 - T_{...}^2 / (rcn) \qquad 4.43b$$

$$\text{SSC} = \sum_{i=1}^{c} (T_{i..}^2 / rn) - T_{...}^2 / (rcn), \qquad 4.43c$$

$$\text{SSR} = \sum_{j=1}^{r} T_{.j.}^2 / (cn) - T_{...}^2 / (rcn), \qquad 4.43d$$

$$SSE = SST - \sum_{i=1}^{c} \sum_{j=1}^{r} T_{ij.}^2 / n - T_{...}^2 /(rcn),$$ 4.43e

where $T_{i..}$, $T_{.j.}$, $T_{ij.}$ and $T_{...}$ are the totals along the columns, along the rows, in each cell and the grand total, respectively. These last formulas are useful for manual computation (or when using EXCEL).

Example 4.19

Q: Consider the 3×2 experiment shown in Figure 4.18, with $n = 4$ cases per cell. Determine all interesting sums of squares, variances and ANOVA results.

A: In order to analyse the data with SPSS and STATISTICA, one must first create a table with two variables corresponding to the columns and row factors and one variable corresponding to the data values (see Figure 4.18).

Table 4.21 shows the results obtained with SPSS. We see that only Factor 2 is found to be significant at a 5% level. Notice also that the interaction effect is only slightly above the 5% level; therefore, it can be suspected to have some influence on the cell means. In order to elucidate this issue, we inspect Figure 4.19, which is a plot of the estimated marginal means for all combinations of categories. If no interaction exists, we expect that the evolution of both curves is similar. This is not the case, however, in this example. We see that the category value of Factor 2 has an influence on how the estimated means depend on Factor 1.

The sums of squares can also be computed manually using the formulas 4.43. For instance, SSC is computed as:

$$SSC = 374^2/8 + 342^2/8 + 335^2/8 - 1051^2/24 = 108.0833.$$

	Factor 1			
Factor 2	1	2	3	Totals
	42	40	33	
	39	37	46	
	33	28	40	
1	43	38	45	464
	56	41	39	
	56	43	40	
	47	55	49	
2	58	60	43	587
Totals	374	342	335	1051

	f1	f2	x
1	1	1	42.00
2	1	1	39.00
3	1	1	33.00
4	1	1	43.00
5	1	2	56.00
6	1	2	56.00
7	1	2	47.00
8	1	2	58.00
9	2	1	40.00
10	2	1	37.00
11	2	1	28.00
12	2	1	38.00
13	2	2	41.00

Figure 4.18. Dataset for Example 4.19 two-way ANOVA test ($c=3$, $r=2$, $n=4$). On the left, the original table is shown. On the right, a partial view of the corresponding SPSS datasheet (f1 and f2 are the factors).

Notice that in Table 4.21 the total sum of squares and the model sum of squares are computed using formulas 4.43a and 4.43b, respectively, without the last term of these formulas. Therefore, the degrees of freedom are crn and cr, respectively. ⬜

Table 4.21. Two-way ANOVA test results, obtained with SPSS, for Example 4.19.

Source	Type III Sum of Squares	df	Mean Square	F	Sig.
Model	46981.250	6	7830.208	220.311	0.000
F1	108.083	2	54.042	1.521	0.245
F2	630.375	1	630.375	17.736	0.001
F1 * F2 [a]	217.750	2	108.875	3.063	0.072
Error	639.750	18	35.542		
Total	47621.000	24			

a Interaction term.

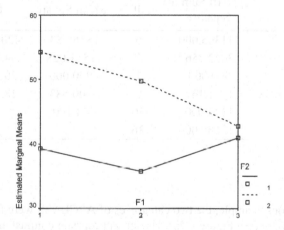

Figure 4.19. Plot of estimated marginal means for Example 4.19. Factor 2 (F2) interacts with Factor 1 (F1).

Example 4.20

Q: Consider the FHR-Apgar dataset, relating variability indices of foetal heart rate (FHR, given in percentage) with the responsiveness of the new-born (Apgar) measured on a 0-10 scale (see Appendix E). The dataset includes observations collected in three hospitals. Perform a factorial model analysis on this dataset, for the variable ASTV (FHR variability index), using two factors: Hospital (3 categories, HUC ≡ 1, HGSA ≡ 2 and HSJ ≡ 3); Apgar 1 class (2 categories: 0 ≡ [0, 8], 1 ≡ [9,10]). In order to use an orthogonal model, select a random sample of $n = 6$ cases for each combination of the categories.

A: Using specific tests described in the following chapter, it is possible to show that variable ASTV can be assumed to approximately follow a normal distribution for most combinations of the factor levels. We use the subset of cases marked with yellow colour in the FHR-Apgar.xls file. For these cases Levene's test yields an observed significance of $p = 0.48$; therefore, the equality of variance assumption is not rejected. We are then entitled to apply the two-way ANOVA test to the dataset.

The two-way ANOVA test results, obtained with SPSS, are shown in Table 4.22 (factors HOSP $\equiv$ Hospital; APCLASS $\equiv$ Apgar 1 class). We see that the null hypothesis is rejected for the effects and their interaction (HOSP * APCLASS). Thus, the test provides evidence that the heart rate variability index ASTV has different means according to the Hospital and to the Apgar 1 category.

Figure 4.20 illustrates the interaction effect on the means. Category 3 of HOSP has quite different means depending on the APCLASS category.

$\square$

Table 4.22. Two-way ANOVA test results, obtained with SPSS, for Example 4.20.

Source	Type III Sum of Squares	df	Mean Square	F	Sig.
Model	111365.000	6	18560.833	420.881	0.000
HOSP	3022.056	2	1511.028	34.264	0.000
APCLASS	900.000	1	900.000	20.408	0.000
HOSP * APCLASS	1601.167	2	800.583	18.154	0.000
Error	1323.000	30	44.100		
Total	112688.000	36			

Example 4.21

Q: In the previous example, the two categories of APCLASS were found to exhibit distinct behaviours (see Figure 4.20). Use an appropriate contrast analysis in order to elucidate this behaviour. Also analyse the following comparisons: hospital 2 vs. 3; hospital 3 vs. the others; all hospitals among them for category 1 of APCLASS.

A: Contrasts in two-way ANOVA are carried out in a similar manner as to what was explained in section 4.5.2.2. The only difference is that in two-way ANOVA, one can specify contrast coefficients that do not sum up to zero. Table 4.23 shows the contrast coefficients used for the several comparisons:

a. The comparison between both categories of APCLASS uses symmetric coefficients for this variable, as in 4.5.2.2. Since this comparison treats all levels of HOSP in the same way, we assign to this variable equal coefficients.

b. The comparison between hospitals 2 and 3 uses symmetric coefficients for these categories. Hospital 1 is removed from the analysis by assigning a zero coefficient to it.

c. The comparison between hospital 3 versus the others uses the assignment rule for merged groups already explained in 4.5.2.2.

d. The comparison between all hospitals, for category 1 of APCLASS, uses two independent contrasts. These are tested simultaneously, representing an exhaustive set of contrasts that compare all levels of HOSP. Category 0 of APCLASS is removed from the analysis by assigning a zero coefficient to it.

Table 4.23. Contrast coefficients and significance for the comparisons described in Example 4.21.

Contrast	(a)			(b)			(c)			(d)		
Description	APCLASS 0 vs. APCLASS 1			HOSP 2 vs. HOSP 3			HOSP 3 vs. {HOSP 1, HOSP 2}			HOSP for APCLASS 1		
HOSP coef.	1	1	1	0	1	−1	1	1	−2	$\begin{matrix} 1 & 0 & -1 \\ 0 & 1 & -1 \end{matrix}$		
APCLASS coef.	1	−1		1	1		1	1		0	1	
p	0.00			0.00			0.29			0.00		

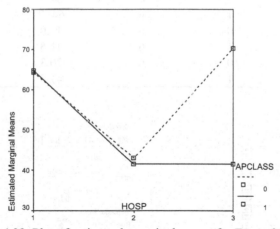

Figure 4.20. Plot of estimated marginal means for Example 4.20.

SPSS and STATISTICA provide the possibility of testing contrasts in multi-way ANOVA analysis. With STATISTICA, the user fills in at will the contrast coefficients in a specific window (e.g. click `Specify contrasts for LS means` in the `Planned comps` tab of the `ANOVA` command, with `HOSP*APCLASS` interaction effect selected). SPSS follows the approach of computing an exhaustive set of contrasts.

The observed significance values in the last row of Table 4.23 lead to the rejection of the null hypothesis for all contrasts except contrast (c).

☐

Example 4.22

Q: Determine the power for the two-way ANOVA test of previous Example 4.20 and the minimum number of cases per group that affords a row effect power above 95%.

A: Power computations for the two-way ANOVA follow the approach explained in section 4.5.2.3.

First, one has to determine the cell statistics in order to be able to compute the standardised effects of the columns, rows and interaction. The cell statistics can be easily computed with SPSS, STATISTICA MATLAB or R. The values for this example are shown in Table 4.24. With STATISTICA one can fill in these values in order to compute the standardised effects as shown in Figure 4.21b. The other specifications are entered in the power specification window, as shown in Figure 4.21a.

Table 4.24. Cell statistics for the FHR-Apgar dataset used in Example 4.20.

HOSP	APCLASS	N	Mean	Std. Dev.
1	0	6	64.3	4.18
1	1	6	64. 7	5.57
2	0	6	43.0	6.81
2	1	6	41.5	7.50
3	0	6	70.3	5.75
3	1	6	41.5	8.96

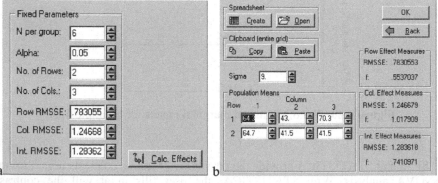

Figure 4.21. Specifying the parameters for the power computation with STATISTICA in Example 4.22: a) Fixed parameters; b) Standardised effects computed with the values of Table 4.24.

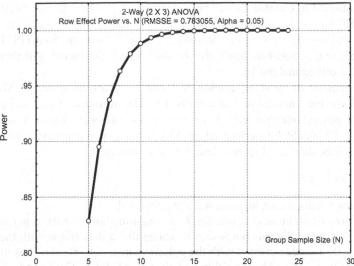

Figure 4.22. Power curve for the row effect of Example 4.22.

The power values computed by STATISTICA are 0.90, 1.00 and 0.97 for the rows, columns and interaction, respectively.

The power curve for the row effects, dependent on n is shown in Figure 4.22. We see that we need at least 8 cases per cell in order to achieve a row effect power above 95%.   ☐

Commands 4.6. SPSS, STATISTICA, MATLAB and R commands used to perform the two-way ANOVA test.

SPSS	`Analyze; General Linear Model;` `Univariate|Multivariate`
	`Statistics; ANOVA; Factorial ANOVA`

STATISTICA	`Statistics; Advanced Linear/Nonlinear`
	`Models; General Linear Models; Main`
	`effects ANOVA | Factorial ANOVA`

MATLAB	`[p,table]=anova2(x,reps,'dispopt')`
R	`anova(lm(X~f1f*f2f))`

The easiest commands to perform the two-way ANOVA test with SPSS and STATISTICA are `General Linear Model; Univariate` and `ANOVA`, respectively. Contrasts in STATISTICA can be specified using the `Planned comps` tab.

As mentioned in Commands 4.5 be sure to check the `No intercept` box in STATISTICA (`Options` tab) and uncheck `Include intercept in model` in SPSS (`General Linear Model, Model` tab). In STATISTICA the `Sigma-restricted` box must also be unchecked; the model will then be the `Type III` orthogonal model.

The meanings of most arguments and return values of `anova2` MATLAB command are the same as in Commands 4.5. The argument `reps` indicates the number of observations per cell. For instance, the two-way ANOVA analysis of Example 4.19 would be performed in MATLAB using a matrix `x` containing exactly the data shown in Figure 4.18a, with the command:

```
» anova2(x,4)
```

The same results shown in Table 4.21 are obtained.

Let us now illustrate how to use the R `anova` function in order to perform two-way ANOVA tests. For this purpose we assume that a data frame with the data of Example 4.19 has been created with the column names `f1`, `f2` and `X` as in the left picture of Figure 4.18. The first thing to do (as we did in Commands 4.5) is to convert `f1` and `f2` into factors with:

```
> f1f <- factor(f1,labels = c("1","2","3"))
> f2f <- factor(f2,labels = c("1","2"))
```

We now obtain the two-way ANOVA similar to Table 4.21 using:

```
> anova(lm(X~f1f*f2f))
```

A model without interaction effects can be obtained with `anova(lm(X~f1f+f2f))` (for details see the help on `lm`)

■

Exercises

4.1 Consider the meteorological dataset used in Example 4.1. Test whether 1980 and 1982 were atypical years with respect to the average maximum temperature. Use the same test value as in Example 4.1.

4.2 Show that the alternative hypothesis $\mu_{T81} = 39.8$ for Example 4.3 has a high power. Determine the smallest deviation from the test value that provides at least a 90% protection against Type II Errors.

4.3 Perform the computations of the powers and critical region thresholds for the one-sided test examples used to illustrate the RS and AS situations in section 4.2.

4.4 Compute the power curve corresponding to Example 4.3 and compare it with the curve obtained with STATISTICA or SPSS. Determine for which deviation of the null hypothesis "typical" temperature one obtains a reasonable protection (power > 80%) against alternative hypothesis.

4.5 Consider the Programming dataset containing student scores during the period 1986-88. Test at 5% level of significance whether or not the mean score is 10. Study the power of the test.

4.6 Determine, at 5% level of significance, whether the standard deviations of variables CG and EG of the Moulds dataset are larger than 0.005 mm.

4.7 Check whether the correlations studied in Exercises 2.9, 2.10. 2.17, 2.18 and 2.19 are significant at 5% level.

4.8 Study the correlation of HFS with $IOA = |IO - 1235| + 0.1$, where HFS and IO are variables of the Breast Tissue dataset. Is this correlation more significant than the one between HFS and IOS in Example 2.18?

4.9 The CFU datasheet of the Cells dataset contains bacterial counts in three organs of sacrificed mice at three different times. Counts are performed in the same conditions in two groups of mice: a protein-deficient group (KO) and a normal, control group (C). Assess at 5% level whether the spleen bacterial count in the two groups are different after two weeks of infection. Which type of test must be used?

4.10 Assume one wishes to compare the measurement sets CG and EG of the Moulds dataset.
    a) Which type of test must be used?
    b) Perform the two-sample mean test at 5% level and study the respective power.
    c) Assess the equality of variance of the sets.

4.11 Consider the CTG dataset. Apply a two-sample mean test comparing the measurements of the foetal heart rate baseline (LB variable) performed in 1996 against those performed in other years. Discuss the results and pertinence of the test.

4.12 Assume we want to discriminate carcinoma from other tissue types, using one of the characteristics of the Breast Tissue dataset.
    a) Assess, at 5% significance level, whether such discrimination can be achieved with one of the characteristics I0, AREA and PERIM.
    b) Assess the equality of variance issue.
    c) Assess whether the rejection of the alternative hypothesis corresponding to the sample means is made with a power over 80%.

4.13 Consider the Infarct dataset containing the measurements EF, IAD and GRD and a score variable (SCR), categorising the severeness of left ventricle necrosis. Determine which of those three variables discriminates at 5% level of significance the score group 2 from the group with scores 0 and 1. Discuss the methods used checking the equality of variance assumption.

4.14 Consider the comparison between the mean neonatal mortality rate at home (MH) and in Health Centres (MI) based on the samples of the Neonatal dataset. What kind of test should be applied in order to assess this two-sample mean comparison and which conclusion is drawn from the test at 5% significance level?

4.15 The FHR-Apgar dataset contains measurements, ASTV, of the percentage of time that foetal heart rate tracings exhibit abnormal short-term variability. Use a two-sample t test in order to compare ASTV means for pairs of Hospitals HSJ, HGSA and HUC. State the conclusions at a 5% level of significance and study the power of the tests.

4.16 The distinction between white and red wines was analysed in Example 4.9 using variables ASP and PHE from the Wines dataset. Perform the two-sample mean test for all variables of this dataset in order to obtain the list of the variables that are capable of achieving the white vs. red discrimination with 95% confidence level. Also determine the variables for which the equality of variance assumption can be accepted.

4.17 For the variable with lowest p in the previous Exercise 4.15 check that the power of the test is 100% and that the test guarantees the discrimination of a 1.3 mg/l mean deviation with power at least 80%.

4.18 Perform the comparison of white vs. red wines using the GLY variable of the Wines dataset. Also depict the situations of an RS and an AS test, computing the respective power for $\alpha = 0.05$ and a deviation of the means as large as the sample mean deviation. Hint: Represent the test as a single mean test with $\mu = \mu_1 - \mu_2$ and pooled standard deviation.

4.19 Determine how large the sample sizes in the previous exercise should be in order to reach a power of at least 80%.

4.20 Using the Programming dataset, compare at 5% significance level the scores obtained by university freshmen in a programming course, for the following two groups: "No pre-university knowledge of programming"; "Some degree of pre-university knowledge of programming".

4.21 Consider the comparison of the six tissue classes of the Breast Tissue dataset studied in Example 4.15. Perform the following analyses:
   a)   Verify that PA500 is the only suitable variable to be used in one-way ANOVA, according to Levene's test of equality of variance.
   b)   Use adequate contrasts in order to assess the following class discriminations: {car}, {con, adi}, {mas, fad, gla}; {car} vs. all other classes.

4.22 Assuming that in the previous exercise one wanted to compare classes {fad}, {mas} and {con}, answer the following questions:
   a)   Does the one-way ANOVA test reject the null hypothesis at $\alpha = 0.005$ significance level?
   b)   Assuming that one would perform all possible two-sample t tests at the same $\alpha = 0.005$ significance level, would one reach the same conclusion as in a)?
   c)   What value should one set for the significance level of the two-sample t tests in order to reject the null hypothesis in the same conditions as the one-way ANOVA does?

4.23 Determine whether or not one should accept with 95% confidence that pre-university knowledge of programming has no influence on the scores obtained by university freshmen in a programming course (Porto University), based on the Programming dataset.

Use the Levene test to check the equality of variance assumption and determine the power of the test.

4.24 Perform the following post-hoc comparisons for the previous exercise:
a)   Scheffé test.
b)   "No previous knowledge" vs. "Some previous knowledge" contrast. Compare the results with those obtained in Exercise 4.19

4.25 Consider the comparison of the bacterial counts as described in the CFU datasheet of the Cells dataset (see Exercise 4.9) for the spleen and the liver at two weeks and at one and two months ("time of count" categories). Using two-way ANOVA performed on the first 5 counts of each group ("knock-out" and "control"), check the following results:
a)   In what concerns the spleen, there are no significant differences at 5% level either for the group categories or for the "time of count" categories. There is also no interaction between both factors.
b)   For the liver there are significant differences at 5% level, both for the group categories and for the "time of count" categories. There is also a significant interaction between these factors as can also be inferred from the respective marginal mean plot.
c)   The test power in this last case is above 80% for the main effects.

4.26 The SPLEEN datasheet of the Cells dataset contains percent counts of bacterial load in the spleen of two groups of mice ("knock-out" and "control") measured by two biochemical markers (CD4 and CD8). Using two-way ANOVA, check the following results:
a)   Percent counts after two weeks of bacterial infection exhibit significant differences at 5% level for the group categories, the biochemical marker categories and the interaction of these factors. However, these results are not reliable since the observed significance of the Levene test is low ($p = 0.014$).
b)   Percent counts after two months of bacterial infection exhibit a significant difference ($p = 0$) only for the biochemical marker. This is a reliable result since the observed significance of the Levene test is larger than 5% ($p = 0.092$).
c)   The power in this last case is very large ($p \approx 1$).

4.27 Using appropriate contrasts check the following results for the ANOVA study of Exercise 4.24 b:
a)   The difference of means for the group categories is significant with $p = 0.006$.
b)   The difference of means for "two weeks" vs "one or two months" is significant with $p = 0.001$.
c)   The difference of means of the time categories for the "knock-out" group alone is significant with $p = 0.027$.

5 Non-Parametric Tests of Hypotheses

The tests of hypotheses presented in the previous chapter were "parametric tests", that is, they concerned parameters of distributions. In order to apply these tests, certain conditions about the distributions must be verified. In practice, these tests are applied when the sampling distributions of the data variables reasonably satisfy the normal model.

Non-parametric tests make *no assumptions regarding the distributions* of the data variables; only a few mild conditions must be satisfied when using most of these tests. Since non-parametric tests make no assumptions about the distributions of the data variables, they are adequate to *small samples*, which would demand the distributions to be known precisely in order for a parametric test to be applied. Furthermore, non-parametric tests often concern different hypotheses about populations than do parametric tests. Finally, unlike parametric tests, there are non-parametric tests that can be applied to *ordinal and/or nominal data*.

The use of fewer or milder conditions imposed on the distributions comes with a price. The non-parametric tests are, in general, less powerful than their parametric counterparts, when such a counterpart exists and is applied in identical conditions. In order to compare the power of a test B with a test A, we can determine the sample size needed by B, n_B, in order to attain the same power as test A, using sample size n_A, and with the same level of significance. The following *power-efficiency* measure of test B compared with A, η_{BA}, is then defined:

$$\eta_{BA} = \frac{n_A}{n_B}.$$ 5.1

For many non-parametric tests (B) the power efficiency, η_{BA}, relative to a parametric counterpart (A) has been studied and the respective results divulged in the literature. Surprisingly enough, the non-parametric tests often have a high power-efficiency when compared with their parametric counterparts. For instance, as we shall see in a later section, the Mann-Whitney test of central location, for two independent samples, has a power-efficiency that is usually larger than 95%, when compared with its parametric counterpart, the t test. This means that when applying the Mann-Whitney test we usually attain the same power as the t test using a sample size that is only 1/0.95 bigger (i.e., about 5% bigger).

5.1   Inference on One Population

5.1.1 The Runs Test

The *runs test* assesses whether or not a sequence of observations can be accepted as a random sequence, that is, with independent successive observations. Note that most tests of hypotheses do not care about the order of the observations. Consider, for instance, the meteorological data used in Example 4.1. In this example, when testing the mean based on a sample of maximum temperatures, the order of the observations is immaterial. The maximum temperatures could be ordered by increasing or decreasing order, or could be randomly shuffled, still giving us exactly the same results.

Sometimes, however, when analysing sequences of observations, one has to decide whether a given sequence of values can be assumed as exhibiting a random behaviour.

Consider the following sequences of $n = 12$ trials of a dichotomous experiment, as one could possibly obtain when tossing a coin:

Sequence 1:	0	0	0	0	0	0	1	1	1	1	1	1
Sequence 2:	0	1	0	1	0	1	0	1	0	1	0	1
Sequence 3:	0	0	1	0	1	1	1	0	1	1	0	0

Sequences 1 and 2 would be rejected as random since a dependency pattern is clearly present[1]. Such sequences raise a reasonable suspicion concerning either the "fairness" of the coin-tossing experiment or the absence of some kind of data manipulation (e.g. sorting) of the experimental results. Sequence 3, on the other hand, seems a good candidate of a sequence with a random pattern.

The runs test analyses the randomness of a sequence of dichotomous trials. Note that all the tests described in the previous chapter (and others to be described next as well) are insensitive to data sorting. For instance, when testing the mean of the three sequences above, with H_0: $\mu = 6/12 = \frac{1}{2}$, one obtains the same results.

The test procedure uses the values of the number of occurrences of each category, say n_1 and n_2 for 1 and 0 respectively, and the number of *runs*, i.e., the number of occurrences of an equal value subsequence delimited by a different value. For sequence 3, the number of runs, r, is equal to 7, as seen below:

| Sequence 3: | 0 0 | 1 | 0 | 1 1 1 | 0 | 1 1 | 0 0 |
| Runs: | 1 | 2 | 3 | 4 | | 5 6 | 7 |

---

[1] Note that we are assessing the randomness of the sequence, not of the process that generated it.

The runs test assesses the null hypothesis of sequence randomness, using the sampling distribution of r, given n_1 and n_2. Tables of this sampling distribution can be found in the literature. For large n_1 or n_2 (say above 20) the sampling distribution of r is well approximated by the normal distribution with the following parameters:

$$\mu_r = \frac{2n_1 n_2}{(n_1 + n_2)} + 1; \quad \sigma_r^2 = \frac{2n_1 n_2 (2n_1 n_2 - n_1 - n_2)}{(n_1 + n_2)^2 (n_1 + n_2 - 1)}. \qquad 5.2$$

Notice that the number of runs always satisfies, $1 \le r \le n$, with $n = n_1 + n_2$. The null hypothesis is rejected when there are either too few runs (as in Sequence 1) or too many runs (as in Sequence 2). For the previous sequences, at a 5% level the critical values of r for $n_1 = n_2 = 6$ are 3 and 11, i.e. the non-critical region of r is [4, 10]. We, therefore, reject at 5% level the null hypothesis of randomness for Sequence 1 ($r = 2$) and Sequence 2 ($r = 12$), and do not reject the null hypothesis for Sequence 3 ($r = 7$).

The runs test can be used with any sequence of values and not necessarily dichotomous, if previously the values are dichotomised, e.g. using the mean or the median.

Example 5.1

Q: Consider the noise sequence in the Signal & Noise dataset (first column) generated with the "normal random number" routine of EXCEL with zero mean. The sequence has $n = 100$ noise values. Use the runs test to assess the randomness of the sequence.

A: We apply the SPSS runs test command, using an imposed (Custom) dichotomization around zero, obtaining an observed two-tailed significance of $p = 0.048$. At a 5% level of significance the randomness of the sequence is not rejected. We may also use the MATLAB or R runs function. We obtain the values of Table 5.1. The interval [n_{low}, n_{up}] represents the non critical region. We see that the observed number of runs coincides with one of the interval ends.

Table 5.1. Results obtained with MATLAB or R runs test for the noise data.

n_1	n_2	r	n_{low}	n_{up}
53	47	41	41	61

□

Example 5.2

Q: Consider the Forest Fires dataset (see Appendix E), which contains the area (ha) of burnt forest in Portugal during the period 1943-1978. Is there evidence from this sample, at a 5% significance level, that the area of burnt forest behaves as a random sequence?

A: The area of burnt forest depending on the year is shown in Figure 5.1. Notice that there is a clear trend we must remove before attempting the runs test. Figure 5.1 also shows the regression line with a null intercept, i.e. passing through the point (0,0), obtained with the methods that will be explained later in Chapter 7.

We now compute the deviations from the linear trend and use them for the runs test. When analysed with SPSS, we find an observed two-tailed significance of $p = 0.335$. Therefore, we do not reject the null hypothesis that the area of burnt forest behaves as a random sequence superimposed on a linear trend.

□

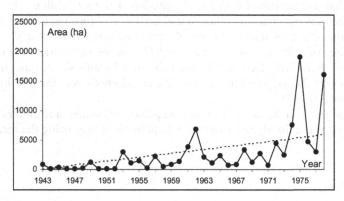

Figure 5.1. Area of burnt forest in Portugal during the years 1943-1978. The dotted line is a linear fit with null intercept.

Commands 5.1. SPSS, MATLAB and R commands used to perform the runs test.

SPSS	`Analyze; Nonparametric Tests; Runs`
MATLAB	`runs(x,alpha)`
R	`runs(x,alpha=0.05)`

STATISTICA, MATLAB statistical toolbox and R `stats` package do not have the runs test. We provide the `runs` function for MATLAB and R (see appendix F) returning the values of Table 5.1. The function should only be used when n_1 or n_2 are large (say, above 20).

■

5.1.2 The Binomial Test

The *binomial or proportion test* is used to assess whether there is evidence from the sample that one category of a dichotomised population occurs in a certain proportion of times. Let us denote the categories or classes of the population by ω, coded 1 for the category of interest and 0 for the complement. The two-tailed test can be then formalised as:

H_0: $P(\omega=1) = p$ (and $P(\omega=0) = 1 - p = q$);
H_1: $P(\omega=1) \neq p$ (and $P(\omega=0) \neq q$).

Given a data sample with n i.i.d. cases, k of which correspond to $\omega=1$, we know from Chapter 3 (see also Appendix C) that the point estimate of p is $\hat{p} = k/n$. In order to establish the critical region of the test, we take into account that the probability of obtaining k events of $\omega =1$ in n trials is given by the binomial law. Let K denote the random variable associated to the number of times that $\omega = 1$ occurs in a sample of size n. We then have the binomial sampling distribution (section A.7.1):

$$P(K = k) = \binom{n}{k} p^k q^{n-k}; \qquad k = 0, 1, \ldots, n.$$

When n is small (say, below 25), the non-critical region is usually quite large and the power of the test quite low. We have also found useless large confidence intervals for small samples in section 3.3, when estimating a proportion. The test yields useful results only for *large samples* (say, above 25). In this case (especially when np or nq are larger than 25, see A.7.3), we use the normal approximation of the standardised sampling distribution:

$$Z = \frac{K - np}{\sqrt{npq}} \quad \sim \quad N_{0,1} \tag{5.3}$$

Notice that denoting by P the random variable corresponding to the proportion of successes in the sample (with observed value $\hat{p} = k/n$), we may write 5.3 as:

$$Z = \frac{K - np}{\sqrt{npq}} = \frac{K/n - p}{\sqrt{pq/n}} = \frac{P - p}{\sqrt{pq/n}}. \tag{5.4}$$

The binomial test is then performed in the same manner as the test of a single mean described in section 4.3.1. The approximation to the normal distribution becomes better if a *continuity correction* is used, reducing by 0.5 the difference between the observed mean ($n\hat{p}$) and the expected mean (np).

As shown in Commands 5.3, SPSS and R have a specific command for carrying out the binomial test. SPSS uses the normal approximation with continuity correction for $n > 25$. R uses a similar procedure. In order to perform the binomial test with STATISTICA or MATLAB, one uses the single sample t test command.

Example 5.3

Q: According to Mendel's Heredity Theory, a cross breeding of yellow and green peas should produce them in a proportion of three times more yellow peas than green peas. A cross breeding of yellow and green peas was performed and produced 176 yellow peas and 48 green peas. Are these experimental results explainable by the Theory?

A: Given the theoretically expected values of the proportion of yellow peas, the test is formalised as:

H_0: $P(\omega=1) = \frac{3}{4}$;
H_1: $P(\omega=1) \neq \frac{3}{4}$.

In order to apply the binomial test to this example, using SPSS, we start by filling in a datasheet as shown in Table 5.2.

Next, in order to specify that category 1 of pea-type occurs 176 times and the category 0 occurs 48 times, we use the "weight cases" option of SPSS, as shown in Commands 5.2. In the Weight Cases window we specify that the *weight variable* is *n*.

Finally, with the binomial command of SPSS, we obtain the results shown in Table 5.3, using 0.75 (¾) as the tested proportion. Note the "Based on Z Approximation" foot message displayed by SPSS. The two-tailed significance is 0.248, so therefore, we do not reject the null hypothesis $P(\omega=1) = 0.75$.

Table 5.2. Datasheet for Example 5.3.

group	pea-type	n
1	1	176
2	0	48

Table 5.3. Binomial test results obtained with SPSS for the Example 5.3.

		Category	n	Observed Prop.	Test Prop.	Asymp. Sig. (1-tailed)
PEA_TYPE	Group 1	1	176	0.79	0.75	0.124[a]
	Group 2	0	48	0.21		
	Total		224	1.00		

a Based on Z approximation.

Let us now carry out this test using the values of the standardised normal distribution. The important values to be computed are:

$np = 224 \times 0.75 = 168$;

$$s = \sqrt{npq} = \sqrt{224 \times 0.75 \times 0.25} = 6.48.$$

Hence, using the continuity correction, we obtain $z = (168 - 176 + 0.5)/6.48 = -1.157$, to which corresponds a one-tailed probability of 0.124 as reported in Table 5.3.

□

Example 5.4

Q: Consider the Freshmen dataset, relative to the Porto Engineering College. Assume that this dataset represents a random sample of the population of freshmen in the College. Does this sample support the hypothesis that there is an even chance that a freshman in this College can be either male or female?

A: We formalise the test as:

H_0: $P(\omega=1) = \frac{1}{2}$;
H_1: $P(\omega=1) \neq \frac{1}{2}$.

The results obtained with SPSS are shown in Table 5.4. Based on these results, we reject the null hypothesis with high confidence.

Note that SPSS always computes a two-tailed significance for a test proportion of 0.5 and a one-tailed significance otherwise. □

Table 5.4. Binomial test results, obtained with SPSS, for the freshmen dataset.

		Category	n	Observed Prop.	Test Prop.	Asymp. Sig. (2-tailed)
SEX	Group 1	female	35	0.27	0.50	0.000
	Group 2	male	97	0.73		
	Total		132	1.00		

Commands 5.2. SPSS and STATISTICA commands used to specify case weighing.

SPSS	Data; Weight Cases
STATISTICA	Tools; Weight

These commands pop up a window where one specifies which variable to use as weight variable and whether weighing is "On" or "Off". Many STATISTICA commands also include a weight button (⚙ w) in connection with the weight specification window. Case weighing is useful whenever the datasheet presents the data in a compact way, with a specific column containing the number of occurrences of each case.

∎

Commands 5.3. SPSS, STATISTICA, MATLAB and R commands used to perform the binomial test.

SPSS	`Analyze; Nonparametric Tests; Binomial`
STATISTICA	`Statistics; Basic Statistics and Tables; t-test, single sample`
MATLAB	`[h,sig,ci]=ttest(x,m,alpha,tail)`
R	`binom.test(x,n,p,conf.level=0.95)`

When performing the binomial test with STATISTICA or MATLAB using the single sample t test, a somewhat different value is obtained because no continuity correction is used and the standard deviation is estimated from $\hat{p}$. This difference is frequently of no importance. With MATLAB the test is performed as follows:

```
» x = [ones(176,1); zeros(48,1)];
» [h, sig, ci]=ttest(x,0.75,0.05,0)
h =
      0
sig =
     0.195
ci =
     0.7316     0.8399
```

Note that x is defined as a column vector filled in with 176 ones followed by 48 zeros. The commands `ones(m,n)` and `zeros(m,n)` define matrices with m rows and n columns filled with ones and zeros, respectively. The notation `[A; B]` defines a matrix by juxtaposition of the matrices A and B side by side along the columns (along the rows when omitting the semicolon).

The results of the test indicate that the null hypothesis cannot be rejected (h=0). The two-tailed significance is 0.195, somewhat lower than previously found (0.248), for the above mentioned reasons.

The arguments x, n and p of the R `binom.test` function represent the number of successes, the number of trials and the tested value of p, respectively. Other details can be found with `help(binom.test)`. For the Example 5.3 we run `binom.test(176,176+48,0.75)`, obtaining a two-tailed significance of 0.247, nearly the double of the value published in Table 5.3 as it should. A 95% confidence interval of [0.726, 0.838] is also published, containing the observed proportion of 0.786.

∎

5.1.3 The Chi-Square Goodness of Fit Test

The previous binomial test applied to a dichotomised population. When there are more than two categories, one often wishes to assess whether the observed frequencies of occurrence in each category are in accordance to what should be expected. Let us start with the random variable 5.4 and square it:

$$Z^2 = \frac{(P-p)^2}{pq/n} = n(P-p)^2\left(\frac{1}{p}+\frac{1}{q}\right) = \frac{(X_1-np)^2}{np} + \frac{(X_2-nq)^2}{nq}, \qquad 5.5$$

where X_1 and X_2 are the random variables associated with the number of "successes" and "failures" in the n-sized sample, respectively. In the above derivation note that denoting $Q = 1 - P$ we have $(nP - np)^2 = (nQ - nq)^2$. Formula 5.5 conveniently expresses the fitting of $X_1 = nP$ and $X_2 = nQ$ to the theoretical values in terms of square deviations. Square deviation is a popular distance measure given its many useful properties, and will be extensively used in Chapter 7.

Let us now consider k categories of events, each one represented by a random variable X_i, and, furthermore, let us denote by p_i the probability of occurrence of each category. Note that the joint distribution of the X_i is a *multinomial distribution*, described in B.1.6. The result 5.5 is generalised for this multinomial distribution, as follows (see property 5 of B.2.7):

$$\chi^{*2} = \sum_{i=1}^{k} \frac{(X_i-np_i)^2}{np_i} \quad \sim \quad \chi^2_{k-1}, \qquad 5.6$$

where the number of degrees of freedom, $df = k - 1$, is imposed by the restriction:

$$\sum_{i=1}^{k} x_i = n. \qquad 5.7$$

As a matter of fact, the chi-square law is only an approximation for the sampling distribution of χ^{*2}, given the dependency expressed by 5.7.

In order to test the goodness of fit of the observed counts O_i to the expected counts E_i, that is, to test whether or not the following null hypothesis is rejected:

H_0: The population has absolute frequencies E_i for each of the $i =1, .., k$ categories,

we then use test the statistic:

$$\chi^{*2} = \sum_{i=1}^{k} \frac{(O_i-E_i)^2}{E_i}, \qquad 5.8$$

which, according to formula 5.6, has approximately a chi-square distribution with $df = k - 1$ degrees of freedom. The approximation is considered acceptable if the following conditions are met:

i. For $df = 1$, no E_i must be smaller than 5;

ii. For $df > 1$, no E_i must be smaller than 1 and no more than 20% of the E_i must be smaller than 5.

Expected absolute frequencies can sometimes be increased, in order to meet the above conditions, by merging adjacent categories.

When the difference between observed (O_i) and expected counts (E_i) is large, the value of χ^{*2} will also be large and the respective tail probability small. For a 0.95 confidence level, the critical region is above $\chi^2_{k-1,0.95}$.

Example 5.5

Q: A die was thrown 40 times with the observed number of occurrences 8, 6, 3, 10, 7, 6, respectively for the face value running from 1 through 6. Does this sample provide evidence that the die is not honest?

A: Table 5.5 shows the chi-square test results obtained with SPSS. Based on the high value of the observed significance, we do not reject the null hypothesis that the die is honest. Applying the R function chisq.test(c(8,6,3,10,7,6)) one obtains the same results as in Table 5.5b. This function can have a second argument with a vector of expected probabilities, which when omitted, as we did, assigns equal probability to all categories.   □

Table 5.5. Dataset (a) and results (b), obtained with SPSS, of the chi-square test for the die-throwing experiment (Example 5.5). The residual column represents the differences between observed and expected frequencies.

FACE	Observed N	Expected N	Residual		FACE
1	8	6.7	1.3	Chi-Square	4.100
2	6	6.7	−0.7		
3	3	6.7	−3.7	df	5
4	10	6.7	3.3		
5	7	6.7	0.3	Asymp. Sig.	0.535
6	6	6.7	−0.7		

a   b

Example 5.6

Q: It is a common belief that the best academic freshmen students usually participate in freshmen initiation rites only because they feel compelled to do so.

Does the Freshmen dataset confirm that belief for the Porto Engineering College?

A: We use the categories of answers obtained for Question 6, "I felt compelled to participate in the Initiation", of the freshmen dataset (see Appendix E). The respective EXCEL file contains the computations of the frequencies of occurrence of each category and for each question, assuming a specified threshold for the average results in the examinations. Using, for instance, the threshold = 10, we see that there are 102 "best" students, with average examination score not less than the threshold. From these 102, there are varied counts for the five categories of Question 6, ranging from 16 students that "fully disagree" to 5 students that "fully agree".

Under the null hypothesis, the answers to Question 6 have no relation with the freshmen performance and we would expect equal frequencies for all categories.

The chi-square test results obtained with SPSS are shown in Table 5.6. Based on these results, we reject the null hypothesis: there is evidence that the answer to Question 6 of the freshmen enquiry bears some relation with the student performance.   ☐

Table 5.6. Dataset (a) and results (b), obtained with SPSS, for Question 6 of the freshmen enquiry and 102 students with average score ≥ 10.

CAT	Observed N	Expected N	Residual		CAT
1	16	20.4	−4.4	Chi-Square	32.020
2	26	20.4	5.6		
3	39	20.4	18.6	df	4
4	16	20.4	−4.4		
5	5	20.4	−15.4	Asymp. Sig.	0.000
a				b	

Example 5.7

Q: Consider the variable ART representing the total area of defects of the Cork Stoppers' dataset, for the class 1 (Super) of corks. Does the sample data provide evidence that this variable can be accepted as being normally distributed in that class?

A: This example illustrates the application of the chi-square test for assessing the goodness of fit to a known distribution. In this case, the chi-square test uses the deviations of the observed absolute frequencies vs. the expected absolute frequencies under the condition of the stated null hypothesis, i.e., that the variable ART is normally distributed.

In order to compute the absolute frequencies, we have to establish a set of intervals based on the percentiles of the normal distribution. Since the number of cases is $n = 50$, and we want the conditions for using the chi-square distribution to be fulfilled, we use intervals corresponding to 20% of the cases. Table 5.7 shows these intervals, under the "z-Interval" heading, which can be obtained from the tables of the standard normal distribution or using software functions, such as the ones already described for SPSS, STATISTICA, MATLAB and R.

The corresponding interval cutpoints, x_{cut}, for the random variable under analysis, X, can now be easily determined, using:

$$x_{cut} = \bar{x} + z_{cut} s_X ,\qquad\qquad 5.9$$

where we use the sample mean and standard deviation as well as the cutpoints determined for the normal distribution, z_{cut}. In the present case, the mean and standard deviation are 137 and 43, respectively, which leads to the intervals under the "ART-Interval" heading.

The absolute frequency columns are now easily computed. With SPSS, STATISTICA and R we now obtain the value of $\chi^{*2} = 2.2$. We must be careful, however, when obtaining the corresponding significance in this application of the chi-square test. The problem is that now we do not have $df = k - 1$ degrees of freedom, but $df = k - 1 - n_p$, where n_p is the number of parameters computed from the sample. In our case, we derived the interval boundaries using the sample mean and sample standard deviation, i.e., we lost two degrees of freedom. Therefore, we have to compute the probability using $df = 5 - 1 - 2 = 2$ degrees of freedom, or equivalently, compute the critical region boundary as:

$$\chi^2_{2,0.95} = 5.99 .$$

Since the computed value of the χ^{*2} is smaller than this critical region boundary, we do not reject at 5% significance level the null hypothesis of variable ART being normally distributed.   □

Table 5.7. Observed and expected (under the normality assumption) absolute frequencies, for variable ART of the cork-stopper dataset.

Cat.	z-Interval	Cumulative p	ART-Interval	Expected Frequencies	Observed Frequencies
1	]− ∞, −0.8416]	0.20	[0, 101]	10	10
2	]−0.8416, −0.2533]	0.40	]101, 126]	10	8
3	]−0.2533, 0.2533]	0.60	]126, 148]	10	14
4	] 0.2533, 0.8416]	0.80	]148, 173]	10	9
5	] 0.8416, +∞ [	1.00	> 173	10	9

Commands 5.4. SPSS, STATISTICA, MATLAB and R commands used to perform the chi-square goodness of fit test.

SPSS	`Analyze; Nonparametric Tests; Chi-Square`
STATISTICA	`Statistics; Nonparametrics; Observed versus expected X`^{2}.
MATLAB	`[c,df,sig] = chi2test(x)`
R	`chisq.test(x,p)`

MATLAB does not have a specific function for the chi-square test. We provide in the book CD the `chi2test` function for that purpose.   ∎

5.1.4 The Kolmogorov-Smirnov Goodness of Fit Test

The Kolmogorov-Smirnov goodness of fit test is a one-sample test designed to assess the goodness of fit of a data sample to a hypothesised continuous distribution, $F_X(x)$. The null hypothesis is formalised as:

H_0: Data variable X has a cumulative probability distribution $F_X(x) \equiv F(x)$.

Let $S_n(x)$ be the observed cumulative distribution of the random sample, x_1, $x_2,..., x_n$, also called *empirical distribution*. Assuming the sample data is sorted in increasing order, the values of $S_n(x)$ are obtained by adding the successive frequencies of occurrence, k_i/n, for each distinct x_i.

Under the null hypothesis one expects to obtain small deviations of $S_n(x)$ from $F(x)$. The Kolmogorov-Smirnov test uses the largest of such deviations as a goodness of fit measure:

$$D_n = \max | F(x) - S_n(x) |, \text{ for every distinct } x_i.$$ 5.10

The sampling distribution of D_n is given in the literature. Unless n is very small the following asymptotic result can be used:

$$\lim_{n \to \infty} P\left(\sqrt{n}D_n \le t \right) = 1 - 2\sum_{i=1}^{\infty} (-1)^{i-1} e^{-2i^2 t^2}.$$ 5.11

The Kolmogorov-Smirnov test rejects the null hypothesis at level α if $D_n > d_{n,\alpha}$, where $d_{n,\alpha}$ is such that:

$$P_{H_0}(D_n > d_{n,\alpha}) = \alpha.$$ 5.12

Using formula 5.11 the following critical points are obtained:

$$d_{n,0.01} = 1.63 / \sqrt{n}; \qquad d_{n,0.05} = 1.36 / \sqrt{n}; \qquad d_{n,0.10} = 1.22 / \sqrt{n}.$$ 5.13

Note that when applying the Kolmogorov-Smirnov test, one often uses the distribution parameters computed from the actual data. For instance, in the case of assessing the normality of an empirical distribution, one often uses the sample mean and sample standard deviation. This is a source of uncertainty in the interpretation of the results.

Example 5.8

Q: Redo the previous Example 5.7 (assessing the normality of ART for class 1 of the cork-stopper data), using the Kolmogorov-Smirnov test.

A: Running the test with SPSS we obtain the results displayed in Table 5.8, showing no evidence ($p = 0.8$) supporting the rejection of the null hypothesis (normal distribution). In R the test would be run as:

```
> x <- ART[1:50]
> ks.test(x, "pnorm", mean(x), sd(x))
```

The following results are obtained confirming the ones in Table 5.8:

```
D = 0.0922, p-value = 0.7891                              ☐
```

Table 5.8. Kolmogorov-Smirnov test results for variable ART obtained with SPSS in the goodness of fit assessment of normal distribution.

		ART
N		50
Normal Parameters	Mean	137.0000
	Std. Deviation	42.9969
Most Extreme Differences	Absolute	0.092
	Positive	0.063
	Negative	−0.092
Kolmogorov-Smirnov Z		0.652
Asymp. Sig. (2-tailed)		0.789

In the goodness of fit assessment of a normal distribution it may be convenient to inspect *cumulative distribution plots* and *normal probability plots*. Figure 5.2 exemplifies these plots for the ART variable of Example 5.8. The cumulative distribution plot helps to detect the regions where the empirical distribution mostly deviates from the theoretical distribution, and can also be used to measure the statistic D_n (formula 5.10). The normal probability plot displays z-scores for the data and for the standard normal distribution along the vertical axis. These last ones lie on a straight line. Large deviations of the observed z-scores, from the straight line corresponding to the normal distribution, are a symptom of poor normal approximation.

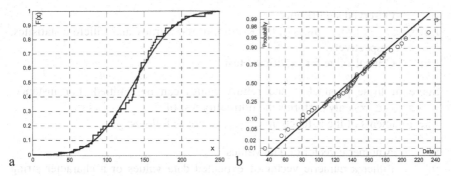

Figure 5.2. Visually assessing the normality of the ART variable (cork stopper dataset) with MATLAB: a) Empirical cumulative distribution plot with superimposed normal distribution (smooth line); b) Normal probability plot.

Commands 5.5. SPSS, STATISTICA, MATLAB and R commands used to perform goodness of fit tests.

SPSS	`Analyze; Nonparametric Tests; 1-Sample K-S`
	`Analyze; Descriptive Statistics; Explore;`
	`Plots; Normality plots with tests`

STATISTICA	`Statistics; Basic Statistics/Tables;`
	`Histograms`
	`Graphs; Histograms`

MATLAB	`[h,p,ksstat,cv]= kstest(x,cdf,alpha,tail)` `[h,p,lstat,cv]= lillietest(x,alpha)`
R	`ks.test(x, y, ...)`

With STATISTICA the one-sample Kolmogorov-Smirnov test is not available as a separate test. It can, however, be performed together with other goodness of fit tests when displaying a histogram (`Advanced` option). SPSS also affords the goodness of fit tests with the normality plots that can be obtained with the `Explore` command.

With the MATLAB commands `kstest` and `lillietest`, the meaning of the parameters and return values when testing the data sample x at level `alpha`, is as follows:

`cdf`:   Two-column matrix, with the first column containing the random sample x and the second column containing the hypothesised cumulative distribution.

`tail`: Type of test with values 0, −1, 1 corresponding to the alternative hypothesis $F(x) \neq S_n(x)$, $F(x) > S_n(x)$ and $F(x) < S_n(x)$, respectively.

`h`:   Test result, equal to 1 if H_0 can be rejected, 0 otherwise.

p: Observed significance.

ksstat, lstat: Values of the Kolmogorov-Smirnov and Liliefors statistics, respectively.

cv: Critical value for significant test.

Some of these parameters and return values can be omitted. For instance, h = kstest(x) only performs the normality test of x.

The arguments of the R function ks.test are as follows:

x: A numeric vector of data values.

y: Either a numeric vector of expected data values or a character string naming a distribution function.

...   Parameters of the distribution specified by y.   ∎

Commands 5.6. SPSS, STATISTICA, MATLAB and R commands used to obtain cumulative distribution plots and normal probability plots.

SPSS	Graphs; Interactive; Histogram; Cumulative histogram
	Analyze; Descriptive Statistics; Explore; Plots; Normality plots with tests │ Graphs; P-P
STATISTICA	Graphs; Histograms; Showing Type; Cumulative
	Graphs; 2D Graphs; Probability-Probability Plots
MATLAB	cdfplot(x) ; normplot(x)
R	plot.ecdf(x) ; qqnorm(x)

The cumulative distribution plot shown in Figure 5.2a was obtained with MATLAB using the following sequence of commands:

```
» art = corkstoppers(1:50,3);
» cdfplot(art)
» hold on
» xaxis = 0:1:250;
» plot(xaxis,normcdf(xaxis,mean(art),std(art)))
```

Note the hold on command used to superimpose the standard normal distribution over the previous empirical distribution of the data. This facility is disabled with hold off. The normcdf command is used to obtain the normal cumulative distribution in the interval specified by xaxis with the mean and standard deviation also specified.   ∎

5.1.5 The Lilliefors Test for Normality

The Lilliefors test resembles the Kolmogorov-Smirnov but it is especially tailored to assess the normality of a distribution, with the null hypothesis formalised as:

$$H_0: \quad F(x) = N_{\mu,\sigma}(x) . \tag{5.14}$$

For this purpose, the test standardises the data using the sample estimates of μ and σ. Let Z represent the standardised data, i.e., $z_i = (x_i - \bar{x})/s$. The Lilliefors' test statistic is:

$$D_n = \max | F(z) - S_n(z) | . \tag{5.15}$$

The test is, therefore, performed like the Kolmogorov-Smirnov test (see formula 5.12), but with the advantage that the sampling distribution of D_n takes into account the fact that the sample mean and sample standard deviation are used. The asymptotic critical points are:

$$d_{n,0.01} = 1.031/\sqrt{n}; \quad d_{n,0.05} = 0.886/\sqrt{n}; \quad d_{n,0.10} = 0.805/\sqrt{n} . \tag{5.16}$$

Critical values and extensive tables of the sampling distribution of D_n can be found in the literature (see e.g. Conover, 1980).

The Lilliefors test can be performed with SPSS and STATISTICA as described in Commands 5.5. When applied to Example 5.8 it produces a lower bound for the significance ($p = 0.2$), therefore not providing evidence allowing us to reject the null hypothesis.

5.1.6 The Shapiro-Wilk Test for Normality

The Shapiro-Wilk test is also tailored to assess the goodness of fit to the normal distribution. It is based on the observed distance between symmetrically positioned data values. Let us assume that the sample size is n and the successive values x_1, $x_2,..., x_n$, were preliminarily sorted by increasing value:

$$x_1 \leq x_2 \leq ... \leq x_n.$$

The distance of symmetrically positioned data values, around the middle value, is measured by:

$$(x_{n-i+1} - x_i), \quad \text{for} \quad i = 1, 2, ..., k,$$

where $k = (n + 1)/2$ if n is odd and $k = n/2$ otherwise.
The Shapiro-Wilk statistic is given by:

$$W = \left[\sum_{i=1}^{k} a_i (x_{n-i+1} - x_i) \right]^2 / \sum_{i=1}^{n} (x_i - \bar{x})^2 . \tag{5.17}$$

The coefficients a_i in formula 5.17 and the critical values of the sampling distribution of W, for several confidence levels, can be obtained from table look-up (see e.g. Conover, 1980).

The Shapiro-Wilk test is considered a better test than the previous ones, especially when the sample size is small. It is available in SPSS and STATISTICA as a complement of histograms and normality plots, respectively (see Commands 5.5). It is also available in R as the function `shapiro.test(x)`. When applied to Example 5.8, it produces an observed significance of $p = 0.88$. With this high significance, it is safe to accept the null hypothesis.

Table 5.9 illustrates the behaviour of the goodness of fit tests in an experiment using small to moderate sample sizes ($n = 10$, 25 and 50), generated according to a known law. The lognormal distribution corresponds to a random variable whose logarithm is normally distributed. The "Bimodal" samples were generated using the sum of two Gaussian functions separated by 4σ. For each value of n a large number of samples were generated (see top of Table 5.9), and the percentage of correct decisions at a 5% level of significance was computed.

Table 5.9. Percentages of correct decisions in the assessment at 5% level of the goodness of fit to the normal distribution, for several empirical distributions (see text).

	$n = 10$ (200 samples)			$n = 25$ (80 samples)			$n = 50$ (40 samples)		
	KS	L	SW	KS	L	SW	KS	L	SW
Normal, $N_{0,1}$	100	95	98	100	100	98	100	100	100
Lognormal	2	42	62	32	94	100	92	100	100
Exponential, ε_1	1	33	43	9	74	91	32	100	100
Student t_2	2	28	27	11	55	66	38	88	95
Uniform, $U_{0,1}$	0	8	6	0	6	24	0	32	88
Bimodal	0	16	15	0	46	51	5	82	92

KS: Kolmogorov-Smirnov; L: Lilliefors; SW: Shapiro-Wilk.

As can be seen in Table 5.9, when the sample size is very small ($n = 10$), all the three tests make numerous mistakes. For larger sample sizes the Shapiro-Wilk test performs somewhat better than the Lilliefors test, which in turn, performs better than the Kolmogorov-Smirnov test. This test is only suitable for very large samples (say $n \gg 50$). It also has the advantage of allowing an assessment of the goodness of fit to other distributions, whereas the Liliefors and Shapiro-Wilk tests can only assess the normality of a distribution.

Also note that most of the test errors in the assessment of the normal distribution occurred for symmetric distributions (three last rows of Table 5.9). The tests made fewer mistakes when the data was generated by asymmetric distributions, namely the lognormal or exponential distribution. Taking into account these observations the reader should keep in mind that the statements "a data sample can be well modelled by the normal distribution" and a "data sample comes from a population with a normal distribution" mean entirely different things.

5.2 Contingency Tables

Contingency tables were introduced in section 2.2.3 as a means of representing multivariate data. In sections 2.3.5 and 2.3.6, some measures of association computed from these tables were also presented. In this section, we describe tests of hypotheses concerning these tables.

5.2.1 The 2×2 Contingency Table

The 2×2 contingency table is a convenient formalism whenever one has two random and independent samples obtained from two distinct populations whose cases can be categorised into two classes, as shown in Figure 5.3. The sample sizes are n_1 and n_2 and the observed occurrence counts are the O_{ij}.

This formalism is used when one wants to assess whether, based on the samples, one can conclude that the probability of occurrence of one of the classes is different for the two populations. It is a quite useful formalism, namely in clinical research, when one wants to assess whether a specific treatment is beneficial; then, the populations correspond to "without" and "with" the treatment.

	Class 1	Class 2	
Population 1	O_{11}	O_{12}	n_1
Population 2	O_{21}	O_{22}	n_2

Figure 5.3. The 2×2 contingency table with the sample sizes (n_1 and n_2) and the observed absolute frequencies (counts O_{ij}).

Let p_1 and p_2 denote the probabilities of occurrence of one of the classes, e.g. class 1, for the populations 1 and 2, respectively. For the two-sided test, the hypotheses are:

H_0: $p_1 = p_2$;
H_1: $p_1 \neq p_2$.

The one-sided test is formalised as:

H_0: $p_1 \leq p_2$, H_1: $p_1 > p$;   or
H_0: $p_1 \geq p_2$; H_1: $p_1 < p_2$.

In order to assess the null hypothesis, we use the same goodness of fit measure as in formula 5.8, now reflecting the sum of the squared deviations for all four cells in the contingency table:

$$T = \sum_{i=1}^{2} \sum_{j=1}^{2} \frac{\left(O_{ij} - E_{ij}\right)^2}{E_{ij}}, \qquad 5.18$$

where the expected absolute frequencies E_{ij} are estimated as:

$$E_{ij} = \frac{n_i \sum_{i=1}^{2} O_{ij}}{n} = \frac{n_i (O_{1j} + O_{2j})}{n}, \qquad 5.19$$

with $n = n_1 + n_2$ (total number of cases).

Thus, we estimate the expected counts in each cell as the ratio of the observed marginal counts. With these estimates, one can rewrite 5.18 as:

$$T = \frac{n(O_{11}O_{22} - O_{12}O_{21})^2}{n_1 n_2 (O_{11} + O_{21})(O_{12} + O_{22})}. \qquad 5.20$$

The sampling distribution of T, assuming that the null hypothesis is true, $p_1 = p_2 = p$, can be computed by first noticing that the probability of obtaining O_{11} cases of class 1 in a sample of n_1 cases from population 1, is given by the binomial law (see A.7):

$$P(O_{11}) = \binom{n_1}{O_{11}} p^{O_{11}} q^{n_1 - O_{11}}.$$

Similarly, for the probability of obtaining O_{21} cases of class 1 in a sample of n_2 cases from population 2:

$$P(O_{21}) = \binom{n_2}{O_{21}} p^{O_{21}} q^{n_2 - O_{21}}.$$

Because the two samples are independent the probability of the joint event is given by:

$$P(O_{11}, O_{21}) = \binom{n_1}{O_{11}} \binom{n_2}{O_{21}} p^{O_{11} + O_{21}} q^{n - O_{11} - O_{21}}, \qquad 5.21$$

The exact values of $P(O_{11}, O_{21})$ are, however, very difficult to compute, except for very small n_1 and n_2 (see e.g. Conover, 1980). Fortunately, the asymptotic distribution of T is well approximated by the chi-square distribution with one degree of freedom (χ_1^2). We then use the critical values of the chi-square distribution in order to test the null hypothesis in the usual way. When dealing with a one-sided test we face the difficulty that the T statistic does not reflect the direction of the deviation between observed and expected frequencies. In this situation, it is simpler to use the sampling distribution of the signed square root of T (with the sign of $O_{11}O_{22} - O_{12}O_{21}$), which is approximated by the standard normal distribution. Denoting by T_1 the signed square root of T, the one-sided test is performed as:

H_0: $p_1 \leq p_2$:   reject at level α if $T_1 > z_{1-\alpha}$;
H_0: $p_1 \geq p_2$:   reject at level α if $T_1 < z_{\alpha}$.

A "continuity correction", known as "Yates' correction", is sometimes used in the chi-square test of 2×2 contingency tables. This correction attempts to compensate for the inaccuracy introduced by using the continuous chi-square distribution, instead of the discrete distribution of T, as follows:

$$T = \frac{n[|O_{11}O_{22} - O_{12}O_{21}| - (n/2)]^2}{n_1 n_2 (O_{11} + O_{21})(O_{12} + O_{22})}.$$ 5.22

Example 5.9

Q: Consider the male and female populations related to the Freshmen dataset. Based on the evidence provided by the respective samples, is it possible to conclude that the proportion of male students that are "initiated" differs from the proportion of female students?

A: We apply the chi-square test to the 2×2 contingency table whose rows are the populations (variable SEX) and whose columns are the counts of initiated freshmen (column INIT).
    The contingency table is shown in Table 5.10. The chi-square test results are shown in Table 5.11. Since the observed significance, with and without the continuity correction, is above the 5% significance level, we do not reject the null hypothesis at that level.

□

Table 5.10. Contingency table obtained with SPSS for the SEX and INIT variables of the freshmen dataset. Note that a missing case for INIT (case #118) is not included.

		INIT		Total
		yes	no	
SEX	male	91	5	96
	female	30	5	35
Total		121	10	131

Table 5.11. Partial list of the chi-square test results obtained with SPSS for the SEX and INIT variables of the freshmen dataset.

	Value	df	Asymp. Sig. (2-sided)
Chi-Square	2.997	1	0.083
Continuity Correction	1.848	1	0.174

Example 5.10

Q: Redo the previous example assuming that the null hypothesis is "the proportion of male students that are 'initiated' is higher than that of female students".

A: We now perform a one-sided chi-square test. For this purpose we notice that the sign of $O_{11}O_{22} - O_{12}O_{21}$ is positive, therefore $T_1 = +\sqrt{2.997} = 1.73$. Since $T_1 > z_\alpha = -1.64$, we also do not reject the null hypothesis for this one-sided test.

□

Commands 5.7. SPSS, STATISTICA, MATLAB and R commands used to perform tests on contingency tables.

SPSS	`Analyze; Descriptive Statistics; Crosstabs`
STATISTICA	`Statistics; Basic Statistics/Tables; Tables and banners`
MATLAB	`[table,chi2,p]=crosstab(col1,col2)`
R	`chisq.test(x, correct=TRUE)`

The meaning of the MATLAB `crosstab` parameters and return values is as follows:

`col1, col2`: vectors containing integer data used for the cross-tabulation.
`table`:   cross-tabulation matrix.
`chi2, p`:   value and significance of the chi-square test.

The R function `chisq.test` can be applied to contingency tables. The x parameter represents then a matrix (the contingency table). The correct parameter corresponds to the Yates' correction for 2×2 contingency tables. Let us illustrate with Example 5.9 data. The contingency table can be built as follows:

```
> ct <- array(0,dim=c(2,2)) ## building the matrix
> ct[1,1] <- sum(SEX==1 & INIT==1) ## & means AND
> ct[1,2] <- sum(SEX==1 & INIT==2)
> ct[2,1] <- sum(SEX==2 & INIT==1)
> ct[2,2] <- sum(SEX==2 & INIT==2)
```

An alternative and easier way to build the contingency table is by using the table function mentioned in Commands 2.1:

```
> ct <- table(SEX,INIT,exclude=c(9))
```

Note the exclude=c(9) argument which excludes non-valid data (corresponding to missing data) coded with 9. Finally, we apply:

```
> chisq.test(ct,correct=FALSE)
X-squared = 2.9323, df = 1, p-value = 0.08682
```

These values agree quite well with those published in Table 5.11.

In order to solve the Example 5.12 we first recode Q7 by merging the values 1 and 2 as follows:

```
> Q7_12<-as.numeric(Q7<=2)+as.numeric(Q7>2)*Q7
```

This creates a new vector with only 4 categorical values: 1, 3, 4 and 5. The as.numeric function converts FALSE and TRUE into 0 and 1, respectively. We then proceed as above:

```
> ct<-table(SEX,Q7_12,exclude=c(9))
> chisq.test(ct)
X-squared = 5.3334, df = 3, p-value = 0.1490
```

∎

5.2.2 The *rxc* Contingency Table

The *rxc* contingency table is an obvious extension of the 2×2 contingency table, when there are more than two categories of the nominal (or ordinal) variable involved. However, some aspects described in the previous section, namely the Yates' correction and the computation of exact probabilities, are only applicable to 2×2 tables.

	Class 1	Class 2	. . .	Class c	
Population 1	O_{11}	O_{12}	. . .	O_{1c}	n_1
Population 2	O_{21}	O_{22}	. . .	O_{2c}	n_2
. . .	. . .	. . .	. . .	. . .	. . .
Population r	O_{r1}	O_{r2}	. . .	O_{rc}	n_r
	c_1	c_2	. . .	c_c	

Figure 5.4. The *rxc* contingency table with the sample sizes (n_i) and the observed absolute frequencies (counts O_{ij}).

The $r \times c$ contingency table is shown in Figure 5.4. All samples from the r populations are assumed to be independent and randomly drawn. All observations are assumedly categorised into exactly one of c categories. The total number of cases is:

$$n = n_1 + n_2 + ... + n_r = c_1 + c_2 + ... + c_c,$$

where the c_j are the column counts, i.e., the total number of observations in the jth class:

$$c_j = \sum_{i=1}^{r} O_{ij}.$$

Let p_{ij} denote the probability that a randomly selected case of population i is from class j. The hypotheses formalised for the $r \times c$ contingency table are a generalisation of the two-sided hypotheses for the 2×2 contingency table (see 5.2.1):

H_0: For any class, the probabilities are the same for all populations: $p_{1j} = p_{2j} = ... = p_{rj}, \forall j$.

H_1: There are at least two populations with different probabilities in one class: $\exists\, i,j,\ p_{ij} \neq p_{kj}$.

The test statistic is also a generalisation of 5.18:

$$T = \sum_{i=1}^{r} \sum_{j=1}^{c} \frac{\left(O_{ij} - E_{ij}\right)^2}{E_{ij}}, \text{ with } E_{ij} = \frac{n_i c_j}{n}. \qquad 5.23$$

If H_0 is true, we expect the observed counts O_{ij} to be near the expected counts E_{ij}, estimated as in the above formula 5.23, using the row and column marginal counts. The asymptotic distribution of T is the chi-square distribution with $df = (r - 1)(c - 1)$ degrees of freedom. As with the chi-square goodness of fit test described in section 5.1.3, the approximation is considered acceptable if the following conditions are met:

i. For $df = 1$, i.e. for 2×2 contingency tables, no E_{ij} must be smaller than 5;
ii. For $df > 1$, no E_{ij} must be smaller than 1 and no more than 20% of the E_{ij} must be smaller than 5.

The SPSS STATISTICA, MATLAB and R commands for testing $r \times c$ contingency tables are indicated in Commands 5.7.

Example 5.11

Q: Consider the male and female populations of the Freshmen dataset. Based on the evidence provided by the respective samples, is it possible to conclude that male and female students have different behaviour participating in the "initiation" on their own will?

A: Question 7 (column Q7) of the freshmen dataset addresses the issue of participating in the initiation on their own will. The 2×5 contingency table, using variables SEX and Q7, has more than 20% of the cells with expected counts below 5 because of the reduced number of cases ranked 1 and 2. We, therefore, create a new variable Q7_12 where the ranks 1 and 2 are merged into a new rank, coded 12.

The contingency table for the variables SEX and Q7_12 is shown in Table 5.11. The chi-square value for this table has an observed significance $p = 0.15$; therefore, we do not reject the null hypothesis of equal behaviour of male and female students at the 5% level.

Since one of the variables, SEX, is nominal, we can determine the association measures suitable to nominal variables, as we did in section 2.3.6. In this example the phi and uncertainty coefficients both have significances (0.15 and 0.08, respectively) that do not support the rejection of the null hypothesis (no association between the variables) at the 5% level.

□

Table 5.12. Contingency table obtained with SPSS for the SEX and Q7_12 variables of the freshmen dataset. Q7_12 is created with the SPSS recode command, using Q7. Note that three missing cases are not included.

			Q7_12				Total
			3	4	5	12	
SEX	male	Count	18	36	29	12	95
		Expected Count	14.0	36.8	30.9	13.3	95.0
	female	Count	1	14	13	6	34
		Expected Count	5.0	13.2	11.1	4.7	34.0
Total		Count	19	50	42	18	129
		Expected Count	19.0	50.0	42.0	18.0	129.0

5.2.3 The Chi-Square Test of Independence

When performing tests of hypotheses one often faces the situation in which a decision must be made as to whether or not two or more variables pertaining to the same population can be considered independent. In order to assess the independency of two variables we use the contingency table formalism, which now, however, is applied to only one population whose variables can be categorised into two or more categories. The variables can either be discrete (nominal or ordinal) or continuous. In this latter case, one must choose suitable categorisations for the continuous variables.

The $r \times c$ contingency table for this situation is the same as shown in Figure 5.4. The only differences being that whereas in the previous section the rows represented different populations and the row totals were assumed to be fixed, now the rows represent categories of a second variable and the row totals can vary arbitrarily, constrained only by the fact that their sum is the total number of cases.

The test is formalised as:

H_0: The event "an observation is in row i" is independent of the event "the same observation is in column j", i.e.:

$$P(\text{row } i, \text{column } j) = P(\text{row } i) \times P(\text{column } j), \forall i, j.$$

H_1: The events "an observation is in row i" and "the same observation is in column j", are dependent, i.e.:

$$\exists\, i, j, P(\text{row } i, \text{column } j) \neq P(\text{row } i) \times P(\text{column } j).$$

Let r_i denote the row totals as in Figure 2.18, such that:

$$r_i = \sum_{j=1}^{c} O_{ij} \text{ and } n = r_1 + r_2 + \ldots + r_r = c_1 + c_2 + \ldots + c_c.$$

As before, we use the test statistic:

$$T = \sum_{i=1}^{r} \sum_{j=1}^{c} \frac{\left(O_{ij} - E_{ij}\right)^2}{E_{ij}}, \text{ with } E_{ij} = \frac{r_i c_j}{n}, \tag{5.24}$$

which has the asymptotic chi-square distribution with $df = (r - 1)(c - 1)$ degrees of freedom. Note, however, that since the row totals can vary in this situation, the exact probability associated to a certain value of T is even more difficult to compute than before because there are a greater number of possible tables with the same T.

Example 5.12

Q: Consider the Programming dataset, containing results of pedagogical enquiries made during the period 1986-1988, of freshmen attending the course "Programming and Computers" in the Electrotechnical Engineering Department of Porto University. Based on the evidence provided by the respective samples, is it possible to conclude that the performance obtained by the students at the final examination is independent of their previous knowledge on programming?

A: Note that we have a single population with two attributes: "previous knowledge on programming" (variable PROG), and "final examination score" (variable SCORE). In order to test the independence hypothesis of these two attributes, we first categorise the SCORE variable into four categories. These can be classified as: "Poor" corresponding to a final examination score below 10; "Fair" corresponding to a score between 10 and 13; "Good" corresponding to a score between 14 and 16; "Very Good" corresponding to a score above 16. Let us call PERF (performance) this new categorised variable.

The 3×4 contingency table, using variables PROG and PERF, is shown in Table 5.13. Only two (16.7%) cells have expected counts below 5; therefore, the recommended conditions, mentioned in the previous section, for using the asymptotic distribution of T, are met.

The value of T is 43.044. The asymptotic chi-square distribution of T has $(3 - 1)(4 - 1) = 6$ degrees of freedom. At a 5% level, the critical region is above 12.59 and therefore the null hypothesis is rejected at that level. As a matter of fact, the observed significance of T is $p \approx 0$.

□

Table 5.13. The 3×4 contingency table obtained with SPSS for the independence test of Example 5.12.

				PERF			Total
			Poor	Fair	Good	Very Good	
PROG	0	Count	76	78	16	7	177
		Expected Count	63.4	73.8	21.6	18.3	177.0
	1	Count	19	29	10	13	71
		Expected Count	25.4	29.6	8.6	7.3	71.0
	2	Count	2	6	7	8	23
		Expected Count	8.2	9.6	2.8	2.4	23.0
Total		Count	97	113	33	28	271
		Expected Count	97.0	113.0	33.0	28.0	271.0

The chi-square test of independence can also be applied to assess whether two or more groups of data are independent or can be considered as sampled from the same population. For instance, the results obtained for Example 5.7 can also be interpreted as supporting, at a 5% level, that the male and female groups are not independent for variable Q7; they can be considered samples from the same population.

5.2.4 Measures of Association Revisited

When analysing contingency tables, it is also convenient to assess the degree of association between the variables, using the ordinal and nominal association measures described in sections 2.3.5 and 2.3.6, respectively. As in 4.4.1, the hypotheses in a two-sided test concerning any measure of association γ are formalised as:

H_0: $\gamma = 0$;
H_1: $\gamma \neq 0$.

5.2.4.1 Measures for Ordinal Data

Let X and Y denote the variables whose association is being assessed. The exact values of the sampling distribution of the *Spearman's rank correlation*, when H_0 is true, can be derived if we note that for any given ranking of Y, any rank order of X is equally likely, and vice-versa. Therefore, any particular ranking has a probability of occurrence of $1/n!$. As an example, let us consider the situation of $n = 3$, with X and Y having ranks 1, 2 and 3. As shown in Table 5.14, there are $3! = 6$ possible permutations of the X ranks. Applying formula 2.21, one then obtains the r_s values shown in the last row. Therefore, under H_0, the ± 1 values have a 1/6 probability and the $\pm\frac{1}{2}$ values have a 1/3 probability. When n is large (say, above 20), the significance of r_s under H_0 can be obtained using the test statistic:

$$z^* = r_s \sqrt{n-1}\,, \qquad\qquad\qquad 5.25$$

which is approximately distributed as the standard normal distribution.

Table 5.14. Possible rankings and Spearman correlation for $n = 3$.

X	Y	Y	Y	Y	Y	Y
1	1	1	2	2	3	3
2	2	3	1	3	1	2
3	3	2	3	1	2	1
r_s	1	0.5	0.5	−0.5	−0.5	−1

In order to test the significance of the *gamma statistic* a large sample (say, above 25) is required. We then use the test statistic:

$$z^* = (G - \gamma)\sqrt{\frac{P+Q}{n(1 - G^2)}}\,, \qquad\qquad 5.26$$

which, under H_0 ($\gamma = 0$), is approximately distributed as the standard normal distribution. The values of P and Q were defined in section 2.3.5.

The Spearman correlation and the gamma statistic were computed for Example 5.12, with the results shown in Table 5.15. We see that the observed significance is very low, leading to the conclusion that there is an association between both variables (PERF, PROG).

Table 5.15. Measures of association for ordinal data computed with SPSS for Example 5.12.

	Value	Asymp. Std. Error	Approx. T	Approx. Sig.
Gamma	0.486	0.076	5.458	0.000
Spearman Correlation	0.332	0.058	5.766	0.000

5.2.4.2   Measures for Nominal Data

In Chapter 2, the following measures of association were described: the *index of association* (phi coefficient), the *proportional reduction of error* (Goodman and Kruskal lambda), and the κ statistic for the *degree of agreement*.

Note that taking into account formulas 2.24 and 5.20, the phi coefficient can be computed as:

$$\phi = \sqrt{\frac{T}{n}} = \frac{|T_1|}{\sqrt{n}} , \qquad 5.27$$

with the phi coefficient now lying in the interval [0, 1]. Since the asymptotic distribution of T_1 is the standard normal distribution, one can then use this distribution in order to evaluate the significance of the signed phi coefficient (using the sign of $O_{11}O_{22} - O_{12}O_{21}$) multiplied by $\sqrt{n}$.

Table 5.16 displays the value and significance of the phi coefficient for Example 5.9. The computed two-sided significance of phi is 0.083; therefore, at a 5% significance level, we do not reject the hypothesis that there is no association between SEX and INIT.

Table 5.16. Phi coefficient computed with SPSS for the Example 5.9 with the two-sided significance.

	Value	Approx. Sig.
Phi	0.151	0.083

The proportional reduction of error has a complex sampling distribution that we will not discuss. For Example 5.9 the only situation of interest for this measure of association is: INIT depending on SEX. Its value computed with SPSS is 0.038. This means that variable SEX will only reduce by about 4% the error of predicting

INIT. As a matter of fact, when using INIT alone, the prediction error is $(131 - 121)/131 = 0.076$. With the contribution of variable SEX, the prediction error is the same $(5/131 + 5/131)$. However, since there is a tie in the row modes, the contribution of INIT is computed as half of the previous error.

In order to test the significance of the κ statistic measuring the agreement among several variables, the following statistic, approximately normally distributed for large n with zero mean and unit standard deviation, is used:

$$z = \kappa / \sqrt{\operatorname{var}(\kappa)} \text{ , with} \tag{5.28}$$

$$\operatorname{var}(\kappa) \approx \frac{2}{n\kappa(\kappa-1)} \frac{P(E) - (2\kappa - 3)[P(E)]^2 + 2(\kappa - 2)\sum p_j^3}{[1 - P(E)]^2}. \tag{5.28a}$$

As described in 2.3.6.3, the κ statistic can be computed with function kappa implemented in MATLAB or R; kappa(x,alpha) computes for a matrix x, (formatted as columns N, S and P in Table 2.13), the row vector denoted [ko,z,zc] in MATLAB containing the observed value of κ, ko, the z value of formula 5.28 and the respective critical value, zc, at alpha level. The meaning of the returned values for the R kappa function is the same. The results of the κ statistic significance for Example 2.11 are obtained as shown below. We see that the null hypothesis (disagreement among all four classifiers) is rejected at a 5% level of significance, since z > zc.

```
[ko,z,zc]=kappa(x,0.05)
ko =
    0.2130
z =
    3.9436
zc =
    3.2897
```

5.3   Inference on Two Populations

In this section, we describe non-parametric tests that have parametric counterparts described in section 4.4.3. As discussed in 4.4.3.1, when testing two populations, one must first assess whether or not the available samples are independent. Tests for two paired or matched samples are used to assess whether two treatments are different or whether one treatment is better than the other. Either treatment is applied to the same group of cases (the "before" and "after" experiments), or applied to pairs of cases which are as much alike as possible, the so-called "matched pairs". When it is impossible to design a study with paired samples, we resort to tests for independent samples. Note that some of the tests described for contingency tables also apply to two independent samples.

5.3.1 Tests for Two Independent Samples

Commands 5.8. SPSS, STATISTICA, MATLAB and R commands used to perform non-parametric tests on two independent samples.

SPSS	Analyze; Nonparametric Tests; 2 Independent Samples
STATISTICA	Statistics; Nonparametrics; Comparing two independent samples (groups)
MATLAB	[p,h,stats]=ranksum(x,y,alpha)
R	ks.test(x,y) ; wilcox.test(x,y) | wilcox.test(x~y)

∎

5.3.1.1 The Kolmogorov-Smirnov Two-Sample Test

The Kolmogorov-Smirnov test is used to assess whether two independent samples were drawn from the same population or from populations with the same distribution, for the variable X being tested, which is assumed to be continuous. Let $F(x)$ and $G(x)$ represent the unknown distributions for the two independent samples. The null hypothesis is formalised as:

H_0: Data variable X has equal cumulative probability distributions for the two samples: $F(x) = G(x)$.

The test is conducted similarly to the way described in section 5.1.4. Let $S_m(x)$ and $S_n(x)$ represent the empirical distributions of the two samples, with sizes m and n, respectively. We then use as test statistic, the maximum deviation of these empirical distributions:

$$D_{m,n} = \max | S_n(x) - S_m(x) |.$$ 5.29

For large samples (say, m and n above 25) and two-tailed tests (the most usual), the significance of $D_{m,n}$ can be evaluated using the critical values obtained with the expression:

$$c\sqrt{\frac{m+n}{mn}},$$ 5.30

where c is a coefficient that depends on the significance level, namely $c = 1.36$ for $\alpha = 0.05$ (for details, see e.g. Siegel S, Castellan Jr NJ, 1998).

When compared with its parametric counterpart, the t test, the Kolmogorov-Smirnov test has a high power-efficiency of about 95%, even for small samples.

Example 5.13

Q: Consider the variable ART, the total area of defects, of the cork-stopper dataset. Can one assume that the distributions of ART for the first two classes of cork-stoppers are the same?

A: Variable ART can be considered a continuous variable, and the samples are independent. Table 5.17 shows the Kolmogorov test results, from where we conclude that the null hypothesis is rejected, i.e., for variable ART, the first two classes have different distributions. The test is performed in R with ks.test (ART[1:50],ART[51:100]).   □

Table 5.17. Two sample Kolmogorov-Smirnov test results obtained with SPSS for variable ART of the cork-stopper dataset.

		ART
Most Extreme Differences	Absolute	0.800
	Positive	0.800
	Negative	0.000
Kolmogorov-Smirnov Z		4.000
Asymp. Sig. (2-tailed)		0.000

5.3.1.2   The Mann-Whitney Test

The *Mann-Whitney test*, also known as *Wilcoxon-Mann-Whitney* or *rank-sum test*, is used like the previous test to assess whether two independent samples were drawn from the same population, or from populations with the same distribution, for the variable being tested, which is assumed to be at least ordinal.

Let $F_X(x)$ and $G_Y(x)$ represent the unknown distributions of the two independent populations, where we explicitly denote by X and Y the corresponding random variables. The null hypothesis can be formalised as in the previous section ($F_X(x) = G_Y(x)$). However, when the distributions are different, it often happens that the probability associated to the event "$X > Y$" is not ½, as should be expected for equal distributions. Following this approach, the hypotheses for the Mann-Whitney test are formalised as:

H_0: $P(X > Y) = ½$;
H_1: $P(X > Y) \neq ½$,

for the two-sided test, and

H_0: $P(X > Y) \geq ½$; H_1: $P(X > Y) < ½$,   or
H_0: $P(X > Y) \leq ½$; H_1: $P(X > Y) > ½$,

for the one-sided test.

In order to assess these hypotheses, the Mann-Whitney test starts by assigning ranks to the samples. Let the samples be denoted $x_1, x_2, ..., x_n$ and $y_1, y_2, ..., y_m$. The ranking of the x_i and y_i assigns ranks in $1, 2, ..., n + m$. As an example, let us consider the following situation:

x_i :   12 21 15 8
y_i :   9 13 19

The ranking of x_i and y_i would then yield the result:

Variable:	X	Y	X	Y	X	Y	X
Data:	8	9	12	13	15	19	21
Rank:	1	2	3	4	5	6	7

The test statistic is the sum of the ranks for one of the variables, say X:

$$W_X = \sum_{i=1}^{n} R(x_i),$$ 5.31

where $R(x_i)$ are the ranks assigned to the x_i. For the example above, $W_X = 16$. Similarly, $W_Y = 12$ with:

$$W_X + W_Y = \frac{N(N+1)}{2}, \quad \text{total sum of the ranks from 1 through } N = n + m.$$

The rationale for using W_X as a test statistic is that under the null hypothesis, $P(X > Y) = \frac{1}{2}$, one expects the ranks to be randomly distributed between the x_i and y_i, therefore resulting in approximately equal average ranks in each of the two samples. For small samples, there are tables with the exact probabilities of W_X. For large samples (say m or n above 10), the sampling distribution of W_X rapidly approaches the normal distribution with the following parameters:

$$\mu_{W_X} = \frac{n(N+1)}{2}; \qquad \sigma^2_{W_X} = \frac{nm(N+1)}{12}.$$ 5.32

Therefore, for large samples, the following test statistic with standard normal distribution is used:

$$z^* = \frac{W_X \pm 0.5 - \mu_{W_X}}{\sigma_{W_X}}.$$ 5.33

The 0.5 continuity correction factor is added when one wants to determine critical points in the left tail of the distribution, and subtracted to determine critical points in the right tail of the distribution.

When compared with its parametric counterpart, the t test, the Mann-Whitney test has a high power-efficiency, of about 95.5%, for moderate to large n. In some cases, it was even shown that the Mann-Whitney test is more powerful than the t test! There is also evidence that it should be preferred over the previous Kolmogorov-Smirnov test for large samples.

Example 5.14

Q: Consider the Programming dataset. Does this data support the hypothesis that freshmen and non-freshmen have different distributions of their scores?

A: The Mann-Whitney test results are summarised in Table 5.18. From this table one concludes that the null hypothesis (equal distributions) cannot be rejected at the 5% level. In R this test would be solved with wilcox.test (Score~F) yielding the same results for the "Mann-Whitney U" and "Asymp. Sig." as in Table 5.18.

□

Table 5.18. Mann-Whitney test results obtained with SPSS for Example 5.14: a) Ranks; b) Test statistic and significance. F=1 for freshmen; 0, otherwise.

F	N	Mean Rank	Sum of Ranks		SCORE
				Mann-Whitney U	3916
0	34	132.68	4511	Wilcoxon W	4511
1	237	136.48	32345	Z	−0.265
Total	271			Asymp. Sig. (2-tailed)	0.791
a				b	

Table 5.19. Ranks for variables ASP and PHE (Example 5.15), obtained with SPSS.

	TYPE	N	Mean Rank	Sum of Ranks
ASP	1	30	40.12	1203.5
	2	37	29.04	1074.5
	Total	67		
PHE	1	30	42.03	1261.0
	2	37	27.49	1017.0
	Total	67		

Example 5.15

Q: Consider the t test performed in Example 4.9, for variables ASP and PHE of the wine dataset. Apply the Mann-Whitney test to these continuous variables and compare the results with those previously obtained.

A: Tables 5.19 and 5.20 show the results with identical conclusions (and p values!) to those presented in Example 4.9.

Note that at a 1% level, we do not reject the null hypothesis for the ASP variable. This example constitutes a good illustration of the power-efficiency of the Mann-Whitney test when compared with its parametric counterpart, the t test.
□

Table 5.20. Mann-Whitney test results for variables ASP and PHE (Example 5.15) with grouping variable TYPE, obtained with SPSS.

	ASP	PHE
Mann-Whitney U	371.5	314
Wilcoxon W	1074.5	1017
Z	−2.314	−3.039
Asymp. Sig. (2-tailed)	0.021	0.002

5.3.2 Tests for Two Paired Samples

Commands 5.9. SPSS, STATISTICA, MATLAB and R commands used to perform non-parametric tests on two paired samples.

STATISTICA	`Statistics; Nonparametrics; Comparing two dependent samples (variables)`
SPSS	`Analyze; Nonparametric Tests; 2 Related Samples`
MATLAB	`[p,h,stats]=signrank(x,y,alpha)` `[p,h,stats]=signtest(x,y,alpha)`
R	`mcnemar.test(x)	mcnemar.test(x,y)` `wilcox.test(x,y,paired=TRUE)`

■

5.3.2.1   The McNemar Change Test

The McNemar change test is particularly suitable to "before and after" experiments, in which each case can be in either of two categories or *responses* and is used as its own control. The test addresses the issue of deciding whether or not the change of response is due to hazard. Let the responses be denoted by the + and − signs and a change denoted by an arrow, →. The test is formalised as:

H_0: After the treatment, $P(+ \rightarrow -) = P(- \rightarrow +)$;
H_1: After the treatment, $P(+ \rightarrow -) \neq P(- \rightarrow +)$.

Let us use a 2×2 table for recording the before and after situations, as shown in Figure 5.5. We see that a change occurs in situations A and D, i.e., the number of cases which change of response is $A + D$. If both changes of response are equally likely, the expected count in both cells is $(A + D)/2$.

The McNemar test uses the following test statistic:

$$\chi^{*2} = \sum_{i=1}^{2} \frac{(O_i - E_i)^2}{E_i} = \frac{\left[A - \dfrac{A+D}{2}\right]^2}{\dfrac{A+D}{2}} + \frac{\left[D - \dfrac{A+D}{2}\right]^2}{\dfrac{A+D}{2}} = \frac{(A-D)^2}{A+D}.$$ 5.34

The sampling distribution of this test statistic, when the null hypothesis is true, is asymptotically the chi-square distribution with $df = 1$. A continuity correction is often used, especially for small absolute frequencies, in order to make the computation of significances more accurate.

An alternative to using the chi-square test is to use the binomial test. One would then consider the sample with $n = A + D$ cases, and assess the null hypothesis that the probabilities of both changes are equal to ½.

		After	
		−	+
	+	A	B
Before			
	−	C	D

Figure 5.5. Table for the McNemar change test, where A, B, C and D are cell counts.

Example 5.16

Q: Consider that in an enquiry into consumer preferences of two products A and B, a group of 57 out of 160 persons preferred product A, before reading a study of a consumer protection organisation. After reading the study, 8 persons that had preferred product A and 21 persons that had preferred product B changed opinion. Is it possible to accept, at a 5% level, that the change of opinion was due to hazard?

A: Table 5.21a shows the respective data in a convenient format for analysis with STATISTICA or SPSS. The column "Number" should be used for weighing the cases corresponding to the cells of Figure 5.5 with "1" denoting product A and "2" denoting product B. Case weighing was already used in section 5.1.2.

Table 5.21b shows the results of the test; at a 5% significance level, we reject the null hypothesis that the change of opinion was due to hazard.

In R the test is run (with the same results) as follows:

```
> x <- array(c(49,21,8,82),dim=c(2,2))
> mcnemar.test(x)                                          ⧠
```

Table 5.21. (a) Data of Example 5.16 in an adequate format for running the McNmear test with STATISTICA or SPSS, (b) Results of the test obtained with SPSS.

Before	After	Number			BEFORE & AFTER
1	1	49			
1	2	8	N		160
2	2	82	Chi-Square		4.966
2	1	21	Asymp. Sig.		0.026
a				b	

5.3.2.2   The Sign Test

The sign test compares two paired samples (x_1, y_1), (x_2, y_2), ... , (x_n, y_n), using the *sign* of the respective differences: $(x_1 - y_1)$, $(x_2 - y_2)$, ... , $(x_n - y_n)$, i.e., using a set of dichotomous values (+ and − signs), to which the binomial test described in section 5.1.2 can be applied in order to assess the truth of the null hypothesis:

$$H_0: \ P(x_i > y_i) = P(x_i < y_i) = \tfrac{1}{2}. \qquad\qquad 5.35$$

Note that the null hypothesis can also be stated in terms of the sign of the differences $x_i - y_i$, by setting their median to zero.

Previous to applying the binomial test, all cases with tied decisions, $x_i = y_i$, are removed from the analysis, and the sample size, n, adjusted accordingly. The null hypothesis is rejected if too few differences of one sign occur.

The power-efficiency of the test is about 95% for $n = 6$, decreasing towards 63% for very large n. Although there are more powerful tests for paired data, an important advantage of the sign test is its broad applicability to ordinal data. Namely, when the magnitude of the differences cannot be expressed as a number, the sign test is the only possible alternative.

Example 5.17

Q: Consider the Metal Firms' dataset containing several performance indices of a sample of eight metallurgic firms (see Appendix E). Use the sign test in order to analyse the following comparisons: a) leadership teamwork (TW) vs. leadership commitment to quality improvement (CI), b) management of critical processes (MC) vs. management of alterations (MA). Discuss the results.

A: All variables are ordinal type, measured on a 1 to 5 scale. One must note, however, that the numeric values of the variables cannot be taken to the letter. One could as well use a scale of A to E or use "very poor", "poor", "fair", "good" and "very good". Thus, the sign test is the only two-sample comparison test appropriate here.

Running the test with STATISTICA, SPSS or MATLAB yields observed one-tailed significances of 0.0625 and 0.5 for comparisons (a) and (b), respectively. Thus, at a 5% significance level, we do not reject the null hypothesis of comparable distributions for pair TW and CI nor for pair MC and MA.

Let us analyse in detail the sign test results for the TW-CI pair of variables. The respective ranks are:

TW: 4   4   3   2   4   3   3   3
CI :   3   2   3   2   4   3   2   2
Difference: +   +   0   0   0   0   +   +

We see that there are 4 ties (marked with 0) and 4 positive differences TW – CI. Figure 5.6a shows the binomial distribution of the number k of negative differences for $n = 4$ and $p = \frac{1}{2}$. The probability of obtaining as few as zero negative differences TW – CI, under H_0, is $(\frac{1}{2})^4 = 0.0625$.

We now consider the MC-MA comparison. The respective ranks are:

MC: 2   2   2   2   1   2   3   2
MA: 1   3   1   1   1   4   2   4
Difference: +   –   +   +   0   –   +   –

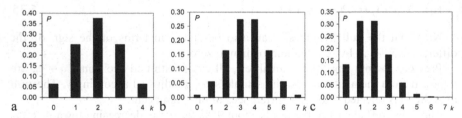

a   b   c

Figure 5.6. Binomial distributions for the sign tests in Example 5.18: a) TW-CI pair, under H_0; b) MC-MA pair, under H_0; c) MC-MA pair for the alternative hypothesis H_1: $P(MC < MA) = \frac{1}{4}$.

Figure 5.6b shows the binomial distribution of the number of negative differences for $n = 7$ and $p = \frac{1}{2}$. The probability of obtaining at most 3 negative differences MC – MA, under H_0, is $\frac{1}{2}$, given the symmetry of the distribution. The critical value of the negative differences, $k = 1$, corresponds to a Type I Error of $\alpha = 0.0625$.

Let us now determine the Type II Error for the alternative hypothesis "positive differences occur three times more often than negative differences". In this case, the distributions of MC and MA are not identical; the distribution of MC favours higher ranks than the distribution of MA. Figure 5.6c shows the binomial distribution for this situation, with $p = P(MC < MA) = \frac{1}{4}$. We clearly see that, in this case, the probability of obtaining at most 3 negative differences MC – MA increases. The Type II Error for the critical value $k = 1$ is the sum of all probabilities for $k \geq 2$, which amounts to $\beta = 0.56$. Even if we relax the α level to 0.23 for a critical value $k = 2$, we still obtain a high Type II Error, $\beta = 0.24$. This low power of the binomial test, already mentioned in 5.1.2, renders the conclusions for small sample sizes quite uncertain.

□

Example 5.18

Q: Consider the FHR dataset containing measurements of basal heart rate frequency (beats per minute) made on 51 foetuses (see Appendix E). Use the sign test in order to assess whether the measurements performed by an automatic system (SPB) are comparable to the computed average (denoted AEB) of the measurements performed by three human experts.

A: There is a clear lack of fit of the distributions of SPB and AEB to the normal distribution. A non-parametric test has, therefore, to be used here. The sign test results, obtained with STATISTICA are shown in Table 5.22. At a 5% significance level, we do not reject the null hypothesis of equal measurement performance of the automatic system and the "average" human expert.

□

Table 5.22. Sign test results obtained with STATISTICA for the SPB-AEB comparison (FHR dataset).

No. of Non-Ties	Percent v < V	Z	p-level
49	63.26531	1.714286	0.086476

5.3.2.3   The Wilcoxon Signed Ranks Test

The Wilcoxon signed ranks test uses the magnitude of the differences $d_i = x_i - y_i$, which the sign test disregards. One can, therefore, expect an enhanced power-efficiency of this test, which is in fact asymptotically 95.5%, when compared with its parametric counterpart, the t test. The test ranks the d_i's according to their magnitude, assigning a rank of 1 to the d_i with smallest magnitude, the rank of 2 to the next smallest magnitude, etc. As with the sign test, x_i and y_i ties ($d_i = 0$) are removed from the dataset. If there are ties in the magnitude of the differences, these are assigned the average of the ranks that would have been assigned without ties. Finally, each rank gets the sign of the respective difference. For the MC and MA variables of Example 5.17, the ranks are computed as:

MC:	2	2	2	2	1	2	3	2
MA:	1	3	1	1	1	4	2	4
MC – MA:	+1	–1	+1	+1	0	–2	+1	–2

Ranks:	1	2	3	4		6	5	7
Signed Ranks:	3	–3	3	3		–6.5	3	–6.5

Note that all the magnitude 1 differences are tied; we, therefore, assign the average of the ranks from 1 to 5, i.e., 3. Magnitude 2 differences are assigned the average rank $(6+7)/2 = 6.5$.

The Wilcoxon test uses the test statistic:

$$T^+ = \text{sum of the ranks of the positive } d_i. \qquad\qquad 5.36$$

The rationale is that under the null hypothesis – samples are from the same population or from populations with the same median – one expects that the sum of the ranks for positive d_i will balance the sum of the ranks for negative d_i. Tables of the sampling distribution of T^+ for small samples can be found in the literature. For large samples (say, $n > 15$), the sampling distribution of T^+ converges asymptotically, under the null hypothesis, to a normal distribution with the following parameters:

$$\mu_{T^+} = \frac{n(n+1)}{4}; \qquad\qquad \sigma^2_{T^+} = \frac{n(n+1)(2n+1)}{24}. \qquad\qquad 5.37$$

A test procedure similar to the t test can then be applied in the large sample case. Note that instead of T^+ the test can also use T^- the sum of the negative ranks.

Table 5.23. Wilcoxon test results obtained with SPSS for the SPB-AEB comparison (FHR dataset) in Example 5.19: a) ranks, b) significance based on negative ranks.

	N	Mean Rank	Sum of Ranks		AE – SP
Negative Ranks	18	20.86	375.5		
Positive Ranks	31	27.40	849.5	Z	–2.358
Ties	2				
Total	51			Asymp. Sig. (2-tailed)	0.018

a   b

Example 5.19

Q: Redo the two-sample comparison of Example 5.18, using the Wilcoxon signed ranks test.

A: The Wilcoxon test results obtained with SPSS are shown in Table 5.23. At a 5% significance level, we reject the null hypothesis of equal measurement performance of the automatic system and the "average" human expert. Note that the conclusion is different from the one reached using the sign test in Example 5.18.

In R the command wilcox.test(SPB, AEB, paired = TRUE) yields the same "p-value".

$\square$

Example 5.20

Q: Estimate the power of the Wilcoxon test performed in Example 5.19 and the needed value of n for reaching a power of at least 90%.

A: We estimate the power of the Wilcoxon test using the concept of power-efficiency (see formula 5.1). Since Example 5.19 involves a large sample ($n = 51$), the power-efficiency of the Wilcoxon test is of about 95.5%.

Figure 5.7a shows the STATISTICA specification window for the dependent samples t test. The values filled in are the sample means and sample standard deviations of the two samples, as well as the correlation between them. The "Alpha" value is the previous two-tailed observed significance (see Table 5.22). The value of n, using formula 5.1, is $n = n_A = 0.955 \times 51 \approx 49$. STATISTICA computes a power of 76% for these specifications.

The power curve shown in Figure 5.7b indicates that the parametric test reaches a power of 90% for $n_A = 70$. Therefore, for the Wilcoxon test we need a number of samples of $n_B = 70/0.955 \approx 73$ for the same power.

$\square$

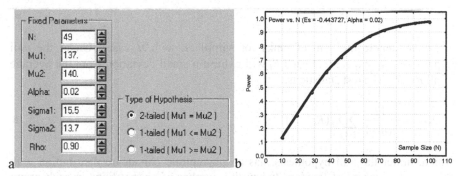

Figure 5.7. Determining the power for a two-paired samples t test, with STATISTICA: a) Specification window, b) Power curve dependent on n.

5.4   Inference on More Than Two Populations

In the present section, we describe non-parametric tests that have parametric counterparts already described in section 4.5. Note that some of the tests described for contingency tables also apply to more than two independent samples.

5.4.1 The Kruskal-Wallis Test for Independent Samples

The *Kruskal-Wallis test* is the non-parametric counterpart of the one-way ANOVA test described in section 4.5.2. The test assesses whether c independent samples are from the same population or from populations with continuous distribution and the same median for the variable being tested. The variable being tested must be at least of ordinal type. The test procedure is a direct generalisation of the Mann-Whitney rank sum test described in section 5.3.1.2. Thus, one starts by assigning natural ordered ranks to the sample values, from the smallest to the largest. Tied ranks are substituted by their average.

Commands 5.10. SPSS, STATISTICA, MATLAB and R commands used to perform the Kruskal-Wallis test.

SPSS	Analyze; Nonparametric Tests; K Independent Samples
STATISTICA	Statistics; Nonparametrics; Comparing multiple indep. samples (groups)
MATLAB	p=kruskalwallis(x)
R	kruskal.test(X~CLASS)

■

Let R_i denote the sum of ranks for sample i, with n_i cases. Under the null hypothesis, we expect that each R_i will exhibit a small deviation from the average of all R_i, $\overline{R}$. The test statistic is:

$$KW = \frac{12}{n(n+1)} \sum_{i=1}^{c} n_i (R_i - \overline{R})^2 \, , \qquad\qquad 5.38$$

which, under the null hypothesis, has an asymptotic chi-square distribution with $df = c - 1$ degrees of freedom (when the number of observations in each group exceeds 5).

When there are tied ranks, a correction is inserted in formula 5.38, dividing the KW value by:

$$1 - \left(\sum_{i=1}^{g} (t_i^3 - t_i) \right) / \left(N^3 - N \right), \tag{5.39}$$

where t_i is the number of ties in group i of g tied groups, and N is the total number of cases in the c samples (sum of the n_i).

The power-efficiency of the Kruskal-Wallis test, referred to the one-way ANOVA, is asymptotically 95.5%.

Example 5.21

Q: Consider the Clays' dataset (see Appendix E). Assume that at a certain stage of the data collection process, only the first 15 cases were available and the Kruskal-Wallis test was used to assess which clay features best discriminated the three types of clays (variable AGE). Perform this test and analyse its results for the alumina content (Al_2O_3) measured with only 3 significant digits.

A: Table 5.24 shows the 15 cases sorted and ranked. Notice the tied values for $Al_2O_3 = 17.3$, corresponding to ranks 6 and 7, which are assigned the mean rank (6+7)/2.

The sum of the ranks is 57, 41 and 22 for the groups 1, 2 and 3, respectively; therefore, we obtain the mean ranks shown in Table 5.25. The asymptotic significance of 0.046 leads us to reject the null hypothesis of equality of medians for the three groups at a 5% level.

□

Table 5.24. The first fifteen cases of the Clays' dataset, sorted and ranked.

AGE	1	1	1	1	1	2	2	2	2	2	3	3	3	3	3
Al_2O_3	23.0	21.4	16.6	22.1	18.8	17.3	17.8	18.4	17.3	19.1	11.5	14.9	11.6	15.8	19.5
Rank	15	13	5	14	10	6.5	8	9	6.5	11	1	3	2	4	12

Table 5.25. Results, obtained with SPSS, for the Kruskal-Wallis test of alumina in the Clays' dataset: a) ranks, b) significance.

AGE	N	Mean Rank
pliocenic good clay	5	11.40
pliocenic bad clay	5	8.20
holocenic clay	5	4.40
Total	15	

a

AL2O3	
Chi-Square	6.151
df	2
Asymp. Sig.	0.046

b

Example 5.22

Q: Consider the Freshmen dataset and use the Kruskal-Wallis test in order to assess whether the freshmen performance (EXAMAVG) differs according to their attitude towards skipping the Initiation (Question 8).

A: The mean ranks and results of the test are shown in Table 5.26. Based on the observed asymptotic significance, we reject the null hypothesis at a 5% level, i.e., we have evidence that the freshmen answer Question 8 of the enquiry differently, depending on their average performance on the examinations.

Table 5.26. Results, obtained with SPSS, for the Kruskal-Wallis test of average freshmen performance in 5 categories of answers to Question 8: a) ranks; b) significance.

Q8	N	Mean Rank	EXAMAVG	
1	10	104.45		
2	22	75.16	Chi-Square	14.081
3	48	60.08		
4	39	59.04	df	4
5	12	63.46		
Total	131		Asymp. Sig.	0.007
a			b	

□

Example 5.23

Q: The variable ART of the Cork Stoppers' dataset was analysed in section 4.5.2.1 using the one-way ANOVA test. Perform the same analysis using the Kruskal-Wallis test and estimate its power for the alternative hypothesis corresponding to the sample means.

A: We saw in 4.5.2.1 that a logarithmic transformation of ART was needed in order to be able to apply the ANOVA test. This transformation is not needed with the Kruskal-Wallist test, whose only assumption is the independency of the samples.

Table 5.27 shows the results, from which we conclude that the null hypothesis of median equality of the three populations is rejected at a 5% significance level (or even at a smaller level).

In order to estimate the power of this Kruskal-Wallis test, we notice that the sample size is large, and therefore, we expect the power to be the same as for the one-way ANOVA test using a number of cases equal to $n = 50 \times 0.955 \approx 48$. The power of the one-way ANOVA, for the alternative hypothesis corresponding to the sample means and with $n = 48$, is 1.

□

Table 5.27. Results, obtained with SPSS, for the Kruskal-Wallis test of variable ART of the `Cork Stoppers'` dataset: a) ranks, b) significance.

C	N	Mean Rank		ART
1	50	28.18	Chi-Square	121.590
2	50	74.35	df	2
3	50	123.97	Asymp.	
Total	150		Sig.	0.000
a			b	

5.4.2   The Friedmann Test for Paired Samples

The *Friedman test* can be considered the non-parametric counterpart of the two-way ANOVA test described in section 4.5.3. The test assesses whether c-paired samples, each with n cases, are from the same population or from populations with continuous distributions and the same median. The variable being tested must be at least of ordinal type. The test procedure starts by assigning natural ordered ranks from 1 to c to the matched case values in each row, from the smallest to the largest. Tied ranks are substituted by their average.

Commands 5.11. SPSS, STATISTICA, MATLAB and R commands used to perform the Friedmann test.

SPSS	`Analyze; Nonparametric Tests; K Related Samples`
STATISTICA	`Statistics; Nonparametrics; Comparing multiple dep. samples (groups)`
MATLAB	`[p,table,stats]=friedman(x,reps)`
R	`friedman.test(x, group) | friedman.test(x~group)`

∎

Let R_i denote the sum of ranks for sample i. Under the null hypothesis, we expect that each R_i will exhibit a small deviation from the value that would be obtained by chance, i.e., $n(c + 1)/2$. The test statistic is:

$$F_r = \frac{12\sum_{i=1}^{c} R_i^2 - 3n^2c(c+1)^2}{nc(c+1)}. \qquad 5.40$$

Tables with the exact probabilities of F_r, under the null hypothesis, can be found in the literature. For $c > 5$ or for $n > 15$ F_r has an asymptotic chi-square distribution with $df = c - 1$ degrees of freedom.

When there are tied ranks, a correction is inserted in formula 5.40, subtracting from $nc(c + 1)$ in the denominator the following term:

$$\frac{nc - \sum_{i=1}^{n} \sum_{j=1}^{g_i} t_{i.j}^3}{c - 1}, \qquad\qquad 5.41$$

where $t_{i.j}$ is the number of ties in group j of g_i tied groups in the ith row.

The power-efficiency of the Friedman test, when compared with its parametric counterpart, the two-way ANOVA, is 64% for $c = 2$ and increases with c, namely to 80% for $c = 5$.

Example 5.24

Q: Consider the evaluation of a sample of eight metallurgic firms (Metal Firms' dataset), in what concerns social impact, with variables: CEI = "commitment to environmental issues"; IRM = "incentive towards using recyclable materials"; EMS = "environmental management system"; CLC = "co-operation with local community"; OEL = "obedience to environmental legislation". Is there evidence at a 5% level that all variables have distributions with the same median?

Table 5.28. Scores and ranks of the variables related to "social impact" in the Metal Firms dataset (Example 5.24).

	Data					Ranks				
	CEI	IRM	EMS	CLC	OEL	CEI	IRM	EMS	CLC	OEL
Firm #1	2	1	1	1	2	4.5	2	2	2	4.5
Firm #2	2	1	1	1	2	4.5	2	2	2	4.5
Firm #3	2	1	1	2	2	4	1.5	1.5	4	4
Firm #4	2	1	1	1	2	4.5	2	2	2	4.5
Firm #5	2	2	1	1	1	4.5	4.5	2	2	2
Firm #6	2	2	2	3	2	2.5	2.5	2.5	5	2.5
Firm #7	2	1	1	2	2	4	1.5	1.5	4	4
Firm #8	3	3	1	2	2	4.5	4.5	1	2.5	2.5
Total						33	20.5	14.5	23.5	28.5

A: Table 5.28 lists the scores assigned to the eight firms. From the scores, the ranks are computed as previously described. Note particularly how ranks are assigned in the case of ties. For instance, Firm #1 IRM, EMS and CLC are tied for rank 1 through 3; thus they get the average rank 2. Firm #1 CEI and OEL are tied for ranks 4 and 5; thus they get the average rank 4.5. Table 5.29 lists the results of the Friedman test, obtained with SPSS. Based on these results, the null hypothesis is rejected at 5% level (or even at a smaller level).

□

Table 5.29. Results obtained with SPSS for the Friedman test of social impact scores of the `Metal Firms'` dataset: a) mean ranks, b) significance.

		Mean Rank			
	CEI	4.13		N	8
	IRM	2.56		Chi-Square	13.831
	EMS	1.81		df	4
	CLC	2.94		Asymp. Sig.	0.008
	OEL	3.56			
a			b		

5.4.3 The Cochran Q test

The *Cochran Q test* is particularly suitable to dichotomous data of k related samples with n items, e.g., when k judges evaluate the presence or absence of an event in the same n cases. The null hypothesis is that there is no difference of probability of one of the events (say, a "success") for the k judges. If the null hypothesis is true, the statistic:

$$Q = \frac{k(k-1)\sum_{j=1}^{k}(G_j - \overline{G})^2}{k\sum_{i=1}^{n}L_i - \sum_{i=1}^{n}L_i^2}, \qquad 5.42$$

is distributed approximately as χ^2 with $df = k - 1$, for not too small n ($n > 4$ and $nk > 24$), where G_j is the total number of successes in the jth column, $\overline{G}$ is the mean of G_j and L_i is the total number of successes in the ith row.

Example 5.25

Q: Consider the FHR dataset, which includes 51 foetal heart rate cases classified by three human experts (E1C, E2C, E3C) and an automatic diagnostic system (SPC) into three categories: normal, suspect and pathologic. Apply the Cochran Q test for the dichotomy normal (0) vs. not normal (1).

A: Table 5.30 shows the frequencies and the value and significance of the Q statistic. Based on these results, we reject with p ≈ 0 the null hypothesis of equal classification of the "normal" event for the three human experts and the automatic system. As a matter of fact, the same conclusion is obtained for the three human experts group (left as an exercise).

□

Table 5.30. Frequencies (a) and Cochran Q test results (b) obtained with SPSS for the FHR dataset in the classification of the normal event.

	Value				
	0	1	N		51
SPCB	41	10	Cochran's Q		61.615
E1CB	20	31	df		3
E2CB	12	39			
E3CB	35	16	Asymp. Sig.		0.000
a			b		

Exercises

5.1 Consider the three sets of measurements, RC, CG and EG, of the Moulds dataset. Assess their randomness with the Runs test, dichotomising the data with the mean, median and mode. Check with a data plot why the random hypothesis is always rejected for the RC measurements (see Exercise 3.2).

5.2 In Statistical Quality Control a process variable is considered out of control if the respective data sequence exhibits a non-random pattern. Assuming that the Cork Stoppers dataset is a valid sample of a cork stopper manufacture process, apply the Runs test to Example 3.4 data, in order to verify that the process is not out of control.

5.3 Consider the Culture dataset, containing a sample of budget expenditure in cultural and sport activities, given in percentage of the total budget. Based on this sample, one could state that more than 50% of the budget is spent on sport activities. Test the validity of this statement with 95% confidence.

5.4 The Flow Rate dataset contains measurements of water flow rate at two dams, denoted AC and T. Assuming the data is a valid sample of the flow rates at those two dams, assess at a 5% level of significance whether or not the flow rate at AC is half of the time higher than at T. Compute the power of the test.

5.5 Redo Example 5.5 for Questions Q1, Q4 and Q7 (Freshmen dataset).

5.6 Redo Example 5.7 for variable PRT (Cork Stoppers dataset).

5.7 Several previous Examples and Exercises assumed a normal distribution for the variables being tested. Using the Lilliefors and Shapiro-Wilk tests, check this assumption for variables used in:

a) Examples 3.6, 3.7, 4.1, 4.5, 4.13, 4.14 and 4.20.

b) Exercises 3.2, 3.8, 4.9, 4.12 and 4.13.

5.8 The Signal & Noise dataset contains amplitude values of a noisy signal for consecutive time instants, and a "detection" variable indicating when the amplitude is above a specified threshold, Δ. For $\Delta = 1$, compute the number of time instants between successive detections and use the chi-square test to assess the goodness of fit of the geometric, Poisson and Gamma distributions to the empirical inter-detection time. The geometric, Poisson and Gamma distributions are described in Appendix B.

5.9 Consider the temperature data, T, of the Weather dataset (Data 1) and assume that it is a valid sample of the yearly temperature at 12H00 in the respective locality. Determine whether one can, with 95% confidence, accept the Beta distribution model with $p = q = 3$ for the empirical distribution of T. The Beta distribution is described in Appendix B.

5.10 Consider the ASTV measurement data sample of the FHR-Apgar dataset. Check the following statements:

a) Variable ASTV cannot have a normal distribution.

b) The distribution of ASTV in hospital HUC can be well modelled by the normal distribution.

c) The distribution of ASTV in hospital HSJ cannot be modelled by the normal distribution.

d) If variable ASTV has a normal distribution in the three hospitals, HUC, HGSA and HSJ, then ASTV has a normal distribution in the Portuguese population.

e) If variable ASTV has a non-normal distribution in one of the three hospitals, HUC, HGSA and HSJ, then ASTV cannot be well modelled by a normal distribution in the Portuguese population.

5.11 Some authors consider Yates' correction overly conservative. Using the Freshmen dataset (see Example 5.9), assess whether or not "the proportion of male students that are 'initiated' is smaller than that of female students" with and without Yates' correction and comment on the results.

5.12 Consider the "Commitment to quality improvement" and "Time dedicated to improvement" variables of the Metal Firms' dataset. Assume that they have binary ranks: 1 if the score is below 3, and 0 otherwise. Can one accept the association of these two variables with 95% confidence?

5.13 Redo the previous exercise using the original scores. Can one use the chi-square statistic in this case?

5.14 Consider the data describing the number of students passing (SCORE $\geq$ 10) or flunking (SCORE < 10) the Programming examination in the Programming dataset. Assess whether or not one can be 95% confident that the pass/flunk variable is independent of previous knowledge in Programming (variable PROG). Also assess whether or not the variables describing the previous knowledge of Boole's Algebra and binary arithmetic are independent.

5.15 Redo Example 5.14 for the variable AB.

5.16 The FHR dataset contains measurements of foetal heart rate baseline performed by three human experts and an automatic system. Is there evidence at the 5% level of significance that there is no difference among the four measurement methods? Is there evidence, at 5% level, of no agreement among the human experts?

5.17 The Culture dataset contains budget percentages spent on promoting sport activities in samples of Portuguese boroughs randomly drawn from three regions. Based on the sample evidence is it possible to conclude that there are no significant differences among those three regions on how the respective boroughs assign budget percentages to sport activities? Also perform the budget percentage comparison for pairs of regions.

5.18 Consider the flow rate data measured at Cávado and Toco Dams included in the Flow Rate dataset. Assume that the December samples are valid random samples for that period of the year and, furthermore, assume that one wishes to compare the flow rate distributions at the two samples.
   a) Can the comparison be performed using a parametric test?
   b) Show that the conclusions of the sign test and of the Wilcoxon signed ranks test are contradictory at 5% level of significance.
   c) Estimate the power of the Wilcoxon signed ranks test.
   d) Repeat the previous analyses for the January samples.

5.19 Using the McNemar Change test compare the pre and post-functional class of patients having undergone heart valve implant using the data sample of the Heart Valve dataset.

5.20 Determine which variables are important in the discrimination of carcinoma from other tissue types using the Breast Tissue dataset, as well as in the discrimination among all tissue types.

5.21 Consider the bacterial counts in the spleen contained in the Cells' dataset and check the following statements:
   a) In general, the CD4 marker is more efficacious than the CD8 marker in the discrimination of the knock-out vs. the control group.
   b) However, in the first two weeks the CD8 marker is by far the most efficacious in the discrimination of the knock-out vs. the control group.
   c) Two months after the infection the biochemical markers CD4 and CD8 are unable to discriminate the knock-out from the control group.

5.22 Based on the sample data included in the Clays' dataset, compare the holocenic with pliocenic clays according to the content of chemical oxides and show that the main difference is in terms of alumina, Al_2O_3. Estimate what is the needed difference in alumina that will correspond to an approximate power of 90%.

5.23 Run the non-parametric counterparts of the tests used in Exercises 4.9, 4.10 and 4.20. Compare the results and the power of the tests with those obtained using parametric tests.

5.24 Using appropriate non-parametric tests, determine which variables of the Wines' dataset are most discriminative of the white from the red wines.

5.25 The Neonatal dataset contains mortality data for delivery taking place at home (MH) and at a Health Centre (MI). Assess whether there are significant differences at 5% level between delivery conditions, using the sign and the Wilcoxon tests.

5.26 Consider the Firms' dataset containing productivity figures (P) for a sample of Portuguese firms in four branches of activity (BRANCH). Study the dataset in order to:
a)   Assess with 5% level of significance whether there are significant differences among the productivity medians of the four branches.
b)   Assess with 1% level of significance whether Commerce and Industry have significantly different medians.

5.27 Apply the appropriate non-parametric test in order to rank the discriminative capability of the features used to characterise the tissue types in the Breast Tissue dataset.

5.28 Redo the previous Exercise 5.27 for the CTG dataset and the three-class discrimination expressed by the grouping variable NSP.

5.29 Consider the discrimination of the three clay types based on the sample data of the Clays' dataset. Show that the null hypothesis of equal medians for the three clay types is:
a)   Rejected with more than 95% confidence for all grading variables (LG, MG, HG).
b)   Not rejected for the iron oxide features.
c)   Rejected with higher confidence for the lime (CaO) than for the silica (SiO_2).

5.30 The FHR dataset contains measurements of basal heart rate performed by three human experts and an automatic diagnostic system. Assess whether the null hypothesis of equal median measurements can be accepted with 5% significance for the three human experts and the automatic diagnostic system.

5.31 When analysing the contents of questions Q4, Q5, Q6 and Q7, someone said that "these questions are essentially evaluating the same thing". Assess whether this statement can be accepted at a 5% significance level. Compute the coefficient of agreement κ and discuss its significance.

5.32 The Programming dataset contains results of an enquiry regarding freshman previous knowledge on programming (PROG), Boole's Algebra (AB), binary arithmetic (BA) and computer hardware (H). Consider the variables PROG, AB, BA and H dichotomised in a "yes/no" fashion. Can one reject with 99% confidence the hypothesis that the four dichotomised variables essentially evaluate the same thing?

5.33 Consider the share values of the firms BRISA, CIMPOR, EDP and SONAE of the Stock Exchange dataset. Assess whether or not the distribution of the daily increase and decrease of the share values can be assumed to be similar for all the firms. Hint: Create new variables with the daily "increase/decrease" information and use an appropriate test for this dichotomous information.

6 Statistical Classification

Statistical classification deals with rules of case assignment to categories or classes. The classification, or *decision rule*, is expressed in terms of a set of random variables – the case *features*. In order to derive the decision rule, one assumes that a *training set* of pre-classified cases – the data sample – is available, and can be used to determine the sought after rule applicable to new cases. The decision rule can be derived in a model-based approach, whenever a joint distribution of the random variables can be assumed, or in a model-free approach, otherwise.

6.1 Decision Regions and Functions

Consider a data sample constituted by n cases, depending on d features. The central idea in statistical classification is to use the data sample, represented by vectors in an $\Re^d$ feature space, in order to derive a decision rule that partitions the feature space into regions assigned to the classification classes. These regions are called *decision regions*. If a feature vector falls into a certain decision region, the associated case is assigned to the corresponding class.

Let us assume two classes, ω_1 and ω_2, of cases described by two-dimensional feature vectors (coordinates x_1 and x_2) as shown in Figure 6.1. The features are random variables, X_1 and X_2, respectively.

Each case is represented by a vector $\mathbf{x} = \begin{bmatrix} x_1 & x_2 \end{bmatrix}' \in \Re^2$. In Figure 6.1, we used "o" to denote class ω_1 cases and "×" to denote class ω_2 cases. In general, the cases of each class will be characterised by random distributions of the corresponding feature vectors, as illustrated in Figure 6.1, where the ellipses represent equal-probability density curves that enclose most of the cases.

Figure 6.1 also shows a straight line separating the two classes. We can easily write the equation of the straight line in terms of the features X_1, X_2 using coefficients or *weights* w_1, w_2 and a *bias* term w_0 as shown in equation 6.1. The weights determine the slope of the straight line; the bias determines the straight line intersect with the axes.

$$d_{X_1,X_2}(\mathbf{x}) \equiv d(\mathbf{x}) = w_1 x_1 + w_2 x_2 + w_0 = 0. \qquad 6.1$$

Equation 6.1 also allows interpretation of the straight line as the root set of a linear function $d(\mathbf{x})$. We say that $d(\mathbf{x})$ is a *linear decision function* that divides (categorises) $\Re^2$ into two decision regions: the upper half plane corresponding to $d(\mathbf{x}) > 0$ where feature vectors are assigned to ω_1; the lower half plane corresponding to $d(\mathbf{x}) < 0$ where feature vectors are assigned to ω_2. The classification is arbitrary for $d(\mathbf{x}) = 0$.

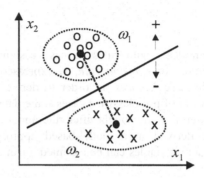

Figure 6.1. Two classes of cases described by two-dimensional feature vectors (random variables X_1 and X_2). The black dots are class means.

The generalisation of the linear decision function for a d-dimensional feature space in $\Re^d$ is straightforward:

$$d(\mathbf{x}) = \mathbf{w'}\,\mathbf{x} + w_0 , \qquad\qquad\qquad 6.2$$

where $\mathbf{w'x}$ represents the dot product[1] of the weight vector and the d-dimensional feature vector.

The root set of $d(\mathbf{x}) = 0$, the *decision surface*, or *discriminant*, is now a linear d-dimensional surface called a *linear discriminant* or *hyperplane*.

Besides the simple linear discriminants, one can also consider using more complex decision functions. For instance, Figure 6.2 illustrates an example of two-dimensional classes separated by a decision boundary obtained with a quadratic decision function:

$$d(\mathbf{x}) = w_5 x_1^2 + w_4 x_2^2 + w_3 x_1 x_2 + w_2 x_2 + w_1 x_1 + w_0 . \qquad\qquad 6.3$$

Linear decision functions are quite popular, as they are easier to compute and have simpler statistical analysis. For this reason in the following we will only deal with linear discriminants.

---

[1] The dot product $\mathbf{x'y}$ is obtained by adding the products of corresponding elements of the two vectors $\mathbf{x}$ and $\mathbf{y}$.

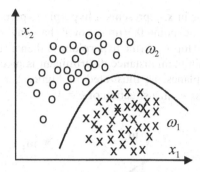

Figure 6.2. Decision regions and boundary for a quadratic decision function.

6.2 Linear Discriminants

6.2.1 Minimum Euclidian Distance Discriminant

The *minimum Euclidian distance discriminant* classifies cases according to their distance to class prototypes, represented by vectors $\mathbf{m}_k$. Usually, these prototypes are class means. We consider the distance taken in the "natural" Euclidian sense. For any d-dimensional feature vector $\mathbf{x}$ and any number of classes, ω_k ($k = 1, ..., c$), represented by their prototypes $\mathbf{m}_k$, the square of the Euclidian distance between the feature vector $\mathbf{x}$ and a prototype $\mathbf{m}_k$ is expressed as follows:

$$d_k^2(\mathbf{x}) = \sum_{i=1}^{d}(x_i - m_{ik})^2 .$$

(6.4)

This can be written compactly in vector form, using the vector dot product:

$$d_k^2(\mathbf{x}) = (\mathbf{x} - \mathbf{m}_k)'(\mathbf{x} - \mathbf{m}_k) = \mathbf{x}'\mathbf{x} - \mathbf{m}_k'\mathbf{x} - \mathbf{x}'\mathbf{m}_k + \mathbf{m}_k'\mathbf{m}_k .$$

(6.5)

Grouping together the terms dependent on $\mathbf{m}_k$, we obtain:

$$d_k^2(\mathbf{x}) = -2(\mathbf{m}_k'\mathbf{x} - 0.5\mathbf{m}_k'\mathbf{m}_k) + \mathbf{x}'\mathbf{x} .$$

(6.6a)

We choose class ω_k, therefore the $\mathbf{m}_k$, which minimises $d_k^2(\mathbf{x})$. Let us assume $c = 2$. The decision boundary between the two classes corresponds to:

$$d_1^2(\mathbf{x}) = d_2^2(\mathbf{x}) .$$

(6.6b)

Thus, using 6.6a, one obtains:

$$(\mathbf{m}_1 - \mathbf{m}_2)'[\mathbf{x} - 0.5(\mathbf{m}_1 + \mathbf{m}_2)] = 0 .$$

(6.6c)

Equation 6.6c, linear in $\mathbf{x}$, represents a hyperplane perpendicular to $(\mathbf{m}_1 - \mathbf{m}_2)$' and passing through the point $0.5(\mathbf{m}_1 + \mathbf{m}_2)$' halfway between the means, as illustrated in Figure 6.1 for $d = 2$ (the hyperplane is then a straight line).

For c classes, the minimum distance discriminant is piecewise linear, composed of segments of hyperplanes, as illustrated in Figure 6.3 with an example of a decision region for class ω_1 in a situation of $c = 4$.

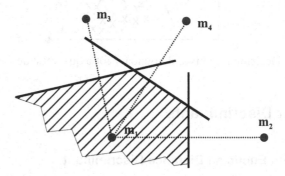

Figure 6.3. Decision region for ω_1 (hatched area) showing linear discriminants relative to three other classes.

Example 6.1

Q: Consider the `Cork Stoppers'` dataset (see Appendix E). Design and evaluate a minimum Euclidian distance classifier for classes 1 (ω_1) and 2 (ω_2), using only feature N (number of defects).

A: In this case, a feature vector with only one element represents each case: $\mathbf{x} = [N]$. Let us first inspect the case distributions in the feature space ($d = 1$) represented by the histograms of Figure 6.4. The distributions have a similar shape with some amount of overlap. The sample means are $m_1 = 55.3$ for ω_1 and $m_2 = 79.7$ for ω_2.

Using equation 6.6c, the linear discriminant is the point at half distance from the means, i.e., the classification rule is:

If $\mathbf{x} < (m_1 + m_2)/2 = 67.5$ then $\mathbf{x} \in \omega_1$ else $\mathbf{x} \in \omega_2$.   6.7

The separating "hyperplane" is simply point 68^2. Note that in the equality case ($\mathbf{x} = 68$), the class assignment is arbitrary.

The classifier performance evaluated in the whole dataset can be computed by counting the wrongly classified cases, i.e., falling into the wrong decision region (a half-line in this case). This amounts to 23% of the cases.

□

---

[2] We assume an underlying real domain for the ordinal feature N. Conversion to an ordinal is performed when needed.

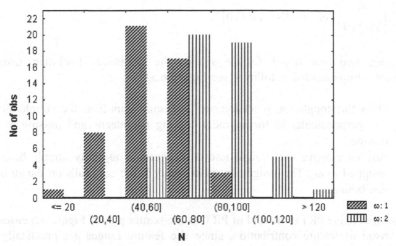

Figure 6.4. Feature N histograms obtained with STATISTICA for the first two classes of the cork-stopper data.

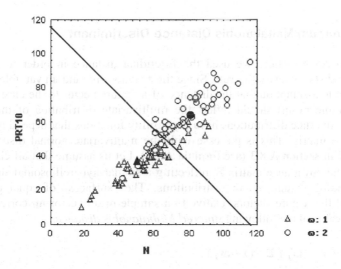

Figure 6.5. Scatter diagram, obtained with STATISTICA, for two classes of cork stoppers (features N, PRT10) with the linear discriminant (solid line) at half distance from the means (solid marks).

Example 6.2

Q: Redo the previous example, using one more feature: PRT10 = PRT/10.

A: The feature vector is:

$$\mathbf{x} = \begin{bmatrix} N \\ PRT10 \end{bmatrix} \quad \text{or} \quad \mathbf{x} = \begin{bmatrix} N & PRT10 \end{bmatrix}'. \tag{6.8}$$

In this two-dimensional feature space, the minimum Euclidian distance classifier is implemented as follows (see Figure 6.5):

1. Draw the straight line (decision surface) equidistant from the sample means, i.e., perpendicular to the segment linking the means and passing at half distance.
2. Any case above the straight line is assigned to ω_2. Any sample below is assigned to ω_1. The assignment is arbitrary if the case falls on the straight-line boundary.

Note that using PRT10 instead of PRT in the scatter plot of Figure 6.5 eases the comparison of feature contribution, since the feature ranges are practically the same.

Counting the number of wrongly classified cases, we notice that the overall error falls to 18%. The addition of PRT10 to the classifier seems beneficial.

□

6.2.2 Minimum Mahalanobis Distance Discriminant

In the previous section, we used the Euclidian distance in order to derive the minimum distance, classifier rule. Since the features are random variables, it seems a reasonable assumption that the distance of a feature vector to the class prototype (class sample mean) should reflect the multivariate distribution of the features. Many multivariate distributions have probability functions that depend on the joint covariance matrix. This is the case with the multivariate normal distribution, as described in section A.8.3 (see formula A.53). Let us assume that all classes have an identical covariance matrix Σ, reflecting a similar hyperellipsoidal shape of the corresponding feature vector distributions. The "surfaces" of equal probability density of the feature vectors relative to a sample mean vector $\mathbf{m}_k$ correspond to a constant value of the following *squared Mahalanobis distance*:

$$d_k^2(\mathbf{x}) = (\mathbf{x} - \mathbf{m}_k)' \Sigma^{-1} (\mathbf{x} - \mathbf{m}_k), \tag{6.9}$$

When the covariance matrix is the unit matrix, we obtain:

$$d_k^2(\mathbf{x}) = (\mathbf{x} - \mathbf{m}_k)' \mathbf{I}^{-1} (\mathbf{x} - \mathbf{m}_k) = (\mathbf{x} - \mathbf{m}_k)' (\mathbf{x} - \mathbf{m}_k),$$

which is the squared Euclidian distance of formula 6.7.

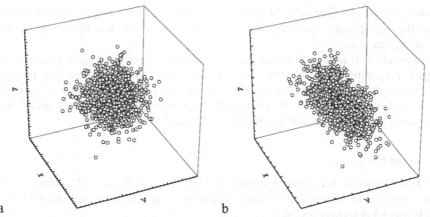

a   b

Figure 6.6. 3D plots of 1000 points with normal distribution: a) Uncorrelated variables with equal variance; b) Correlated variables with unequal variance.

Let us now interpret these results. When all the features are uncorrelated and have equal variance, the covariance matrix is the unit matrix multiplied by the equal variance factor. In the three-dimensional space, the clouds of points are distributed as spheres, illustrated in Figure 6.6a, and the usual Euclidian distance to the mean is used in order to estimate the probability density at any point. The Mahalanobis distance is a generalisation of the Euclidian distance applicable to the general case of correlated features with unequal variance. In this case, the points of equal probability density lie on an ellipsoid and the data points cluster in the shape of an ellipsoid, as illustrated in Figure 6.6b. The orientations of the ellipsoid axes correspond to the correlations among the features. The lengths of straight lines passing through the centre and intersecting the ellipsoid correspond to the variances along the lines. The probability density is now estimated using the squared Mahalanobis distance 6.9.

Formula 6.9 can also be written as:

$$d_k^2(\mathbf{x}) = \mathbf{x'\Sigma^{-1}x} - \mathbf{m}_k{}'\mathbf{\Sigma^{-1}x} - \mathbf{x'\Sigma^{-1}m}_k + \mathbf{m}_k{}'\mathbf{\Sigma^{-1}m}_k .$$ 6.10a

Grouping, as we have done before, the terms dependent on $\mathbf{m}_k$, we obtain:

$$d_k^2(\mathbf{x}) = -2\left((\mathbf{\Sigma^{-1}m}_k)'\mathbf{x} - 0.5\mathbf{m}_k{}'\mathbf{\Sigma^{-1}m}_k\right) + \mathbf{x'\Sigma^{-1}x} .$$ 6.10b

Since $\mathbf{x'\Sigma^{-1}x}$ is independent of k, minimising $d_k(\mathbf{x})$ is equivalent to *maximising* the following decision functions:

$$g_k(\mathbf{x}) = \mathbf{w}_k{}'\mathbf{x} + w_{k,0} ,$$ 6.10c

with $\mathbf{w}_k = \mathbf{\Sigma^{-1}m}_k$; $w_{k,0} = -0.5\mathbf{m}_k{}'\mathbf{\Sigma^{-1}m}_k$. 6.10d

Using these decision functions, we again obtain linear discriminant functions in the form of hyperplanes passing through the middle point of the line segment linking the means. The only difference from the results of the previous section is that the hyperplanes separating class ω_i from class ω_j are now orthogonal to the vector $\Sigma^{-1}(\mathbf{m}_i - \mathbf{m}_j)$.

In practice, it is impossible to guarantee that all class covariance matrices are equal. Fortunately, the decision surfaces are usually not very sensitive to mild deviations from this condition; therefore, in normal practice, one uses an estimate of a pooled covariance matrix, computed as an average of the sample covariance matrices. This is the practice followed by SPSS and STATISTICA.

Example 6.3

Q: Redo Example 6.1, using a minimum Mahalanobis distance classifier. Check the computation of the discriminant parameters and determine to which class a cork with 65 defects is assigned.

A: Given the similarity of both distributions, the Mahalanobis classifier produces the same classification results as the Euclidian classifier. Table 6.1 shows the *classification matrix* (obtained with SPSS) with the predicted classifications along the columns and the true (observed) classifications along the rows. We see that for this simple classifier, the overall percentage of correct classification in the data sample (training set) is 77%, or equivalently, the overall training set error is 23% (18% for ω_1 and 28% for ω_2). For the moment, we will not assess how the classifier performs with independent cases, i.e., we will not assess its *test set* error.

The decision function coefficients (also known as *Fisher's coefficients*), as computed by SPSS, are shown in Table 6.2.

Table 6.1. Classification matrix obtained with SPSS of two classes of cork stoppers using only one feature, N.

			Predicted Group Membership		Total
		Class	1	2	
Original	Count	1	41	9	50
Group		2	14	36	50
	%	1	82.0	18.0	100
		2	28.0	72.0	100

77.0% of original grouped cases correctly classified.

Table 6.2. Decision function coefficients obtained with SPSS for two classes of cork stoppers and one feature, N.

	Class 1	Class 2
N	0.192	0.277
(Constant)	−6.005	−11.746

Let us check these results. The class means are $\mathbf{m}_1 = [55.28]$ and $\mathbf{m}_2 = [79.74]$. The average variance is $s^2 = 287.63$. Applying formula 6.10d we obtain:

$$\mathbf{w}_1 = \mathbf{m}_1 / s^2 = [0.192] \quad ; \quad w_{1,0} = -0.5\|\mathbf{m}_1\|^2 / s^2 = -6.005 . \qquad 6.11a$$

$$\mathbf{w}_2 = \mathbf{m}_2 / s^2 = [0.277] \quad ; \quad w_{2,0} = -0.5\|\mathbf{m}_2\|^2 / s^2 = -11.746 . \qquad 6.11b$$

These results confirm the ones shown in Table 6.2. Let us determine the class assignment of a cork-stopper with 65 defects. As $g_1([65]) = 0.192 \times 65 - 6.005 = 6.48$ is greater than $g_2([65]) = 0.227 \times 65 - 11.746 = 6.26$ it is assigned to class ω_1.

□

Example 6.4

Q: Redo Example 6.2, using a minimum Mahalanobis distance classifier. Check the computation of the discriminant parameters and determine to which class a cork with 65 defects and with a total perimeter of 520 pixels (PRT10 = 52) is assigned.

A: The training set classification matrix is shown in Table 6.3. A significant improvement was obtained in comparison with the Euclidian classifier results mentioned in section 6.2.1; namely, an overall training set error of 10% instead of 18%. The Mahalanobis distance, taking into account the shape of the data clusters, not surprisingly, performed better. The decision function coefficients are shown in Table 6.4. Using these coefficients, we write the decision functions as:

$$g_1(\mathbf{x}) = \mathbf{w}_1'\mathbf{x} + w_{1,0} = [0.262 \quad -0.09783]\mathbf{x} - 6.138 . \qquad 6.12a$$

$$g_2(\mathbf{x}) = \mathbf{w}_2'\mathbf{x} + w_{2,0} = [0.0803 \quad 0.2776]\mathbf{x} - 12.817 . \qquad 6.12b$$

The point estimate of the pooled covariance matrix of the data is:

$$\mathbf{S} = \begin{bmatrix} 287.63 & 204.070 \\ 204.070 & 172.553 \end{bmatrix} \quad \Rightarrow \quad \mathbf{S}^{-1} = \begin{bmatrix} 0.0216 & -0.0255 \\ -0.0255 & 0.036 \end{bmatrix} . \qquad 6.13$$

Substituting $\mathbf{S}^{-1}$ in formula 6.10d, the results shown in Table 6.4 are obtained.

Table 6.3. Classification matrix obtained with SPSS for two classes of cork stoppers with two features, N and PRT10.

		Predicted Group Membership		Total
	Class	1	2	
Original	Count			
Count	1	49	1	50
Group	2	9	41	50
%	1	98.0	2.0	100
	2	18.0	82.0	100

90.0% of original grouped cases correctly classified.

It is also straightforward to compute $S^{-1}(m_1 - m_2) = [0.18 \; -0.376]'$. The orthogonal line to this vector with slope 0.4787 and passing through the middle point between the means is shown with a solid line in Figure 6.7. As expected, the "hyperplane" leans along the regression direction of the features (see Figure 6.5 for comparison).

As to the classification of $x = [65 \; 52]'$, since $g_1([65 \; 52]') = 5.80$ is smaller than $g_2([65 \; 52]') = 6.86$, it is assigned to class ω_2. This cork stopper has a total perimeter of the defects that is too big to be assigned to class ω_1.

□

Table 6.4. Decision function coefficients, obtained with SPSS, for the two classes of cork stoppers with features N and PRT10.

	Class 1	Class 2
N	0.262	0.0803
PRT10	-0.09783	0.278
(Constant)	-6.138	-12.817

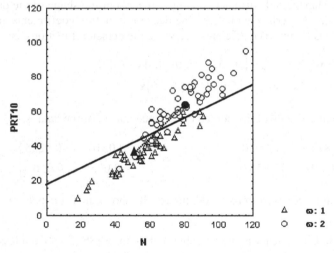

Figure 6.7. Mahalanobis linear discriminant (solid line) for the two classes of cork stoppers. Scatter plot obtained with STATISTICA.

Notice that if the distributions of the feature vectors in the classes correspond to different hyperellipsoidal shapes, they will be characterised by unequal covariance matrices. The distance formula 6.10 will then be influenced by these different shapes in such a way that we obtain quadratic decision boundaries. Table 6.5 summarises the different types of minimum distance classifiers, depending on the covariance matrix.

Table 6.5. Summary of minimum distance classifier types.

Covariance	Classifier	Equal-density surfaces	Discriminants
$\Sigma_i = s^2 \mathbf{I}$	Linear, Euclidian	Hyperspheres	Hyperplanes orthogonal to the segment linking the means
$\Sigma_i = \Sigma$	Linear, Mahalanobis	Hyperellipsoids	Hyperplanes leaning along the regression lines
Σ_i	Quadratic, Mahalanobis	Hyperellipsoids	Quadratic surfaces

Commands 6.1. SPSS, STATISTICA, MATLAB and R commands used to perform discriminant analysis.

SPSS	`Analyze; Classify; Discriminant`
STATISTICA	`Statistics; Multivariate Exploratory Techniques; Discriminant Analysis`
MATLAB	`classify(sample,training,group)` *classmatrix(x,y)*
R	*classify(sample,training,group)* *classmatrix(x,y)*

A large number of statistical analyses are available with SPSS and STATISTICA discriminant analysis commands. For instance, the pooled covariance matrix exemplified in 6.13 can be obtained with SPSS by checking the Pooled Within-Groups Matrices of the Statistics tab. There is also the possibility of obtaining several types of results, such as listings of decision function coefficients, classification matrices, graphical plots illustrating the separability of the classes, etc. The discriminant classifier can also be configured and evaluated in several ways. Many of these possibilities are described in the following sections.

The R stats package does not include discriminant analysis functions. However, it includes a function for computing Mahalanobis distances. We provide in the book CD two functions for performing discriminant analysis. The first function, classify(sample,training,group), returns a vector containing the integer classification labels of a sample matrix based on a training data matrix with a corresponding group vector of supervised classifications (integers starting from 1). The returned classification labels correspond to the minimum Mahalanobis distance using the pooled covariance matrix. The second function, classmatrix(x,y), generates a classification matrix based on two vectors, x and y, of integer classification labels. The classification matrix of Table 6.3 can be obtained as follows, assuming the cork data frame has been attached with columns ND, PRT and CL corresponding to variables N, PRT and CLASS, respectively:

```
> y <- cbind(ND[1:100],PRT[1:100]/10)
> co <- classify(y,y,CL[1:100])
> classmatrix(CL[1:100],co)
```

The meanings of MATLAB's classify arguments are the same as in R. MATLAB does not provide a function for obtaining the classification matrix. We include in the book CD the classmatrix function for this purpose, working in the same way as in R.

We didn't obtain the same values in MATLAB as we did with the other software products. The reason may be attributed to the fact that MATLAB apparently does not use pooled covariances (therefore, is not providing linear discriminants). ∎

6.3   Bayesian Classification

In the previous sections, we presented linear classifiers based solely on the notion of distance to class means. We did not assume anything specific regarding the data distributions. In this section, we will take into account the specific probability distributions of the cases in each class, thereby being able to adjust the classifier to the specific risks of a classification.

6.3.1 Bayes Rule for Minimum Risk

Let us again consider the cork stopper problem and imagine that factory production was restricted to the two classes we have been considering, denoted as: ω_1 = Super and ω_2 = Average. Let us assume further that the factory had a record of production stocks for a reasonably long period, summarised as:

Number of produced cork stoppers of class ω_1: $n_1 =$ 901 420
Number of produced cork stoppers of class ω_2: $n_2 =$ 1 352 130
Total number of produced cork stoppers: n = 2 253 550

With this information, we can readily obtain good estimates of the probabilities of producing a cork stopper from either of the two classes, the so-called *prior probabilities* or *prevalences*:

$$P(\omega_1) = n_1/n = 0.4; \qquad P(\omega_2) = n_2/n = 0.6. \qquad\qquad 6.14$$

Note that the prevalences are not entirely controlled by the factory, and that they depend mainly on the quality of the raw material. Just as, likewise, a cardiologist cannot control how prevalent myocardial infarction is in a given population. Prevalences can, therefore, be regarded as "states of nature".

Suppose we are asked to make a blind decision as to which class a cork stopper belongs without looking at it. If the only available information is the prevalences, the sensible choice is class ω_2. This way, we expect to be wrong only 40% of the times.

Assume now that we were allowed to measure the feature vector x of the presented cork stopper. Let $P(\omega_i \mid x)$ be the conditional probability of the cork stopper represented by x belonging to class ω_i. If we are able to determine the probabilities $P(\omega_1 \mid x)$ and $P(\omega_2 \mid x)$, the sensible decision is now:

If $P(\omega_1 \mid x) > P(\omega_2 \mid x)$ we decide $x \in \omega_1$;

If $P(\omega_1 \mid x) < P(\omega_2 \mid x)$ we decide $x \in \omega_2$;   6.15

If $P(\omega_1 \mid x) = P(\omega_2 \mid x)$ the decision is arbitrary.

We can condense 6.15 as:

If $P(\omega_1 \mid x) > P(\omega_2 \mid x)$ then $x \in \omega_1$ else $x \in \omega_2$.   6.15a

The *posterior probabilities* $P(\omega_i \mid x)$ can be computed if we know the pdfs of the distributions of the feature vectors in both classes, $p(x \mid \omega_i)$, the so-called *likelihood* of x. As a matter of fact, the Bayes law (see Appendix A) states that:

$$P(\omega_i \mid x) = \frac{p(x \mid \omega_i)P(\omega_i)}{p(x)}, \qquad 6.16$$

with $p(x) = \sum_{i=1}^{c} p(x \mid \omega_i)P(\omega_i)$,   the *total probability* of x.

Note that $P(\omega_i)$ and $P(\omega_i \mid x)$ are discrete probabilities (symbolised by a capital letter), whereas $p(x \mid \omega_i)$ and $p(x)$ are values of pdf functions. Note also that the term $p(x)$ is a common term in the comparison expressed by 6.15a, therefore, we may rewrite for two classes:

If $p(x \mid \omega_1)P(\omega_1) > p(x \mid \omega_2)P(\omega_2)$ then $x \in \omega_1$ else $x \in \omega_2$,   6.17

Example 6.5

Q: Consider the classification of cork stoppers based on the number of defects, N, and restricted to the first two classes, "Super" and "Average". Estimate the posterior probabilities and classification of a cork stopper with 65 defects, using prevalences 6.14.

A: The feature vector is $x = [N]$, and we seek the classification of $x = [65]$. Figure 6.8 shows the histograms of both classes with a superimposed normal curve.

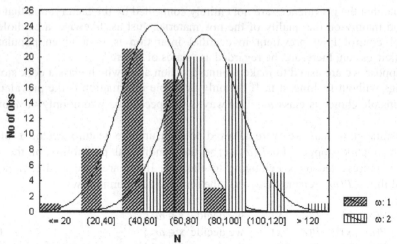

Figure 6.8. Histograms of feature N for two classes of cork stoppers, obtained with STATISTICA. The threshold value N = 65 is marked with a vertical line.

From this graphic display, we can estimate the likelihoods[3] and the posterior probabilities:

$$p(\mathbf{x} \mid \omega_1) = 20 / 24 = 0.833 \quad \Rightarrow \quad P(\omega_1)p(\mathbf{x} \mid \omega_1) = 0.4 \times 0.833 = 0.333 ; \quad \text{6.18a}$$

$$p(\mathbf{x} \mid \omega_2) = 16 / 23 = 0.696 \quad \Rightarrow \quad P(\omega_2)p(\mathbf{x} \mid \omega_2) = 0.6 \times 0.696 = 0.418 . \quad \text{6.18b}$$

We then decide class ω_2, although the likelihood of ω_1 is bigger than that of ω_2. Notice how the statistical model prevalences changed the conclusions derived by the minimum distance classification (see Example 6.3).

           □

Figure 6.9 illustrates the effect of adjusting the prevalence threshold assuming equal and normal pdfs:

- Equal prevalences. With equal pdfs, the decision threshold is at half distance from the means. The number of cases incorrectly classified, proportional to the shaded areas, is equal for both classes. This situation is identical to the minimum distance classifier.

- Prevalence of ω_1 bigger than that of ω_2. The decision threshold is displaced towards the class with smaller prevalence, therefore decreasing the number of wrongly classified cases of the class with higher prevalence, as seems convenient.

---

[3] The normal curve fitted by STATISTICA is multiplied by the factor "number of cases" × "histogram interval width", which is 1000 in the present case. This constant factor is of no importance and is neglected in the computations of 6.18.

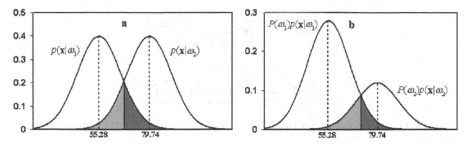

Figure 6.9. Influence of the prevalence threshold on the classification errors, represented by the shaded areas (dark grey represents the errors for class ω_1). (a) Equal prevalences; (b) Unequal prevalences.

DISCRIM. ANALYSIS	Rows: Observed classifications Columns: Predicted classifications		
Group	Percent Correct	G_1:1 p=.40000	G_2:2 p=.60000
G_1:1	64.00000	32	18
G_2:2	82.00000	9	41
Total	73.00000	41	59

Figure 6.10. Classification results, obtained with STATISTICA, of the cork stoppers with unequal prevalences: 0.4 for class ω_1 and 0.6 for class ω_2.

Example 6.6

Q: Compute the classification matrix for all the cork stoppers of Example 6.5 and comment the results.

A: Figure 6.10 shows the classification matrix obtained with the prevalences computed in 6.14, which are indicated in the Group row. We see that indeed the decision threshold deviation led to a better performance for class ω_2 than for class ω_1. This seems reasonable since class ω_2 now occurs more often. Since the overall error has increased, one may wonder if this influence of the prevalences was beneficial after all. The answer to this question is related to the topic of *classification risks*, presented below.

□

Let us assume that the cost of a ω_1 ("super") cork stopper is 0.025 € and the cost of a ω_2 ("average") cork stopper is 0.015 €. Suppose that the ω_1 cork stoppers are to be used in special bottles whereas the ω_2 cork stoppers are to be used in normal bottles.

Let us further consider that the wrong classification of an average cork stopper leads to its rejection with a loss of 0.015 € and the wrong classification of a super quality cork stopper amounts to a loss of 0.025 − 0.015 = 0.01 € (see Figure 6.11).

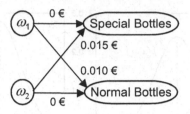

Figure 6.11. Loss diagram for two classes of cork stoppers. Correct decisions have zero loss.

Denote:

SB – Action of using a cork stopper in *special bottles*.
NB – Action of using a cork stopper in *normal bottles*.
$\omega_1=S$ (class super); $\omega_2=A$ (class average)

Define: $\lambda_{ij} = \lambda(\alpha_i \mid \omega_j)$ as the *loss* associated with an action α_i when the correct class is ω_j. In the present case, $\alpha_i \in \{SB, NB\}$.
We can arrange the λ_{ij} in a loss matrix Λ, which in the present case is:

$$\Lambda = \begin{bmatrix} 0 & 0.015 \\ 0.01 & 0 \end{bmatrix}.$$

6.19

Therefore, the *risk* (expected value of the loss) associated with the action of using a cork, characterised by feature vector **x**, in special bottles, can be expressed as:

$$R(SB \mid \mathbf{x}) = \lambda(SB \mid S)P(S \mid \mathbf{x}) + \lambda(NB \mid M)P(A \mid \mathbf{x}) = 0.015 \times P(A \mid \mathbf{x});$$

6.20a

And likewise for normal bottles:

$$R(NB \mid \mathbf{x}) = \lambda(NB \mid S)P(S \mid \mathbf{x}) + \lambda(NB \mid A)P(A \mid \mathbf{x}) = 0.01 \times P(S \mid \mathbf{x});$$

6.20b

We are assuming that in the risk evaluation, the only influence is from wrong decisions. Therefore, correct decisions have zero loss, $\lambda_{ii} = 0$, as in 6.19. If instead of two classes, we have c classes, the risk associated with a certain action α_i is expressed as follows:

$$R(\alpha_i \mid \mathbf{x}) = \sum_{j=1}^{c} \lambda(\alpha_i \mid \omega_j)P(\omega_j \mid \mathbf{x}).$$

6.21

We are obviously interested in minimising an average risk computed for an arbitrarily large number of cork stoppers. The *Bayes rule for minimum risk* achieves this through the minimisation of the individual conditional risks $R(\alpha_i \mid \mathbf{x})$.

Let us assume, first, that wrong decisions imply the same loss, which can be scaled to a unitary loss:

$$\lambda_{ij} = \lambda(\alpha_i \mid \omega_j) = \begin{cases} 0 & \text{if } i = j \\ 1 & \text{if } i \neq j \end{cases}.$$ 6.22a

In this situation, since all posterior probabilities add up to one, we have to minimise:

$$R(\alpha_i \mid \mathbf{x}) = \sum_{j \neq i} P(\omega_j \mid \mathbf{x}) = 1 - P(\omega_i \mid \mathbf{x}).$$ 6.22b

This corresponds to maximising $P(\omega_i \mid \mathbf{x})$, i.e., the Bayes decision rule for minimum risk corresponds to the generalised version of 6.15a:

Decide ω_i if $P(\omega_i \mid \mathbf{x}) > P(\omega_j \mid \mathbf{x})$, $\forall j \neq i$.   6.22c

Thus, the decision function for class ω_i is the posterior probability, $g_i(\mathbf{x}) = P(\omega_i \mid \mathbf{x})$, and the classification rule amounts to selecting the class with maximum posterior probability.

Let us now consider the situation of different losses for wrong decisions, assuming, for the sake of simplicity, that $c = 2$. Taking into account expressions 6.20a and 6.20b, it is readily concluded that we will decide ω_1 if:

$$\lambda_{21} P(\omega_1 \mid \mathbf{x}) > \lambda_{12} P(\omega_2 \mid \mathbf{x}) \Rightarrow p(\mathbf{x} \mid \omega_1) \lambda_{21} P(\omega_1) > p(\mathbf{x} \mid \omega_2) \lambda_{12} P(\omega_2).$$ 6.23

This is equivalent to formula 6.17 using the following *adjusted prevalences*:

$$P^*(\omega_1) = \frac{\lambda_{21} P(\omega_1)}{\lambda_{21} P(\omega_1) + \lambda_{12} P(\omega_2)} \;;\; P^*(\omega_2) = \frac{\lambda_{12} P(\omega_2)}{\lambda_{21} P(\omega_1) + \lambda_{12} P(\omega_2)}.$$ 6.23a

STATISTICA and SPSS allow specifying the priors as estimates of the sample composition (as in 6.14) or by user assignment of specific values. In the latter the user can adjust the priors in order to cope with specific classification risks.

Example 6.7

Q: Redo Example 6.6 using adjusted prevalences that take into account 6.14 and the loss matrix 6.19. Compare the classification risks with and without prevalence adjustment.

A: The losses are $\lambda_{12} = 0.015$ and $\lambda_{21} = 0.01$. Using the prevalences 6.14, one obtains $P^*(\omega_1) = 0.308$ and $P^*(\omega_2) = 0.692$. The higher loss associated with a wrong classification of a ω_2 cork stopper leads to an increase of $P^*(\omega_2)$ compared with $P^*(\omega_1)$. The consequence of this adjustment is the decrease of the number of

ω_2 cork stoppers wrongly classified as ω_1. This is shown in the classification matrix of Table 6.6.

We can now compute the average risk for this two-class situation, as follows:

$$R = \lambda_{12} Pe_{12} + \lambda_{21} Pe_{21},$$

where Pe_{ij} is the error probability of deciding class ω_i when the true class is ω_j.

Using the training set estimates of these errors, $Pe_{12} = 0.1$ and $Pe_{21} = 0.46$ (see Table 6.6), the estimated average risk per cork stopper is computed as $R = 0.015 \times Pe_{12} + 0.01 \times Pe_{21} = 0.015 \times 0.01 + 0.01 \times 0.46 = 0.0061$ €. If we had not used the adjusted prevalences, we would have obtained the higher risk estimate of 0.0063 € (use the Pe_{ij} estimates from Figure 6.10).   ☐

Table 6.6. Classification matrix obtained with STATISTICA of two classes of cork stoppers with adjusted prevalences (Class 1 $\equiv \omega_1$; Class 2 $\equiv \omega_2$). The column values are the predicted classifications.

	Percent Correct	Class 1	Class 2
Class 1	54	27	23
Class 2	90	5	45
Total	72	32	68

6.3.2 Normal Bayesian Classification

Up to now, we have assumed no particular distribution model for the likelihoods. Frequently, however, the normal distribution model is a reasonable assumption. SPSS and STATISTICA make this assumption when computing posterior probabilities.

A normal likelihood for class ω_i is expressed by the following pdf (see Appendix A):

$$p(\mathbf{x} \mid \omega_i) = \frac{1}{(2\pi)^{d/2} |\mathbf{\Sigma}_i|^{1/2}} \exp\left(-\frac{1}{2}(\mathbf{x} - \mathbf{\mu}_i)' \mathbf{\Sigma}_i^{-1}(\mathbf{x} - \mathbf{\mu}_i)\right),\qquad 6.24$$

with:

$$\mathbf{\mu}_i = E_i[\mathbf{x}], \text{ mean vector for class } \omega_i ;\qquad 6.24a$$

$$\mathbf{\Sigma}_i = E_i[(\mathbf{x} - \mathbf{\mu}_i)(\mathbf{x} - \mathbf{\mu}_i)'], \text{ covariance for class } \omega_i .\qquad 6.24b$$

Since the likelihood 6.24 depends on the Mahalanobis distance of a feature vector to the respective class mean, we obtain the same types of classifiers shown in Table 6.5.

Note that even when the data distributions are not normal, as long as they are symmetric and in correspondence to ellipsoidal shaped clusters of points, we obtain the same decision surfaces as for a normal classifier, although with different error rates and posterior probabilities.

As previously mentioned SPSS and STATISTICA use a pooled covariance matrix when performing linear discriminant analysis. The influence of this practice on the obtained error, compared with the theoretical optimal Bayesian error corresponding to a quadratic classifier, is discussed in detail in (Fukunaga, 1990). Experimental results show that when the covariance matrices exhibit mild deviations from the pooled covariance matrix, the designed classifier has a performance similar to the optimal performance with equal covariances. This makes sense since for covariance matrices that are not very distinct, the difference between the optimum quadratic solution and the sub-optimum linear solution should only be noticeable for cases that are far away from the prototypes, as illustrated in Figure 6.12.

As already mentioned in section 6.2.3, using decision functions based on the individual covariance matrices, instead of a pooled covariance matrix, will produce quadratic decision boundaries. SPSS affords the possibility of computing such quadratic discriminants, using the `Separate-groups` option of the `Classify` tab. However, a quadratic classifier is less robust (more sensitive to parameter deviations) than a linear one, especially in high dimensional spaces, and needs a much larger training set for adequate design (see e.g. Fukunaga and Hayes, 1989).

SPSS and STATISTICA provide complete listings of the posterior probabilities 6.18 for the normal Bayesian classifier, i.e., using the likelihoods 6.24.

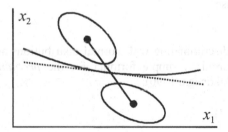

Figure 6.12. Discrimination of two classes with optimum quadratic classifier (solid line) and sub-optimum linear classifier (dotted line).

Example 6.8

Q: Determine the posterior probabilities corresponding to the classification of two classes of cork stoppers with equal prevalences as in Example 6.4 and comment the results.

A: Table 6.7 shows a partial listing of the computed posterior probabilities, obtained with SPSS. Notice that case #55 is marked with **, indicating a misclassified case, with a posterior probability that is higher for class 1 (0.782) than for class 2 (0.218). Case #61 is also misclassified, but with a small difference of posterior probabilities. Borderline cases as case #61 could be re-analysed, e.g. using more features.   □

Table 6.7. Partial listing of the posterior probabilities, obtained with SPSS, for the classification of two classes of cork stoppers with equal prevalences. The columns headed by "P(G=g | D=d)" are posterior probabilities.

Case Number	Actual Group	Highest Group Predicted Group	P(G=g | D=d)	Second Highest Group Group	P(G=g | D=d)
...					
50	1	1	0.964	2	0.036
51	2	2	0.872	1	0.128
52	2	2	0.728	1	0.272
53	2	2	0.887	1	0.113
54	2	2	0.843	1	0.157
55	2	1**	0.782	2	0.218
56	2	2	0.905	1	0.095
57	2	2	0.935	1	0.065
...					
61	2	1**	0.522	2	0.478
...					

** Misclassified case

For a two-class discrimination with normal distributions and equal prevalences and covariance, there is a simple formula for the *probability of error* of the classifier (see e.g. Fukunaga, 1990):

$$Pe = 1 - N_{0,1}(\delta / 2),$$ 6.25

with:

$$\delta^2 = (\mathbf{\mu}_1 - \mathbf{\mu}_2)' \Sigma^{-1} (\mathbf{\mu}_1 - \mathbf{\mu}_2),$$ 6.25a

the square of the so-called *Bhattacharyya distance*, a Mahalanobis distance of the means, reflecting the class separability.

Figure 6.13 shows the behaviour of *Pe* with increasing squared Bhattacharyya distance. After an initial quick, exponential-like decay, *Pe* converges asymptotically to zero. It is, therefore, increasingly difficult to lower a classifier error when it is already small.

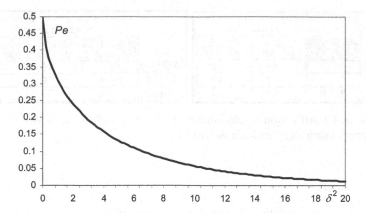

Figure 6.13. Error probability of a Bayesian two-class discrimination with normal distributions and equal prevalences and covariance.

6.3.3 Dimensionality Ratio and Error Estimation

The Mahalanobis and the Bhattacharyya distances can only increase when adding more features, since for every added feature a non-negative distance contribution is also added. This would certainly be the case if we had the true values of the means and the covariances available, which, in practical applications, we do not.

When using a large number of features we get numeric difficulties in obtaining a good estimate of Σ^{-1}, given the finiteness of the training set. Surprising results can then be expected; for instance, the performance of the classifier can degrade when more features are added, instead of improving.

Figure 6.14 shows the classification matrix for the two-class, cork-stopper problem, using the whole ten-feature set and equal prevalences. The training set performance did not increase significantly compared with the two-feature solution presented previously, and is worse than the solution using the four-feature vector [ART PRM NG RAAR]', as shown in Figure 6.14b.

There are, however, further compelling reasons for not using a large number of features. In fact, when using estimates of means and covariance derived from a training set, we are designing a biased classifier, fitted to the training set. Therefore, we should expect that our training set error estimates are, on average, optimistic. On the other hand, error estimates obtained in independent test sets are expected to be, on average, pessimistic. It is only when the number of cases, n, is sufficiently larger than the number of features, d, that we can expect that our classifier will generalise, that is it will perform equally well when presented with new cases. The n/d ratio is called the *dimensionality ratio*.

The choice of an adequate dimensionality ratio has been studied by several authors (see References). Here, we present some important results as an aid for the designer to choose sensible values for the n/d ratio. Later, when we discuss the topic of classifier evaluation, we will come back to this issue from another perspective.

a	Rows: Observ. classif. Cols: Pred. classif.			b	Rows: Observ. classif. Cols: Pred. classif.		
Group	% Correct	G_1:1 p=.50	G_2:2 p=.50	Group	% Correct	G_1:1 p=.50	G_2:2 p=.50
G_1:1	98.0	49	1	G_1:1	98.0	49	1
G_2:2	86.0	7	43	G_2:2	88.0	6	44
Total	92.0	56	44	Total	93.0	55	45

Figure 6.14. Classification results obtained with STATISTICA, of two classes of cork stoppers using: (a) Ten features; (b) Four features.

Let us denote:

Pe –   Probability of error of a given classifier;

Pe^* –   Probability of error of the optimum Bayesian classifier;

$Pe_d(n)$ –   Training (design) set estimate of Pe based on a classifier designed on n cases;

$Pe_t(n)$ –   Test set estimate of Pe based on a set of n test cases.

The quantity $Pe_d(n)$ represents an estimate of Pe influenced only by the finite size of the design set, i.e., the classifier error is measured exactly, and its deviation from Pe is due solely to the finiteness of the design set. The quantity $Pe_t(n)$ represents an estimate of Pe influenced only by the finite size of the test set, i.e., it is the expected error of the classifier when evaluated using n-sized test sets. These quantities verify $Pe_d(\infty) = Pe$ and $Pe_t(\infty) = Pe$, i.e., they converge to the theoretical value Pe with increasing values of n. If the classifier happens to be designed as an optimum Bayesian classifier Pe_d and Pe_t converge to Pe^*.

In normal practice, these error probabilities are not known exactly. Instead, we compute estimates of these probabilities, $\hat{Pe}_d$ and $\hat{Pe}_t$, as percentages of misclassified cases, in exactly the same way as we have done in the classification matrices presented so far. The probability of obtaining k misclassified cases out of n for a classifier with a theoretical error Pe, is given by the binomial law:

$$P(k) = \binom{n}{k} Pe^k (1 - Pe)^{n-k} .$$ 6.26

The maximum likelihood estimation of Pe under this binomial law is precisely (see Appendix C):

$$\hat{Pe} = k / n ,$$ 6.27

with standard deviation:

$$\sigma = \sqrt{\frac{Pe(1 - Pe)}{n}} .$$ 6.28

Formula 6.28 allows the computation of confidence interval estimates for $\hat{P}e$, by substituting $\hat{P}e$ in place of Pe and using the normal distribution approximation for sufficiently large n (say, $n \geq 25$). Note that formula 6.28 yields zero for the extreme cases of $Pe = 0$ or $Pe = 1$.

In normal practice, we first compute $\hat{P}e_d$ by designing and evaluating the classifier in the same set with n cases, $\hat{P}e_d(n)$. This is what we have done so far. As for $\hat{P}e_t$, we may compute it using an independent set of n cases, $\hat{P}e_t(n)$. In order to have some guidance on how to choose an appropriate dimensionality ratio, we would like to know the deviation of the expected values of these estimates from the Bayes error. Here the expectation is computed on a population of classifiers of the same type and trained in the same conditions. Formulas for these expectations, $E[\hat{P}e_d(n)]$ and $E[\hat{P}e_t(n)]$, are quite intricate and can only be computed numerically. Like formula 6.25, they depend on the Bhattacharyya distance. A software tool, SC Size, computing these formulas for two classes with normally distributed features and equal covariance matrices, separated by a linear discriminant, is included with on the book CD. SC Size also allows the computation of confidence intervals of these estimates, using formula 6.28.

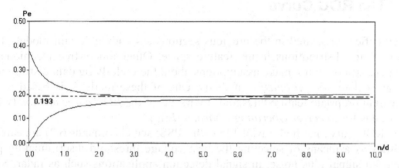

Figure 6.15. Two-class linear discriminant $E[\hat{P}e_d(n)]$ and $E[\hat{P}e_t(n)]$ curves, for $d = 7$ and $\delta^2 = 3$, below and above the dotted line, respectively. The dotted line represents the Bayes error (0.193).

Figure 6.15 is obtained with SC Size and illustrates how the expected values of the error estimates evolve with the n/d ratio, where n is assumed to be the number of cases in each class. The feature set dimension id $d = 7$. Both curves have an asymptotic behaviour with $n \rightarrow \infty$ [4], with the average design set error estimate converging to the Bayes error from below and the average test set error estimate converging from above.

---

[4] Numerical approximations in the computation of the average test set error may sometimes result in a slight deviation from the asymptotic behaviour, for large n.

Both standard deviations, which can be inspected in text boxes for a selected value of n/d, are initially high for low values of n and converge slowly to zero with $n \to \infty$. For the situation shown in Figure 6.15, the standard deviation of $\hat{Pe}_d(n)$ changes from 0.089 for $n = d$ (14 cases, 7 per class) to 0.033 for $n = 10d$ (140 cases, 70 per class).

Based on the behaviour of the $E[\hat{Pe}_d(n)]$ and $E[\hat{Pe}_t(n)]$ curves, some criteria can be established for the dimensionality ratio. As a general rule of thumb, using dimensionality ratios well above 3 is recommended.

If the cases are not equally distributed by the classes, it is advisable to use the smaller number of cases per class as value of n. Notice also that a multi-class problem can be seen as a generalisation of a two-class problem if every class is well separated from all the others. Then, the total number of needed training samples for a given deviation of the expected error estimates from the Bayes error can be estimated as cn^*, where n^* is the particular value of n that achieves such a deviation in the most unfavourable, two-class dichotomy of the multi-class problem.

6.4   The ROC Curve

The classifiers presented in the previous sections assumed a certain model of the feature vector distributions in the feature space. Other model-free techniques to design classifiers do not make assumptions about the underlying data distributions. They are called *non-parametric* methods. One of these methods is based on the choice of appropriate feature thresholds by means of the *ROC curve method* (where ROC stands for *Receiver Operating Characteristic*).

The ROC curve method (available with SPSS; see Commands 6.2) appeared in the fifties as a means of selecting the best voltage threshold discriminating pure noise from signal plus noise, in signal detection applications such as radar. Since the seventies, the concept has been used in the areas of medicine and psychology, namely for test assessment purposes.

The ROC curve is an interesting analysis tool for two-class problems, especially in situations where one wants to detect rarely occurring events such as a special signal, a disease, etc., based on the choice of feature thresholds. Let us call the absence of the event the *normal* situation (N) and the occurrence of the rare event the *abnormal* situation (A). Figure 6.16 shows the classification matrix for this situation, based on a given decision rule, with true classes along the rows and decided (predicted) classifications along the columns[5].

---

[5]   The reader may notice the similarity of the canonical two-class classification matrix with the hypothesis decision matrix in chapter 4 (Figure 4.2).

Decision

		A	N
Reality	A	a	b
	N	c	d

Figure 6.16. The canonical classification matrix for two-class discrimination of an abnormal event (A) from the normal event (N).

From the classification matrix of Figure 6.16, the following parameters are defined:

- True Positive Ratio $\equiv$ TPR $= a/(a+b)$. Also known as *sensitivity*, this parameter tells us how sensitive our decision method is in the detection of the abnormal event. A classification method with high sensitivity will rarely miss the abnormal event when it occurs.

- True Negative Ratio $\equiv$ TNR $= d/(c+d)$. Also known as *specificity*, this parameter tells us how specific our decision method is in the detection of the abnormal event. A classification method with a high specificity will have a very low rate of false alarms, caused by classifying a normal event as abnormal.

- False Positive Ratio $\equiv$ FPR $= c/(c+d) = 1 -$ specificity.

- False Negative Ratio $\equiv$ FNR $= b/(a+b) = 1 -$ sensitivity.

Both the sensitivity and specificity are usually given in percentages. A decision method is considered good if it simultaneously has a high sensitivity (rarely misses the abnormal event when it occurs) and a high specificity (has a low false alarm rate). The ROC curve depicts the sensitivity versus the FPR (complement of the specificity) for every possible decision threshold.

Example 6.9

Q: Consider the Programming dataset (see Appendix E). Determine whether a threshold-based decision rule using attribute AB, "previous learning of Boolean Algebra", has a significant influence deciding the student passing (SCORE $\geq$ 10) or flunking (SCORE $<$ 10) the Programming course, by visual inspection of the respective ROC curve.

A: Using the Programming dataset we first establish the following Table 6.8. Next, we set the following decision rule for the attribute (feature) AB:

Decide "Pass the Programming examination" if AB $\geq \Delta$.

We then proceed to determine for every possible threshold value, Δ, the sensitivity and specificity of the decision rule in the classification of the students. These computations are summarised in Table 6.9.

Note that when $\Delta = 0$ the decision rule assigns all students to the "Pass" group (all students have AB $\geq$ 0). For $0 < \Delta \leq 1$ the decision rule assigns to the "Pass" group 135 students that have indeed "passed" and 60 students that have "flunked" (these 195 students have AB $\geq$ 1). Likewise for other values of Δ up to $\Delta > 2$ where the decision rule assigns all students to the flunk group since no students have $\Delta > 2$. Based on the classification matrices for each value of Δ the sensitivities and specificities are computed as shown in Table 6.9.

The ROC curve can be directly drawn using these computations, or using SPSS as shown in Figure 6.17c. Figures 6.17a and 6.17b show how the data must be specified. From visual inspection, we see that the ROC curve is only moderately off the diagonal, corresponding to a non-informative decision rule (more details, later).

□

Table 6.8. Number of students passing and flunking the "Programming" examination for three categories of AB (see the `Programming` dataset).

Previous learning of AB = Boolean Algebra	1 = Pass	0 = Flunk
0 = None	39	37
1 = Scarcely	86	46
2 = A lot	49	14
Total	174	97

Table 6.9. Computation of the sensitivity (TPR) and 1–specificity (FPR) for Example 6.9.

Pass / Flunk Reality	Total Cases	Pass/Flunk Decision Based on AB $\geq \Delta$							
		$\Delta = 0$		$0 < \Delta \leq 1$		$1 < \Delta \leq 2$		$\Delta > 2$	
		1	0	1	0	1	0	1	0
1	174	174	0	135	39	49	125	0	174
0	97	97	0	60	37	14	83	0	97
TPR		1		0.78		0.28		0	
FPR		1		0.62		0.14		0	

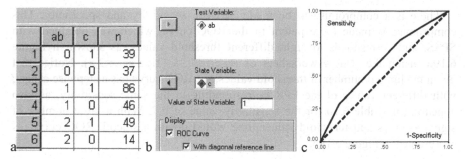

Figure 6.17. ROC curve for Example 6.9, solved with SPSS: a) Datasheet with column "n" used as weight variable; b) ROC curve specification window; c) ROC curve.

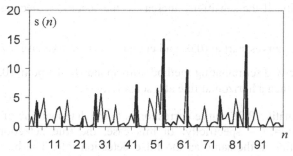

Figure 6.18. One hundred samples of a signal consisting of noise plus signal impulses (bold lines) occurring at random times.

Example 6.10

Q: Consider the Signal & Noise dataset (see Appendix E). This set presents 100 signal plus noise values $s(n)$ (Signal+Noise variable), consisting of random noise plus signal impulses with random amplitude, occurring at random times according to the Poisson law. The Signal & Noise data is shown in Figure 6.18. Determine the ROC curve corresponding to the detection of signal impulses using several threshold values to separate signal from noise.

A: The signal plus noise amplitude shown in Figure 6.18 is often greater than the average noise amplitude, therefore revealing the presence of the signal impulses (e.g. at time instants 53 and 85). The discrimination between signal and noise is made setting an amplitude threshold, Δ, such that we decide "impulse" (our rare event) if $s(n) > \Delta$, and "noise" (the normal event) otherwise. For each threshold value, it's then possible to establish the signal vs. noise classification matrix and compute the sensitivity and specificity values. By varying the threshold (easily done in the Signal & Noise.xls file), the corresponding sensitivity and specificity values can be obtained, as shown in Table 6.10.

There is a compromise to be made between sensitivity and specificity. This compromise is made more patent in the ROC curve, which was obtained with SPSS, and corresponds to eight different threshold values, as shown in Figure 6.19a (using the Data worksheet of Signal & Noise.xls). Notice that given the limited number of threshold values, the ROC curve has a stepwise aspect, with different values of the FPR corresponding to the same sensitivity, as also appearing in Table 6.10 for the sensitivity value of 0.7. With a large number of signal samples and threshold values, one would obtain a smooth ROC curve, as represented in Figure 6.19b.　　　　　　　　　　　　　　　　　　　　　　　□

Looking at the ROC curves shown in Figure 6.19 the following characteristic aspects are clearly visible:

- The ROC curve graphically depicts the compromise between sensitivity and specificity. If the sensitivity increases, the specificity decreases, and vice-versa.

- All ROC curves start at (0,0) and end at (1,1) (see Exercise 6.7).

- A perfectly discriminating method corresponds to the point (0,1). The ROC curve is then a horizontal line at a sensitivity =1.

A non-informative ROC curve corresponds to the diagonal line of Figures 6.19, with sensitivity = 1 − specificity. In this case, the true detection rate of the abnormal situation is the same as the false detection rate. The best compromise decision of sensitivity = specificity = 0.5 is then just as good as flipping a coin.

Table 6.10. Sensitivity and specificity in impulse detection (100 signal values).

Threshold	Sensitivity	Specificity
1	0.90	0.66
2	0.80	0.80
3	0.70	0.87
4	0.70	0.93

One of the uses of the ROC curve is related to the issue of choosing the best decision threshold that can differentiate both situations; in the case of Example 6.10, the presence of the impulses from the presence of the noise alone. Let us address this discriminating issue as a cost decision issue as we have done in section 6.3.1. Representing the sensitivity and specificity of the method for a threshold Δ by $s(\Delta)$ and $f(\Delta)$ respectively, and using the same notation as in formula 6.20, we can write the total risk as:

$$R = \lambda_{aa}P(A)s(\Delta) + \lambda_{an}P(A)(1 - s(\Delta)) + \lambda_{na}P(N)f(\Delta) + \lambda_{nn}P(N)(1 - f(\Delta)),$$
or, $R = s(\Delta)(\lambda_{aa}P(A) - \lambda_{an}P(A)) + f(\Delta)(\lambda_{na}P(N) - \lambda_{nn}P(N)) + \text{constant}.$

In order to obtain the best threshold, we minimise the risk R by differentiating and equalling to zero, obtaining then:

$$\frac{ds(\Delta)}{df(\Delta)} = \frac{(\lambda_{nn} - \lambda_{na})P(N)}{(\lambda_{aa} - \lambda_{an})P(A)}.$$ 6.29

The point of the ROC curve where the slope has the value given by formula 6.29 represents the optimum operating point or, in other words, corresponds to the best threshold for the two-class problem. Notice that this is a model-free technique of choosing a feature threshold for discriminating two classes, with no assumptions concerning the specific distributions of the cases.

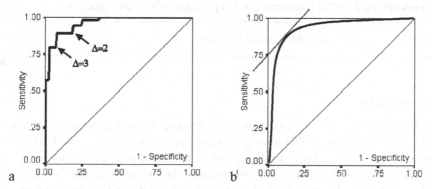

Figure 6.19. ROC curve (bold line), obtained with SPSS, for the signal + noise data: (a) Eight threshold values (the values for $\Delta = 2$ and $\Delta = 3$ are indicated); b) A large number of threshold values (expected curve) with the 45° slope point.

Let us now assume that, in a given situation, we assign zero cost to correct decisions, and a cost that is inversely proportional to the prevalences to a wrong decision. Then, the slope of the optimum operating point is at 45°, as shown in Figure 6.19b. For the impulse detection example, the best threshold would be somewhere between 2 and 3.

Another application of the ROC curve is in the comparison of classification performance, namely for feature selection purposes. We have already seen in 6.3.1 how prevalences influence classification decisions. As illustrated in Figure 6.9, for a two-class situation, the decision threshold is displaced towards the class with the smaller prevalence. Consider that the classifier is applied to a population where the prevalence of the abnormal situation is low. Then, for the previously mentioned reason, the decision maker should operate in the lower left part of the ROC curve in order to keep FPR as small as possible. Otherwise, given the high prevalence of the normal situation, a high rate of false alarms would be obtained. Conversely, if the classifier is applied to a population with a high prevalence of the abnormal situation, the decision-maker should adjust the decision threshold to operate on the FPR high part of the curve.

Briefly, in order for our classification method to perform optimally for a large range of prevalence situations, we would like to have an ROC curve very near the perfect curve, i.e., with an underlying area of 1. It seems, therefore, reasonable to select from among the candidate classification methods (or features) the one that has an ROC curve with the highest underlying area.

The area under the ROC curve is computed by the SPSS with a 95% confidence interval.

Despite some shortcomings, the ROC curve area method is a popular method of assessing classifier or feature performance. This and an alternative method based on information theory are described in Metz *et al.* (1973).

Commands 6.2. SPSS command used to perform ROC curve analysis.

SPSS	Graphs; ROC Curve

■

Example 6.11

Q: Consider the FHR-Apgar dataset, containing several parameters computed from foetal heart rate (FHR) tracings obtained previous to birth, as well as the so-called Apgar index. This is a ranking index, measured on a one-to-ten scale, and evaluated by obstetricians taking into account clinical observations of a newborn baby. Consider the two FHR features, ALTV and ASTV, representing the percentages of abnormal long term and abnormal short-term heart rate variability, respectively. Use the ROC curve in order to elucidate which of these parameters is better in the clinical practice for discriminating an Apgar > 6 (normal situation) from an Apgar ≤ 6 (abnormal or suspect situation).

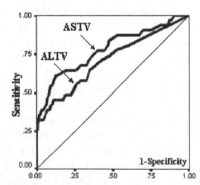

Figure 6.20. ROC curves for the FHR Apgar dataset, obtained with SPSS, corresponding to features ALTV and ASTV.

A: The ROC curves for ALTV and ASTV are shown in Figure 6.20. The areas under the ROC curve, computed by SPSS with a 95% confidence interval, are 0.709 ± 0.11 and 0.781 ± 0.10 for ALTV and ASTV, respectively. We, therefore, select the ASTV parameter as the best diagnostic feature.

□

6.5 Feature Selection

As already discussed in section 6.3.3, great care must be exercised in reducing the number of features used by a classifier, in order to maintain a high dimensionality ratio and, therefore, reproducible performance, with error estimates sufficiently near the theoretical value. For this purpose, one may use the hypothesis test methods described in chapters 4 and 5 with the aim of discarding features that are clearly non-useful at an initial stage of the classifier design. This feature assessment task, while assuring that an information-carrying feature set is indeed used in the classifier, does not guarantee it will need the whole set. Consider, for instance, that we are presented with a classification problem described by 4 features, x_1, x_2, x_3 and x_4, with x_1 and x_2 perfectly discriminating the classes, and x_3 and x_4 being linearly dependent of x_1 and x_2. The hypothesis tests will then find that all features contribute to class discrimination. However, this discrimination could be performed equally well using the alternative sets $\{x_1, x_2\}$ or $\{x_3, x_4\}$. Briefly, discarding features with no aptitude for class discrimination is no guarantee against redundant features.

There is abundant literature on the topic of feature selection (see References). Feature selection uses a search procedure of a feature subset (*model*) obeying a stipulated merit criterion. A possible choice for this criterion is minimising *Pe*, with the disadvantage of the search process depending on the classifier type. More often, a class separability criterion such as the Bhattacharyya distance or the ANOVA *F* statistic is used. The *Wilks' lambda*, defined as the ratio of the determinant of the pooled covariance over the determinant of the total covariance, is also a popular criterion. Physically, it can be interpreted as the ratio between the average class volume and the total volume of all cases. Its value will range from 0 (complete class separation) to 1 (complete class fusion).

As for the search method, the following are popular ones and available in STATISTICA and SPSS:

1. Sequential search (direct)

The direct sequential search corresponds to performing successive feature additions or eliminations to the target set, based on a separability criterion.

In a *forward search*, one starts with the feature of most merit and, at each step, all the features not yet included in the subset are revised; the one that contributes the most to class discrimination is evaluated through the merit criterion. This feature is then included in the subset and the procedure advances to the next search step. The process goes on until the merit criterion for any candidate feature is below a specified threshold.

In a *backward search*, the process starts with the whole feature set and, at each step, the feature that contributes the least to class discrimination is removed. The process goes on until the merit criterion for any candidate feature is above a specified threshold.

2. Sequential search (dynamic)

The problem with the previous search methods is the possible existence of "nested" feature subsets that are not detected by direct sequential search. This problem is tackled in a dynamic search by performing a combination of forward and backward searches at each level, known as "plus *l*-take away *r*" selection.

Direct sequential search methods can be applied using STATISTICA and SPSS, the latter affording a dynamic search procedure that is in fact a "plus 1-take away 1" selection. As merit criterion, STATISTICA uses the ANOVA F (for all selected features at a given step) with default value of one. SPSS allows the use of other merit criteria such as the squared Bhattacharyya distance (i.e., the squared Mahalanobis distance of the means).

It is also common to set a lower limit to the so-called *tolerance level*, $T = 1 - r^2$, which must be satisfied by all features, where r is the multiple correlation factor of one candidate feature with all the others. Highly correlated features are therefore removed. One must be quite conservative, however, in the specification of the tolerance. A value at least as low as 1% is common practice.

Example 6.12

Q: Consider the first two classes of the Cork Stoppers' dataset. Perform forward and backward searches on the available 10-feature set, using default values for the tolerance (0.01) and the ANOVA F (1.0). Evaluate the training set errors of both solutions.

A: Figure 6.21 shows the summary listing of a forward search for the first two classes of the cork-stopper data obtained with STATISTICA. Equal priors are assumed. Note that variable ART, with the highest F, entered in the model in "Step 1". The Wilk's lambda, initially 1, decreased to 0.42 due to the contribution of ART. Next, in "Step 2", the variable with highest F contribution for the model containing ART, enters in the model, decreasing the Wilks' lambda to 0.4. The process continues until there are no variables with F contribution higher than 1. In the listing an approximate F for the model, based on the Wilk's lambda, is also indicated. Figure 6.21 shows that the selection process stopped with a highly significant ($p \approx 0$) Wilks' lambda. The four-feature solution {ART, PRM, NG, RAAR} corresponds to the classification matrix shown before in Figure 6.14b.

Using a backward search, a solution with only two features (N and PRT) is obtained. It has the performance presented in Example 6.2. Notice that the backward search usually needs to start with a very low tolerance value (in the present case T = 0.002 is sufficient). The dimensionality ratio of this solution is comfortably high: $n/d = 25$. One can therefore be confident that this classifier performs in a nearly optimal way.

□

Example 6.13

Q: Redo the previous Example 6.12 for a three-class classifier, using dynamic search.

A: Figure 6.22 shows the listing produced by SPSS in a dynamic search performed on the cork-stopper data (three classes), using the squared Bhattacharyya distance (D squared) of the two closest classes as a merit criterion. Furthermore, features were only entered or removed from the selected set if they contributed significantly to the ANOVA F. The solution corresponding to Figure 6.22 used a 5% level for the statistical significance of a candidate feature to enter the model, and a 10% level to remove it. Notice that PRT, which had entered at step 1, was later removed, at step 5. The nested solution {PRM, N, ARTG, RAAR} would not have been found by a direct forward search.

```
Stepwise Analysis - Step 0

Number of variables in the model: 0
Wilks' Lambda: 1.000000

Stepwise Analysis - Step 1

Number of variables in the model: 1
Last variable entered:      ART   F (  1,   99) = 136.5565    p < .0000
Wilks' Lambda: .4178098   approx. F (  1,   98) = 136.5565    p < .0000

Stepwise Analysis - Step 2

Number of variables in the model: 2
Last variable entered:      PRM   F (  1,   98) = 3.880044    p < .0517
Wilks' Lambda: .4017400   approx. F (  2,   97) = 72.22485    p < .0000

Stepwise Analysis - Step 3

Number of variables in the model: 3
Last variable entered:      NG    F (  1,   97) = 2.561449    p < .1128
Wilks' Lambda: .3912994   approx. F (  3,   96) = 49.77880    p < .0000

Stepwise Analysis - Step 4

Number of variables in the model: 4
Last variable entered:      RAAR  F (  1,   96) = 1.619636    p < .2062
Wilks' Lambda: .3847401   approx. F (  4,   95) = 37.97999    p < .0000

Stepwise Analysis - Step 4 (Final Step)

Number of variables in the model: 4
Last variable entered:      RAAR  F (  1,   95) = .3201987    p < .5728
```

Figure 6.21. Feature selection listing, obtained with STATISTICA, using a forward search for two classes of the cork-stopper data.

Step	Entered	Removed	Min. D Squared		Exact F			
			Statistic	Between Groups	Statistic	df1	df2	Sig.
1	PRT		2.401	1.00and 2.00	60.015	1	147.000	1.176E-12
2	PRM		3.083	1.00and 2.00	38.279	2	146.000	4.330E-14
3	N		4.944	1.00and 2.00	40.638	3	145.000	.000
4	ARTG		5.267	1.00and 2.00	32.248	4	144.000	7.438E-15
5		PRT	5.098	1.00and 2.00	41.903	3	145.000	.000
6	RAAR		6.473	1.00and 2.00	39.629	4	144.000	2.316E-22

Figure 6.22. Feature selection listing, obtained with SPSS (Stepwise Method; Mahalanobis), using a dynamic search on the cork stopper data (three classes).

6.6   Classifier Evaluation

The determination of reliable estimates of a classifier error rate is obviously an essential task in order to assess its usefulness and to compare it with alternative solutions.

As explained in section 6.3.3, design set estimates are on average optimistic and the same can be said about using an error formula such as 6.25, when true means and covariance are replaced by their sample estimates. It is, therefore, mandatory that the classifier be empirically tested, using a test set of independent cases. As previously mentioned in section 6.3.3, these test set estimates are, on average, pessimistic.

The influence of the finite sample sizes can be summarised as follows (for details, consult Fukunaga K, 1990):

–   The bias – deviation of the error estimate from the true error – is predominantly influenced by the finiteness of the design set;

–   The variance of the error estimate is predominantly influenced by the finiteness of the test set.

In normal practice, we only have a data set S with n samples available. The problem arises of how to divide the available cases into design set and test set. Among a vast number of methods (see e.g. Fukunaga K, Hayes RR, 1989b) the following ones are easily implemented in SPSS and/or STATISTICA:

Resubstitution method

The whole set S is used for design, and for testing the classifier. As a consequence of the non-independence of design and test sets, the method yields, on average, an optimistic estimate of the error, $E[\hat{Pe}_d(n)]$, mentioned in section 6.3.3. For the two-class linear discriminant with normal distributions an example of such an estimate for various values of n is plotted in Figure 6.15 (lower curve).

Holdout method

The available n samples of S are randomly divided into two disjointed sets (traditionally with 50% of the samples each), S_d and S_t used for design and test, respectively. The error estimate is obtained from the test set, and therefore, suffers from the bias and variance effects previously described. By taking the average over many partitions of the same size, a reliable estimate of the test set error, $E[\hat{Pe}_t(n)]$, is obtained (see section 6.3.3). For the two-class linear discriminant with normal distributions an example of such an estimate for various values of n is plotted in Figure 6.15 (upper curve).

Partition methods

Partition methods, also called *cross-validation* methods divide the available set S into a certain number of subsets, which rotate in their use of design and test, as follows:

1. Divide S into $k > 1$ subsets of randomly chosen cases, with each subset having n/k cases.

2. Design the classifier using the cases of $k - 1$ subsets and test it on the remaining one. A test set estimate Pe_{ti} is thereby obtained.

3. Repeat the previous step rotating the position of the test set, obtaining thereby k estimates Pe_{ti}.

4. Compute the average test set estimate $Pe_t = \sum_{i=1}^{k} Pe_{ti} / k$ and the variance of the Pe_{ti}.

This is the so-called *k-fold cross-validation*. For $k = 2$, the method is similar to the traditional holdout method. For $k = n$, the method is called the *leave-one-out method*, with the classifier designed with $n - 1$ samples and tested on the one remaining sample. Since only one sample is being used for testing, the variance of the error estimate is large. However, the samples are being used independently for design in the best possible way. Therefore the average test set error estimate will be a good estimate of the classifier error for sufficiently high n, since the bias contributed by the finiteness of the design set will be low. For other values of k, there is a compromise between the high bias-low variance of the holdout method, and the low bias-high variance of the leave-one-out method, with less computational effort.

Statistical software products such as SPSS and STATISTICA allow the selection of the cases used for training and for testing linear discriminant classifiers. With SPSS, it is possible to use a selection variable, easing the task of specifying randomly selected samples. SPSS also affords performing a leave-one-out classification. With STATISTICA, one can initially select the cases used for training (Selection Conditions option in the Tools menu), and once the classifier is designed, specify test cases (Select Cases button in the Classification tab of the command window). In MATLAB and R one may create a case-selecting vector, called a filter, with random 0s and 1s.

Example 6.14

Q: Consider the two-class cork-stopper classifier, with two features, presented in section 6.2.2 (see classification matrix in Table 6.3). Evaluate the performance of this classifier using the partition method with $k = 3$, and the leave-one-out method.

A: Using the partition method with $k = 3$, a test set estimate of $Pe_t = 9.9$ % was obtained, which is near the training set error estimate of 10%. The leave-one-out method also produces $Pe_t = 10$ % (see Table 6.11; the "Original" matrix is the training set estimate, the "Cross-validated" matrix is the test set estimate). The closeness of these figures is an indication of reliable error estimation for this high dimensionality ratio classification problem ($n/d = 25$). Using formula 6.28 the 95% confidence limits for these error estimates are: $s = 0.03 \Rightarrow Pe = 10\% \pm 5.9\%$.

□

Table 6.11. Listing of the classification matrices obtained with SPSS, using the leave-one-out method in the classification of the first two classes of the cork-stopper data with two features.

		C	Predicted Group Membership		Total
			1	2	
Original	Count	1	49	1	50
		2	9	41	50
	%	1	98.0	2.0	100
		2	18.0	82.0	100
Cross-validated	Count	1	49	1	50
		2	9	41	50
	%	1	98.0	2.0	100
		2	18.0	82.0	100

Example 6.15

Q: Consider the three-class, cork-stopper classifier, with four features, determined in Example 6.13. Evaluate the performance of this classifier using the leave-one-out method.

A: Table 6.12 shows the leave-one-out results, obtained with SPSS, in the classification of the three cork-stopper classes, using the four features selected by dynamic search in Example 6.13. The training set error is 10.7%; the test set error estimate is 12%. Therefore, we still have a reliable error estimate of about $(10.7 + 12)/2 = 11.4\%$ for this classifier, which is not surprising since the dimensionality ratio is high ($n/d = 12.5$). For the estimate $Pe = 11.4\%$ the 95% confidence interval corresponds to an error tolerance of 5%.   ☐

Table 6.12. Listing of the classification matrices obtained with SPSS, using the leave-one-out method in the classification of the three classes of the cork-stopper data with four features.

		C	Predicted Group Membership			Total
			1	2	3	
Original	Count	1	43	7	0	50
		2	5	45	0	50
		3	0	4	46	50
	%	1	86.0	14.0	0.0	100
		2	10.0	90.0	.0	100
		3	0.0	8.0	92.0	100
Cross-validated	Count	1	43	7	0	50
		2	5	44	1	50
		3	0	5	45	50
	%	1	86.0	14.0	0.0	100
		2	10.0	88.0	2.0	100
		3	0.0	10.0	90.0	100

6.7   Tree Classifiers

In multi-group classification, one is often confronted with the problem that reasonable performances can only be achieved using a large number of features. This requires a very large design set for proper training, probably much larger than what we have available. Also, the feature subset that is the most discriminating set for some classes can perform rather poorly for other classes. In an attempt to overcome these difficulties, a "divide and conquer" principle using multistage classification can be employed. This is the approach of *decision tree* classifiers, also known as *hierarchical classifiers*, in which an unknown case is classified into a class using decision functions in successive stages.

At each stage of the tree classifier, a simpler problem with a smaller number of features is solved. This is an additional benefit, namely in practical multi-class problems where it is rather difficult to guarantee normal or even symmetric distributions with similar covariance matrices for all classes, but it may be possible, with the multistage approach, that those conditions are approximately met at each stage, affording then optimal classifiers.

Example 6.16

Q: Consider the `Breast Tissue` dataset (electric impedance measurements of freshly excised breast tissue) with 6 classes denoted CAR (carcinoma), FAD (fibro-adenoma), GLA (glandular), MAS (mastopathy), CON (connective) and ADI (adipose). Derive a decision tree solution for this classification problem.

A: Performing a Kruskal-Wallis analysis, it is readily seen that all the features have discriminative capabilities, namely I0 and PA500, and that it is practically impossible to discriminate between classes GLA, FAD and MAS. The low dimensionality ratio of this dataset for the individual classes (e.g. only 14 cases for class CON) strongly recommends a decision tree approach, with the use of merged classes and a greatly reduced number of features at each node.

As I0 and PA500 are promising features, it is worthwhile to look at the respective scatter diagram shown in Figure 6.23. Two case clusters are visually identified: one corresponding to {CON, ADI}, the other to {MAS, GLA, FAD, CAR}. At the first stage of the tree we then use I0 alone, with a threshold of I0 = 600, achieving zero errors.

At stage two, we attempt the most useful discrimination from the medical point of view: class CAR (carcinoma) vs. {FAD, MAS, GLA}. Using discriminant analysis, this can be performed with an overall training set error of about 8%, using features AREA_DA and IPMAX, whose distributions are well modelled by the normal distribution.

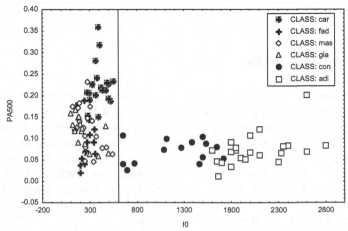

Figure 6.23. Scatter plot of six classes of breast tissue with features I0 and PA500.

Figure 6.24 shows the corresponding linear discriminant. Performing two randomised runs using the partition method in halves (i.e., the 2-fold cross-validation with half of the samples for design and the other half for testing), an average test set error of 8.6% was obtained, quite near the design set error. At stage two, the discrimination CON vs. ADI can also be performed with feature I0 (threshold I0 =1550), with zero errors for ADI and 14% errors for CON.

With these results, we can establish the decision tree shown in Figure 6.25. At each level of the decision tree, a decision function is used, shown in Figure 6.25 as a *decision rule* to be satisfied. The left descendent tree branch corresponds to compliance with a rule, i.e., to a "Yes" answer; the right descendent tree branch corresponds to a "No" answer.

Since a small number of features is used at each level, one for the first level and two for the second level, respectively, we maintain a reasonably high dimensionality ratio at both levels; therefore, we obtain reliable estimates of the errors with narrow 95% confidence intervals (less than 2% for the first level and about 3% for the CAR vs. {FAD, MAS, GLA} level).

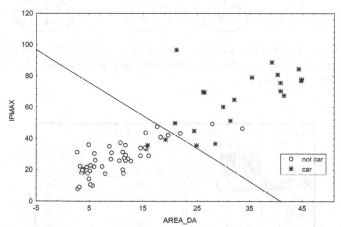

Figure 6.24. Scatter plot of breast tissue classes CAR and {MAS, GLA, FAD} (denoted not car) using features AREA_DA and IPMAX, showing the linear discriminant separating the two classes.

For comparison purposes, the same four-class discrimination was carried out with only one linear classifier using the same three features I0, AREA_DA and IPMAX as in the hierarchical approach. Figure 6.26 shows the classification matrix. Given that the distributions are roughly symmetric, although with some deviations in the covariance matrices, the optimal error achieved with linear discriminants should be close to what is shown in the classification matrix. The degraded performance compared with the decision tree approach is evident.

On the other hand, if our only interest is to discriminate class *car* from all other ones, a linear classifier with only one feature can achieve this discrimination with a performance of about 86% (see Exercise 6.5). This is a comparable result to the one obtained with the tree classifier.

□

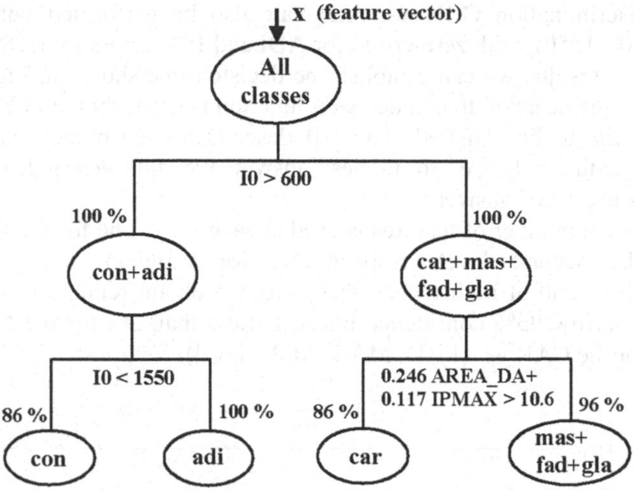

Figure 6.25. Hierarchical tree classifier for the breast tissue data with percentages of correct classifications and decision functions used at each node. Left branch = "Yes"; right branch = "No".

DISCR.	Rows: Observed classific.			
ANAL.	Columns: Predicted classific.			
	%	car	con	adi	fad+
Group	Corr.	p=.198	p=.132	p=.208	p=.462
car	52.4	11	0	0	10
con	64.3	0	9	2	3
adi	95.5	0	1	21	0
fad+	98.0	1	0	0	48
Total	84.0	12	10	23	61

Figure 6.26. Classification matrix obtained with STATISTICA, of four classes of breast tissue using three features and linear discriminants. Class fad+ is actually the class set {FAD, MAS, GLA}.

The decision tree used for the Breast Tissue dataset is an example of a *binary tree*: at each node, a dichotomic decision is made. Binary trees are the most popular type of trees, namely when a single feature is used at each node, resulting in linear discriminants that are parallel to the feature axes, and easily interpreted by human experts. Binary trees also allow categorical features to be easily incorporated with node splits based on a "yes/no" answer to the question whether or not a given case belongs to a set of categories. For instance, this type of trees is frequently used in medical applications, and often built as a result of statistical studies of the influence of individual health factors in a given population.

The design of decision trees can be automated in many ways, depending on the *split criterion* used at each node, and the type of search used for best group discrimination. A split criterion has the form:

$$d(\mathbf{x}) \geq \Delta,$$

where $d(\mathbf{x})$ is a decision function of the feature vector $\mathbf{x}$ and Δ is a threshold. Usually, linear decision functions are used. In many applications, the split criteria are expressed in terms of the individual features alone (the so-called *univariate splits*).

An important concept regarding split criteria is the concept of *node impurity*. The node impurity is a function of the fraction of cases belonging to a specific class at that node.

Consider the two-class situation shown in Figure 6.27. Initially, we have a node with equal proportions of cases belonging to the two classes (white and black circles). We say that its impurity is maximal. The right split results in nodes with zero impurity, since they contain cases from only one of the classes. The left split, on the contrary, increases the proportion of cases from one of the classes, therefore decreasing the impurity, although some impurity remains present.

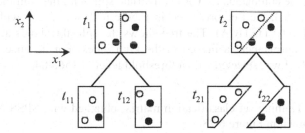

Figure 6.27. Splitting a node with maximum impurity. The left split ($x_1 \geq \Delta$) decreases the impurity, which is still non-zero; the right split ($w_1 x_1 + w_2 x_2 \geq \Delta$) achieves pure nodes.

A popular measure of impurity, expressed in the [0, 1] interval, is the *Gini index of diversity*:

$$i(t) = \sum_{\substack{j,k=1 \\ j \neq k}}^{c} P(j \mid t) P(k \mid t).$$

6.30

For the situation shown in Figure 6.27, we have:

$i(t_1) = i(t_2) = 1 \times 1 = 1;$

$i(t_{11}) = i(t_{12}) = \dfrac{2}{3}\dfrac{1}{3} = \dfrac{2}{9};$

$i(t_{21}) = i(t_{22}) = 1 \times 0 = 0.$

In the automatic generation of binary trees the tree starts at the root node, which corresponds to the whole training set. Then, it progresses by searching for each variable the threshold level achieving the maximum decrease of the impurity at each node. The generation of splits stops when no significant decrease of the impurity is achieved. It is common practice to use the individual feature values of the training set cases as candidate threshold values. Sometimes, after generating a tree automatically, some sort of *tree pruning* should be performed in order to remove branches of no interest.

SPSS and STATISTICA have specific commands for designing tree classifiers, based on univariate splits. The method of exhaustive search for the best univariate splits is usually called the CRT (also CART or C&RT) method, pioneered by Breiman, Friedman, Olshen and Stone (see Breiman *et al.*, 1993).

Example 6.17

Q: Use the CRT approach with univariate splits and the Gini index as splitting criterion in order to derive a decision tree for the Breast Tissue dataset. Assume equal priors of the classes.

A: Applying the commands for CRT univariate split with the Gini index, described in Commands 6.3, the tree presented in Figure 6.28 was found with SPSS (same solution with STATISTICA). The tree shows the split thresholds at each node as well as the improvement achieved in the Gini index. For instance, the first split variable PERIM was selected with a threshold level of 1563.84.

Table 6.13. Training set classification matrix, obtained with SPSS, corresponding to the tree shown in Figure 6.28.

Observed	Predicted						
	car	fad	mas	gla	con	adi	Percent Correct
car	20	0	1	0	0	0	95.2%
fad	0	0	12	3	0	0	0.0%
mas	2	0	15	1	0	0	83.3%
gla	1	0	4	11	0	0	68.8%
con	0	0	0	0	14	0	100.0%
adi	0	0	0	0	1	21	95.5%

The classification matrix corresponding to this classification tree is shown in Table 6.13. The overall percent correct is 76.4% (overall error of 23.6%). Note the good classification results for the classes CAR, CON and ADI and the difficult splitting of {FAD,MAS,GLA} that we had already observed. Also note the gradual error increase as one progresses through the tree. Node splitting stops when no significant improvement is found.

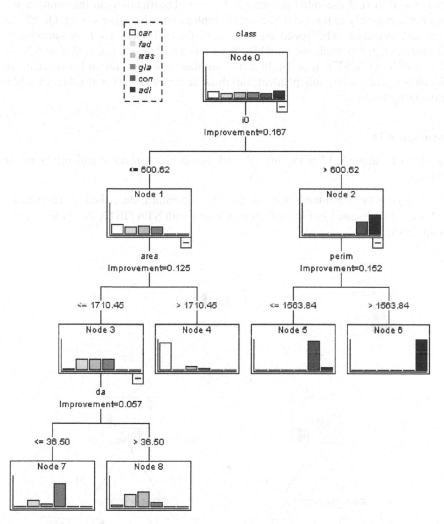

Figure 6.28. CRT tree using the Gini index as impurity criterion, designed with SPSS.

The CRT algorithm based on exhaustive search tends to be biased towards selecting variables that afford more splits. It is also quite time consuming. Other approaches have been proposed in order to remedy these shortcomings, namely the approach followed by the algorithm known as QUEST ("Quick, Unbiased, Efficient Statistical Trees"), proposed by Loh, WY and Shih, YS (1997), that employs a sort of recursive quadratic discriminant analysis for improving the reliability and efficiency of the classification trees that it computes.

It is often interesting to compare the CRT and QUEST solutions, since they tend to exhibit complementary characteristics. CRT, besides its shortcomings, is guaranteed to find the splits producing the best classification (in the training set, but not necessarily in test sets) because it employs an exhaustive search. QUEST is fast and unbiased. The speed advantage of QUEST over CRT is particularly dramatic when the predictor variables have dozens of levels (Loh, WY and Shih, YS, 1997). QUEST's lack of bias in variable selection for splits is also an advantage when some independent variables have few levels and other variables have many levels.

Example 6.18

Q: Redo Example 6.17 using the QUEST approach. Assume equal priors of the classes.

A: Applying the commands for the QUEST algorithm, described in Commands 6.3, the tree presented in Figure 6.29 was found with STATISTICA (same solution with SPSS).

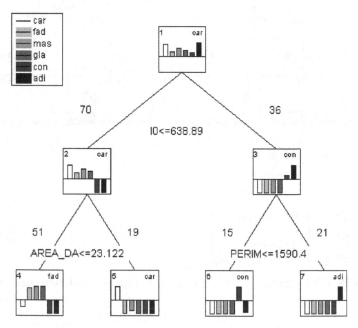

Figure 6.29. Tree plot, obtained with STATISTICA for the breast-tissue, using the QUEST approach.

The classification matrix corresponding to this classification tree is shown in Table 6.14. The overall percent correct is 63.2% (overall error of 36.8%). Note the good classification results for the classes CON and ADI and the splitting off of {FAD,MAS,GLA} as a whole. This solution is similar to the solution we had derived "manually" and represented in Figure 6.25.

□

Table 6.14. Training set classification matrix corresponding to the tree shown in Figure 6.29.

Observed	Predicted						
	car	fad	mas	gla	con	adi	Percent Correct
car	17	4	0	0	0	0	81.0%
fad	0	15	0	0	0	0	100.0%
mas	2	16	0	0	0	0	0.0%
gla	0	16	0	0	0	0	0.0%
con	0	0	0	0	14	0	100.0%
adi	0	0	0	0	1	21	95.5%

The tree solutions should be validated as with any other classifier type. SPSS and STATISTICA afford the possibility of cross-validating the designed trees using the partition method described in section 6.6. In the present case, since the dimensionality ratios are small, one has to perform the cross-validation with very small test samples. Using a 14-fold cross-validation for the CRT and QUEST solutions of Examples 6.17 and 6.18 we obtained the results shown in Table 6.13. We see that although CRT yielded a lower training set error compared with QUEST, this last method provided a solution with better generalization capability (smaller difference between training set and test set errors). Note that 14-fold cross-validation is equivalent to the leave-one-out method for the smaller sized class of this dataset.

Table 6.15. Overall errors and respective standard deviations (obtained with STATISTICA) in 14-fold cross-validation of the tree solutions found in Examples 6.17 and 6.18.

Method	Overall Error	Stand. Deviation
CRT	0.406	0.043
QUEST	0.349	0.040

Commands 6.3. SPSS and STATISTICA commands used to design tree classifiers.

SPSS	`Analyze; Classify; Tree...`
STATISTICA	`Statistics; Multivariate Exploratory Techniques; Classification Trees`

When performing tree classification with SPSS it is advisable to first assign appropriate labels to the categorical variable. This can be done in a "Define Variable Properties..." window. The Tree window allows one to specify the dependent (categorical) and independent variables and the type of Output one wishes to obtain (usually, Chart – a display as in Figure 6.28 – and Classification Table from Statistics). One then proceeds to choosing a growing method (CRT, QUEST), the maximum number of cases per node at input and output (in Criteria), the priors (in Options) and the cross-validation method (in Validation).

In STATISTICA the independent variables are called "predictors". Real-valued variables as the ones used in the previous examples are called "ordered predictors". One must not forget to set the codes for the dependent variable. The CRT and QUEST methods appear in the Methods window denominated as "CR&T-style exhaustive search for univariate splits" and "Discriminant-based univariate splits for categ. and ordered predictors", respectively.

The classification matrices in STATISTICA have a different configuration of the ones shown in Tables 6.13 and 6.14: the observations are along the columns and the predictions along the rows. Cross-validation in STATISTICA provides the average misclassification matrix which can be useful to individually analyse class behaviour.   ∎

Exercises

6.1 Consider the first two classes of the Cork Stoppers' dataset described by features ART and PRT.

  a) Determine the Euclidian and Mahalanobis classifiers using feature ART alone, then using both ART and PRT.

  b) Compute the Bayes error using a pooled covariance estimate as the true covariance for both classes.

  c) Determine whether the Mahalanobis classifiers are expected to be near the optimal Bayesian classifier.

  d) Using SC Size, determine the average deviation of the training set error estimate from the Bayes error, and the 95% confidence interval of the error estimate.

6.2 Repeat the previous exercise for the three classes of the Cork Stoppers' dataset, using features N, PRM and ARTG.

6.3 Consider the problem of classifying cardiotocograms (CTG dataset) into three classes: N (normal), S (suspect) and P (pathological).
    a)   Determine which features are most discriminative and appropriate for a Mahalanobis classifier approach for this problem.
    b)   Design the classifier and estimate its performance using a partition method for the test set error estimation.

6.4 Repeat the previous exercise using the Rocks' dataset and two classes: {granites} vs. {limestones, marbles}.

6.5 A physician would like to have a very simple rule available for screening out carcinoma situations from all other situations using the same diagnostic means and measurements as in the Breast Tissue dataset.
    a)   Using the Breast Tissue dataset, find a linear Bayesian classifier with only one feature for the discrimination of carcinoma versus all other cases (relax the normality and equal variance requirements). Use forward and backward search and estimate the priors from the training set sizes of the classes.
    b)   Obtain training set and test set error estimates of this classifier, and 95% confidence intervals.
    c)   Using the SC Size program, assess the deviation of the error estimate from the true Bayesian error, assuming that the normality and equal variance requirements were satisfied.
    d)   Suppose that the risk of missing a carcinoma is three times higher than the risk of misclassifying a non-carcinoma. How should the classifying rule be reformulated in order to reflect these risks, and what is the performance of the new rule?

6.6 Design a linear discriminant classifier for the three classes of the Clays' dataset and evaluate its performance.

6.7 Explain why all ROC curves start at (0,0) and finish at (1,1) by analysing what kind of situations these points correspond to.

6.8 Consider the Breast Tissue dataset. Use the ROC curve approach to determine single features that will discriminate carcinoma cases from all other cases. Compare the alternative methods using the ROC curve areas.

6.9 Repeat the ROC curve experiments illustrated in Figure 6.20 for the FHR Apgar dataset, using combinations of features.

6.10 Increase the amplitude of the signal impulses by 20% in the Signal & Noise dataset. Consider the following impulse detection rule:

An impulse is detected at time n when $s(n)$ is bigger than $\alpha \sum_{i=1}^{2} \left(s(n-i) + s(n+i) \right)$.

Determine the ROC curve corresponding to several α values, and determine the best α for the impulse/noise discrimination. How does this method compare with the amplitude threshold method described in section 6.4?

6.11 Consider the Infarct dataset, containing four continuous-type measurements of physiological variables of the heart (EF, CK, IAD, GRD), and one ordinal-type variable (SCR: 0 through 5) assessing the severity of left ventricle necrosis. Use ROC curves of the four continuous-type measurements in order to determine the best threshold discriminating "low" necrosis (SCR < 2) from "medium-high" necrosis (SCR ≥ 2), as well as the best discriminating measurement.

6.12 Repeat Exercises 6.3 and 6.4 performing sequential feature selection (direct and dynamic).

6.13 Perform a *resubstitution* and *leave-one-out* estimation of the classification errors for the three classes of cork stoppers, using the features obtained by dynamic selection (Example 6.13). Comment on the reliability of these estimates.

6.14 Compute the 95% confidence interval of the error for the classifier designed in Exercise 6.3 using the standard formula. Perform a partition method evaluation of the classifier, with 10 partitions, obtaining another estimate of the 95% confidence interval of the error.

6.15 Compute the decrease of impurity in the trees shown in Figure 6.25 and Figure 6.29, using the Gini index.

6.16 Compute the classification matrix CAR vs. {MAS, GLA, FAD} for the Breast Tissue dataset in the tree shown in Figure 6.25. Observe its dependence on the prevalences. Compute the linear discriminant shown in the same figure.

6.17 Using the CRT and QUEST approaches, find decision trees that discriminate the three classes of the CTG dataset, N, S and P, using several initial feature sets that contain the four variability indexes ASTV, ALTV, MSTV, MLTV. Compare the classification performances for the several initial feature sets.

6.18 Consider the four variability indexes of foetal heart rate (MLTV, MSTV, ALTV, ASTV) included in the CTG dataset. Using the CRT approach, find a decision tree that discriminates the pathological foetal state responsible for a "flat-sinusoidal" (FS) tracing from all the other classes.

6.19 Design tree classifiers for the three classes of the Clays' dataset using the CRT and QUEST approaches, and compare their performance with the classifier of Exercise 6.6.

6.20 Design a tree classifier for Exercise 6.11 and evaluate its performance comparatively.

6.21 Redesign the tree solutions found in Examples 6.17 and 6.18 using priors estimated from the training set (empirical priors) instead of equal priors. Compare the solutions with those obtained in the mentioned examples and comment the found differences.

7 Data Regression

An important objective in scientific research and in more mundane data analysis tasks concerns the possibility of predicting the value of a *dependent* random variable based on the values of other *independent* variables, establishing a functional relation of a statistical nature. The study of such functional relations, known for historical reasons as *regressions*, goes back to pioneering works in Statistics.

Let us consider a functional relation of one random variable Y depending on a single *predictor* variable X, which may or may not be random:

$$Y = g(X).$$

We study such a functional relation, based on a dataset of observed values $\{(x_1,y_1), (x_1,y_1), \ldots, (x_n,y_n)\}$, by means of a *regression model*, $\hat{Y} = \hat{g}(X)$, which is a formal way of expressing the statistical nature of the unknown functional relation, as illustrated in Figure 7.1. We see that for every predictor value x_i, we must take into account the probability distribution of Y as expressed by the density function $f_Y(y)$. Given certain conditions the stochastic means of these probability distributions determine the sought for functional relation, as illustrated in Figure 7.1. In the following we always assume X to be a deterministic variable.

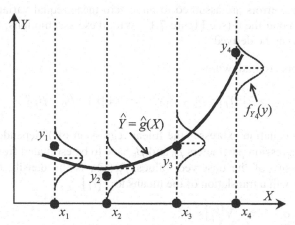

Figure 7.1. Statistical functional model in single predictor regression. The y_i are the observations of the dependent variable for the predictor values x_i.

Correlation differs from regression since in correlation analysis all variables are assumed to be random and play a symmetrical role, with no dependency assignment. As it happens with correlation, one must also be cautious when trying to infer causality relations from regression. As a matter of fact, the existence of a statistical relation between the response Y and the predictor variable X does not necessarily imply that Y depends causally on X (see also 4.4.1).

7.1   Simple Linear Regression

7.1.1 Simple Linear Regression Model

In simple linear regression, one has a single predictor variable, X, and the functional relation is assumed to be linear. The only random variable is Y and the regression model is expressed as:

$$Y_i = \beta_0 + \beta_1 x_i + \mathcal{E}_i,$$
7.1

where:

i.   The Y_i are random variables representing the observed values y_i for the predictor values x_i. The Y_i are distributed as $f_{Y_i}(y)$. The linear regression parameters, β_0 and β_1, are known as *intercept* and *slope*, respectively.

ii.   The $\mathcal{E}_i$ are random *error* terms (variables), with:

$$E[\mathcal{E}_i] = 0; \quad V[\mathcal{E}_i] = \sigma^2; \quad V[\mathcal{E}_i \mathcal{E}_j] = 0, \quad \forall i \neq j.$$

Therefore, the errors are assumed to have zero mean, equal variance and to be uncorrelated among them (see Figure 7.1). With these assumptions, the following model features can be derived:

i.   The errors are i.i.d. with:

$$E[\mathcal{E}_i] = 0 \implies E[Y_i] = \beta_0 + \beta_1 x_i \implies E[Y] = \beta_0 + \beta_1 X.$$

The last equation expresses the linear regression of Y dependent on X. The linear regression parameters β_0 and β_1 have to be estimated from the dataset. The density of the observed values, $f_{Y_i}(y)$, is the density of the errors, $f_{\mathcal{E}}(\varepsilon)$, with a translation of the means to $E[Y_i]$.

ii. $V[\mathcal{E}_i] = \sigma^2 \implies V[Y_i] = \sigma^2.$

iii.   The Y_i and Y_j are uncorrelated.

7.1.2 Estimating the Regression Function

A popular method of estimating the regression function parameters is to use a *least square error* (LSE) approach, by minimising the total sum of the squares of the errors (deviations) between the observed values y_i and the estimated values $b_0 + b_1 x_i$:

$$E = \sum_{i=1}^{n} \varepsilon_i^2 = \sum_{i=1}^{n} (y_i - b_0 - b_1 x_i)^2 .$$ 7.2

where b_0 and b_1 are estimates of β_0 and β_1, respectively.

In order to apply the LSE method one starts by differentiating E in order to b_0 and b_1 and equalising to zero, obtaining the so-called *normal equations*:

$$\begin{cases} \sum y_i = n b_0 + b_1 \sum x_i \\ \sum x_i y_i = b_0 \sum x_i + b_1 \sum x_i^2 \end{cases},$$ 7.3

where the summations, from now on, are always assumed to be for the n predictor values. By solving the normal equations, the following parameter estimates, b_0 and b_1, are derived:

$$b_1 = \frac{\sum (x_i - \bar{x})(y_i - \bar{y})}{\sum (x_i - \bar{x})^2} .$$ 7.4

$$b_0 = \bar{y} - b_1 \bar{x} .$$ 7.5

The least square estimates of the linear regression parameters enjoy a number of desirable properties:

i.   The parameters b_0 and b_1 are *unbiased estimates* of the true parameters β_0 and β_1 ($E[b_0] = \beta_0$, $E[b_1] = \beta_1$), and have *minimum variance* among all unbiased linear estimates.

ii. The *predicted* (or *fitted*) *values* $\hat{y}_i = b_0 + b_1 x_i$ are point estimates of the true, *observed* values, y_i. The same is valid for the whole relation $\hat{Y} = b_0 + b_1 X$, which is the point estimate of the *mean response* $E[Y]$.

iii. The regression line always goes through the point ($\bar{x}, \bar{y}$).

iv. The computed errors $e_i = y_i - \hat{y}_i = y_i - b_0 - b_1 x_i$, called the *residuals*, are point estimates of the error values ε_i. The sum of the residuals is zero: $\sum e_i = 0$.

v.   The residuals are uncorrelated with the predictor and the predicted values: $\sum e_i x_i = 0$; $\sum e_i \hat{y}_i = 0$.

vi. $\sum y_i = \sum \hat{y}_i$ $\Rightarrow$ $\bar{y} = \bar{\hat{y}}$, i.e., the predicted values have the same mean as the observed values.

These properties are a main reason of the popularity of the LSE method. However, the reader must bear in mind that other error measures could be used. For instance, instead of minimising the sum of the squares of the errors one could minimise the sum of the absolute values of the errors: $E = \sum |\varepsilon_i|$. Another linear regression would then be obtained with other properties. In the following we only deal with the LSE method.

Example 7.1

Q: Consider the variables ART and PRT of the Cork Stoppers' dataset. Imagine that we wanted to predict the total area of the defects of a cork stopper (ART) based on their total perimeter (PRT), using a linear regression approach. Determine the regression parameters and represent the regression line.

A: Figure 7.2 shows the scatter plot obtained with STATISTICA of these two variables with the linear regression fit (Linear Fit box in Scatterplot), using equations 7.4 and 7.5. Figure 7.3 shows the summary of the regression analysis obtained with STATISTICA (see Commands 7.1). Using the values of the linear parameters (Column B in Figure 7.3) we conclude that the fitted regression line is:

ART = −64.5 + 0.547×PRT.

Note that the regression line passes through the point of the means of ART and PRT: ($\overline{\text{ART}}$, $\overline{\text{PRT}}$) = (324, 710).

⊓

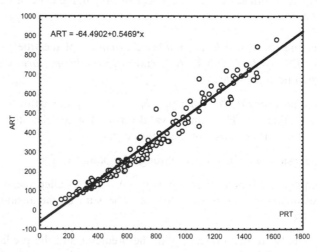

Figure 7.2. Scatter plot of variables ART and PRT (cork-stopper dataset), obtained with STATISTICA, with the fitted regression line.

N=150	Beta	Std. Err. of Beta	B	Std. Err. of B	t(148)	p-level
Intercept			-64.4902	7.053354	-9.14320	0.000000
PRT	0.981142	0.015888	0.5469	0.008857	61.75316	0.000000

R= .98114218 R²= .96263997 Adjusted R²= .96238754
F(1,148)=3813.5 p<0.0000 Std.Error of estimate: 39.050

Figure 7.3. Table obtained with STATISTICA containing the results of the simple linear regression for the Example 7.1.

The value of Beta, mentioned in Figure 7.3, is related to the so-called *standardised regression model*:

$$Y_i^* = \beta_1^* x_i^* + \mathcal{E}_i .$$
7.6

In equation 7.6 only one parameter is used, since Y_i^* and x_i^* are standardised variables (mean = 0, standard deviation = 1) of the observed and predictor variables, respectively. (By equation 7.5, $\beta_0 = E[Y] - \beta_1 \bar{x}$ implies $(Y_i - E[Y]) / \sigma_Y = \beta_1^* (x_i - \bar{x}) / s_X + \mathcal{E}_i^*$.)

It can be shown that:

$$\beta_1 = \left(\frac{\sigma_Y}{s_X}\right)\beta_1^* .$$
7.7

The standardised β_1^* is the so-called *beta coefficient*, which has the point estimate value $b_1^* = 0.98$ in the table shown in Figure 7.3.

Figure 7.3 also mentions the values of R, R^2 and Adjusted R^2. These are measures of association useful to assess the goodness of fit of the model. In order to understand their meanings we start with the estimation of the error variance, by computing the *error sum of squares* or *residual sum of squares* (SSE)[1], i.e. the quantity E in equation 7.2, as follows:

$$SSE = \sum (y_i - \hat{y}_i)^2 = \sum e_i^2 .$$
7.8

Note that the deviations are referred to each predicted value; therefore, SSE has $n - 2$ degrees of freedom since two degrees of freedom are lost: b_0 and b_1. The following quantities can also be computed:

- *Mean square error*: $MSE = \dfrac{SSE}{n-2}$.

- *Root mean square error*, or *standard error*: $RMS = \sqrt{MSE}$.

---

[1] Note the analogy of SSE and SST with the corresponding ANOVA sums of squares, formulas 4.25b and 4.22, respectively.

This last quantity corresponds to the "Std. Error of estimate" in Figure 7.3.

The total variance of the observed values is related to the *total sum of squares* (SST)[1]:

$$SST \equiv SSY = \sum (y_i - \bar{y})^2 .$$

7.9

The contribution of X to the prediction of Y can be evaluated using the following association measure, known as *coefficient of determination* or *R-square*:

$$r^2 = \frac{SST - SSE}{SST} \in [0, 1].$$

7.10

Therefore, "R-square", which can also be shown to be the square of the Pearson correlation between x_i and y_i, measures the contribution of X in reducing the variation of Y, i.e., in reducing the uncertainty in predicting Y. Notice that:

1. If all observations fall on the regression line (perfect regression, complete certainty), then SSE = 0, $r^2 = 1$.

2. If the regression line is horizontal (no contribution of X in predicting Y), then SSE = SST, $r^2 = 0$.

However, as we have seen in 2.3.4 when discussing the Pearson correlation, "R-square" does not assess the appropriateness of the linear regression model.

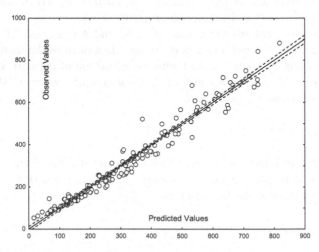

Figure 7.4. Scatter plot, obtained with STATISTICA, of the observed values versus predicted values of the ART variable (cork-stopper data) with the fitted line and the 95% confidence interval (dotted line).

Often the value of "R-square" is found to be slightly optimistic. Several authors propose using the following "Adjusted R-square" instead:

$$r_a^2 = r^2 - (1 - r^2)/(n-2).$$ 7.11

For the cork-stopper example the value of the "R square" is quite high, $r^2 = 0.96$, as shown in Figure 7.3. STATISTICA highlights the summary table when this value is found to be significant (same test as in 4.4.1), therefore showing evidence of a tight fit. Figure 7.4 shows the observed versus predicted values for the Example 7.1. A perfect model would correspond to a unit slope straight line.

Commands 7.1. SPSS, STATISTICA, MATLAB and R commands used to perform simple linear regression.

SPSS	`Analyze; Regression; Linear`
STATISTICA	`Statistics; Multiple regression | Advanced Linear/Nonlinear Models; General Linear Models`
MATLAB	`[b,bint,r,rint,stats]=regress(y,X,alpha)`
R	`lm(y~X)`

SPSS and STATISTICA commands for regression analysis have a large number of options that the reader should explore in the following examples. With SPSS and STATISTICA, there is also the possibility of obtaining a variety of detailed listings of predicted values and residuals as well as graphic help, such as specialised scatter plots. For instance, Figure 7.4 shows the scatter plot of the observed versus the predicted values of variable ART (cork-stopper example), together with the 95% confidence interval for the linear fit.

Regression analysis is made in MATLAB with the `regress` function, which computes the LSE coefficient estimates b of the equation $y = Xb$ where y is the dependent data vector and X is the matrix whose columns are the predictor data vectors. We will use more than one predictor variable in section 7.3 and will then adopt the matrix notation. The meaning of the other return values is as follows:

`r`: residuals;   `rint`: alpha confidence intervals for `r`;
`stats`: r^2 and other statistics   `bint`: alpha confidence interval for `b`;

Let us use Example 7.1 to illustrate the use of the `regress` function. We start by defining the ART and PRT data vectors using the `cork` matrix containing the whole dataset. These variables correspond to columns 2 and 4, respectively (see the EXCEL data file):

```
>> ART = cork(:,2); PRT = cork(:,4);
```

Next, we create the X matrix by binding a column of ones, corresponding to the intercept term in equation 7.1, to the PRT vector:

```
>> X = [PRT ones(size(PRT,1),1)]
```

We are now ready to apply the regress function:

```
>> [b,bint,r,rint,stats] = regress(ART,X,0.05);
```

The values of b, bint and stats are as follows:

```
>> b
b =
    0.5469
  -64.4902
>> bint
bint =
    0.5294    0.5644
  -78.4285  -50.5519
>> stats
stats =
  1.0e+003 *
    0.0010    3.8135         0
```

The values of b coincide with those in Figure 7.3. The intercept coefficient is here the second element of b in correspondence with the (second) column of ones of X. The values of bint are the 95% confidence intervals of b agreeing with the values computed in Example 7.2 and Example 7.4, respectively. Finally, the first value of stats is the R-square statistic; the second and third values are respectively the ANOVA F and p discussed in section 7.1.4 and reported in Table 7.1. The exact value of the R-square statistic (without the four-digit rounding effect of the above representation) can be obtained by previously issuing the format long command.

Let us now illustrate the use of the R lm function for the same problem as in Example 7.1. We have already used the lm function in Chapter 4 when computing the ANOVA tests (see Commands 4.5 and 4.6). This function fits a linear model describing the y data as a function of the X data. In chapter 4 the X data was a categorical data vector (an R factor). Here, the X data correspond to the real-valued predictors. Using the cork data frame we may run the lm function as follows:

```
> load("e:cork")
> attach(cork)
> summary(lm(ART~PRT))

Call:
lm(formula = ART ~ PRT)

Residuals:
     Min       1Q   Median       3Q      Max
 -95.651  -22.727   -1.016   19.012  152.143
```

```
Coefficients:
            Estimate Std. Error t value Pr(>|t|)
(Intercept) -64.49021    7.05335  -9.143 4.38e-16 ***
PRT          0.54691    0.00885  61.753  < 2e-16 ***
---
Signif. codes:  0 '***' 0.001 '**' 0.01 '*' 0.05 '.'
0.1 ' ' 1

Residual standard error: 39.05 on 148 degrees of
freedom
Multiple R-Squared: 0.9626,Adjusted R-squared: 0.9624
F-statistic:  3813 on 1 and 148 DF,p-value: < 2.2e-16
```

We thus obtain the same results published in Figure 7.3 and Table 7.1 plus some information on the residuals. The lm function returns an object of class "lm" with several components, such as coefficients and residuals (for more details use the help). Returning objects with components is a general feature of R. We found it already when describing how to obtain the density estimate of a histogram object in Commands 2.3 and the histogram of a bootstrap object in Commands 3.7. The summary function when applied to an object produces a summary display of the most important object components, as exemplified above. If one needs to obtain a particular component one uses the "$" notation. For instance, the residuals of the above regression model can be stored in a vector x with:

```
r <- lm(ART~PRT)
x <- r$residuals
```

The fitted values can be obtained with r$fitted.   ∎

7.1.3 Inferences in Regression Analysis

In order to make inferences about the regression model, the errors ε_i are assumed to be independent and normally distributed, $N_{0,\sigma}$. This constitutes the so-called *normal regression model*. It can then be shown that the unbiased estimate of σ is the RMS.

The inference tests described in the following sections continue to be valid in the case of mild deviations from normality. Even if the distributions of Y_i are far from normal, the estimators of b_0 and b_1 have the property of *asymptotic normality*: their distributions approach normality under very general conditions, as the sample size increases.

7.1.3.1 Inferences About b_1

The point estimate of b_1 is given by formula 7.4. This formula can also be expressed as:

$$b_1 = \sum k_i y_i \qquad \text{with} \qquad k_i = \frac{(x_i - \bar{x})}{\sum (x_i - \bar{x})^2}.$$ 7.12

The sampling distribution of b_1 for the normal regression model is also normal (since b_1 is a linear combination of the y_i), with:

- Mean: $E[b_1] = E\left[\sum k_i Y_i\right] = \beta_0 \sum k_i + \beta_1 \sum k_i x_i = \beta_1.$

- Variance: $V[b_1] = V\left[\sum k_i Y_i\right] = \sum k_i^2 V[Y_i] = \sigma^2 \sum k_i^2 = \dfrac{\sigma^2}{\sum (x_i - \bar{x})^2}.$

If instead of σ, we use its estimate $RMS = \sqrt{MSE}$, we then have:

$$s_{b_1} = \sqrt{MSE / \sum (x_i - \bar{x})^2}.$$ 7.13

Thus, in order to make inferences about b_1, we take into account that:

$$t^* = \frac{b_1 - \beta_1}{s_{b_1}} \sim t_{n-2}.$$ 7.14

The sampling distribution of the studentised statistic t^* allows us to compute confidence intervals for β_1 as well as to perform tests of hypotheses, in order to, for example, assess if there is no linear association: H_0: $\beta_1 = 0$.

Example 7.2

Q: Determine the 95% confidence interval of b_1 for the ART(PRT) linear regression in Example 7.1.

A: The MSE value can be found in the SPSS or STATISTICA ANOVA table (see Commands 7.2). The Model Summary of SPSS or STATISTICA also publishes the value of RMS (Standard Error of Estimate). When using MATLAB, the values of MSE and RMS can also be easily computed using the vector r of the residuals (see Commands 7.1). The value of $\sum (x_i - \bar{x})^2$ is computed from the variance of the predictor values. Thus, in the present case we have:

$$MSE = 1525, \; s_{PRT} = 361.2 \; \Rightarrow \; s_{b_1} = \sqrt{MSE / ((n-1)s_{PRT}^2)} = 0.00886.$$

Since $t_{148,0.975} = 1.976$ the 95% confidence interval of b_1 is [0.5469 − 0.0175, 0.5469 + 0.0175] = [0.5294, 0.5644], which agrees with the values published by SPSS (confidence intervals option), STATISTICA (Advanced Linear/Nonlinear Models), MATLAB and R.

□

Example 7.3

Q: Consider the ART(PRT) linear regression in Example 7.1. Is it valid to reject the null hypothesis of no linear association, at a 5% level of significance?

A: The results of the respective t test are shown in the last two columns of Figure 7.3. Taking into account the value of p ($p \approx 0$ for $t^* = 61.8$), the null hypothesis is rejected.   □

7.1.3.2   Inferences About b_o

The point estimate of b_0 is given by formula 7.5. The sampling distribution of b_0 for the normal regression model is also normal (since b_0 is a linear combination of the y_i), with:

–   Mean: $\mathrm{E}[b_0] = \beta_0$;

–   Variance $\mathrm{V}[b_0] = \sigma^2 \left(\dfrac{1}{n} + \dfrac{\bar{x}^2}{\sum (x_i - \bar{x})^2} \right)$.

Since σ is usually unknown we use the point estimate of the variance:

$$s_{b_0}^2 = \mathrm{MSE} \left(\frac{1}{n} + \frac{\bar{x}^2}{\sum (x_i - \bar{x})^2} \right). \qquad\qquad 7.15$$

Therefore, in order to make inferences about b_0, we take into account that:

$$t^* = \frac{b_0 - \beta_0}{s_{b_0}} \sim t_{n-2}. \qquad\qquad 7.16$$

This allows us to compute confidence intervals for β_0, as well as to perform tests of hypotheses, namely in order to assess whether or not the regression line passes through the origin: H_0: $\beta_0 = 0$.

Example 7.4

Q: Determine the 95% confidence interval of b_0 for the ART(PRT) linear regression in Example 7.1.

A: Using the MSE and s_{PRT} values as described in Example 7.2, we obtain:

$$s_{b_0}^2 = \mathrm{MSE} \left(1/n + \bar{x}^2 / \sum (x_i - \bar{x})^2 \right) = 49.76.$$

Since $t_{148,0.975} = 1.976$ we thus have $s_{b_0} \in [-64.49 - 13.9, -64.49 + 13.9] = [-78.39, -50.59]$ with 95% confidence level. This interval agrees with previously mentioned SPSS, STATISTICA, MATLAB and R results.

□

Example 7.5

Q: Consider the ART(PRT) linear regression in Example 7.1. Is it valid to reject the null hypothesis of a linear fit through the origin at a 5% level of significance?

A: The results of the respective t test are shown in the last two columns of Figure 7.3. Taking into account the value of p ($p \approx 0$ for $t^* = -9.1$), the null hypothesis is rejected. This is a somewhat strange result, since one expects a null area corresponding to a null perimeter. As a matter of fact an ART(PRT) linear regression without intercept is also a valid data model (see Exercise 7.3).   □

7.1.3.3   Inferences About Predicted Values

Let us assume that one wants to derive interval estimators of $E[\hat{Y}_k]$, i.e., one wants to determine which value would be obtained, on average, for a predictor variable level x_k, and if *repeated samples* (or *trials*) were used.

The point estimate of $E[\hat{Y}_k]$, corresponding to a certain value x_k, is the computed predicted value:

$$\hat{y}_k = b_0 + b_1 x_k.$$

The $\hat{y}_k$ value is a possible value of the random variable $\hat{Y}_k$ which represents all possible predicted values. The sampling distribution for the normal regression model is also normal (since it is a linear combination of observations), with:

–   Mean: $E[\hat{Y}_k] = E[b_0 + b_1 x_k] = E[b_0] + x_k E[b_1] = \beta_0 + \beta_1 x_k$;

–   Variance: $V[\hat{Y}_k] = \sigma^2 \left(\dfrac{1}{n} + \dfrac{(x_k - \bar{x})^2}{\sum (x_i - \bar{x})^2} \right)$.

Note that the variance is affected by how far x_k is from the sample mean $\bar{x}$. This is a consequence of the fact that all regression estimates must pass through $(\bar{x}, \bar{y})$. Therefore, values x_k far away from the mean lead to higher variability in the estimates.

Since σ is usually unknown we use the estimated variance:

$$s[\hat{Y}_k] = \text{MSE} \left(\frac{1}{n} + \frac{(x_k - \bar{x})^2}{\sum (x_i - \bar{x})^2} \right). \qquad\qquad 7.17$$

Thus, in order to make inferences about $\hat{Y}_k$, we use the studentised statistic:

$$t^* = \frac{\hat{y}_k - E[\hat{Y}_k]}{s[\hat{Y}_k]} \sim t_{n-2}. \qquad\qquad 7.18$$

This sampling distribution allows us to compute confidence intervals for the predicted values. Figure 7.4 shows with dotted lines the 95% confidence interval for the cork-stopper Example 7.1. Notice how the confidence interval widens as we move away from $(\bar{x}, \bar{y})$.

Example 7.6

Q: The observed value of ART for PRT = 1612 is 882. Determine the 95% confidence interval of the predicted ART value using the ART(PRT) linear regression model derived in Example 7.1.

A: Using the MSE and s_{PRT} values as described in Example 7.2, and taking into account that $\overline{PRT} = 710.4$, we compute:

$$(x_k - \bar{x})^2 = (1612-710.4)^2 = 812882.6; \quad \sum(x_i - \bar{x})^2 = 19439351;$$

$$s[\hat{Y}_k] = MSE\left(\frac{1}{n} + \frac{(x_k - \bar{x})^2}{\sum(x_i - \bar{x})^2}\right) = 73.94.$$

Since $t_{148,0.975} = 1.976$ we obtain $\hat{y}_k \in [882 - 17, 882 + 17]$ with 95% confidence level. This corresponds to the 95% confidence interval depicted in Figure 7.4. $\square$

7.1.3.4 *Prediction of New Observations*

Imagine that we want to predict a new observation, that is an observation for new predictor values independent of the original n cases. The new observation on y is viewed as the *result of a new trial*. To stress this point we call it:

$$Y_{k(new)}.$$

If the regression parameters were perfectly known, one would easily find the confidence interval for the prediction of a new value. Since the parameters are usually unknown, we have to take into account two sources of variation:

- The location of $E[Y_{k(new)}]$, i.e., where one would locate, on average, the new observation. This was discussed in the previous section.

- The distribution of $Y_{k(new)}$, i.e., how to assess the expected deviation of the new observation from its average value. For the normal regression model, the variance of the prediction error for the new prediction can be obtained as follows, assuming that the new observation is independent of the original n cases:

$$V_{pred} = V[Y_{k(new)} - \hat{Y}_k] = \sigma^2 + V[\hat{Y}_k].$$

The sampling distribution of $Y_{k(\text{new})}$ for the normal regression model takes into account the above sources of variation, as follows:

$$t^* = \frac{y_{k(\text{new})} - \hat{y}_k}{s_{\text{pred}}} \sim t_{n-2},$$ 7.19

where s_{pred}^2 is the unbiased estimate of V_{pred} :

$$s_{\text{pred}}^2 = \text{MSE} + s^2[\hat{Y}_k] = \text{MSE}\left(1 + \frac{1}{n} + \frac{(x_k - \bar{x})^2}{\sum(x_i - \bar{x})^2}\right).$$ 7.20

Thus, the $1 - \alpha$ confidence interval for the new observation, $y_{k(\text{new})}$, is:

$$\hat{y}_k \pm t_{n-2,1-\alpha/2}\, s_{\text{pred}}.$$ 7.20a

Example 7.7

Q: Compute the estimate of the total area of defects of a cork stopper with a total perimeter of the defects of 800 pixels, using Example 7.1 regression model.

A: Using formula 7.20 with the MSE, s_{PRT}, $\overline{\text{PRT}}$ and $t_{148,0.975}$ values presented in Examples 7.2 and 7.6, as well as the coefficient values displayed in Figure 7.3, we compute:

$$\hat{y}_{k(\text{new})} \in [437.5 - 77.4, 437.5 + 77.4] \approx [360, 515], \text{ with 95\% confidence level.}$$

Figure 7.5 shows the table obtained with STATISTICA (using the Predict dependent variable button of the Multiple regression command), displaying the predicted value of variable ART for the predictor value PRT = 800, together with the 95% confidence interval. Notice that the 95% confidence interval is quite smaller than we have computed above, since STATISTICA is using formula 7.17 instead of formula 7.20, i.e., is considering the predictor value as making part of the dataset.

In R the same results are obtained with:

```
x <- c(800,0)    ## 0 is just a dummy value
z <- rbind(cork,x)
predict(r,z,interval=c("confidence"),type=c("response
"))
```

The second command line adds the predictor value to the data frame. The predict function lists all the predicted values with the 95% confidence interval. In this case we are interested in the last listed values, which agree with those of Figure 7.5.

□

Variable	B-Weight	Value	B-Weight * Value
PRT	0.546918	800.0000	437.5347
Intercept			-64.4902
Predicted			373.0445
-95.0%CL			366.5515
+95.0%CL			379.5375

Figure 7.5. Prediction of the new observation of ART for PRT = 800 (cork-stopper dataset), using STATISTICA.

7.1.4 ANOVA Tests

The analysis of variance tests are quite popular in regression analysis since they can be used to evaluate the regression model in several aspects. We start with a basic ANOVA test for evaluating the following hypotheses:

H_0: $\beta_1 = 0$;   7.21a

H_1: $\beta_1 \neq 0$.   7.21b

For this purpose, we break down the total deviation of the observations around the mean, given in equation 7.9, into two components:

$$SST = \sum (y_i - \bar{y})^2 = \sum (\hat{y}_i - \bar{y})^2 + \sum (y_i - \hat{y}_i)^2 .$$ 7.22

The first component represents the deviations of the fitted values around the mean, and is known as *regression sum of squares*, SSR:

$$SSR = \sum (\hat{y}_i - \bar{y})^2 .$$ 7.23

The second component was presented previously as the error sum of squares, SSE (see equation 7.8). It represents the deviations of the observations around the regression line. We, therefore, have:

$$SST = SSR + SSE.$$ 7.24

The number of degrees of freedom of SST is $n - 1$ and it breaks down into one degree of freedom for SSR and $n - 2$ for SSE. Thus, we define the *regression mean square*:

$$MSR = \frac{SSR}{1} = SSR .$$

The mean square error was already defined in section 7.1.2. In order to test the null hypothesis 7.21a, we then use the following ratio:

$$F^* = \frac{MSR}{MSE} \sim F_{1,n-2} .$$ 7.25

From the definitions of MSR and MSE we expect that large values of F support H_1 and values of F near 1 support H_0. Therefore, the appropriate test is an upper-tail F test.

Example 7.8

Q: Apply the ANOVA test to the regression Example 7.1 and discuss its results.

A: For the cork-stopper Example 7.1, the ANOVA array shown in Table 7.1 can be obtained using either SPSS or STATISTICA. The MATLAB and R functions listed in Commands 7.1 return the same F and p values as in Table 7.1. The complete ANOVA table can be obtained in R with the anova function (see Commands 7.2). Based on the observed significance of the test, we reject H_0, i.e., we conclude the existence of the linear component ($\beta_1 \neq 0$).　　　　　　　　□

Table 7.1. ANOVA test for the simple linear regression example of predicting ART based on the values of PRT (cork-stopper data).

	Sum of Squares	df	Mean Squares	F	p
SSR	5815203	1	5815203	3813.453	0.00
SSE	225688	148	1525		
SST	6040891				

Commands 7.2. SPSS, STATISTICA, MATLAB and R commands used to perform the ANOVA test in simple linear regression.

SPSS	`Analyze; Regression; Linear; Statistics; Model Fit`
STATISTICA	`Statistics; Multiple regression; Advanced; ANOVA`
MATLAB	`[b,bint,r,rint,stats]=regress(y,X,alpha)`
R	`anova(lm(y~X))`

■

There are also specific ANOVA tests for assessing whether a certain regression function adequately fits the data. We will now describe the ANOVA *test for lack of fit*, which assumes that the observations of Y are independent, normally distributed and with the same variance. The test takes into account what happens to repeat observations at one or more X levels, the so-called *replicates*.

Let us assume that there are c distinct values of X, replicates or not, each with n_j replicates:

$$n = \sum_{j=1}^{c} n_j .$$
7.26

The ith replicate for the j level is denoted y_{ij}. Let us first assume that the replicate variables Y_{ij} are not constrained by the regression line; in other words, they obey the so-called *full model*, with:

$$Y_{ij} = \mu_j + \varepsilon_{ij} , \text{ with i.i.d. } \varepsilon_{ij} \sim N_{0,\sigma} \Rightarrow E[Y_{ij}] = \mu_j.$$
7.27

The full model does not impose any restriction on the μ_j, whereas in the linear regression model the mean responses are linearly related.

To fit the full model to the data, we require:

$$\hat{\mu}_j = \bar{y}_j .$$
7.28

Thus, we have the following error sum of squares for the full model (F denotes the full model):

$$SSE(F) = \sum_{j}\sum_{i} (y_{ij} - \bar{y}_j)^2 , \text{ with } df_F = \sum_{j}(n_j - 1) = n - c .$$
7.29

In the above summations any X level with no replicates makes no contribution to SSE(F). SSE(F) is also called *pure error sum of squares* and denoted SSPE.

Under the linear regression assumption, the μ_j are linearly related with x_j. They correspond to a *reduced model*, with:

$$Y_{ij} = \beta_0 + \beta_1 x_j + \varepsilon_{ij} .$$

The error sum of squares for the reduced model is the usual error sum (R denotes the reduced model):

$$SSE(R) \equiv SSE, \text{ with } df_R = n - 2 .$$

The difference SSLF = SSE − SSPE is called the *lack of fit sum of squares* and has $(n - 2) - (n - c) = c - 2$ degrees of freedom. The decomposition SSE = SSPE + SSLF corresponds to:

$$\underbrace{y_{ij} - \hat{y}_{ij}}_{\text{error deviation}} = \underbrace{(y_{ij} - \bar{y}_j)}_{\text{pure error deviation}} + \underbrace{(\bar{y}_j - \hat{y}_{ij})}_{\text{lack of fit deviation}} .$$
7.30

If there is a lack of fit, SSLF will dominate SSE, compared with SSPE. Therefore, the ANOVA test, assuming that the null hypothesis is the lack of fit, is performed using the following statistic:

$$F^* = \frac{\text{SSLF}}{c-2} \div \frac{\text{SSPE}}{n-c} = \frac{\text{MSLF}}{\text{MSPE}} \sim F_{c-2,n-c}$$ 7.30a

The test for lack of fit is formalised as:

H_0: $E[Y] = \beta_0 + \beta_1 X$.   7.31a

H_1: $E[Y] \neq \beta_0 + \beta_1 X$.   7.31b

Let $F_{1-\alpha}$ represent the $1 - \alpha$ percentile of $F_{c-2,n-c}$. Then, if $F^* \leq F_{1-\alpha}$ we accept the null hypothesis, otherwise (significant test), we conclude for the lack of fit.

Repeat observations at only one or some levels of X are usually deemed sufficient for the test. When no replications are present in a data set, an approximate test for lack of fit can be conducted if there are some cases, at adjacent X levels, for which the mean responses are quite close to each other. These adjacent cases are grouped together and treated as pseudo-replicates.

Example 7.9

Q: Apply the ANOVA lack of fit test for the regression in Example 7.1 and discuss its results.

A: First, we know from the previous results of Table 7.1, that:

SSE = 225688; $df = n - 2 = 148$; MSE = 1525 .   7.32

In order to obtain the value of SSPE, using STATISTICA, we must run the General Linear Models command and in the Options tab of Quick Specs Dialog, we must check the Lack of fit box. After conducting a Whole Model R (whole model regression) with the variable ART depending on PRT, the following results are obtained:

SSPE = 65784.3; $df = n - c = 20$; MSPE = 3289.24 .   7.33

Notice, from the value of df, that there are 130 distinct values of PRT. Using the results 7.32 and 7.33, we are now able to compute:

SSLF = SSE − SSPE = 159903.7; $df = c - 2 = 128$; MSLF = 1249.25 .

Therefore, $F^* = \text{MSLF}/\text{MSPE} = 0.38$. For a 5% level of significance, we determine the 95% percentile of $F_{128,20}$, which is $F_{0.95} = 1.89$. Since $F^* < F_{0.95}$, we then conclude for the goodness of fit of the simple linear model.   □

7.2   Multiple Regression

7.2.1 General Linear Regression Model

Assuming the existence of $p - 1$ predictor variables, the general linear regression model is the direct generalisation of 7.1:

$$Y_i = \beta_0 + \beta_1 x_{i1} + \beta_2 x_{i2} + \ldots + \beta_{p-1} x_{i,p-1} + \varepsilon_i = \sum_{k=0}^{p-1} \beta_k x_{ik} + \varepsilon_i, \qquad 7.34$$

with $x_{i0} = 1$. In the following we always consider normal regression models with i.i.d. errors $\varepsilon_i \sim N_{0,\sigma}$.

Note that:

–   The general linear regression model implies that the observations are independent normal variables.

–   When the x_i represent values of different predictor variables the model is called a first-order model, in which there are no interaction effects between the predictor variables.

–   The general linear regression model encompasses also qualitative predictors. For example:

$$Y_i = \beta_0 + \beta_1 x_{i1} + \beta_2 x_{i2} + \varepsilon_i. \qquad 7.35$$

$x_{i1} =$ patient's weight

$$x_{i2} = \begin{cases} 1 & \text{if patient female} \\ 0 & \text{if patient male} \end{cases}$$

Patient is male: $Y_i = \beta_0 + \beta_1 x_{i1} + \varepsilon_i.$

Patient is female: $Y_i = (\beta_0 + \beta_2) + \beta_1 x_{i1} + \varepsilon_i.$

Multiple linear regression can be performed with SPSS, STATISTICA, MATLAB and R with the same commands and functions listed in Commands 7.1.

7.2.2 General Linear Regression in Matrix Terms

In order to understand the computations performed to fit the general linear regression model to the data, it is convenient to study the normal equations 7.3 in matrix form.

We start by expressing the general linear model (generalisation of 7.1) in matrix terms as:

$$\mathbf{y} = \mathbf{X}\boldsymbol{\beta} + \boldsymbol{\varepsilon}, \qquad 7.36$$

where:

- $\mathbf{y}$ is an $n \times 1$ matrix (i.e., a column vector) of the predictions;
- $\mathbf{X}$ is an $n \times p$ matrix of the $p - 1$ predictor values plus a bias (of value 1) for the n predictor levels;
- $\boldsymbol{\beta}$ is a $p \times 1$ matrix of the coefficients;
- $\boldsymbol{\varepsilon}$ is an $n \times 1$ matrix of the errors.

For instance, the multiple regression expressed by formula 7.35 is represented as follows in matrix form, assuming $n = 3$ predictor levels:

$$
\begin{bmatrix} y_1 \\ y_2 \\ y_3 \end{bmatrix} = \begin{bmatrix} 1 & x_{11} & x_{12} \\ 1 & x_{21} & x_{22} \\ 1 & x_{31} & x_{32} \end{bmatrix} \begin{bmatrix} \beta_0 \\ \beta_1 \\ \beta_2 \end{bmatrix} + \begin{bmatrix} \varepsilon_1 \\ \varepsilon_2 \\ \varepsilon_3 \end{bmatrix}.
$$

We assume, as for the simple regression, that the errors are i.i.d. with zero mean and equal variance:

$$ E[\boldsymbol{\varepsilon}] = 0; \qquad V[\boldsymbol{\varepsilon}] = \sigma^2 \mathbf{I} . $$

Thus: $E[\mathbf{y}] = \mathbf{X}\boldsymbol{\beta}$.

The least square estimation of the coefficients starts by computing the total error:

$$ E = \sum \varepsilon_i^2 = \boldsymbol{\varepsilon}'\boldsymbol{\varepsilon} = (\mathbf{y} - \mathbf{Xb})'(\mathbf{y} - \mathbf{Xb}) = \mathbf{y}'\mathbf{y} - (\mathbf{X}'\mathbf{y})'\mathbf{b} - \mathbf{b}'\mathbf{X}\,\mathbf{y} + \mathbf{b}'\mathbf{X}'\mathbf{Xb} . \quad 7.37 $$

Next, the error is minimised by setting to zero the derivatives in order to the coefficients, obtaining the normal equations in matrix terms:

$$ \frac{\partial E}{\partial b_i} = 0 \quad \Rightarrow \quad -2\mathbf{X}'\mathbf{y} + 2\mathbf{X}'\mathbf{Xb} = 0 \quad \Rightarrow \quad \mathbf{X}'\mathbf{Xb} = \mathbf{X}'\mathbf{y} . $$

Hence:

$$ \mathbf{b} = (\mathbf{X}'\mathbf{X})^{-1}\mathbf{X}'\mathbf{y} = \mathbf{X}^{*}\mathbf{y} , \qquad\qquad 7.38 $$

where $\mathbf{X}^{*}$ is the so-called *pseudo-inverse* matrix of $\mathbf{X}$.

The fitted values can now be computed as:

$$ \hat{\mathbf{y}} = \mathbf{Xb} . $$

Note that this formula, using the predictors and the estimated coefficients, can also be expressed in terms of the predictors and the observations, substituting the vector of the coefficients given in 7.38.

Let us consider the normal equations:

$$ \mathbf{b} = (\mathbf{X}'\mathbf{X})^{-1}\mathbf{X}'\mathbf{Y} . $$

For the *standardised model* (i.e., using standardised variables) we have:

$$\mathbf{X'X} = \mathbf{r_{xx}} = \begin{bmatrix} 1 & r_{12} & \cdots & r_{1,p-1} \\ r_{21} & 1 & \cdots & r_{2,p-1} \\ \cdots & \cdots & \cdots & \cdots \\ r_{p-1,1} & r_{p-1,2} & \cdots & 1 \end{bmatrix};$$

7.39

$$\mathbf{X'Y} = \mathbf{r_{yx}} = \begin{bmatrix} r_{y1} \\ r_{y2} \\ \vdots \\ r_{y,p-1} \end{bmatrix}.$$

7.40

Hence:

$$\mathbf{b} = \begin{bmatrix} b_1' \\ b_2' \\ \vdots \\ b_{p-1}' \end{bmatrix} = \mathbf{r_{xx}^{-1} r_{yx}},$$

7.41

where **b** is the vector containing the point estimates of the beta coefficients (compare with formula 7.7 in section 7.1.2), $\mathbf{r_{xx}}$ is the symmetric matrix of the *predictor correlations* (see A.8.2) and $\mathbf{r_{yx}}$ is the vector of the correlations between Y and each of the predictor variables.

Example 7.10

Q: Consider the following six cases of the Foetal Weight dataset:

Variable	Case #1	Case #2	Case #3	Case #4	Case #5	Case #6
CP	30.1	31.1	32.4	32	32.4	35.9
AP	28.8	31.3	33.1	34.4	32.8	39.3
FW	2045	2505	3000	3520	4000	4515

Determine the beta coefficients of the linear regression of FW (foetal weight in grams) depending on CP (cephalic perimeter in mm) and AP (abdominal perimeter in mm) and performing the computations expressed by formula 7.41.

A: We can use MATLAB function corrcoef or appropriate SPSS, STATISTICA and R commands to compute the correlation coefficients. Using MATLAB and denoting by fw the matrix containing the above data with cases along the rows and variables along the columns, we obtain:

```
» c=corrcoef(fw(:,:))
» c =
```

```
1.0000       0.9692       0.8840
0.9692       1.0000       0.8880
0.8840       0.8880       1.0000
```

We now apply formula 7.41 as follows:

```
» rxx = c(1:2,1:2); ryx = c(1:2,3);
» b = inv(rxx)*ryx
b =
    0.3847
    0.5151
```

These are also the values obtained with SPSS, STATISTICA and R. It is interesting to note that the beta coefficients for the 414 cases of the Foetal Weight dataset are 0.3 and 0.64 respectively.   □

Example 7.11

Q: Determine the multiple linear regression coefficients of the previous example.

A: Since the beta coefficients are the regression coefficients of the standardised model, we have:

$$\frac{FW - \overline{FW}}{s_{FW}} = 0.3847 \frac{CP - \overline{CP}}{s_{CP}} + 0.5151 \frac{AP - \overline{AP}}{s_{AP}}.$$

Thus:

$$b_0 = \overline{FW} + s_{FW}\left(-0.3847\frac{\overline{CP}}{s_{CP}} - 0.5151\frac{\overline{AP}}{s_{AP}}\right) = -7125.7.$$

$$b_1 = 0.3847\frac{s_{FW}}{s_{CP}} = 181.44.$$

$$b_2 = 0.5151\frac{s_{FW}}{s_{AP}} = 135.99.$$

These computations can be easily carried out in MATLAB or R. For instance, in MATLAB b_2 is computed as b2=0.5151*std(fw(:,3))/std(fw(:,2)). The same values can of course be obtained with the commands listed in Commands 7.1   □

7.2.3 Multiple Correlation

Let us go back to the R-square statistic described in section 7.1, which represented the square of the correlation between the independent and the dependent variables. It also happens that it represents the square of the correlation between the dependent and predicted variable, i.e., the square of:

$$r_{Y\hat{Y}} = \frac{\sum (y_i - \bar{y})(\hat{y}_i - \bar{\hat{y}})}{\sqrt{\sum (y_i - \bar{y})^2 \sum (\hat{y}_i - \bar{\hat{y}})^2}} \qquad 7.42$$

In multiple regression this quantity represents the correlation between the dependent variable and the predicted variable *explained by all the predictors*; it is therefore appropriately called *multiple correlation* coefficient. For $p-1$ predictors we will denote this quantity as $r_{Y|X_1,...,X_{p-1}}$.

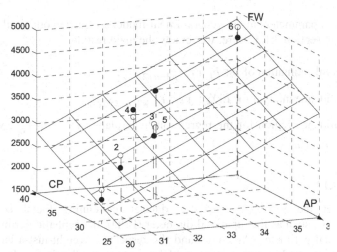

Figure 7.6. The regression linear model (plane) describing FW as a function of (CP,AP) using the dataset of Example 7.10. The observations are the solid balls. The predicted values are the open balls (lying on the plane). The multiple correlation corresponds to the correlation between the observed and predicted values.

Example 7.12

Q: Compute the multiple correlation coefficient for the Example 7.10 regression, using formula 7.42.

A: In MATLAB the computations can be carried out with the matrix fw of Example 7.10 as follows:

```
» fw = [fw(:,1) ones(1,6) fw(:,2:3)];
» [b,bint,r,rint,stats] = regress(fw(:,1),fw(:,2:4));
» y = fw(:,1); ystar = y-r;
» corrcoef(y,ytar)
ans =
      1.0000      0.8930
      0.8930      1.0000
```

The first line includes the independent terms in the fw matrix in order to compute a linear regression model with intercept. The third line computes the predicted values in the ystar vector. The square of the multiple correlation coefficient, $r_{FW|CP,AP} = 0.893$ computed in the fourth line coincides with the value of R-square computed in the second line (r) as it should be. Figure 7.6 illustrates this multiple correlation situation.

□

7.2.4 Inferences on Regression Parameters

Inferences on parameters in the general linear model are carried out similarly to the inferences in section 7.1.3. Here, we review the main results:

- Interval estimation of β_k: $b_k \pm t_{n-p,1-\alpha/2} \, s_{b_k}$.

- Confidence interval for $E[Y_k]$: $\hat{y}_k \pm t_{n-p,1-\alpha/2} s_{\hat{y}_k}$.

- Confidence region for the regression hyperplane: $\hat{y}_k \pm W s_{\hat{y}_k}$, with $W^2 = pF_{p,n-p,1-\alpha}$.

Example 7.13

Q: Consider the Foetal Weight dataset, containing foetal echographic measurements, such as the biparietal diameter (BPD), the cephalic perimeter (CP), the abdominal perimeter (AP), etc., and the respective weight-just-after-delivery, FW. Determine the linear regression model needed to predict the newborn weight, FW, using the three variables BPD, CP and AP. Discuss the results.

A: Having filled in the three variables BPD, CP and AP as predictor or independent variables and the variable FW as the dependent variable, one can obtain with STATISTICA the result summary table shown in Figure 7.7.

The standardised beta coefficients have the same meaning as in 7.1.2. Since these reflect the contribution of standardised variables, they are useful for comparing the relative contribution of each variable. In this case, variable AP has the highest contribution and variable CP the lowest. Notice the high coefficient of multiple determination, R^2 and that in the last column of the table, all t tests are found significant. Similar results are obtained with the commands listed in Commands 7.1 for SPSS, MATLAB and R.

Figure 7.8 shows line plots of the true (observed) values and predicted values of the foetal weight using the multiple linear regression model. The horizontal axis of these line plots is the case number. The true foetal weights were previously sorted by increasing order. Figure 7.9 shows the scatter plot of the observed and predicted values obtained with the Multiple Regression command of STATISTICA.

□

N=414	Beta	Std.Err. of Beta	B	Std.Err. of B	t(410)	p-level
R= .88655938 R²= .78598754 Adjusted R²= .78442160						
F(3,410)=501.93 p<0.0000 Std.Error of estimate: 291.84						
Intercept			-4765.66	261.9039	-18.1962	0.000000
BPD	0.262660	0.040504	292.28	45.0721	6.4848	0.000000
CP	0.105382	0.043764	36.00	14.9485	2.4079	0.016483
AP	0.609095	0.031987	124.72	6.5499	19.0421	0.000000

Figure 7.7. Estimation results obtained with STATISTICA of the trivariate linear regression of the foetal weight data.

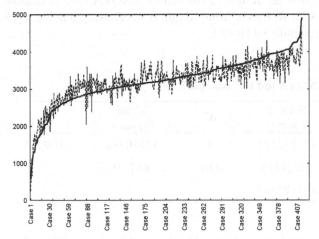

Figure 7.8. Plot obtained with STATISTICA of the predicted (dotted line) and observed (solid line) foetal weights with a trivariate (BPD, CP, AP) linear regression model.

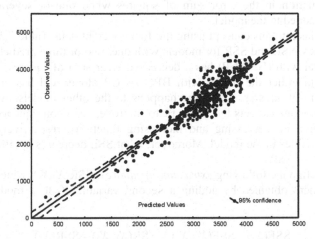

Figure 7.9. Plot obtained with STATISTICA of the observed versus predicted foetal weight values with fitted line and 95% confidence interval.

7.2.5 ANOVA and Extra Sums of Squares

The simple ANOVA test presented in 7.1.4, corresponding to the decomposition of the total sum of squares as expressed by formula 7.24, can be generalised in a straightforward way to the multiple regression model.

Example 7.14

Q: Apply the simple ANOVA test to the foetal weight regression in Example 7.13.

A: Table 7.2 lists the results of the simple ANOVA test, obtainable with SPSS STATISTICA, or R, for the foetal weight data, showing that the regression model is statistically significant ($p \approx 0$).

□

Table 7.2. ANOVA test for Example 7.13.

	Sum of Squares	df	Mean Squares	F	p
SSR	128252147	3	42750716	501.9254	0.00
SSE	34921110	410	85173		
SST	163173257				

It is also possible to apply the ANOVA test for lack of fit in the same way as was done in 7.1.4. However, when there are several predictor values playing their influence in the regression model, it is useful to assess their contribution by means of the so-called *extra sums of squares*. An extra sum of squares measures the marginal reduction in the error sum of squares when one or several predictor variables are added to the model.

We now illustrate this concept using the foetal weight data. Table 7.3 shows the regression lines, SSE and SSR for models with one, two or three predictors. Notice how the model with (BPD,CP) has a decreased error sum of squares, SSE, when compared with either the model with BPD or CP alone, and has an increased regression sum of squares. The same happens to the other models. As one adds more predictors one expects the linear fit to improve. As a consequence, SSE and SSR are monotonic decreasing and increasing functions, respectively, with the number of variables in the model. Moreover, what SSE decreases is reflected by an equal increase of SSR.

We now define the following *extra sum of squares*, $SSR(X_2|X_1)$, which measures the improvement obtained by adding a second variable X_2 to a model that has already X_1:

$$SSR(X_2 \mid X_1) = SSE(X_1) - SSE(X_1, X_2) = SSR(X_1, X_2) - SSR(X_1).$$

Table 7.3. Computed models with SSE, SSR and respective degrees of freedom for the foetal weight data (sums of squares divided by 10^6).

Abstract Model	Computed model	SSE	df	SSR	df
$Y = g(X_1)$	FW(BPD) = $-4229.1 + 813.3$ BPD	76.0	412	87.1	1
$Y = g(X_2)$	FW(CP) = $-5096.2 + 253.8$ CP	73.1	412	90.1	1
$Y = g(X_3)$	FW(AP) = $-2518.5 + 173.6$ AP	46.2	412	117.1	1
$Y = g(X_1, X_2)$	FW(BPD,CP) = $-5464.7 + 412.0$ BPD + 149.9 CP	65.8	411	97.4	2
$Y = g(X_1, X_3)$	FW(BPD,AP) = $-4481.1 + 367.2$ BPD + 131.0 AP	35.4	411	127.8	2
$Y = g(X_2, X_3)$	FW(CP,AP) = $-4476.2 + 102.9$ CP + 130.7 AP	38.5	411	124.7	2
$Y = g(X_1, X_2, X_3)$	FW(BPD,CP,AP) = $-4765.7 + 292.3$ BPD + 36.0 CP + 127.7 AP	34.9	410	128.3	3

$X_1 \equiv$ BPD; $X_2 \equiv$ CP; $X_3 \equiv$ AP

For the data of Table 7.3 we have: SSR(CP | BPD) – SSE(BPD) – SSE(BPD, CP) = 76 – 65.8 = 10.2, which is practically the same as SSR(BPD, CP) – SSR(BPD) = 97.4 – 87.1 = 10.3 (difference only due to numerical roundings).

Similarly, one can define:

$$\text{SSR}(X_3 \mid X_1, X_2) = \text{SSE}(X_1, X_2) - \text{SSE}(X_1, X_2, X_3)$$
$$= \text{SSR}(X_1, X_2, X_3) - \text{SSR}(X_1, X_2).$$
$$\text{SSR}(X_2, X_3 \mid X_1) = \text{SSE}(X_1) - \text{SSE}(X_1, X_2, X_3) = \text{SSR}(X_1, X_2, X_3) - \text{SSR}(X_1).$$

The first extra sum of squares, SSR($X_3 \mid X_1, X_2$), represents the improvement obtained when adding a third variable, X_3, to a model that has already two variables, X_1 and X_2. The second extra sum of squares, SSR($X_2, X_3 \mid X_1$), represents the improvement obtained when adding two variables, X_2 and X_3, to a model that has only one variable, X_1.

The extra sums of squares are especially useful for performing tests on the regression coefficients and for detecting multicollinearity situations, as explained in the following sections.

With the extra sums of squares it is also possible to easily compute the so-called *partial correlations*, measuring the degree of linear relationship between two variables after including other variables in a regression model. Let us illustrate this topic with the foetal weight data. Imagine that the only predictors were BPD, CP and AP as in Table 7.3, and that we wanted to build a regression model of FW by successively entering in the model the predictor which is most correlated with the predicted variable. In the beginning there are no variables in the model and we choose the predictor with higher correlation with the independent variable FW. Looking at Table 7.4 we see that, based on this rule, AP enters the model. Now we must ask which of the remaining variables, BPD or CP, has a higher correlation with the predicted variable of the model that has already AP. The answer to this question amounts to computing the partial correlation of a candidate variable, say X_2, with the predicted variable of a model that has already X_1, $r_{Y,X_2|X_1}$. The respective formula is:

$$r_{Y,X_2|X_1}^2 = \frac{SSR(X_2 \mid X_1)}{SSE(X_1)} = \frac{SSE(X_1) - SSE(X_1, X_2)}{SSE(X_1)}$$

For the foetal weight dataset the computations with the values in Table 7.3 are as follows:

$$r_{FW,BPD|AP}^2 = \frac{SSR(BPD \mid AP)}{SSE(AP)} = 0.305 ,$$

$$r_{FW,CP|AP}^2 = \frac{SSR(CP \mid AP)}{SSE(AP)} = 0.167 ,$$

resulting in the partial correlation values listed in Table 7.4. We therefore select BPD as the next predictor to enter the model. This process could go on had we more predictors. For instance, the partial correlation of the remaining variable CP with the predicted variable of a model that has already AP and BPD is computed as:

$$r_{FW,CP|BPD,AP}^2 = \frac{SSR(CP \mid BPD, AP)}{SSE(BPD, AP)} = 0.014 .$$

Further details on the meaning and statistical significance testing of partial correlations can be found in (Kleinbaum DG *et al.*, 1988).

Table 7.4. Correlations and partial correlations for the foetal weight dataset.

Variables in the model	Variables to enter the model	Correlation	Sample Value
(None)	BPD	$r_{FW,BPD}$	0.731
(None)	CP	$r_{FW,CP}$	0.743
(None)	AP	$r_{FW,AP}$	0.847
AP	BPD	$r_{FW,BPD|AP}$	0.552
AP	CP	$r_{FW,CP|AP}$	0.408
AP, BPD	CP	$r_{FW,CP|BPD,AP}$	0.119

7.2.5.1   Tests for Regression Coefficients

We will only present the test for a single coefficient, formalised as:

H_0: $\beta_k = 0$;
H_1: $\beta_k \neq 0$.

The statistic appropriate for this test, is:

$$t^* = \frac{b_k}{s_{b_k}} \sim t_{n-p}.$$

7.43

We may also use, as in section 7.1.4, the ANOVA test approach. As an illustration, let us consider a model with three variables, X_1, X_2, X_3, and, furthermore, let us assume that we want to test whether "H_0: $\beta_3 = 0$" can be accepted or rejected. For this purpose, we first compute the error sum of squares for the *full model*:

$$SSE(F) = SSE(X_1, X_2, X_3), \quad \text{with} \quad df_F = n - 4.$$

The *reduced model*, corresponding to H_0, has the following error sum of squares:

$$SSE(R) = SSE(X_1, X_2), \quad \text{with} \quad df_R = n - 3.$$

The ANOVA test assessing whether or not any benefit is derived from adding X_3 to the model, is then based on the computation of:

$$F^* = \frac{SSE(R) - SSE(F)}{df_R - df_F} \div \frac{SSE(F)}{df_F} = \frac{SSR(X_3 \mid X_1, X_2)}{1} \div \frac{SSE(X_1, X_2, X_3)}{n-4}$$

$$= \frac{MSR(X_3 \mid X_1, X_2)}{MSE(X_1, X_2, X_3)}$$

In general, we have:

$$F^* = \frac{MSR(X_k \mid X_1 \ldots X_{k-1} X_{k+1} \ldots X_{p-1})}{MSE} \sim F_{1, n-p}.$$

7.44

The F test using this sampling distribution is equivalent to the t test expressed by 7.43. This F test is known as *partial F test*.

7.2.5.2   Multicollinearity and its Effects

If the predictor variables are uncorrelated, the regression coefficients remain constant, irrespective of whether or not another predictor variable is added to the model. Similarly, the same applies for the sum of squares. For instance, for a model with two uncorrelated predictor variables, the following should hold:

$$SSR(X_1 \mid X_2) = SSE(X_2) - SSE(X_1, X_2) = SSR(X_1); \qquad\qquad 7.45a$$
$$SSR(X_2 \mid X_1) = SSE(X_1) - SSE(X_1, X_2) = SSR(X_2). \qquad\qquad 7.45b$$

On the other hand, if there is a perfect correlation between X_1 and X_2 – in other words, X_1 and X_2 are collinear – we would be able to determine an infinite number of regression solutions (planes) intersecting at the straight line relating X_1 and X_2. Multicollinearity leads to imprecise determination coefficients, imprecise fitted values and imprecise tests on the regression coefficients.

In practice, when predictor variables are correlated, the marginal contribution of any predictor variable in reducing the error sum of squares varies, depending on which variables are already in the regression model.

Example 7.15

Q: Consider the trivariate regression of the foetal weight in Example 7.13. Use formulas 7.45 to assess the collinearity of CP given BPD and of AP given BPD and BPD, CP.

A: Applying formulas 7.45 to the results displayed in Table 7.3, we obtain:

$$SSR(CP) = 90 \times 10^6 .$$
$$SSR(CP \mid BPD) = SSE(BPD) - SSE(CP, BPD) = 76 \times 10^6 - 66 \times 10^6 = 10 \times 10^6 .$$

We see that SSR(CP|BPD) is small compared with SSR(CP), which is a symptom that BPD and CP are highly correlated. Thus, when BPD is already in the model, the marginal contribution of CP in reducing the error sum of squares is small because BPD contains much of the same information as CP.

In the same way, we compute:

$$SSR(AP) = 46 \times 10^6 .$$
$$SSR(AP \mid BPD) = SSE(BPD) - SSE(BPD, AP) = 41 \times 10^6 .$$
$$SSR(AP \mid BPD, CP) = SSE(BPD, CP) - SSE(BPD, CP, AP) = 31 \times 10^6 .$$

We see that AP seems to bring a definite contribution to the regression model by reducing the error sum of squares. □

7.2.6 Polynomial Regression and Other Models

Polynomial regression models may contain squared, cross-terms and higher order terms of the predictor variables. These models can be viewed as a generalisation of the multivariate linear model.

As an example, consider the following second order model:

$$Y_i = \beta_0 + \beta_1 x_i + \beta_2 x_i^2 + \varepsilon_i . \qquad\qquad 7.46$$

The Y_i can also be linearly modelled as:

$$Y_i = \beta_0 + \beta_1 u_{i1} + \beta_2 u_{i2} + \varepsilon_i \quad \text{with} \quad u_{i1} = x_i; \; u_{i2} = x_i^2.$$

As a matter of fact, many complex dependency models can be transformed into the general linear model after suitable transformation of the variables. The general linear model encompasses also the *interaction effects*, as in the following example:

$$Y_i = \beta_0 + \beta_1 x_{i1} + \beta_2 x_{i2} + \beta_3 x_{i1} x_{i2} + \varepsilon_i, \qquad 7.47$$

which can be transformed into the linear model, using the extra variable $x_{i3} = x_{i1} x_{i2}$ for the cross-term $x_{i1} x_{i2}$.

Frequently, when dealing with polynomial models, the predictor variables are previously centred, replacing x_i by $x_i - \bar{x}$. The reason is that, for instance, X and X^2 will often be highly correlated. Using centred variables reduces multi-collinearity and tends to avoid computational difficulties.

Note that in all the previous examples, the model is linear in the parameters β_k. When this condition is not satisfied, we are dealing with a *non-linear model*, as in the following example of the so-called *exponential regression*:

$$Y_i = \beta_0 \exp(\beta_1 x_i) + \varepsilon_i. \qquad 7.48$$

Unlike linear models, it is not generally possible to find analytical expressions for the estimates of the coefficients of non-linear models, similar to the normal equations 7.3. These have to be found using standard numerical search procedures. The statistical analysis of these models is also a lot more complex. For instance, if we linearise the model 7.48 using a logarithmic transformation, the errors will no longer be normal and with equal variance.

Commands 7.3. SPSS, STATISTICA, MATLAB and R commands used to perform polynomial and non-linear regression.

SPSS	`Analyze; Regression; Curve Estimation` `Analyze; Regression; Nonlinear`

STATISTICA	`Statistics; Advanced Linear/Nonlinear`
	`Models; General Linear Models; Polynomial`
	`Regression`
	`Statistics; Advanced Linear/Nonlinear`
	`Models; Non-Linear Estimation`

MATLAB	`[p,S] = polyfit(X,y,n)`
	`[y,delta] = polyconf(p,X,S)`
	`[beta,r,J]= nlinfit(X,y,FUN,beta0)`

R	`lm(formula)	glm(formula)` `nls(formula, start, algorithm, trace)`

The MATLAB `polyfit` function computes a polynomial fit of degree n using the predictor matrix X and the observed data vector y. The function returns a vector p with the polynomial coefficients and a matrix S to be used with the `polyconf` function producing confidence intervals y ± delta at alpha confidence level (95% if alpha is omitted). The `nlinfit` returns the coefficients beta and residuals r of a nonlinear fit $y = f(X, beta)$, whose formula is specified by a string FUN and whose initial coefficient estimates are beta0.

The R `glm` function operates much in the same way as the `lm` function, with the support of extra parameters. The parameter `formula` is used to express a polynomial dependency of the independent variable with respect to the predictors, such as `y ~ x + I(x^2)`, where the function I inhibits the interpretation of "^" as a formula operator, so it is used as an arithmetical operator. The `nls` function for nonlinear regression is used with a `start` vector of initial estimates, an `algorithm` parameter specifying the algorithm to use and a `trace` logical value indicating whether a trace of the iteration progress should be printed. An example is: `nls(y~1/(1+exp((a-log(x))/b)), start=list(a=0, b=1), alg="plinear", trace=TRUE)`.   ∎

Example 7.16

Q: Consider the `Stock Exchange` dataset (see Appendix E). Design and evaluate a second order polynomial model, without interaction effects, for the SONAE share values depending on the predictors EURIBOR and USD.

A: Table 7.5 shows the estimated parameters of this second order model, along with the results of *t* tests. From these results, we conclude that all coefficients have an important contribution to the designed model. The simple ANOVA test gives also significant results. However, Figure 7.10 suggests that there is some trend of the residuals as a function of the observed values. This is a symptom that some lack of fit may be present. In order to investigate this issue we now perform the ANOVA test for lack of fit. We may use STATISTICA for this purpose, in the same way as in the example described in section 7.1.4.

Table 7.5. Results obtained with STATISTICA for a second order model, with predictors EURIBOR and USD, in the regression of SONAE share values (`Stock Exchange` dataset).

Effect	SONAE Param.	SONAE Std.Err	t	p	−95% Cnf.Lmt	+95% Cnf.Lmt
Intercept	−283530	24151	−11.7	0.00	−331053	−236008
EURIBOR	13938	1056	13.2	0.00	11860	16015
$EURIBOR^2$	−1767	139.8	−12.6	0.00	−2042	−1491
USD	560661	49041	11.4	0.00	464164	657159
USD^2	−294445	24411	−12.1	0.00	−342479	−246412

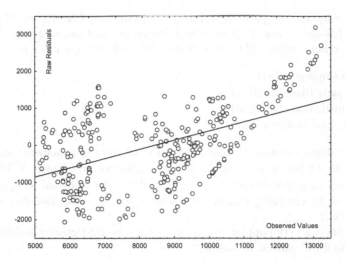

Figure 7.10. Residuals versus observed values in the Stock Exchange example.

First, note that there are $p - 1 = 4$ predictor variables in the model; therefore, $p = 5$. Secondly, in order to have enough replicates for STATISTICA to be able to compute the pure error, we use two new variables derived from EURIBOR and USD by rounding them to two and three significant digits, respectively. We then obtain (removing a 10^3 factor):

SSE = 345062; $df = n - p = 308$;   MSE = 1120.
SSPE = 87970; $df = n - c = 208$;   MSPE = 423.

From these results we compute:

SSLF = SSE − SSPE = 257092; $df = c - p = 100$;   MSLF = 2571.
$F^* = $ MSLF/MSPE − 6.1 .

The 95% percentile of $F_{100,208}$ is 1.3. Since $F^* > 1.3$, we then conclude for the lack of fit of the model.   ▯

7.3   Building and Evaluating the Regression Model

7.3.1 Building the Model

When there are several variables that can be used as candidates for predictor variables in a regression model, it would be fastidious having to try every possible combination of variables. In such situations, one needs a search procedure operating in the variable space in order to build up the regression model much in the same way as we performed feature selection in Chapter 6. The search procedure has also to use an appropriate criterion for predictor selection. There are many such criteria published in the literature. We indicate here just a few:

- SSE (minimisation)
- R square (maximisation)
- t statistic (maximisation)
- F statistic (maximisation)

When building the model, these criteria can be used in a stepwise manner the same way as we performed sequential feature selection in Chapter 6. That is, by either adding consecutive variables to the model – the so-called *forward search method* –, or by removing variables from an initial set – the so-called *backward search method*.

For instance, a very popular method is to use forward stepwise building up the model using the F statistic, as follows:

1. Initially enters the variable, say X_1, that has maximum $F_k = $ MSR(X_k)/MSE(X_k), which must be above a certain specified level.

2. Next is added the variable with maximum $F_k = $ MSR($X_k \mid X_1$) / MSE(X_k, X_1) and above a certain specified level.

3. The Step 2 procedure goes on until no variable has a partial F above the specified level.

Example 7.17

Q: Apply the forward stepwise procedure to the foetal weight data (see Example 7.13), using as initial predictor sets {BPD, CP, AP} and {MW, MH, BPD, CP, AP, FL}.

A: Figure 7.11 shows the evolution of the model using the forward stepwise method to {BPD, CP, AP}. The first variable to be included, with higher F, is the variable AP. The next variables that are included have a decreasing F contribution but still higher than the specified level of "F to Enter", equal to 1. These results confirm the findings on partial correlation coefficients discussed in section 7.2.5 (Table 7.4).

Variable	Step +in/-out	Multiple R	Multiple R-square	R-square change	F - to entr/rem	p-level	Variabls included
AP	1	0.846657	0.716827	0.716827	1042.943	0.000000	1
BPD	2	0.884851	0.782961	0.066134	125.235	0.000000	2
CP	3	0.886559	0.785988	0.003027	5.798	0.016483	3

Figure 7.11. Forward stepwise regression (obtained with STATISTICA) for the foetal weight example, using {BPD, CP, AP} as initial predictor set.

Let us now assume that the initial set of predictors is {MW, MH, BPD, CP, AP, FL}. Figure 7.12 shows the evolution of the model at each step. Notice that one of the variables, MH, was not included in the model, and the last one, CP, has a non-significant F test ($p > 0.05$), and therefore, should also be excluded.

□

Variable	Step +in/-out	Multiple R	Multiple R-square	R-square change	F - to entr/rem	p-level	Variabls included
AP	1	0.846657	0.716827	0.716827	1042.943	0.000000	1
BPD	2	0.884851	0.782961	0.066134	125.235	0.000000	2
FL	3	0.897886	0.806198	0.023237	49.160	0.000000	3
MW	4	0.902938	0.815298	0.009099	20.149	0.000009	4
CP	5	0.903231	0.815827	0.000529	1.172	0.279681	5

Figure 7.12. Forward stepwise regression (obtained with STATISTICA) for the foetal weight example, using {MW, MH, BPD, CP, AP, FL} as initial predictor set.

Commands 7.4. SPSS, STATISTICA, MATLAB and R commands used to perform stepwise linear regression.

SPSS	`Analyze; Regression; Linear; Method Forward`
STATISTICA	`Statistics; Multiple Regression; Advanced; Forward Stepwise`
MATLAB	`stepwise(X,y)`
R	`step(object, direction = c("both", "backward", "forward"), trace)`

With SPSS and STATISTICA the user can specify the level of F in order to enter or remove variables.

The MATLAB `stepwise` function fits a regression model of y depending on X, displaying figure windows for interactively controlling the stepwise addition and removal of model terms.

The R `step` function allows the stepwise selection of a model, represented by the parameter `object` and generated by R `lm` or `glm` functions. The selection is based on a more sophisticated criterion than the ANOVA F. The parameter `direction` specifies the direction (forward, backward or a combination of both) of the stepwise search. The parameter `trace` when left with its default value will force `step` to generate information during its running.

■

7.3.2 Evaluating the Model

7.3.2.1   Identifying Outliers

Outliers correspond to cases exhibiting a strong deviation from the fitted regression curve, which can have a harmful influence in the process of fitting the model to the data. Identification of outliers, for their eventual removal from the dataset, is usually carried out using the so-called *semistudentised residuals* (or *standard residuals*), defined as:

$$e_i^* = \frac{e_i - \bar{e}}{\sqrt{MSE}} = \frac{e_i}{\sqrt{MSE}}.$$ 7.49

Cases whose magnitude of the semistudentised residuals exceeds a certain threshold (usually 2), are considered outliers and are candidates for removal.

Example 7.18

Q: Detect the outliers of the first model designed in Example 7.13, using semistudentised residuals.

A: Figure 7.13 shows the partial listing, obtained with STATISTICA, of the 18 outliers for the foetal weight regression with the three predictors AP, BPD and CP. Notice that the magnitudes of the Standard Residual column are all above 2.

□

	Standard Residuals					Standard Residual: FW (fetalweight.STA) Outliers				
Case	-4.	-3.	±2.	3.	4.	Residual	Standard Residual	Mahalanobis Distance	Deleted Residual	Cook's Distance
62	.	.	*.	.	.	-628.427	-2.15329	0.24594	-630.325	0.003511
74	.	.	*	.	.	-625.597	-2.14360	0.15088	-627.342	0.003212
86	.	.	.	*.	.	807.796	2.76790	18.59577	848.028	0.100142
87	.	.	* .	.	.	-722.758	-2.47652	2.68339	-729.258	0.013913
139	.	.	*	.	.	-592.949	-2.03173	1.61768	-596.728	0.006618
174	.	.	*	.	.	-610.306	-2.09120	3.65717	-617.263	0.012604
325	.	.	. *	.	.	683.513	2.34204	0.56303	686.105	0.005221
329	.	.	. *	.	.	699.999	2.39853	0.90118	703.232	0.006673
358	.	.	. *	.	.	652.537	2.23590	0.88541	655.526	0.005751
359	.	.	. *	.	.	639.499	2.19123	8.33510	654.284	0.028394
371	.	.	. *	.	.	679.481	2.32823	0.04968	681.208	0.003454
377	.	.	. *	.	.	648.622	2.22249	7.17145	661.711	0.025421

Figure 7.13. Outlier list obtained with STATISTICA for the foetal weight example.

There are other ways to detect outliers, such as:

- Use of *deleted residuals*: the residual is computed for the respective case, assuming that it was not included in the regression analysis. If the deleted residual differs greatly from the original residual (i.e., with the case included) then the case is, possibly, an outlier. Note in Figure 7.13 how case 86 has a deleted residual that exhibits a large difference from the original residual, when compared with similar differences for cases with smaller standard residual.

- *Cook's distance*: measures the distance between beta values with and without the respective case. If there are no outlier cases, these distances are of approximately equal amplitude. Note in Figure 7.13 how the Cook's distance for case 86 is quite different from the distances of the other cases.

7.3.2.2   Assessing Multicollinearity

Besides the methods described in 7.2.5.2, multicollinearity can also be assessed using the so-called *variance inflation factors* (VIF), which are defined for each predictor variable as:

$$\text{VIF}_k = (1 - r_k^2)^{-1},$$ 7.50

where r_k^2 is the coefficient of multiple determination when x_k is regressed on the $p - 2$ remaining variables in the model. An r_k^2 near 1, indicating significant correlation with the remaining variables, will result in a large value of VIF. A VIF larger than 10 is usually taken as an indicator of multicollinearity.

For assessing multicollinearity, the mean of the VIF values is also computed:

$$\overline{\text{VIF}} = \sum_{k=1}^{p-1} \text{VIF}_k / (p-1).$$ 7.51

A mean VIF considerably larger than 1 is indicative of serious multicollinearity problems.

Commands 7.5. SPSS, STATISTICA, MATLAB and R commands used to evaluate regression models.

SPSS	Analyze; Regression; Linear; Statistics; Model Fit
STATISTICA	Statistics; Multiple regression; Advanced; ANOVA
MATLAB	regstats(y,X)
R	influence.measures

The MATLAB `regstats` function generates a set of regression diagnostic measures, such as the studentised residuals and the Cook's distance. The function creates a window with check boxes for each diagnostic measure and a `Calculate Now` button. Clicking `Calculate Now` pops up another window where the user can specify names of variables for storing the computed measures.

The R `influence.measures` is a suite of regression diagnostic functions, including those diagnostics that we have described, such as deleted residuals and Cook's distance.   ∎

7.3.3 Case Study

We have already used the foetal weight prediction task in order to illustrate specific topics on regression. We will now consider this task in a more detailed fashion so that the reader can appreciate the application of the several topics that were previously described in a complete worked-out case study.

7.3.3.1 Determining a Linear Model

We start with the solution obtained by forward stepwise search, summarised in Figure 7.11. Table 7.6 shows the coefficients of the model. The values of beta indicate that their contributions are different. All t tests are significant; therefore, no coefficient is discarded at this phase. The ANOVA test, shown in Table 7.7 gives also a good prognostic of the goodness of fit of the model.

Table 7.6. Parameters and t tests of the trivariate linear model for the foetal weight example.

	Beta	Std. Err. of Beta	B	Std. Err. of B	t_{410}	p
Intercept			−4765.7	261.9	−18.2	0.00
AP	0.609	0.032	124.7	6.5	19.0	0.00
BPD	0.263	0.041	292.3	45. 1	6.5	0.00
CP	0.105	0.044	36.0	15.0	2.4	0.02

Table 7.7. ANOVA test of the trivariate linear model for the foetal weight example.

	Sum of Squares	df	Mean Squares	F	p
Regress.	128252147	3	42750716	501.9254	0.00
Residual	34921110	410	85173		
Total	163173257				

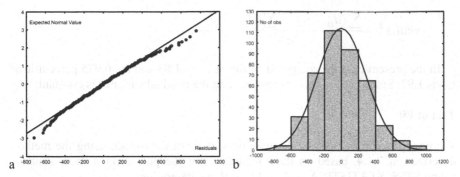

Figure 7.14. Distribution of the residuals for the foetal weight example: a) Normal probability plot; b) Histogram.

7.3.3.2   Evaluating the Linear Model

Distribution of the Residuals

In order to assess whether the errors can be assumed normally distributed, one can use graphical inspection, as in Figure 7.14, and also perform the distribution fitting tests described in chapter 5. In the present case, the assumption of normal distribution for the errors seems a reasonable one.

The constancy of the residual variance can be assessed using the following *modified Levene test*:

1. Divide the data set into two groups: one with the predictor values comparatively low and the other with the predictor values comparatively high. The objective is to compare the residual variance in the two groups. In the present case, we divide the cases into the two groups corresponding to observed weights below and above 3000 g. The sample sizes are $n_1 = 118$ and $n_2 = 296$, respectively.

2. Compute the medians of the residuals e_i in the two groups: med_1 and med_2. In the present case $\mathrm{med}_1 = -182.32$ and $\mathrm{med}_2 = 59.87$.

3. Let $d_{i1} = |e_{i1} - \mathrm{med}_1|$ and $d_{i2} = |e_{i2} - \mathrm{med}_2|$ represent the absolute deviations of the residuals around the medians in each group. We now compute the respective sample means, $\bar{d}_1$ and $\bar{d}_2$, of these absolute deviations, which in our study case are: $\bar{d}_1 = 187.37$, $\bar{d}_2 = 221.42$.

4. Compute:

$$t^* = \frac{\bar{d}_1 - \bar{d}_2}{s\sqrt{\dfrac{1}{n_1} + \dfrac{1}{n_2}}} \quad \sim \quad t_{n-2}, \qquad\qquad 7.52$$

$$\text{with } s^2 = \frac{\sum (d_{i1} - \bar{d}_1)^2 + \sum (d_{i2} - \bar{d}_2)^2}{n-2}.$$

In the present case the computed t value is $t^* = -1.83$ and the 0.975 percentile of t_{412} is 1.97. Since $|t^*| < t_{412,0.975}$, we accept that the residual variance is constant.

Test of Fit

We now proceed to evaluate the goodness of fit of the model, using the method described in 7.1.4, based on the computation of the pure error sum of squares. Using SPSS, STATISTICA, MATLAB or R, we determine:

$n = 414$; $c = 381$; $n - c = 33$; $c - 2 = 379$.
SSPE = 1846345.8; MSPE=SSPE/$(n - c)$ = 55949.9.
SSE = 34921109.

Based on these values, we now compute:

SSLF = SSE − SSPE = 33074763.2; MSLF = SSLF/$(c - 2)$ = 87268.5.

Thus, the computed F^* is: F^* = MSLF/MSPE = 1.56. On the other hand, the 95% percentile of $F_{379, 33}$ is 1.6. Since $F^* < F_{379, 33}$, we do not reject the goodness of fit hypothesis.

Detecting Outliers

The detection of outliers was already performed in 7.3.2.1. Eighteen cases are identified as being outliers. The evaluation of the model without including these outlier cases is usually performed at a later phase. We leave as an exercise the preceding evaluation steps after removing the outliers.

Assessing Multicollinearity

Multicollinearity can be assessed either using the extra sums of squares as described in 7.2.5.2 or using the VIF factors described in 7.3.2.2. This last method is particularly fast and easy to apply.

Using SPSS, STATISTICA, MATLAB or R, one can easily obtain the coefficients of determination for each predictor variable regressed on the other ones. Table 7.8 shows the values obtained for our case study.

Table 7.8. VIF factors obtained for the foetal weight data.

	BPD(CP,AP)	CP(BPD,AP)	AP(BPD,CP)
r^2	0.6818	0.7275	0.4998
VIF	3.14	3.67	2

Although no single VIF is larger than 10, the mean VIF is 2.9, larger than 1 and, therefore, indicative that some degree of multicollinearity may be present.

Cross-Validating the Linear Model

Until now we have assessed the regression performance using the same set that was used for the design. Assessing the performance in the design (training) set yields on average optimistic results, as we have already seen in Chapter 6, when discussing data classification. We need to evaluate the ability of our model to generalise when applied to an independent test set. For that purpose we apply a cross-validation method along the same lines as in section 6.6.

Let us illustrate this procedure by applying a two-fold cross-validation to our FW(AP,BPD,CP) model. For that purpose we randomly select approximately half of the cases for training and the other half for test, and then switch the roles. This can be implemented in SPSS, STATISTICA, MATLAB and R by setting up a filter variable with random 0s and 1s. Denoting the two sets by D_0 and D_1 we obtained the results in Table 7.9 in one experiment. Based on the F tests and on the proximity of the RMS values we conclude the good generalisation of the model.

Table 7.9. Two-fold cross-validation results. The test set results are in italic.

Design with D_0 (204 cases)			Design with D_1 (210 cases)		
D_0 RMS	D_1 RMS	D1 F (p)	D_1 RMS	D_0 RMS	D_0 F (p)
272.6	*312.7*	*706 (0)*	277.1	*308.3*	*613 (0)*

7.3.3.3 *Determining a Polynomial Model*

We now proceed to determine a third order polynomial model for the foetal weight regressed by the same predictors but without interaction terms. As previously mentioned in 7.2.6, in order to avoid numerical problems, we use centred predictors by subtracting the respective mean. We then use the following predictor variables:

$$X_1 = \text{BPD} - \text{mean(BPD)}; \quad X_{11} = X_1^2; \quad X_{111} = X_1^3.$$
$$X_2 = \text{CP} - \text{mean(CP)}; \quad X_{22} = X_2^2; \quad X_{222} = X_2^3.$$
$$X_3 = \text{AP} - \text{mean(AP)}; \quad X_{33} = X_3^2; \quad X_{333} = X_3^3.$$

With SPSS and STATISTICA, in order to perform the forward stepwise search, the predictor variables must first be created before applying the respective regression commands. Table 7.9 shows some results obtained with the forward stepwise search. Note that although six predictors were included in the model using the threshold of 1 for the "F to enter", the three last predictors do not have significant F tests and the predictors X_{222} and X_{11} also do not pass in the respective t tests (at 5% significance level).

Let us now apply the backward search process. Figure 7.15 shows the summary table of this search process, obtained with STATISTICA, using a threshold of "F to remove" = 10 (one more than the number of initial predictors). The variables are removed consecutively by increasing order of their F contribution until reaching the end of the process with two included variables, X_1 and X_3. Notice, however, that variable X_2 is found significant in the F test, and therefore, it should probably be included too.

Table 7.10. Parameters of a third order polynomial regression model found with a forward stepwise search for the foetal weight data (using SPSS or STATISTICA).

	Beta	Std. Err. of Beta	F to Enter	p	t_{410}	p
Intercept					181.7	0.00
X_3	0.6049	0.033	1043	0.00	18.45	0.00
X_1	0.2652	0.041	125.2	0.00	6.492	0.00
X_2	0.1399	0.047	5.798	0.02	2.999	0.00
X_{222}	−0.0942	0.056	1.860	0.17	−1.685	0.09
X_{22}	−0.1341	0.065	2.496	0.12	−2.064	0.04
X_{11}	0.0797	0.0600	1.761	0.185	1.327	0.19

Variable	Step +in/-out	Multiple R	Multiple R-square	R-square change	F - to entr/rem	p-level	Variabls included
x333	-1	0.888697	0.789783	-0.000000	0.000285	0.986540	8
x111	-2	0.888551	0.789522	-0.000261	0.502880	0.478645	7
x33	-3	0.888349	0.789164	-0.000358	0.690113	0.406614	6
x11	-4	0.887836	0.788252	-0.000912	1.761362	0.185199	5
x22	-5	0.887106	0.786957	-0.001295	2.495749	0.114929	4
x222	-6	0.886559	0.785988	-0.000969	1.860499	0.173317	3
x2	-7	0.884851	0.782961	-0.003027	5.798190	0.016483	2

Figure 7.15. Parameters and tests obtained with STATISTICA for the third order polynomial regression model (foetal weight example) using the backward stepwise search procedure.

7.3.3.4 Evaluating the Polynomial Model

We now evaluate the polynomial model found by forward search and including the six predictors X_1, X_2, X_3, X_{11}, X_{22}, X_{222}. This is done for illustration purposes only since we saw in the previous section that the backward search procedure found a simpler linear model. Whenever a simpler (using less predictors) and similarly performing model is found, it should be preferred for the same generalisation reasons that were explained in the previous chapter.

The distribution of the residuals is similar to what is displayed in Figure 7.14. Since the backward search cast some doubts as to whether some of these predictors have a valid contribution, we will now use the methods based on the extra sums of squares. This is done in order to evaluate whether each regression coefficient can be assumed to be zero, and to assess the multicollinearity of the model. As a final result of this evaluation, we will conclude that the polynomial model does not bring about any significant improvement compared to the previous linear model with three predictors.

Table 7.11. Results of the test using extra sums of squares for assessing the contribution of each predictor in the polynomial model (foetal weight example).

Variable	X_1	X_2	X_3	X_{11}	X_{22}	X_{222}
Coefficient	b_1	b_2	b_3	b_{11}	b_{22}	b_{222}
Variables in the Reduced Model	$X_2, X_3, X_{11},$ X_{22}, X_{222}	$X_1, X_3, X_{11},$ X_{22}, X_{222}	$X_1, X_2, X_{11},$ X_{22}, X_{222}	$X_1, X_2, X_3,$ X_{22}, X_{222}	$X_1, X_2, X_3,$ X_{11}, X_{222}	$X_1, X_2, X_3,$ X_{11}, X_{22}
SSE(R) $(/10^3)$	37966	36163	36162	34552	347623	34643
SSR = SSE(R) – SSE(F) $(/10^3)$	3563	1760	1759	149	360	240
$F^* = $ SSR/MSE	42.15	20.82	20.81	1.76	4.26	2.84
Reject H_0	Yes	Yes	Yes	No	Yes	No

Testing whether individual regression coefficients are zero

We use the partial F test described in section 7.2.5.1 as expressed by formula 7.44. As a preliminary step, we determine with SPSS, STATISTICA, MATLAB or R the SSE and MSE of the model:

SSE = 34402739; MSE = 84528.

We now use the 95% percentile of $F_{1,407} = 3.86$ to perform the individual tests as summarised in Table 7.11. According to these tests, variables X_{11} and X_{222} should be removed from the model.

Assessing multicollinearity

We use the test described in section 7.2.5.2 using the same SSE and MSE as before. Table 7.12 summarises the individual computations. According to Table

7.11, the larger differences between SSE(X) and SSE(X | R) occur for variables X_{11}, X_{22} and X_{222}. These variables have a strong influence in the multicollinearity of the model and should, therefore, be removed. In other words, we come up with the first model of Example 7.17.

Table 7.12. Sums of squares for each predictor in the polynomial model (foetal weight example) using the full and reduced models.

Variable	X_1	X_2	X_3	X_{11}	X_{22}	X_{222}
SSE(X) ($/10^3$)	76001	73062	46206	131565	130642	124828
Reduced Model	$X_2, X_3, X_{11},$ X_{22}, X_{222}	$X_1, X_3, X_{11},$ X_{22}, X_{222}	$X_1, X_2, X_{11},$ X_{22}, X_{222}	$X_1, X_2, X_3,$ X_{22}, X_{222}	$X_1, X_2, X_3,$ X_{11}, X_{222}	$X_1, X_2, X_3,$ X_{11}, X_{22}
SSE(R) ($/10^3$)	37966	36163	36162	34552	34763	34643
SSE(X | R) = SSE(R) − SSE ($/10^3$)	3563	1760	1759	149	360	240
Larger Differences				↑	↑	↑

7.4   Regression Through the Origin

In some applications of regression models we may know beforehand that the regression function must pass through the origin. SPSS and STATISTICA have options that allow the user to include or exclude the "intercept" or "constant" term in/from the model. In MATLAB and R one only has to discard a column of ones from the independent data matrix in order to build a model without the intercept term. Let us discuss here the simple linear model with normal errors. Without the "intercept" term the model is written as:

$$Y_i = \beta_1 x_i + \varepsilon_i .$$
7.53

The point estimate of β_1 is:

$$b_1 = \frac{\sum x_i y_i}{\sum x_i^2} .$$
7.54

The unbiased estimate of the error variance is now:

$$MSE = \frac{\sum e_i^2}{n-1}, \text{ with } n-1 \text{ (instead of } n-2\text{) degrees of freedom.}$$
7.55

Example 7.19

Q: Determine the simple linear regression model FW(AP) with and without intercept for the `Foetal Weight` dataset. Compare both solutions.

A: Table 7.13 shows the results of fitting a single linear model to the regression FW(AP) with and without the intercept term. Note that in this last case the magnitude of t for b_1 is much larger than with the intercept term. This would lead us to prefer the without-intercept model, which by the way seems to be the most reasonable model since one expects FW and AP tending jointly to zero.

Figure 7.16 shows the observed versus the predicted cases in both situations. The difference between fitted lines is huge.   □

Table 7.13. Parameters of single linear regression FW(AP), with and without the "intercept" term.

		b	Std. Err. of b	t	p
With Intercept	b_0	−1996.37	188.954	−10.565	0.00
	b_1	157.61	5.677	27.763	0.00
Without Intercept	b_1	97.99	0.60164	162.874	0.00

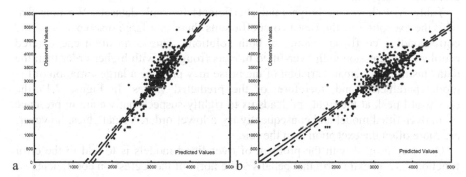

a   b

Figure 7.16. Scatter plots of the observed vs. predicted values for the single linear regression FW(AP): a) with "intercept" term, b) without "intercept" term.

An important aspect to be taken into consideration when regressing through the origin is that the sum of the residuals is not zero. The only constraint on the residuals is:

$$\sum x_i e_i = 0 \, . \tag{7.56}$$

Another problem with this model is that SSE may exceed SST! This can occur when the data has an intercept away from the origin. Hence, the coefficient of determination r^2 may turn out to be negative. As a matter of fact, the coefficient of determination r^2 has no clear meaning for the regression through the origin.

7.5   Ridge Regression

Imagine that we had the dataset shown in Figure 7.17a and that we knew to be the result of some process with an unknown polynomial response function plus some added zero mean and constant standard deviation normal noise. Let us further assume that we didn't know the order of the polynomial function; we only knew that it didn't exceed the 9th order. Searching for a 9th order polynomial fit we would get the regression solution shown with dotted line in Figure 7.17a. The fit is quite good (the R-square is 0.99), but do we really need a 9th order fit? Does the 9th order fit, we have found for the data of Figure 7.17a, generalise for a new dataset generated in the same conditions?

We find here again the same "training set"-"test set" issue that we have found in Chapter 6 when dealing with data classification. It is, therefore, a good idea to get a new dataset and try to fit the found polynomial to it. As an alternative we may also fit a new polynomial to the new dataset and compare both solutions. Figure 7.17b shows a possible instance of a new dataset, generated by the same process for the same predictor values, with the respective 9th order polynomial fit. Again the fit is quite good (R-square is 0.98) although the large downward peak at the right end looks quite suspicious.

Table 7.14 shows the polynomial coefficients for both datasets. We note that with the exception of the first two coefficients there is a large discrepancy of the corresponding coefficient values in both solutions. This is an often encountered problem in regression with over-fitted models (roughly, with higher order than the data "justifies"): a small variation of the noise may produce a large variation of the model parameters and, therefore, of the predicted values. In Figure 7.17 the downward peak at the right end leads us to rightly suspect that we are in presence of an over-fitted model and consequently try a lower order. Visual clues, however, are more often the exception than the rule.

One way to deal with the problem of over-fitted models is to add to the error function 7.37 an extra term that penalises the norm of the regression coefficients:

$$E = (\mathbf{y} - \mathbf{Xb})'(\mathbf{y} - \mathbf{Xb}) + r\mathbf{b}'\mathbf{b} = \text{SSE} + \text{R} .$$ 7.57

When minimising the new error function 7.57 with the added term $\text{R} = r\mathbf{b}'\mathbf{b}$ (called a *regularizer*) we are constraining the regression coefficients to be as small as possible driving the coefficients of unimportant terms towards zero. The parameter r controls the degree of penalisation of the square norm of $\mathbf{b}$ and is called the *ridge* factor. The new regression solution obtained by minimizing 7.57 is known as *ridge regression* and leads to the following ridge parameter vector $\mathbf{b}_R$:

$$\mathbf{b}_R = (\mathbf{X}'\mathbf{X} + r\mathbf{I})^{-1}\mathbf{X}'\mathbf{y} = (\mathbf{r}_{XX} + r\mathbf{I})^{-1}\mathbf{r}_{YX} .$$ 7.58

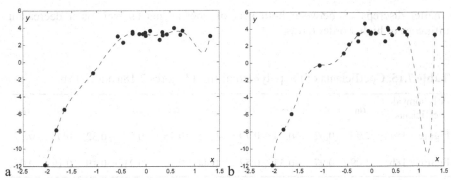

Figure 7.17. A set of 21 points (solid circles) with 9th order polynomial fits (dotted lines). In both cases the x values and the noise statistics are the same; only the y values correspond to different noise instances.

Table 7.14. Coefficients of the polynomial fit of Figures 7.17a and 7.17b.

Polynomyal coefficients	a_0	a_1	a_2	a_3	a_4	a_5	a_6	a_7	a_8	a_9
Figure 7.17a	3.21	−0.93	0.31	8.51	−3.27	−9.27	−0.47	3.05	0.94	0.03
Figure 7.17b	3.72	−1.21	−6.98	20.87	19.98	−30.92	−31.57	6.18	12.48	2.96

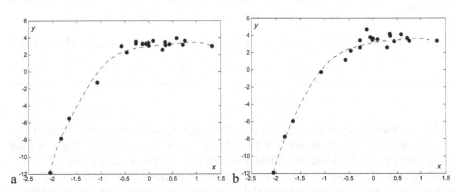

Figure 7.18. Ridge regression solutions with $r = 1$ for the Figure 7.17 datasets.

Figure 7.18 shows the ridge regression solutions for the Figure 7.17 datasets using a ridge factor $r = 1$. We see that the solutions are similar to each other and with a smoother aspect. The downward peak of Figure 7.17 disappeared. Table 7.15 shows the respective polynomial coefficients, where we observe a much smaller discrepancy between both sets of coefficients as well as a decreasing influence of higher order terms.

Table 7.15. Coefficients of the polynomial fit of Figures 7.18a and 7.18b.

Polynomyal coefficients	a_0	a_1	a_2	a_3	a_4	a_5	a_6	a_7	a_8	a_9
Figure 7.18a	2.96	0.62	−0.43	0.79	−0.55	0.36	−0.17	−0.32	0.08	0.07
Figure 7.18b	3.09	0.97	−0.53	0.52	−0.44	0.23	−0.21	−0.19	0.10	0.05

One can also penalise selected coefficients by using in 7.58 an adequate diagonal matrix of penalties, **P**, instead of **I**, leading to:

$$\mathbf{b} = \left(\mathbf{X'X} + r\mathbf{P}\right)^{-1}\mathbf{X'y} .$$
7.59

Figure 7.19 shows the regression solution of Figure 7.17b dataset, using as **P** a matrix with diagonal [1 1 1 1 10 10 1000 1000 1000 1000] and $r = 1$. Table 7.16 shows the computed and the true coefficients. We have now almost retrieved the true coefficients. The idea of "over-fitted" model is now clear.

Table 7.16. Coefficients of the polynomial fit of Figure 7.19 and true coefficients.

Polynomyal coefficients	a_0	a_1	a_2	a_3	a_4	a_5	a_6	a_7	a_8	a_9
Figure 7.19	2.990	0.704	−0.980	0.732	−0.180	0.025	−0.002	−0.001	−0.003	−0.002
True	3.292	0.974	−1.601	0.721	0	0	0	0	0	0

Let us now discuss how to choose the ridge factor when performing ridge regression with 7.58 (regression with 7.59 is much less popular). We can gain some insight into this issue by considering the very simple dataset shown in Figure 7.20, constituted by only 3 points, to which we fit a least square linear model – the dotted line –, and a second-order model – the parabola represented with solid line – using a ridge factor.

The regression line satisfies property iv of section 7.1.2: the sum of the residuals is zero. In Figure 7.20a the ridge factor is zero; therefore, the parabola passes exactly at the 3 points. This will always happen no matter where the observed values are positioned. In other words, the second-order solution is in this case an over-fitted solution tightly attached to the "training set" and unable to generalise to another independent set (think of an addition of i.i.d. noise to the observed values).

The **b** vector is in this case **b** = [0   3.5   −1.5]', with no independent term and a large second-order term.

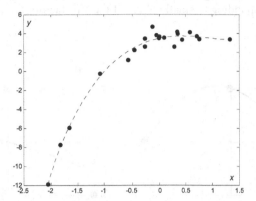

Figure 7.19. Ridge regression solution of Figure 7.17b dataset, using a diagonal matrix of penalties (see text).

Let us now add a regularizer. As we increase the ridge factor the second-order term decreases and the independent term increases. With $r = 0.6$ we get the solution shown in Figure 7.20b with **b** = [0.42   0.74   −0.16]'. We are now quite near the regression line with a large independent term and a reduced second-order term. The addition of i.i.d. noise with small amplitude should not change, on average, this solution. On average we expect some compensation of the errors and a solution that somehow passes half way of the points. In Figure 7.20c the regularizer weighs as much as the classic least squares error. We get **b** = [0.38 0.53   −0.05]' and "almost" a line passing below the "half way". Usually, when performing ridge regression we go as far as $r = 1$. If we go beyond this value the square norm of **b** is driven to small values and we may get strange solutions such as the one shown in Figure 7.20d for $r = 50$ corresponding to **b** = [0.020   0.057 0.078]', i.e., a dominant second-order term.

Figure 7.21 shows for $r \in [0, 2]$ the SSE curve together with the curve of the following error:

$$\text{SSE(L)} = \sum \left(\hat{y}_i - \hat{y}_{iL} \right)^2 ,$$

where the $\hat{y}_i$ are, as usual, the predicted values (second-order model) and the $\hat{y}_{iL}$ are the predicted values of the linear model, which is the preferred model in this case. The minimum of SSE(L) (L from Linear) occurs at $r = 0.6$, where the SSE curve starts to saturate.

We may, therefore choose the best r by graphical inspection of the estimated SSE (or MSE) and the estimated coefficients as functions of r, the so-called *ridge traces*. One usually selects the value of r that corresponds to the beginning of a "stable" evolution of the MSE and coefficients.

Besides its use in the selection of "smooth", non-over-fitted models, ridge regression is also used as a remedy to decrease the effects of multicollinearity as illustrated in the following Example 7.20. In this application one must select a ridge factor corresponding to small values of the VIF factors.

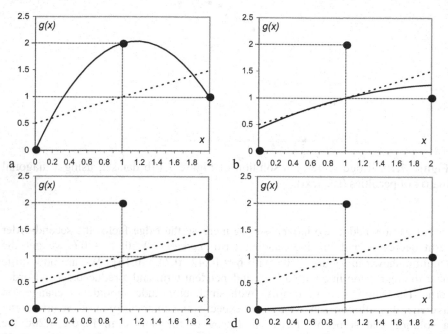

Figure 7.20. Fitting a second-order model to a very simple dataset (3 points represented by solid circles) with ridge factor: a) 0; b) 0.6; c) 1; d) 50.

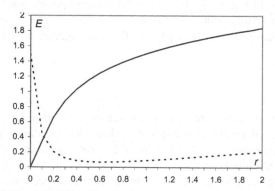

Figure 7.21. SSE (solid line) and SSE(L) (dotted line) curves for the ridge regression solutions of Figure 7.20 dataset.

Example 7.20

Q: Determine the ridge regression solution for the foetal weight prediction model designed in Example 7.13.

A: Table 7.17 shows the evolution with r of the MSE, coefficients and VIF for the linear regression model of the foetal weight data using the predictors BPD, AP and CP. The mean VIF is also included in Table 7.17.

Table 7.17. Values of MSE, coefficients, VIF and mean VIF for several values of the ridge parameter in the multiple linear regression of the foetal weight data.

r		0	0.10	0.20	0.30	0.40	0.50	0.60
MSE		291.8	318.2	338.8	355.8	370.5	383.3	394.8
BPD	b	292.3	269.8	260.7	254.5	248.9	243.4	238.0
	VIF	3.14	2.72	2.45	2.62	2.12	2.00	1.92
CP	b	36.00	54.76	62.58	66.19	67.76	68.21	68.00
	VIF	3.67	3.14	2.80	2.55	3.09	1.82	2.16
AP	b	124.7	108.7	97.8	89.7	83.2	78.0	73.6
	VIF	2.00	1.85	1.77	1.71	1.65	1.61	1.57
Mean VIF		2.90	2.60	2.34	2.17	2.29	1.80	1.88

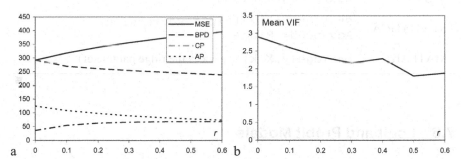

Figure 7.22. a) Plot of the foetal weight regression MSE and coefficients for several values of the ridge parameter; b) Plot of the mean VIF factor for several values of the ridge parameter.

Figure 7.22 shows the ridge traces for the MSE and three coefficients as well as the evolution of the Mean VIF factor. The ridge traces do not give, in this case, a clear indication of the best r value, although the CP curve suggests a "stable" evolution starting at around $r = 0.2$. We don't show the values and the curve corresponding to the intercept term since it is not informative. The evolution of the

VIF and Mean VIF factors (the Mean VIF is shown in Figure 7.22b) suggest the solutions $r = 0.3$ and $r = 0.5$ as the most appropriate.

Figure 7.23 shows the predicted FW values with $r = 0$ and $r = 0.3$. Both solutions are near each other. However, the ridge regression solution has decreased multicollinearity effects (reduced VIF factors) with only a small increase of the MSE.   ☐

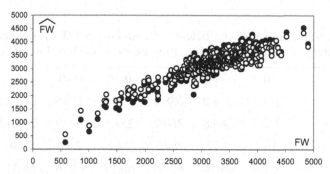

Figure 7.23. Predicted versus observed FW values with $r = 0$ (solid circles) and $r = 0.3$ (open circles).

Commands 7.6. SPSS, STATISTICA and MATLAB commands used to perform ridge regression.

SPSS	`Ridge Regression Macro`
STATISTICA	`Statistics; Multiple Regression; Advanced; Ridge`
MATLAB	`b=ridge(y,X,k)` (k is the ridge parameter)

■

7.6   Logit and Probit Models

Logit and probit regression models are adequate for those situations where the dependent variable of the regression problem is binary, i.e., it has only two possible outcomes, e.g., "success"/"failure" or "normal"/"abnormal". We assume that these binary outcomes are coded as 1 and 0. The application of linear regression models to such problems would not be satisfactory since the fitted predicted response would ignore the restriction of binary values for the observed data.

A simple regression model for this situation is:

$$Y_i = g(x_i) + \varepsilon_i, \text{ with } y_i \in \{0, 1\}. \tag{7.60}$$

Let us consider Y_i to be a Bernoulli random variable with $p_i = P(Y_i = 1)$. Then, as explained in Appendix A and presented in B.1.1, we have:

$$E[Y_i] = p_i.$$
                                                                7.61

On the other hand, assuming that the errors have zero mean, we have from 7.60:

$$E[Y_i] = g(x_i).$$
                                                                7.62

Therefore, no matter which regression model we are using, the mean response for each predictor value represents the probability that the corresponding observed variable is one.

In order to handle the binary valued response we apply a mapping from the predictor domain onto the [0, 1] interval. The logit and probit regression models are precisely popular examples of such a mapping. The *logit model* uses the so-called *logistic* function, which is expressed as:

$$E[Y_i] = \frac{\exp(\beta_0 + \beta_1 x_{i1} + \ldots + \beta_{p-1} x_{ip-1})}{1 + \exp(\beta_0 + \beta_1 x_{i1} + \ldots + \beta_{p-1} x_{ip-1})}.$$
                                                                7.63

The *probit model* uses the normal probability distribution as mapping function:

$$E[Y_i] = N_{0,1}(\beta_0 + \beta_1 x_{i1} + \ldots + \beta_{p-1} x_{ip-1}).$$
                                                                7.64

Note that both mappings are examples of S-shaped functions (see Figure 7.24 and Figure A.7.b), also called *sigmoidal* functions. Both models are examples of non-linear regression.

The logistic response enjoys the interesting property of simple linearization. As a matter of fact, denoting as before the mean response by the probability p_i, and if we apply the *logit transformation*:

$$p_i^* = \ln\left(\frac{p_i}{1 - p_i}\right),$$
                                                                7.65

we obtain:

$$p_i^* = \beta_0 + \beta_1 x_{i1} + \ldots + \beta_{p-1} x_{ip-1}.$$
                                                                7.66

Since the mean binary responses can be interpreted as probabilities, a suitable method to estimate the coefficients for the logit and probit models, is the maximum likelihood method, explained in Appendix C, instead of the previously used least square method. Let us see how this method is applied in the case of the simple logit model. We start by assuming a Bernoulli random variable associated to each observation y_i; therefore, the joint distribution of the n observations is (see B.1.1):

$$p(y_1, \ldots, y_n) = \prod_{i=1}^{n} p_i^{y_i} (1 - p_i)^{1-y_i}.$$
                                                                7.67

Taking the natural logarithm of this likelihood, we obtain:

$$\ln p(y_1,\ldots,y_n) = \sum y_i \ln\left(\frac{p_i}{1-p_i}\right) + \sum \ln(1-p_i).$$ 7.68

Using formulas 7.62, 7.63 and 7.64, the logarithm of the likelihood (*log-likelihood*), which is a function of the coefficients, $L(\beta)$, can be expressed as:

$$L(\beta) = \sum y_i(\beta_0 + \beta_1 x_i) - \sum \ln[1 + \exp(\beta_0 + \beta_1 x_i)].$$ 7.69

The maximization of the $L(\beta)$ function can now be carried out using one of many numerical optimisation methods, such as the quasi-Newton method, which iteratively improves current estimates of function maxima using estimates of its first and second order derivatives.

The estimation of the probit model coefficients follows a similar approach. Both models tend to yield similar solutions, although the probit model is more complex to deal with, namely in what concerns inference procedures and multiple predictor handling.

Example 7.21

Q: Consider the Clays' dataset, which includes 94 samples of analysed clays from a certain region of Portugal. The clays are categorised according to their geological age as being pliocenic ($y_i = 1$; 69 cases) or holocenic ($y_i = 0$; 25 cases). Imagine that one wishes to estimate the probability of a given clay (from that region) to be pliocenic, based on its content in high graded grains (variable HG). Design simple logit and probit models for that purpose. Compare both solutions.

A: Let AgeB represent the binary dependent variable. Using STATISTICA or SPSS (see Commands 7.7), the fitted logistic and probit responses are:

AgeB = exp(−2.646 + 0.23×HG) /[1 + exp(−2.646 + 0.23×HG)];
AgeB = $N_{0,1}$(−1.54 + 0.138×HG).

Figure 7.24 shows the fitted response for the logit model and the observed data. A similar figure is obtained for the probit model. Also shown is the 0.5 threshold line. Any response above this line is assigned the value 1, and below the line, the value 0. One can, therefore, establish a training-set classification matrix for the predicted versus the observed values, as shown in Table 7.18, which can be obtained using either the SPSS or STATISTICA commands. Incidentally, note how the logit and probit models afford a regression solution to classification problems and constitute an alternative to the statistical classification methods described in Chapter 6.   ∎

When dealing with binary responses, we are confronted with the fact that the regression errors can no longer be assumed normal and as having equal variance. Therefore, the statistical tests for model evaluation, described in preceding sections, are no longer applicable. For the logit and probit models, some sort of the chi-square test described in Chapter 5 is usually applied in order to assess the goodness of fit of the model. SPSS and STATISTICA afford another type of chi-square test based on the log-likelihood of the model. Let L_0 represent the log-likelihood for the null model, i.e., where all slope parameters are zero, and L_1 the log-likelihood of the fitted model. In the test used by STATISTICA, the following quantity is computed:

$$L = -2(L_0 - L_1),$$

which, under the null hypothesis that the null model perfectly fits the data, has a chi-square distribution with $p - 1$ degrees of freedom. The test used by SPSS is similar, using only the quantity $-2 L_1$, which, under the null hypothesis, has a chi-square distribution with $n - p$ degrees of freedom.

In Example 7.21, the chi-square test is significant for both the logit and probit models; therefore, we reject the null hypothesis that the null model fits the data perfectly. In other words, the estimated parameters b_1 (0.23 and 0.138 for the logit and probit models, respectively) have a significant contribution for the fitted models.

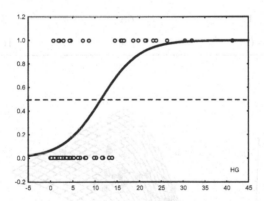

Figure 7.24. Logistic response for the clay classification problem, using variable HG (obtained with STATISTICA). The circles represent the observed data.

Table 7.18. Classification matrix for the clay dataset, using predictor HG in the logit or probit models.

	Predicted Age = 1	Predicted Age = 0	Error rate
Observed Age = 1	65	4	94.2
Observed Age = 0	10	15	60.0

Example 7.22

Q: Redo the previous example using forward search in the set of all original clay features.

A: STATISTICA (Generalized Linear/Nonlinear Models) and SPSS afford forward and backward search in the predictor space when building a logit or probit model. Figure 7.25 shows the response function of a logit bivariate model built with the forward search procedure and using the predictors HG and TiO_2.

In order to derive the predicted Age values, one would have to determine the cases above and below the 0.5 plane. Table 7.19 displays the corresponding classification matrix, which shows some improvement, compared with the situation of using the predictor HG alone. The error rates of Table 7.19, however, are training set estimates. In order to evaluate the performance of the model one would have to compute test set estimates using the same methods as in section 7.3.3.2.

□

Table 7.19. Classification matrix for the clay dataset, using predictors HG and TiO_2 in the logit model.

	Predicted Age = 1	Predicted Age = 0	Error rate
Observed Age = 1	66	3	95.7
Observed Age = 0	9	16	64.0

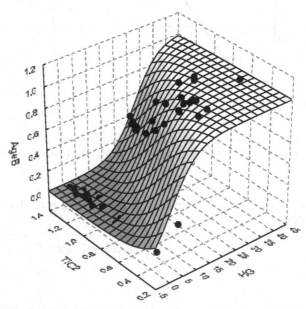

Figure 7.25. 3-D plot of the bivariate logit model for the Clays' dataset. The solid circles are the observed values.

Commands 7.7. SPSS and STATISTICA commands used to perform logit and probit regression.

SPSS	`Analyze; Regression; Binary Logistic |` `Probit`

STATISTICA	`Statistics; Advanced Linear/Nonlinear`
	`Models; Nonlinear Estimation; Quick Logit`
	`| Quick Probit`
	
	`Statistics; Advanced Linear/Nonlinear`
	`Models; Generalized Linear/Nonlinear`
	`Models; Logit | Probit`

∎

Exercises

7.1 The Flow Rate dataset contains daily measurements of flow rates in two Portuguese Dams, denoted AC and T. Consider the estimation of the flow rate at AC by linear regression of the flow rate at T:
   a)   Estimate the regression parameters.
   b)   Assess the normality of the residuals.
   c)   Assess the goodness of fit of the model.
   d)   Predict the flow rate at AC when the flow rate at T is 4 m^3/s.

7.2 Redo the previous Exercise 7.1 using quadratic regression confirming a better fit with higher R^2.

7.3 Redo Example 7.3 without the intercept term, proving the goodness of fit of the model.

7.4 In Exercises 2.18 and 4.8 the correlations between HFS and a transformed variable of I0 were studied. Using polynomial regression, determine a transformed variable of I0 with higher correlation with HFS.

7.5 Using the Clays' dataset, show that the percentage of low grading material depends on their composition of K_2O and Al_2O_3. Use for that purpose a stepwise regression approach with the chemical constituents as predictor candidates. Furthermore, perform the following analyses:
   a)   Assess the contribution of the predictors using appropriate inference tests.
   b)   Assess the goodness of fit of the model.
   c)   Assess the degree of multicollinearity of the predictors.

7.6 Consider the Services' firms of the Firms' dataset. Using stepwise search of a linear regression model estimating the capital revenue, CAPR, of the firms with the predictor candidates {GI, CA, NW, P, A/C, DEPR}, perform the following analyses:
   a)   Show that the best predictor of CAPR is the apparent productivity, P.
   b)   Check the goodness of fit of the model.
   c)   Obtain the regression line plot with the 95% confidence interval.

7.7 Using the `Forest Fires'` dataset, show that, in the conditions of the sample, it is possible to predict the yearly AREA of burnt forest using the number of reported fires as predictor, with an r^2 over 80%. Also, perform the following analyses:
a)   Use ridge regression in order to obtain better parameter estimates.
b)   Cross-validate the obtained model using a partition of even/odd years.

7.8 The search of a prediction model for the foetal weight in section 7.3.3.3 contemplated a third order model. Perform a stepwise search contemplating the interaction effects $X_{12} = X_1X_2$, $X_{13} = X_1X_3$, $X_{23} = X_2X_3$, and show that these interactions have no valid contribution.

7.9 The following Shepard's formula is sometimes used to estimate the foetal weight: $\log_{10}FW = 1.2508 + 0.166BPD + 0.046AP - 0.002646(BPD)(AP)$. Try to obtain this formula using the Foetal Weight dataset and linear regression.

7.10 Variable X_{22}, was found to be a good predictor candidate in the forward search process in section 7.3.3.3. Study in detail the model with predictors X_1, X_2, X_3, X_{22}, assessing namely: the multicollinearity; the goodness of fit; and the detection of outliers.

7.11 Consider the `Wines'` dataset. Design a classifier for the white vs. red wines using features ASP, GLU and PHE and logistic regression. Check if a better subset of features can be found.

7.12 In Example 7.16, the second order regression of the SONAE share values (`Stock Exchange` dataset) was studied. Determine multiple linear regression solutions for the SONAE variable using the other variables of the dataset as predictors and forward and backward search methods. Perform the following analyses:
a)   Compare the goodness of fit of the forward and backward search solutions.
b)   For the best solution found in a), assess the multicollinearity and the contribution of the various predictors and determine an improved model. Test this model using a cross-validation scheme and identify the outliers.

7.13 Determine a multiple linear regression solution that will allow forecasting the temperature one day ahead in the `Weather` dataset (`Data 1` worksheet). Use today's temperature as one of the predictors and evaluate the model.

7.14 Determine and evaluate a logit model for the classification of the CTG dataset in normal vs. non-normal cases using forward and backward searches in the predictor set {LB, AC, UC, ASTV, MSTV, ALTV, MLTV, DL}. Note that variables AC, UC and DL must be converted into time rate (e.g. per minute) variables; for that purpose compute the signal duration based on the start and end instants given in the CTG dataset.

8 Data Structure Analysis

In the previous chapters, several methods of data classification and regression were presented. Reference was made to the dimensionality ratio problem, which led us to describe and use variable selection techniques. The problem with these techniques is that they cannot detect *hidden* variables in the data, responsible for interesting data variability. In the present chapter we describe techniques that allow us to analyse the data structure with the dual objective of dimensional reduction and improved data interpretation.

8.1 Principal Components

In order to illustrate the contribution of data variables to the data variability, let us inspect Figure 8.1 where three datasets with a bivariate normal distribution are shown.

In Figure 8.1a, variables X and Y are uncorrelated and have the same variance, $\sigma^2 = 1$. The circle is the equal density curve for a 2σ deviation from the mean. Any linear combination of X and Y corresponds, in this case, to a radial direction exhibiting the same variance. Thus, in this situation, X and Y are as good in describing the data as any other orthogonal pair of variables.

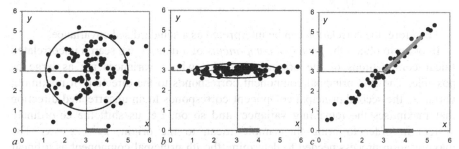

Figure 8.1. Bivariate, normal distributed datasets showing the standard deviations along X and Y with dark grey bars: a) Equal standard deviations (1); b) Very small standard deviation along Y (0.15); and c) Correlated variables of equal standard deviations (1.31) with a light-grey bar showing the standard deviation of the main principal component (3.42).

In Figure 8.1b, X and Y are uncorrelated but have different variances, namely a very small variance along Y, $\sigma_Y^2 = 0.0225$. The importance of Y in describing the data is tenuous. In the limit, with $\sigma_Y^2 \to 0$, Y would be discarded as an interesting variable and the equal density ellipsis would converge to a line segment.

In Figure 8.1c, X and Y are correlated ($\rho = 0.99$) and have the same variance, $\sigma^2 = 1.72$. In this case, as shown in the figure, any equal density ellipsis leans along the regression line at 45°. Based only on the variances of X and Y, we might be led to the idea that two variables are needed in order to explain the variability of the data. However, if we choose an orthogonal co-ordinate system with one axis along the regression line, we immediately see that we have a situation similar to Figure 8.1b, that is, only one hidden variable (absent in the original data), say Z, with high standard deviation (3.42) is needed (light-grey bar in Figure 8.1c). The other orthogonal variable is responsible for only a residual standard deviation (0.02). A variable that maximises a data variance is called a *principal component* of the data. Using only one variable, Z, instead of the two variables X and Y, amounts to a *dimensional reduction* of the data.

Consider a multivariate dataset, with $\mathbf{x} = [X_1\ X_2\ \dots\ X_d]'$, and let $\mathbf{S}$ denote the sample covariance matrix of the data (point estimate of the population covariance Σ), where each element s_{ij} is the covariance between variables X_i and X_j, estimated as follows for n cases (see A.8.2):

$$s_{ij} = \frac{1}{n-1}\sum_{k=1}^{n}(x_{ki} - \bar{x}_i)(x_{kj} - \bar{x}_j).$$

8.1

Notice that covariances are symmetric, $s_{ij} = s_{ji}$, and that s_{ii} is the usual estimate of the variance of X_i, s_i^2. The covariance is related to the correlation, estimated as:

$$r_{ij} = \frac{\sum_{k=1}^{n}(x_{ki} - \bar{x}_i)(x_{kj} - \bar{x}_j)}{(n-1)s_i s_j} = \frac{s_{ij}}{s_i s_j}, \qquad \text{with}\quad r_{ij} \in [-1, 1].$$

8.2

Therefore, the correlation can be interpreted as a standardised covariance.

In order to obtain the *principal components* of a dataset, we search uncorrelated linear combinations of the original variables whose variances are as large as possible. The first principal component corresponds to the direction of maximum variance; the second principal component corresponds to an uncorrelated direction that maximises the remaining variance, and so on. Let us shift the co-ordinate system in order to bring the sample mean to the origin, $\mathbf{x}_c = \mathbf{x} - \bar{\mathbf{x}}$. The maximisation process needed to determine the ith principal component as a linear combination of $\mathbf{x}_c$ co-ordinates, $z_i = \mathbf{u}_i'(\mathbf{x} - \bar{\mathbf{x}})$, is expressed by the following equation (for details see e.g. Fukunaga K, 1990, or Jolliffe IT, 2002):

$$(\mathbf{S} - \lambda_i \mathbf{I})\,\mathbf{u}_i = \mathbf{0},$$

8.3

where $\mathbf{I}$ is the $d{\times}d$ unit matrix, λ_i is a scalar and $\mathbf{u}_i$ is a $d{\times}1$ column vector of the linear combination coefficients.

In order to obtain non-trivial solutions of equation 8.3, one needs to solve the determinant equation $|\mathbf{S} - \lambda \mathbf{I}| = 0$. There are d scalar solutions λ_i of this equation called the *eigenvalues* or *characteristic values* of $\mathbf{S}$, which represent the variances for the new variables z_i. After solving the homogeneous system of equations for the different eigenvalues, one obtains a family of *eigenvectors* or *characteristic vectors* $\mathbf{u}_i$, such that $\forall\ i, j\ \mathbf{u}_i'\mathbf{u}_j = 0$ (*orthogonal system* of uncorrelated variables). Usually, one selects from the family of eigenvectors those that have unit length, $\mathbf{u}_i'\mathbf{u}_i = 1, \forall\ i$ (*orthonormal system*).

We will now illustrate the process of the computation of eigenvalues and eigenvectors for the covariance matrix of Figure 8.1c:

$$\mathbf{S} = \begin{bmatrix} 1.72 & 1.7 \\ 1.7 & 1.72 \end{bmatrix}.$$

The eigenvalues are computed as:

$$|\mathbf{S} - \lambda\mathbf{I}| = \begin{vmatrix} 1.72 - \lambda & 1.7 \\ 1.7 & 1.72 - \lambda \end{vmatrix} = 0 \Rightarrow 1.72 - \lambda = \pm 1.7 \Rightarrow \lambda_1 = 3.42, \lambda_2 = 0.02.$$

For λ_1 the homogeneous system of equations is:

$$\begin{bmatrix} -1.7 & 1.7 \\ 1.7 & -1.7 \end{bmatrix}\begin{bmatrix} u_1 \\ u_2 \end{bmatrix} = 0,$$

from where we derive the unit length eigenvector: $\mathbf{u}_1 = [0.7071\ 0.7071]' \equiv [1/\sqrt{2}\ 1/\sqrt{2}\ ]'$. For λ_2, in the same way we derive the unit length eigenvector orthogonal to $\mathbf{u}_1$: $\mathbf{u}_2 = [-0.7071\ 0.7071]' \equiv [-1/\sqrt{2}\ 1/\sqrt{2}\ ]'$. Thus, the principal components of the co-ordinates are $Z_1 = (X_1 + X_2)/\sqrt{2}$ and $Z_2 = (X_1 + X_2)/\sqrt{2}$ with variances 3.42 and 0.02, respectively.

The unit length eigenvectors make up the column vectors of an *orthonormal* matrix $\mathbf{U}$ (i.e., $\mathbf{U}^{-1} = \mathbf{U}'$) used to determine the co-ordinates of an observation $\mathbf{x}$ in the new uncorrelated system of the principal components:

$$\mathbf{z} = \mathbf{U}'(\mathbf{x} - \bar{\mathbf{x}}). \qquad\qquad 8.4$$

These co-ordinates in the principal component space are often called "*z-scores*". In order to avoid confusion with the previous meaning of z-scores – *standardised data* with zero mean and unit variance – we will use the term *pc-scores* instead.

The extraction of principal components is basically a *variance maximising rotation* of the original variable space. Each principal component corresponds to a certain amount of variance of the whole dataset. For instance, in the example portrayed in Figure 8.1c, the first principal component represents $\lambda_1/(\lambda_1 + \lambda_2) = 99\%$

of the total variance. In short, $\mathbf{u}_1$ alone contains practically all the information about the data; the remaining $\mathbf{u}_2$ is residual "noise".

Let Λ represent the diagonal matrix of the eigenvalues:

$$\Lambda = \begin{bmatrix} \lambda_1 & 0 & \dots & 0 \\ 0 & \lambda_2 & \dots & 0 \\ \dots & \dots & \dots & \dots \\ 0 & 0 & \dots & \lambda_d \end{bmatrix}. \qquad 8.5$$

The following properties are verified:

1. $\mathbf{U'\,S\,U} = \Lambda$ and $\mathbf{S} = \mathbf{U\,\Lambda\,U'}$. \hfill 8.6

2. The determinant of the covariance matrix, $|\mathbf{S}|$, is:

$$|\mathbf{S}| = |\Lambda| = \lambda_1 \lambda_2 \dots \lambda_d. \qquad 8.7$$

$|\mathbf{S}|$ is called the *generalised variance* and its square root is proportional to the area or volume of the data cluster since it is the product of the ellipsoid axes.

3. The traces of $\mathbf{S}$ and Λ are equal to the sum of the variances of the variables:

$$\text{tr}(\mathbf{S}) = \text{tr}(\Lambda) = s_1^2 + s_2^2 + \dots + s_d^2. \qquad 8.8$$

Based on this property, we measure the contribution of a variable X_k by $e = \lambda_k / \sum \lambda_i = \lambda_k/(s_1^2 + s_2^2 + \dots + s_d^2)$, as we did previously.

The contribution of each original variable X_j to each principal component Z_i can be assessed by means of the corresponding *sample correlation* between X_j and Z_i, often called the *loading* of X_j:

$$r_{ij} = (u_{ji}\sqrt{\lambda_i})/s_j. \qquad 8.9$$

Function pccorr implemented in MATLAB and R and supplied in Tools (see Commands 8.1) allows computing the r_{ij} correlations.

Example 8.1

Q: Consider the best class of the Cork Stoppers' dataset (first 50 cases). Compute the covariance matrix and their eigenvalues and engeivectors using the original variables ART and PRT. Determine the algebraic expression and contribution of the main principal component, its correlation with the original variables as well as the new co-ordinates of the first cork-stopper.

A: We use MATLAB to perform the necessary computations (see Commands 8.1). Let cork represent the data matrix with all 10 features. We then use:

```
» % Extract 1st class ART and PRT from cork
» x = [cork(1:50,1) cork(1:50,3)];
» S = cov(x);                % covariance matrix
» [u,lambda,e] = pcacov(S);  % principal components
» r = pccorr(x);             % correlations
```

The results S, u, lambda, e and r are shown in Table 8.1. The scatter plots of the data using the original variables and the principal components are shown in Figure 8.2. The pc-scores can be obtained with:

```
» xc = x-ones(50,1)*mean(x);
» z = (u'*xc')';
```

We see that the first principal component with algebraic expression, $-0.3501\times ART-0.9367\times PRT$, highly correlated with the original variables, explains almost 99% of the total variance. The first cork-stopper, represented by [81 250]' in the ART-PRT plane, maps into:

$$\begin{bmatrix} -0.3501 & -0.9367 \\ -0.9367 & 0.3501 \end{bmatrix}\begin{bmatrix} 81-137 \\ 250-365 \end{bmatrix}=\begin{bmatrix} 127.3 \\ 12.2 \end{bmatrix}.$$

The eigenvector components are the cosines of the angles subtended by the principal components in the ART-PRT plane. In Figure 8.2a, this result can only be visually appreciated after giving equal scales to the axes. □

Table 8.1. Eigenvectors and eigenvalues obtained with MATLAB for the first class of cork-stoppers (variables ART and PRT).

Covariance		Eigenvectors		Eigenvalues	Explained variance	Correlations for z_1
$S\ (\times 10^{-4})$		u_1	u_2	$\lambda\ (\times 10^{-4})$	$e\ (\%)$	r_{1j}
0.1849	0.4482	−0.3501	−0.9367	1.3842	98.76	−0.9579
0.4482	1.2168	−0.9367	0.3501	0.0174	1.24	−0.9991

An interesting application of principal components is in statistical *quality control*. The possibility afforded by principal components of having a much-reduced set of variables explaining the whole data variability is an important advantage. Instead of controlling several variables, with the same type of Error Type I degradation as explained in 4.5.1, sometimes only one variable needs to be controlled.

Furthermore, principal components afford an easy computation of the following *Hotteling's T^2 measure of variability*:

$$T^2 = (\mathbf{x} - \overline{\mathbf{x}})' \mathbf{S}^{-1} (\mathbf{x} - \overline{\mathbf{x}}) = \mathbf{z}' \mathbf{\Lambda}^{-1} \mathbf{z} .$$ 8.10

Critical values of T^2 are computed in terms of the F distribution as follows:

$$T_{d,n,1-\alpha}^2 = \frac{d(n-1)}{n-d} F_{d,n-d,1-\alpha} .$$ 8.11

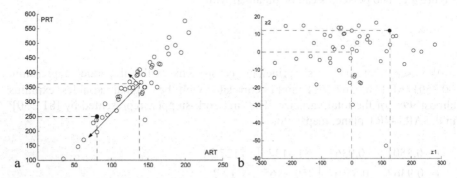

Figure 8.2. Scatter plots obtained with MATLAB of the cork-stopper data (first class) represented in the planes: a) ART-PRT with superimposed principal components; b) Principal components. The first cork is shown with a solid circle.

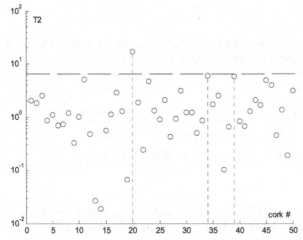

Figure 8.3. T^2 chart for the first class of the cork-stopper data. Case #20 is out of control.

Example 8.2

Q: Determine the Hotteling's T^2 control chart for the previous Example 8.1 and find the corks that are "out of control" at a 95% confidence level.

A: The Hotteling's T^2 values can be determined with MATLAB `princomp` function. The 95% critical value for $F_{2,48}$ is 3.19; hence, the 95% critical value for the Hotteling's T^2, using formula 8.11, is computed as 6.51. Figure 8.3 shows the corresponding control chart. Cork #20 is clearly "out of control", i.e., it should be reclassified. Corks #34 and #39 are borderline cases.

□

Commands 8.1. SPSS, STATISTICA, MATLAB and R commands used to perform principal component and factor analyses.

SPSS	Analyze; Data Reduction; Factor
STATISTICA	Statistics; Multivariate Exploratory Techniques; Factor Analysis

MATLAB	`[u,l]=eig(C); [pc, lat, expl] = pcacov(C)`
	`[pc, score, lat, tsq]= princomp(x)`
	`residuals = pcares(x,ndim)`
	`[ndim,p,chisq] = barttest(x,alpha)`
	`r = pccorr(x) ; f=velcorr(x,icov)`

R	`eigen(C) ; prcomp(x) ; princomp(x)`
	`screeplot(p)`
	`factanal(x,factors,scores,rotation)`
	`pccorr(x) ; velcorr(x,icov)`

SPSS and STATISTICA commands are of straightforward use. SPSS and STATISTICA always use the correlation matrix instead of the covariance matrix for computing the principal components. Figure 8.4 shows STATISTICA specification window for the selection of the two most important components with eigenvalues above 1. If one wishes to obtain all principal components one should set the `Min. eigenvalue` to 0 and the `Max. no. of factors` to the data dimension.

The MATLAB `eig` function returns the eigenvectors, u, and eigenvalues, l, of a covariance matrix C. The `pcacov` function determines the principal components of a covariance matrix C, which are returned in pc. The return vectors lat and expl store the variances and contributions of the principal components to the total variance, respectively. The `princomp` function returns the principal components and eigenvalues of a data matrix x in pc and lat, respectively. The pc-scores and Hotteling's T^2 are returned in score and tsq, respectively. The `pcares` function returns the residuals obtained by retaining the first ndim principal components of x. The `barttest` function returns the number of dimensions to retain together with the Bartlett's test probabilities, p, and χ^2 scores, chisq (see section 8.2).

The MATLAB implemented `pccorr` function computes the partial correlations between the original variables and the principal components of a data matrix x. The `velcorr` function computes the Velicer partial correlations (see section 8.2) using matrix x either as data matrix ($\texttt{icov} \neq 0$) or as covariance matrix ($\texttt{icov} = 0$).

The R eigen function behaves as the MATLAB eig function. For instance, the eigenvalues and eigenvectors of Table 8.1 can be obtained with eigen(cov(cbind(ART[1:50],PRT[1:50]))). The prcomp function computes among other things the principal components (curiously, called "rotation" or "loadings" in R) and their standard deviations (square roots of the eigenvalues). For the dataset of Example 8.1 one would use:

```
> p<-prcomp(cbind(ART[1:50],PRT[1:50]))
> p
Standard deviations:
[1] 117.65407  13.18348

Rotation:
            PC1          PC2
[1,] 0.3500541   0.9367295
[2,] 0.9367295  -0.3500541
```

We thus obtain the same eigenvectors (PC1 and PC2) as in Table 8.1 (with an unimportant change of sign). The standard deviations are the square roots of the eigenvalues listed in Table 8.1. With the R princomp function, besides the principal components and their standard deviations, one can also obtain the data projections onto the eigenvectors (the so-called scores in R).

A scree plot (see section 8.2) can be obtained in R with the screeplot function using as argument an object returned by the princomp function. The R factanal function performs factor analysis (see section 8.4) of the data matrix x returning the number of factors specified by factors with the specified rotation method. Bartlett's test scores can be specified with scores.

The R implemented functions pccorr and velcorr behave in the same way as their MATLAB counterparts.   ∎

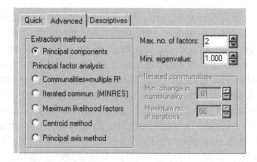

Figure 8.4. Partial view of STATISTICA specification window for principal component analysis with standardised data.

8.2   Dimensional Reduction

When using principal component analysis for dimensional reduction, one must decide how many components (and corresponding variances) to retain. There are several criteria published in the literature to consider. The following are commonly used:

1. Select the principal components that explain a certain percentage (say, 95%) of tr(Λ). This is a very simplistic criterion that is not recommended.

2. The *Guttman-Kaiser criterion* discards eigenvalues below the average tr(Λ)/d (below 1 for standardised data), which amounts to retaining the components responsible for the variance contributed by one variable if the total variance was equally distributed.

3. The so-called *scree test* uses a plot of the eigenvalues (*scree plot*), discarding those starting where the plot levels off.

4. A more elaborate criterion is based on the so-called *broken stick* model. This criterion discards the eigenvalues whose proportion of explained variance is smaller than what should be the expected length l_k of the kth longest segment of a unit length stick randomly broken into d segments:

$$l_k = \frac{1}{d}\sum_{i=k}^{d}\frac{1}{i}.$$ 8.12

A table of l_k values is given in `Tools.xls`.

5. The *Bartlett's test* method is based on the assessment of whether or not the null hypothesis that the last $p - q$ eigenvalues are equal, $\lambda_{q+1} = \lambda_{q+2} = \ldots = \lambda_p$, can be accepted. The mathematics of this test are intricate (see Jolliffe IT, 2002, for a detailed discussion) and its results often unreliable. We pay no further attention to this procedure.

6. The *Velicer partial correlation procedure* uses the partial correlations among the original variables when one or more principal components are removed. Let S_k represent the remaining covariance matrix when the covariance of the first k principal components is removed:

$$\mathbf{S}_k = \mathbf{S} - \sum_{i=1}^{k}\lambda_i\mathbf{u}_i\mathbf{u}_i'; \qquad k = 0, 1, \ldots, d.$$ 8.13

Using the diagonal matrix $\mathbf{D}_k$ of $\mathbf{S}_k$, containing the variances, we compute the correlation matrix:

$$\mathbf{R}_k = \mathbf{D}_k^{-1/2}\mathbf{S}_k\mathbf{D}_k^{-1/2}.$$ 8.14

Finally, with the elements $r_{ij(k)}$ of $\mathbf{R}_k$ we compute the following quantity:

$$f_k = \sum_i \sum_{j \neq i} r_{ij(k)}^2 / [d(d-1)].$$ 8.15

The f_k are the sum of squares of the partial correlations when the first k principal components are removed. As long as f_k decreases, the partial covariances decline faster than the residual variances. Usually, after an initial decrease, f_k will start to increase, reflecting the fact that with the removal of main principal components, we are obtaining increasingly correlated "noise". The k value corresponding to the first f_k minimum is then used as the stopping rule.

The Velicer procedure can be applied using the `velcorr` function implemented in MATLAB and R and available in Tools (see Appendix F).

Example 8.3

Q: Using all the previously described criteria, determine the number of principal components for the `Cork Stoppers'` dataset (150 cases, 10 variables) that should be retained and assess their contribution.

A: Table 8.2 shows the computed eigenvalues of the cork-stopper dataset. Figure 8.5a shows the scree plot and Figure 8.5b shows the evolution of Velicer's f_k. Finally, Table 8.3 compares the number of retained principal components for the several criteria and the respective percentage of explained variance. The highly recommended Velicer's procedure indicates 3 as the appropriate number of principal components to retain.

□

Table 8.2. Eigenvalues of the cork-stopper dataset computed with MATLAB (a scale factor of 10^4 has been removed).

λ_1	λ_2	λ_3	λ_4	λ_5
1.1342	0.1453	0.0278	0.0202	0.0137
λ_6	λ_7	λ_8	λ_9	λ_{10}
0.0087	0.0025	0.0016	0.0006	0.0001

Table 8.3. Comparison of dimensional reduction criteria (Example 8.3).

Criterion	95% variance	Guttman-Kaiser	Scree test	Broken stick	Velicer
k	3	1	3	1	3
Explained variance	96.5%	83.7%	96.5%	83.7%	96.5%

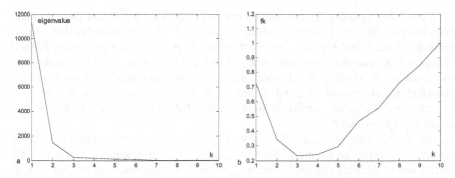

Figure 8.5. Assessing the dimensional reduction to be performed in the cork stopper dataset with: a) Scree plot, b) Velicer partial correlation plot. Both plots obtained with MATLAB.

8.3 Principal Components of Correlation Matrices

Sometimes, instead of computing the principal components out of the original data, they are computed out of the standardised data, i.e., using the z-scores of the data. This is the procedure followed by SPSS and STATISTICA, which is related to the factor analysis approach described in the following section. Using the standardised data has the consequence of eigenvalues and eigenvectors computed from the correlation matrix instead of the covariance matrix (see formula 8.2). The R function `princomp` has a logical argument, `cor`, whose value controls the use of the data correlation or covariance matrix. The results obtained are, in general, different.

Note that since all diagonal elements of a correlation matrix are 1, we have $tr(\Lambda) = d$. Thus, the Guttman-Kaiser criterion amounts, in this case, to selecting the eigenvalues which are greater than 1.

Using standardised data has several benefits, namely imposing equal contribution of the original variables when they have different units or heterogeneous variances.

Example 8.4

Q: Compare the bivariate principal component analysis of the `Rocks` dataset (134 cases, 18 variables), using covariance and correlation matrices.

A: Table 8.4 shows the eigenvectors and correlations (called *factor loadings* in STATISTICA) computed with the original data and the standardised data. The first ones, $\mathbf{u}_1$ and $\mathbf{u}_2$, are computed with MATLAB or R using the covariance matrix; the second ones, $\mathbf{f}_1$ and $\mathbf{f}_2$, are computed with STATISTICA using the correlation matrix. Figure 8.6 shows the corresponding pc scores (called *factor scores* in STATISTICA), that is the data projections onto the principal components.

We see that by using the covariance matrix, only one eigenvector has dominant correlations with the original variables, namely the "compression breaking load" variables RMCS and RCSG. These variables are precisely the ones with highest variance. Note also the dominant values of the first two elements of $\mathbf{u}$. When using the correlation matrix, the $\mathbf{f}$ elements are more balanced and express the contribution of several original features: $\mathbf{f}_1$ highly correlated with chemical features, and $\mathbf{f}_2$ highly correlated with density (MVAP), porosity (PAOA), and water absorption (AAPN).

The scatter plot of Figure 8.6a shows that the pc scores obtained with the covariance matrix are unable to discriminate the several groups of rocks; $\mathbf{u}_1$ only discriminates the rock classes between high and low "compression breaking load" groups. On the other hand, the scatter plot in Figure 8.6b shows that the pc scores obtained with the correlation matrix discriminate the rock classes, both in terms of chemical composition ($\mathbf{f}_1$ basically discriminates Ca vs. SiO_2-rich rocks) and of density-porosity-water absorption features ($\mathbf{f}_2$).

$\square$

Table 8.4. Eigenvectors of the rock dataset computed from the covariance matrix ($\mathbf{u}_1$ and $\mathbf{u}_2$) and from the correlation matrix ($\mathbf{f}_1$ and $\mathbf{f}_2$) with the respective correlations. Correlations above 0.7 are shown in bold.

	$\mathbf{u}_1$	$\mathbf{u}_2$	r_1	r_2	$\mathbf{f}_1$	$\mathbf{f}_2$	r_1	r_2
RMCS	-0.695	0.487	**-0.983**	0.136	-0.079	0.018	-0.569	0.057
RCSG	-0.714	-0.459	**-0.984**	-0.126	-0.069	0.034	-0.499	0.105
RMFX	-0.013	-0.489	-0.078	-0.606	-0.033	0.053	-0.237	0.163
MVAP	-0.015	-0.556	-0.089	-0.664	-0.034	0.271	-0.247	**0.839**
AAPN	0.000	0.003	0.251	0.399	0.046	-0.293	0.331	**-0.905**
PAOA	0.001	0.008	0.241	0.400	0.044	-0.294	0.318	**-0.909**
CDLT	0.001	-0.005	0.240	-0.192	0.001	0.177	0.005	0.547
RDES	0.002	-0.002	0.523	-0.116	0.070	-0.101	0.503	-0.313
RCHQ	-0.002	-0.028	-0.060	-0.200	-0.095	0.042	-0.689	0.131
SiO_2	-0.025	0.046	-0.455	0.169	-0.129	-0.074	**-0.933**	-0.229
Al_2O_3	-0.004	0.001	-0.329	0.016	-0.129	-0.069	**-0.932**	-0.215
Fe_2O_3	-0.001	-0.006	-0.296	-0.282	-0.111	-0.028	**-0.798**	-0.087
MnO	-0.000	-0.000	-0.252	-0.039	-0.090	-0.011	-0.647	-0.034
CaO	0.020	-0.025	0.464	-0.113	0.132	0.073	**0.955**	0.225
MgO	-0.003	-0.007	-0.393	-0.226	-0.024	0.025	-0.175	0.078
Na_2O	-0.001	0.004	-0.428	0.236	-0.119	-0.071	**-0.856**	-0.220
K_2O	-0.001	0.005	-0.320	0.267	-0.117	-0.084	**-0.845**	-0.260
TiO_2	-0.000	-0.000	-0.152	-0.097	-0.088	-0.026	-0.633	-0.079

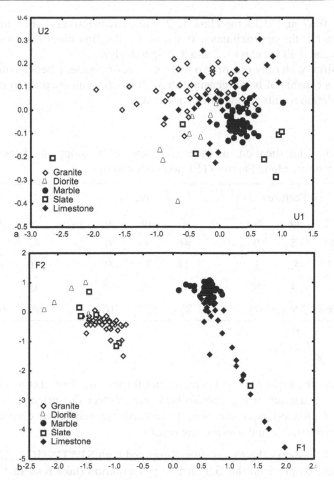

Figure 8.6. The rock dataset analysed with principal components computed from the covariance matrix (a) and from the correlation matrix (b).

Example 8.5

Q: Consider the three classes of the Cork Stoppers' dataset (150 cases). Evaluate the training set error for linear discriminant classifiers using the 10 original features and one or two principal components of the data correlation matrix.

A: The classification matrices, using the linear discriminant procedure described in Chapter 6, are shown in Table 8.5. We see that the dimensional reduction didn't degrade the training set error significantly. The first principal component, F1, alone corresponds to more than 86% of the total variance. Adding the principal component F2, 94.5% of the total data variance is explained. Principal component F1 has a distribution that is well approximated by the normal distribution (Shapiro-Wilk

p = 0.69, 0.67 and 0.33 for class 1, 2 and 3, respectively). For the principal component F2, the approximation is worse for the first class (Shapiro-Wilk p = 0.09, 0.95 and 0.40 for class 1, 2 and 3, respectively).

A classifier with only one or two features has, of course, a better dimensionality ratio and is capable of better generalisation. It is left as an exercise to compare the cross-validation results for the three feature sets.

□

Table 8.5. Classification matrices for the cork stoppers dataset. Correct classifications are along the rows (50 cases per class).

	10 Features			F_1 and F_2			F_1		
	ω_1	ω_2	ω_3	ω_1	ω_2	ω_3	ω_1	ω_2	ω_3
ω_1	45	5	0	46	4	0	47	3	0
ω_2	7	42	1	11	39	0	10	40	0
ω_3	0	4	46	0	5	45	0	5	45
Pe	10%	16%	6%	8%	22%	10%	6%	20%	10%

Example 8.6

Q: Compute the main principal components for the two first classes of the Cork Stoppers' dataset, using standardised data. Select the principal components using the Guttman-Kaiser criterion. Determine the respective correlations with each original variable and interpret the results.

A: Figure 8.7a shows the eigenvalues computed with STATISTICA. The first two eigenvalues comply with the Guttman-Kaiser criterion (take note that the sum of all eigenvalues is 10).

The factor loadings of the two main principal components are shown in Figure 8.8a. Significant values appear in bold. A plot of these factor loadings is shown in Figure 8.8b. It is clearly visible that the first principal component, F_1, is highly correlated with all cork-stopper features except N and the opposite happens with F_2. These observations suggest, therefore, that the description (or classification) of the two cork-stopper classes can be achieved either with F_1 and F_2, or with feature N and one of the other features, namely the highest correlated feature PRTG (total perimeter of the big defects).

Furthermore, we see that the only significant correlation relative to F_2 is smaller than any of the significant correlations relative to F_1. Thus, F_1 or PRTG alone describes most of the data, as suggested by the scatter plot of Figure 8.7b (pc scores). □

When analysing grouped data with principal components, as we did in the previous Examples 8.4 and 8.6, one often wants to determine the most important variables as well as the data groups that best reflect the behaviour of those variables.

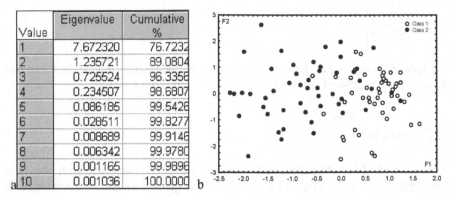

Value	Eigenvalue	Cumulative %
1	7.672320	76.7232
2	1.235721	89.0804
3	0.725524	96.3358
4	0.234507	98.6807
5	0.086185	99.5428
6	0.028511	99.8277
7	0.008689	99.9148
8	0.006342	99.9780
9	0.001165	99.9898
10	0.001036	100.0000

Figure 8.7. Dimensionality reduction of the first two classes of cork-stoppers: a) Eigenvalues; b) Principal component scatter plot (compare with Figure 6.5). (Both graphs obtained with STATISTICA.)

Consider the means of variable F1 in Example 8.6: 0.71 for class 1 and -0.71 for class 2 (see Figure 8.7b). As expected, given the translation $y = x - \bar{x}$, the means are symmetrically located around F1 = 0. Moreover, by visual inspection, we see that the class 1 cases cluster on a high F1 region and class 2 cases cluster on a low F1 region. Notice that since the scatter plot 8.7b uses the projections of the standardised data onto the F1-F2 plane, the cases tend to cluster around the (1, 1) and $(-1, -1)$ points in this plane.

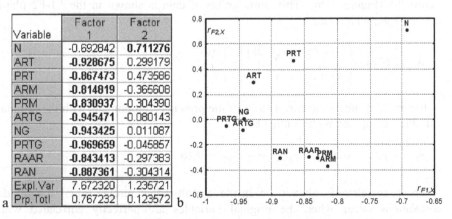

Variable	Factor 1	Factor 2
N	-0.692842	**0.711276**
ART	**-0.928675**	0.299179
PRT	**-0.867473**	0.473586
ARM	**-0.814819**	-0.365608
PRM	**-0.830937**	-0.304390
ARTG	**-0.945471**	-0.080143
NG	**-0.943425**	0.011087
PRTG	**-0.969659**	-0.045857
RAAR	**-0.843413**	-0.297383
RAN	**-0.887361**	-0.304314
Expl.Var	7.672320	1.235721
Prp.Totl	0.767232	0.123572

Figure 8.8. Factor loadings table (a) with significant correlations in bold and graph (b) for the first two classes of cork-stoppers, obtained with STATISTICA.

In order to analyse this issue in further detail, let us consider the simple dataset shown in Figure 8.9a, consisting of normally distributed bivariate data generated with (true) mean $\mu_o = [3 \quad 3]$' and the following (true) covariance matrix:

$$\Sigma_o = \begin{bmatrix} 5 & 3 \\ 3 & 2 \end{bmatrix}.$$

Figure 8.9b shows this dataset after standardisation (subtraction of the mean and division by the standard deviation) with the new covariance matrix:

$$\Sigma = \begin{bmatrix} 1 & 0.9478 \\ 0.9478 & 1 \end{bmatrix}.$$

The standardised data has unit variance along all variables with the new covariance: $\sigma_{12} = \sigma_{21} = 3/(\sqrt{5}\sqrt{2}) = 0.9487$. The eigenvalues and eigenvectors of Σ (computed with MATLAB function eig), are:

$$\Lambda = \begin{bmatrix} 1.9487 & 0 \\ 0 & 0.0513 \end{bmatrix}; \qquad U = \begin{bmatrix} -1/\sqrt{2} & 1/\sqrt{2} \\ 1/\sqrt{2} & 1/\sqrt{2} \end{bmatrix}.$$

Note that $tr(\Lambda) = 2$, the total variance, and that the first principal component explains 97% of the total variance.

Figure 8.9c shows the standardised data projected onto the new system of variables F1 and F2.

Let us now consider a group of data with mean $m_o = [4 \quad 4]$' and a one-standard-deviation boundary corresponding to the ellipsis shown in Figure 8.9a, with $s_x = \sqrt{5}/2$ and $s_y = \sqrt{2}/2$, respectively. The mean vector maps onto $m = m_o - \mu_o = [1 \quad 1]$ '; given the values of the standard deviation, the ellipsis maps onto a circle of radius 0.5 (Figure 8.9b). This same group of data is shown in the F1-F2 plane (Figure 8.9c) with mean:

$$m_p = U'm = \begin{bmatrix} -1/\sqrt{2} & 1/\sqrt{2} \\ 1/\sqrt{2} & 1/\sqrt{2} \end{bmatrix} \begin{bmatrix} 1 \\ 1 \end{bmatrix} = \begin{bmatrix} 0 \\ \sqrt{2} \end{bmatrix}.$$

Figure 8.9d shows the correlations of the principal components with the original variables, computed with formula 8.9:

$$r_{F_1X} = r_{F_1Y} = 0.987; \qquad r_{F_2X} = -r_{F_2Y} = 0.16.$$

These correlations always lie inside a unit-radius circle. Equal magnitude correlations occur when the original variables are perfectly correlated with $\lambda_1 = \lambda_2 = 1$. The correlations are then $|r_{F_1X}| = |r_{F_1Y}| = 1/\sqrt{2}$ (apply formula 8.9).

In the case of Figure 8.9d, we see that F1 is highly correlated with the original variables, whereas F2 is weakly correlated. At the same time, a data group lying in the "high region" of X and Y tends to cluster around the F1 = 1 value after projection of the standardised data. We may superimpose these two different graphs – the pc scores graph and the correlation graph – in order to facilitate the interpretation of the data groups that exhibit some degree of correspondence with high values of the variables involved.

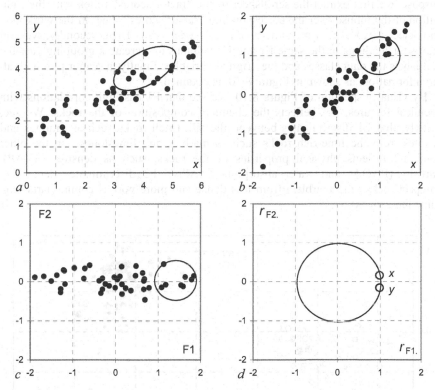

Figure 8.9. Principal component transformation of a bivariate dataset: a) original data with a group delimited by an ellipsis; b) Standardised data with the same group (delimited by a circle); c) Standardised data projection onto the F1-F2 plane; d) Plot of the correlations (circles) of the original variables with F1 and F2.

Example 8.7

Q: Consider the Rocks' dataset, a sample of 134 rocks classified into five classes (1="granite", 2="diorite", 3="marble", 4="slate", 5="limestone") and characterised by 18 features (see Appendix E). Use the two main principal components of the data in order to interpret it.

A: Only the first four eigenvalues satisfy the Kaiser criterion. The first two eigenvalues are responsible for about 58% of the total variance; therefore, when discarding the remaining eigenvalues, we are discarding a substantial amount of the information from the dataset (see Exercise 8.12).

We can conveniently interpret the data by using a graphic display of the standardised data projected onto the plane of the first two principal components, say F1 and F2, superimposed over the correlation plot. In STATISTICA, this overlaid graphic display can be obtained by first creating a datasheet with the projections ("factor scores") and the correlations ("factor loadings"). For this purpose, we first extract the scrollsheet of the "factor scores" (click with the right button of the mouse over the corresponding "factor scores" sheet in the workbook and select `Extract as stand alone window`). Then, secondly, we join the factor loadings in the same F1 and F2 columns and create a grouping variable that labels the data classes and the original variables. Finally, a scatter plot with all the information, as shown in Figure 8.10, is obtained.

By visual inspection of Figure 8.10, we see that F1 has high correlations with chemical features, i.e., reflects the chemical composition of the rocks. We see, namely, that F1 discriminates between the silica-rich rocks such as granites and diorites from the lime-rich rocks such as marbles and limestones. On the other hand, F2 reflects physical properties of the rocks, such as density (MVAP), porosity (PAOA) and water absorption (AAPN). F2 discriminates dense and compact rocks (e.g. marbles) from less dense and more porous counterparts (e.g. some limestones).   ☐

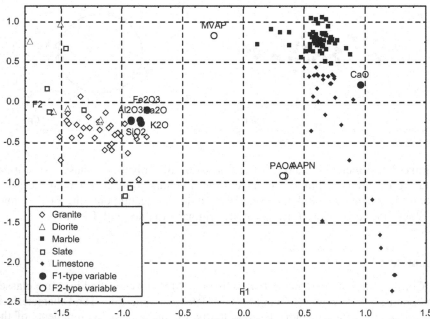

Figure 8.10. Partial view of the standardised rock dataset projected onto the F1-F2 principal component plane, overlaid with the correlation plot.

8.4   Factor Analysis

Let us again consider equation 8.4 which yields the pc-scores of the data using the $d{\times}d$ matrix $\mathbf{U}$ of the eigenvectors:

$$\mathbf{z} = \mathbf{U'}(\mathbf{x} - \overline{\mathbf{x}}).\qquad\qquad 8.16$$

Reversely, with this equation we can obtain the original data from their principal components:

$$\mathbf{x} = \overline{\mathbf{x}} + \mathbf{U}\mathbf{z}.\qquad\qquad 8.17$$

If we discard some principal components, using a reduced $d{\times}k$ matrix $\mathbf{U}_k$, we no longer obtain the original data, but an estimate $\hat{\mathbf{x}}$:

$$\hat{\mathbf{x}} = \overline{\mathbf{x}} + \mathbf{U}_k \mathbf{z}_k.\qquad\qquad 8.18$$

Using 8.17 and 8.18, we can express the original data in terms of the estimation error $\mathbf{e} = \mathbf{x} - \hat{\mathbf{x}}$, as:

$$\mathbf{x} = \overline{\mathbf{x}} + \mathbf{U}_k \mathbf{z}_k + (\mathbf{x} - \hat{\mathbf{x}}) = \overline{\mathbf{x}} + \mathbf{U}_k \mathbf{z}_k + \mathbf{e}.\qquad\qquad 8.19$$

When all principal components are used, the covariance matrix satisfies $\mathbf{S} = \mathbf{U}\,\Lambda\,\mathbf{U'}$ (see formula 8.6 in the properties mentioned in section 8.1). Using the *reduced eigenvector matrix* $\mathbf{U}_k$, and taking 8.19 into account, we can express $\mathbf{S}$ in terms of an approximate covariance matrix $\mathbf{S}_k$ and an error matrix $\mathbf{E}$:

$$\mathbf{S} = \mathbf{U}_k \Lambda \mathbf{U}_k' + \mathbf{E} = \mathbf{S}_k + \mathbf{E}.\qquad\qquad 8.20$$

In factor analysis, the retained principal components are called *common factors*. Their correlations with the original variables are called *factor loadings*. Each common factor $\mathbf{u}_j$ is responsible by a *communality*, h_i^2 , which is the variability associated with the original ith variable:

$$h_i^2 = \sum_{j=1}^{k} \lambda_j u_{ij}^2 .\qquad\qquad 8.21$$

The communalities are the diagonal elements of $\mathbf{S}_k$ and make up a diagonal *communality matrix* $\mathbf{H}$.

Example 8.8

Q: Compute the approximate covariance, communality and error matrices for Example 8.1.

A: Using MATLAB to carry out the computations, we obtain:

$$S_1 = U_1 \Lambda U_1' = \begin{bmatrix} 0.1697 & 0.4539 \\ 0.4539 & 1.2145 \end{bmatrix}; \qquad H = \begin{bmatrix} 0.1697 & 0 \\ 0 & 1.2145 \end{bmatrix};$$

$$E = S - S_1 = \begin{bmatrix} 0.1849 & 0.4482 \\ 0.4482 & 1.2168 \end{bmatrix} - \begin{bmatrix} 0.1697 & 0.4539 \\ 0.4539 & 1.2145 \end{bmatrix} = \begin{bmatrix} 0.0152 & -0.0057 \\ -0.0057 & 0.0023 \end{bmatrix}.$$

□

In the previous example, we can appreciate that the matrix of the diagonal elements of **E** is the difference between the matrix of the diagonal elements of **S** and **H**:

$$\text{diagonal}(S) = \begin{bmatrix} 0.1894 & 0 \\ 0 & 1.2168 \end{bmatrix}$$

$$\text{diagonal}(H) = \begin{bmatrix} 0.1697 & 0 \\ 0 & 1.2145 \end{bmatrix}$$

$$\text{diagonal}(E) = \begin{bmatrix} 0.0152 & 0 \\ 0 & 0.0023 \end{bmatrix} = \text{diagonal}(S) - \text{diagonal}(H)$$

In factor analysis, one searches for a solution for the equation 8.20, such that **E** is a diagonal matrix, i.e., one tries to obtain *uncorrelated errors* from the component estimation process. In this case, representing by **D** the matrix of the diagonal elements of **S**, we have:

$$S = S_k + (D - H). \tag{8.22}$$

In order to cope with different units of the original variables, it is customary to carry out the factor analysis on correlation matrices:

$$R = R_k + (I - H). \tag{8.23}$$

There are several algorithms for finding factor analysis solutions which basically improve current estimates of communalities and factors according to a specific criterion (for details see e.g. Jackson JE, 1991). One such algorithm, known as *principal factor analysis*, starts with an initial estimate of the communalities, e.g. as the multiple R square of the respective variable with all other variables (see formula 7.10). It uses a principal component strategy in order to iteratively obtain improved estimates of communalities and factors.

In principal component analysis, the principal components are directly computed from the data. In factor analysis, the common factors are estimates of unobservable variables, called *latent variables*, which model the data in such a way that the remaining errors are uncorrelated. Equation 8.19 then expresses the observations **x** in terms of the latent variables z_k and uncorrelated errors **e**. The true values of the observations **x**, before any error has been added, are values of the so-called *manifest variables*.

The main benefits of factor analysis when compared with principal component analysis are the non-correlation of the residuals and the invariance of the solutions with respect to scale change.

After finding a factor analysis solution, it is still possible to perform a new transformation that rotates the factors in order to achieve special effects as, for example, to align the factors with maximum variability directions (*varimax* procedure).

Example 8.9

Q: Redo Example 8.8 using principal factor analysis with the communalities computed by the multiple R square method.

A: The correlation matrix is:

$$R = \begin{bmatrix} 1 & 0.945 \\ 0.945 & 1 \end{bmatrix}.$$

Starting with communalities = multiple R^2 square = 0.893, STATISTICA (`Communalities = multiple R`^{2}) converges to solution:

$$H = \begin{bmatrix} 0.919 & 0 \\ 0 & 0.919 \end{bmatrix}; \; \Lambda = \begin{bmatrix} 1.838 & 0 \\ 0 & 0.162 \end{bmatrix}.$$

For unit length eigenvectors, we have:

$$R_1 = U_1 \Lambda U_1' = \begin{bmatrix} 1/\sqrt{2} & 0 \\ 1/\sqrt{2} & 0 \end{bmatrix} \begin{bmatrix} 1.838 & 0 \\ 0 & 0.162 \end{bmatrix} \begin{bmatrix} 1/\sqrt{2} & 1/\sqrt{2} \\ 0 & 0 \end{bmatrix} = \begin{bmatrix} 0.919 & 0.919 \\ 0.919 & 0.919 \end{bmatrix}.$$

Thus: $R_1 + (I - H) = \begin{bmatrix} 1 & 0.919 \\ 0.919 & 1 \end{bmatrix}.$

We see that the residual cross-correlations are only $0.945 - 0.919 = 0.026$. ▯

Example 8.10

Q: Redo Example 8.7 using principal factor analysis and varimax rotation.

A: Using STATISTICA with `Communalities=Multiple R`^{2} checked (see Figure 8.4) in order to apply formula 8.21, we obtain the solution shown in Figure 8.11. The varimax procedure is selected in the `Factor rotation` box included in the `Loadings` tab (after clicking `OK` in the window shown in Figure 8.4).

The rock dataset projected onto the factor plane shown in Figure 8.11 leads us to the same conclusions as in Example 8.7, stressing the opposition SiO_2-CaO and "aligning" the factors in such a way that facilitates the interpretation of the data structure.

▯

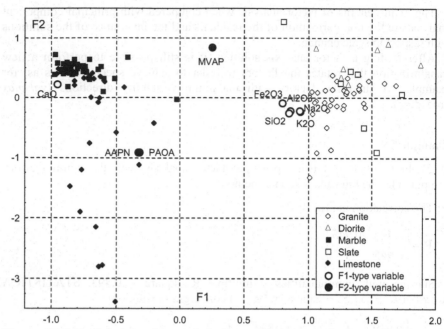

Figure 8.11. Partial view of the rock dataset projected onto the F1-F2 factor plane, after varimax rotation, overlaid with the factor loadings plot.

Exercises

8.1   Consider the standardised electrical impedance features of the Breast Tissue dataset and perform the following principal component analyses:

  a)   Check that only two principal components are needed to explain the data according to the Guttman-Kaiser, broken stick and Velicer criteria.

  b)   Determine which of the original features are highly correlated to the principal components found in a).

  c)   Using a scatter plot of the pc-scores check that the {ADI, CON} class set is separated from all other classes by the first principal component only, whereas the discrimination of the carcinoma class requires the two principal components. (Compare with the results of Examples 6.17 and 6.18.)

  d)   Redo Example 6.16 using the principal components as classifying features. Compare the classification results with those obtained previously.

8.2   Perform a principal component analysis of the correlation matrix of the chemical and grading features of the Clays' dataset, showing that:

  a)   The scree plot has a slow decay after the first eigenvalue. The Velicer criterion indicates that only the first two eigenvalues should be retained.

  b)   The pc correlations show that the first principal component reflects the silica-alumina content of the clays; the second principal component reflects the lime content; and the third principal component reflects the grading.

c)   The scatter plot of the pc-scores of the first two principal components indicates a good discrimination of the two clay types (holocenic and pliocenic).

8.3   Redo the previous Exercise 8.2 using principal factor analysis. Show that only the first factor has a high loading with the original features, namely the alumina content of the clays.

8.4   Design a classifier for the first two classes of the Cork Stoppers' dataset using the main principal components of the data. Compare the classification results with those obtained in Example 6.4.

8.5   Consider the CTG dataset with 2126 cases of foetal heart rate (FHR) features computed in normal, suspect and pathological FHR tracings (variable NSP). Perform a principal component analysis using the feature set {LB, ASTV, MSTV, ALTV, MLTV, WIDTH, MIN, MAX, MODE, MEAN, MEDIAN, V} containing continuous-type features.
a)   Show that the two main principal components computed for the standardised features satisfy the broken-stick criterion.
b)   Obtain a pc correlation plot superimposed onto the pc-scores plot and verify that: first, there is a quite good discrimination of the normal vs. pathological cases with the suspect cases blending in the normal and pathological clusters; and that there are two pathological clusters, one related to a variability feature (MSTV) and the other related to FHR histogram features.

8.6   Using principal factor analysis, determine which original features are the most important explaining the variance of the Firms' dataset. Also compare the principal factor solution with the principal component solution of the standardised features and determine whether either solution is capable to conveniently describe the activity branch of the firms.

8.7   Perform a principal component and a principal factor analysis of the standardised features BASELINE, ACELRATE, ASTV, ALTV, MSTV and MLTV of the FHR-Apgar dataset checking the following results:
a)   The principal factor analysis affords a univariate explanation of the data variance related to the FHR variability features ASTV and ALTV, whereas the principal component analysis affords an explanation requiring three components. Also check the scree plots.
b)   The pc-score plots of the factor analysis solution afford an interpretation of the Apgar index. For this purpose, use the varimax rotation and plot the categorised data using three classes for the Apgar at 1 minute after birth (Apgar1: ≤5; >5 and ≤8; >8) and two classes for the Apgar at 5 minutes after birth (Apgar5: ≤8; >8).

8.8   Redo the previous Exercise 8.7 for the standardised features EF, CK, IAD and GRD of the Infarct dataset showing that the principal component solution affords an explanation of the data based on only one factor highly correlated with the ejection fraction, EF. Check the discrimination capability of this factor for the necrosis severity score SCR > 2 (high) and SCR < 2 (low).

8.9 Consider the Stock Exchange dataset. Using principal factor analysis, determine which economic variable best explains the variance of the whole data.

8.10 Using the Hotteling's T^2 control chart for the wines of the Wines' dataset, determine which wines are "out of control" at 95% confidence level and present an explanation for this fact taking into account the values of the variables highly correlated with the principal components. Use only variables without missing data for the computation of the principal components.

8.11 Perform a principal factor analysis of the wine data studied in the previous Exercise 8.10 showing that there are two main factors, one highly correlated to the GLU-THR variables and the other highly correlated to the PHE-LYS variables. Use varimax rotation and analyse the clustering of the white and red wines in the factor plane superimposed onto the factor loading plane.

8.12 Redo the principal factor analysis of Example 8.10 using three factors and varimax rotation. With the help of a 3D plot interpret the results obtained checking that the three factors are related to the following original variables: SiO2-Al2O3-CaO (silica-lime factor), AAPN-AAOA (porosity factor) and RMCS-RCSG (resistance factor).

9 Survival Analysis

In medical studies one is often interested in studying the expected time until the death of a patient, undergoing a specific treatment. Similarly, in technological tests, one is often interested in studying the amount of time until a device subjected to specified conditions fails. Times until death and times until failure are examples of *survival data*. The statistical analysis of survival data is based on the use of specific data models and probability distributions. In this chapter, we present several introductory topics of survival analysis and their application to survival data using SPSS, STATISTICA, MATLAB and R (survival package).

9.1 Survivor Function and Hazard Function

Consider a random variable $T \in \Re^+$ representing the *lifetime* of a class of objects or individuals, and let $f(t)$ denote the respective pdf. The distribution function of T is:

$$F(t) = P(T < t) = \int_0^t f(u)du. \tag{9.1}$$

In general, $f(t)$ is a positively skewed function, with a long right tail. Continuous distributions such as the exponential or the Weibull distributions (see B.2.3 and B.2.4) are good candidate models for $f(t)$.

The *survivor function* or *reliability function*, $S(t)$, is defined as the probability that the lifetime (survival time) of the object is greater than or equal to t:

$$S(t) = P(T \geq t) = 1 - F(t). \tag{9.2}$$

The *hazard function* (or *failure rate function*) represents the probability that the object ends its lifetime (fails) at time t, conditional on having survived until that time. In order to compute this probability, we first consider the probability that the survival time T lies between t and $t + \Delta t$, conditioned on $T \geq t$: $P(t \leq T < t + \Delta t \mid T \geq t)$. The hazard function is the limit of this probability when $\Delta t \to 0$:

$$h(t) = \lim_{\Delta t \to 0} \frac{P(t \leq T < t + \Delta t \mid T \geq t)}{\Delta t}. \tag{9.3}$$

Given the property A.7 of conditional probabilities, the numerator of 9.3 can be written as:

$$P(t \leq T < t + \Delta t \mid t \geq t) = \frac{P(t \leq T < t + \Delta t)}{P(T \geq t)} = \frac{F(t + \Delta t) - F(t)}{S(t)}.$$ 9.4

Thus:

$$h(t) = \lim_{\Delta t \to 0} \frac{F(t + \Delta t) - F(t)}{\Delta t} \frac{1}{S(t)} = \frac{f(t)}{S(t)},$$ 9.5

since $f(t)$ is the derivative of $F(t)$: $f(t) = dF(t) / dt$.

9.2 Non-Parametric Analysis of Survival Data

9.2.1 The Life Table Analysis

In survival analysis, the survivor and hazard functions are estimated from the observed survival times. Consider a set of ordered survival times t_1, t_2, ..., t_k. One may then estimate any particular value of the survivor function, $S(t_i)$, in the following way:

$S(t_i) = P$(surviving to time t_i) =
    P(surviving to time t_1)
    $\times P$(surviving to time t_1 | survived to time t_2)
    ...
    $\times P$(surviving to time t_i | survived to time t_{i-1}). 9.6

Let us denote by n_j the number of individuals that are alive at the start of the interval $[t_j , t_{j+1}[$, and by d_j the number of individuals that die during that interval. We then derive the following non-parametric estimate:

$$\hat{P}(\text{surviving to } t_{j+1} \mid \text{survived to } t_j) = \frac{n_j - d_j}{n_j},$$ 9.7

from where we estimate $S(t_i)$ using formula 9.6.

Example 9.1

Q: A car stand has a record of the sale date and the date that a complaint was first presented for three different cars (this is a subset of the Car Sale dataset in Appendix E). These dates are shown in Table 9.1. Compute the estimate of the time-to-complaint probability for $t = 300$ days.

A: In this example, the time-to-complaint, "Complaint Date" – "Sale Date", is the survival time. The computed times in days are shown in the last column of Table 9.1. Since there are no complaints occurring between days 261 and 300, we may apply 9.6 and 9.7 as follows:

$\hat{S}(300) = \hat{S}(261) = \hat{P}(\text{surviving to } 240)\, \hat{P}(\text{surviving to } 261 \,|\, \text{survived to } 240)$

$$= \frac{3-1}{2} \frac{2-1}{2} = \frac{1}{3}.$$

Alternatively, one could also compute this estimate as $(3 - 2)/3$, considering the [0, 261] interval.   □

Table 9.1. Time-to-complaint data in car sales (3 cars).

Car	Sale Date	Complaint Date	Time-to-complaint (days)
#1	1-Nov-00	29-Jun-01	240
#2	22-Nov-00	10-Aug-01	261
#3	16-Feb-01	30-Jan-02	348

In a survival study, information concerning the "death" times of one or more cases that entered the study is often not available either because the cases were "lost" during the study or because they are still "alive" at the end of the study. These are the so-called *censored* cases[1].

The information of the censored cases must also be taken into consideration when estimating the survivor function. Let us denote by c_j the number of cases censored in the interval $[t_j, t_{j+1}[$. The *actuarial* or *life-table* estimate of the survivor function is a non-parametric estimate that assumes that the censored survival times occur uniformly throughout that interval, so that the average number of individuals that are at risk of dying during $[t_j, t_{j+1}[$ is:

$$n_j^* = n_j - c_j/2.$$ 9.8

Taking into account formulas 9.6 and 9.7, the life-table estimate of the survivor function is computed as:

$$\hat{S}(t) = \prod_{j=1}^{k} \left(\frac{n_j^* - d_j}{n_j^*} \right), \quad \text{for } t_k \leq t < t_{k+1}.$$ 9.9

The hazard function is an estimate of 9.5, given by:

$$\hat{h}(t) = \frac{d_j}{(n_j^* - d_j/2)\tau_j}, \quad \text{for } t_j \leq t < t_{j+1},$$ 9.10

where τ_j is the length of the jth time interval.

---

[1] The type of censoring described here is the one most frequently encountered, known as right censoring. There are other, less frequent types of censoring.

Example 9.2

Q: Consider that the data records of the car stand (Car Sale dataset), presented in the previous example, was enlarged to 8 cars, as shown in Table 9.2. Determine the survivor and hazard functions using the life-table estimate.

A: We now have two sources of censored data: the three cars that are known to have had no complaints at the end of the study, and one car whose owner could not be contacted at the end of the study, but whose car was known to have had no complaint at a previous date. We can summarise this information as shown in Table 9.3.

Using SPSS, with the time-to-complaint and censored columns of Table 9.3 and a specification of displaying time intervals 0 through 600 days by 75 days, we obtain the life-table estimate results shown in Table 9.4. Figure 9.1 shows the survivor function plot. Note that it is a monotonic decreasing function.

Table 9.2. Time-to-complaint data in car sales (8 cars).

Car	Sale Date	Complaint Date	Without Complaint at the End of the Study	Last Date Known to be Without Complaint
#1	12-Sep-00		31-Mar-02	
#2	26-Oct-00	31-Mar-02		
#3	01-Nov-00	29-Jun-01		
#4	22-Nov-00	10-Aug-01		
#5	18-Jan-01		31-Mar-02	
#6	02-Jul-01			24-Sep-01
#7	16-Feb-01	30-Jan-02		
#8	03-May-01		31-Mar-02	

Table 9.3. Summary table of the time-to-complaint data in car sales (8 cars).

Car	Start Date	Stop Date	Censored	Time-to-complaint (days)
#1	12-Sep-00	31-Mar-02	TRUE	565
#2	26-Oct-00	31-Mar-02	FALSE	521
#3	01-Nov-00	29-Jun-01	FALSE	240
#4	22-Nov-00	10-Aug-01	FALSE	261
#5	18-Jan-01	31-Mar-02	TRUE	437
#6	02-Jul-01	24-Sep-01	TRUE	84
#7	16-Feb-01	30-Jan-02	FALSE	348
#9	03-May-01	31-Mar-02	TRUE	332

Columns 2 through 5 of Table 9.4 list the successive values of n_j, c_j, n_j^*, and d_j, respectively. The "Propn Surviving" column is obtained by applying formula 9.7 with correction for censored data (formula 9.8). The "Cumul Propn Surv at End" column lists the values of $\hat{S}(t)$ obtained with formula 9.9. The "Propn Terminating" column is the complement of the "Propn Surviving" column. Finally, the last two columns list the values of the probability density and hazard functions, computed with the finite difference approximation of $f(t) = \Delta F(t)/\Delta t$ and formula 9.5, respectively.   □

Table 9.4. Life-table of the time-to-complaint data, obtained with SPSS.

Intrvl Start Time	Number Entrng this Intrvl	Number Wdrawn During Intrvl	Number Exposed to Risk	Number of Termnl Events	Propn Termi- nating	Propn Sur- viving	Cumul Propn Surv at End	Proba- bility Density	Hazard Rate
0	8	0	8	0	0	1	1	0	0
75	8	1	7.5	0	0	1	1	0	0
150	7	0	7	0	0	1	1	0	0
225	7	0	7	2	0.2857	0.7143	0.7143	0.0038	0.0044
300	5	1	4.5	1	0.2222	0.7778	0.5556	0.0021	0.0033
375	3	1	2.5	0	0	1	0.5556	0	0
450	2	0	2	1	0.5	0.5	0.2778	0.0037	0.0089
525	1	1	0.5	0	0	1	0.2778	0	0

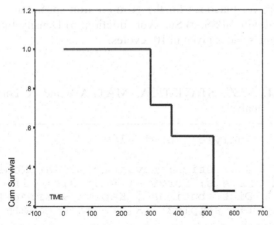

Figure 9.1. Life-table estimate of the survivor function for the time-to-complaint data (first eight cases of the Car Sale dataset) obtained with SPSS.

Example 9.3

Q: Consider the amount of time until breaking of iron specimens, submitted to low amplitude sinusoidal loads (Group 1) in fatigue tests, a sample of which is given in the Fatigue dataset. Determine the survivor, hazard and density functions using the life-table estimate procedure. What is the estimated percentage of specimens breaking beyond 2 million cycles? In addition determine the estimated percentage of specimens that will break at 500000 cycles.

A: We first convert the time data, given in number of 20 Hz cycles, to a lower range of values by dividing it by 10000. Next, we use this data with SPSS, assigning the Break variable as a censored data indicator (Break = 1 if the specimen has broken), and obtain the plots of the requested functions between 0 and 400 with steps of 20, shown in Figure 9.2.

Note the right tailed, positively skewed aspect of the density function, shown in Figure 9.2b, typical of survival data. From Figure 9.2a, we see that the estimated percentage of specimens surviving beyond 2 million cycles (marked 200 in the *t* axis) is over 45%. From Figure 9.2c, we expect a break rate of about 0.4% at 500000 cycles (marked 50 in the *t* axis). □

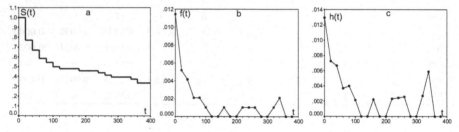

Figure 9.2. Survival functions for the group 1 iron specimens of the Fatigue dataset, obtained with SPSS: a) Survivor function; b) Density function; c) Hazard function. The time scale is given in 10^4 cycles.

Commands 9.1. SPSS, STATISTICA, MATLAB and R commands used to perform survival analysis.

SPSS	Analyze; Survival
STATISTICA	Statistics; Advanced Linear/Nonlinear Models; Survival Analysis; Life tables & Distributions | Kaplan & Meier | Comparing two samples | Regression models
MATLAB	[par, pci] = expfit(x,alpha) [par, pci] = weibfit(x,alpha)
R	Surv(time,event); survfit(survobject) survdif(survobject ~ group, rho) coxph(survobject ~ factor)

SPSS uses as input data in survival analysis the survival time (e.g. last column of Table 9.3) and a censoring variable (Status). STATISTICA allows, as an alternative, the specification of the start and stop dates (e.g., second and third columns of Table 9.3) either in date format or as separate columns for day, month and year. All the features described in the present chapter are easily found in SPSS or STATISTICA windows.

MATLAB stats toolbox does not have specific functions for survival analysis. It has, however, the expfit and weibfit functions which can be used for parametric survival analysis (see section 9.4) since they compute the maximum likelihood estimates of the parameters of the exponential and Weibull distributions, respectively, fitting the data vector x. The parameter estimates are returned in par. The confidence intervals of the parameters, at alpha significance level, are returned in pci.

A suite of R functions for survival analysis, together with functions for operating with dates, is available in the survival package. Be sure to load it first with library(survival). The Surv function is used as a preliminary operation to create an object (a Surv object) that serves as an argument for other functions. The arguments of Surv are a time and event vectors. The event vector contains the censored information. Let us illustrate the use of Surv for the Example 9.2 dataset. We assume that the last two columns of Table 9.3 are stored in t and ev, respectively for "Time-to-complaint" and "Censored", and that the ev values are 1 for "censored" and 0 for "not censored". We then apply Surv as follows:

```
> x <- Surv(t[1:8],ev[1:8]==0)
> x
[1] 565+ 521  240  261  437+  84+ 348  332+
```

The event argument of Surv must specify which value corresponds to the "not censored"; hence, the specification ev[1:8]==0. In the list above the values marked with "+" are the censored observations (any observation with an event label different from 0 is deemed "censored"). We may next proceed, for instance, to create a Kaplan-Meier estimate of the data using survfit(x) (or, if preferred, survfit(Surv(t[1:8],ev[1:8]==0))).

The survdiff function provides tests for comparing groups of survival data. The argument rho can be 0 or 1 depending on whether one wants the log-rank or the Peto-Wilcoxon test, respectively.

The cosxph function fits a Cox regression model for a specified factor. ∎

9.2.2 The Kaplan-Meier Analysis

The Kaplan-Meier estimate, also known as *product-limit* estimate of the survivor function is another type of non-parametric estimate, which uses intervals starting at "death" times. The formula for computing the estimate of the survivor function is similar to formula 9.9, using n_j instead of n_j^*:

$$\hat{S}(t) = \prod_{j=1}^{k} \left(\frac{n_j - d_j}{n_j} \right), \quad \text{for } t_k \leq t < t_{k+1}. \tag{9.11}$$

Since, by construction, there are n_j individuals who are alive just before t_j and d_j deaths occurring at t_j, the probability that an individual dies between $t_j - \delta$ and t_j is estimated by d_j / n_j. Thus, the probability of individuals surviving through $[t_j, t_{j+1}[$ is estimated by $(n_j - d_j)/ n_j$.

The only influence of the censored data is in the computation of the number of individuals, n_j, who are alive just before t_j. If a censored survival time occurs simultaneously with one or more deaths, then the censored survival time is taken to occur immediately after the death time.

The Kaplan-Meier estimate of the hazard function is given by:

$$\hat{h}(t) = \frac{d_j}{n_j \tau_j}, \quad \text{for } t_j \leq t < t_{j+1}, \tag{9.12}$$

where τ_j is the length of the jth time interval. For details, see e.g. (Collet D, 1994) or (Kleinbaum DG, Klein M, 2005).

Example 9.4

Q: Redo Example 9.2 using the Kaplan-Meier estimate.

A: Table 9.5 summarises the computations needed for obtaining the Kaplan-Meier estimate of the "time-to-complaint" data. Figure 9.3 shows the respective survivor function plot obtained with STATISTICA. The computed data in Table 9.5 agrees with the results obtained with either STATISTICA or SPSS.

In R one uses the `survfit` function to obtain the Kaplan-Meier estimate. Assuming one has created the `Surv` object x as explained in Commands 9.1, one proceeds to calling `survfit(x)`. A plot as in Figure 9.3, with Greenwood's confidence interval (see section 9.2.3), can be obtained with `plot(survfit(x))`. Applying `summary` to `survfit(x)` the confidence intervals for $S(t)$ are displayed as follows:

```
time  n.risk n.event survival  std.err lower 95% CI upper 95% CI
240        7       1    0.857    0.132       0.6334            1
261        6       1    0.714    0.171       0.4471            1
348        4       1    0.536    0.201       0.2570            1
521        2       1    0.268    0.214       0.0558            1
```

□

Table 9.5. Kaplan-Meier estimate of the survivor function for the first eight cases of the Car Sale dataset.

Interval Start	Event	n_j	d_j	p_j	S_j
84	Censored	8	0	1	1
240	"Death"	7	1	0.8571	0.8571
261	"Death"	6	1	0.8333	0.7143
332	Censored	5	0	1	0.7143
348	"Death"	4	1	0.75	0.5357
437	Censored	3	0	1	0.5357
521	"Death"	2	1	0.5	0.2679
565	Censored	1	0	1	0.2679

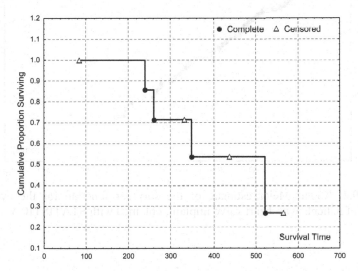

Figure 9.3. Kaplan-Meier estimate of the survivor function for the first eight cases of the Car Sale dataset, obtained with STATISTICA. (The "Complete" cases are the "deaths".)

Example 9.5

Q: Consider the Heart Valve dataset containing operation dates for heart valve implants at São João Hospital, Porto, Portugal, and dates of subsequent event occurrences, namely death, re-operation and endocarditis. Compute the Kaplan-Meier estimate for the event-free survival time, that is, survival time without occurrence of death, re-operation or endocarditis events. What is the percentage of patients surviving 5 years without any event occurring?

A: The `Heart Valve Survival` datasheet contains the computed final date for the study (variable DATE_STOP). This is the date of the first occurring event, if it did occur, or otherwise, the last date the patient was known to be alive and well. The survivor function estimate shown in Figure 9.4 is obtained by using STATISTICA with DATE_OP and DATE_STOP as initial and final dates, and variable EVENT as censored data indicator. From this figure, one can estimate that about 85% of patients survive five years (1825 days) without any event occurring.

□

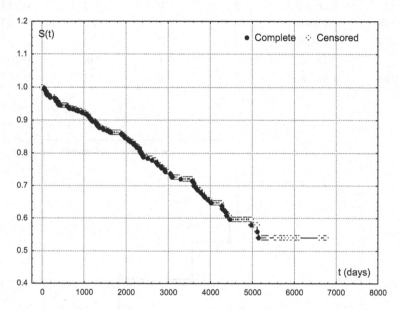

Figure 9.4. Kaplan-Meier estimate of the survivor function for the event-free survival of patients with heart valve implant, obtained with STATISTICA.

9.2.3 Statistics for Non-Parametric Analysis

The following statistics are often needed when analysing survival data:

1. Confidence intervals for $S(t)$.

For the Kaplan-Meier estimate, the confidence interval is computed assuming that the estimate $\hat{S}(t)$ is normally distributed (say for a number of intervals above 30), with mean $S(t)$ and standard error given by the *Greenwood's formula*:

$$s\left[\hat{S}(t)\right] \approx \hat{S}(t)\left\{\sum_{j=1}^{k}\frac{d_j}{n_j(n_j-d_j)}\right\}^2 \;,\; \text{for } t_k \leq t < t_{k+1}. \qquad 9.13$$

2. Median and percentiles of survival time.

Since the density function of the survival times, $f(t)$, is usually a positively skewed function, the median survival time, $t_{0.5}$, is the preferred location measure. The median can be obtained from the survivor function, namely:

$$F(t_{0.5}) = 0.5 \quad \Rightarrow \quad S(t_{0.5}) = 1 - 0.5 = 0.5. \tag{9.14}$$

When using non-parametric estimates of the survivor function, it is usually not possible to determine the exact value of $t_{0.5}$, given the stepwise nature of the estimate $\hat{S}(t)$. Instead, the following estimate is determined:

$$\hat{t}_{0.5} = \min\{t_i; \quad \hat{S}(t_i) \le 0.5\}. \tag{9.15}$$

Percentiles p of the survival time are computed in the same way:

$$\hat{t}_p = \min\{t_i; \quad \hat{S}(t_i) \le 1 - p\}. \tag{9.16}$$

3. Confidence intervals for the median and percentiles.

Confidence intervals for the median and percentiles are usually determined assuming a normal distribution of these statistics for a sufficiently large number of cases (say, above 30), and using the following formula for the standard error of the percentile estimate (for details see e.g. Collet D, 1994 or Kleinbaum DG, Klein M, 2005):

$$s[\hat{t}_p] = \frac{1}{\hat{f}(\hat{t}_p)} s[\hat{S}(\hat{t}_p)], \tag{9.17}$$

where the estimate of the probability density can be obtained by a finite difference approximation of the derivative of $\hat{S}(t)$.

Example 9.6

Q: Determine the 95% confidence interval for the survivor function of Example 9.3, as well as for the median and 60% percentile.

A: SPSS produces an output containing the value of the median and the standard errors of the survivor function. The standard values of the survivor function can be used to determine the 95% confidence interval, assuming a normal distribution. The survivor function with the 95% confidence interval is shown in Figure 9.5.

The median survival time of the specimens is $100 \times 10^4 = 1$ million cycles. The 60% percentile survival time can be estimated as follows:

$$\hat{t}_{0.6} = \min\{t_i; \quad \hat{S}(t_i) \le 1 - 0.6\}.$$

From Figure 9.5 (or from the life table), we then see that $\hat{t}_{0.6} = 280 \times 10^4$ cycles. Let us now compute the standard errors of these estimates:

$$s[100] = \frac{1}{\hat{f}(100)} \, s\left[\hat{S}(100)\right] = \frac{0.0721}{0.001} = 72.1 \,.$$

$$s[280] = \frac{1}{\hat{f}(280)} \, s\left[\hat{S}(280)\right] = \frac{0.0706}{0.001} = 70.6 \,.$$

Thus, under the normality assumption, the 95% confidence intervals for the median and 60% percentile of the survival times are [0, 241.3] and [41.6, 418.4], respectively. We observe that the non-parametric confidence intervals are too large to be useful. Only for a much larger number of cases are the survival functions shown in Figure 9.2 smooth enough to produce more reliable estimates of the confidence intervals.   □

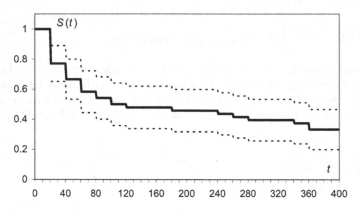

Figure 9.5. Survivor function of the group 1 iron specimens, of the `Fatigue` dataset with the 95% confidence interval (plot obtained with EXCEL using SPSS results). The time scale is given in 10^4 cycles.

9.3   Comparing Two Groups of Survival Data

Let $h_1(t)$ and $h_2(t)$ denote the hazard functions of two independent groups of survival data, often called the *exposed* and *unexposed groups*. Comparison of the two groups of survival data can be performed as a hypothesis test formalised in terms of the *hazard ratio* $\psi = h_1(t)/ h_2(t)$, as follows:

H_0: $\psi = 1$ (survival curves are the same);
H_1: $\psi \neq 1$ (one of the groups will consistently be at a greater risk).

The following two non-parametric tests are of widespread use:

1. The Log-Rank Test.

Suppose that there are r distinct death times, t_1, t_2, ..., t_r, across the two groups, and that at each time t_j, there are d_{1j}, d_{2j} individuals of groups 1 and 2 respectively, that die. Suppose further that just before time t_j, there are n_{1j}, n_{2j} individuals of groups 1 and 2 respectively, at risk of dying. Thus, at time t_j there are $d_j = d_{1j} + d_{2j}$ deaths in a total of $n_j = n_{1j} + n_{2j}$ individuals at risk, as shown in Table 9.6.

Table 9.6. Number of deaths and survivals at time t_j in a two-group comparison.

Group	Deaths at t_j	Survivals beyond t_j	Individuals at risk before $t_j - \delta$
1	d_{1j}	$n_{1j} - d_{1j}$	n_{1j}
2	d_{2j}	$n_{2j} - d_{2j}$	n_{2j}
Total	d_j	$n_j - d_j$	n_j

If the marginal totals along the rows and columns in Table 9.6 are considered fixed, and the null hypothesis is true (survival time is independent of group), the remaining four cells in Table 9.6 only depend on one of the group deaths, say d_{1j}. As described in section B.1.4, the probability of the associated random variable, D_{1j}, taking value in $[0, \min(n_{1j}, d_j)]$, is given by the hypergeometric law:

$$P(D_{1j} = d_{1j}) = h_{n_j, d_j, m_j}(d_{1j}) = \binom{d_j}{d_{1j}} \binom{n_j - d_j}{n_{1j} - d_{1j}} / \binom{n_j}{n_{1j}}.$$
9.18

The mean of D_{1j} is the expected number of group 1 individuals who die at time t_j (see B.1.4):

$$e_{1j} = n_{1j}(d_j / n_j).$$
9.19

The *Log-Rank* test combines the information of all 2×2 contingency tables, similar to Table 9.6 that one can establish for all t_j, using a test based on the χ^2 test (see 5.1.3). The method of combining the information of all 2×2 contingency tables is known as the *Mantel-Haenszel procedure*. The test statistic is:

$$\chi^{*2} = \frac{\left(\left| \sum_{j=1}^{r} d_{1j} - \sum_{j=1}^{r} e_{1j} \right| - 0.5 \right)^2}{\sum_{j=1}^{r} \dfrac{n_{1j} n_{2j} d_j (n_j - d_j)}{n_j^2 (n_j - 1)}} \sim \chi_1^2 \text{ (under } H_0\text{)}.$$
9.20

Note that the numerator, besides the 0.5 continuity correction, is the absolute difference between observed and expected frequencies of deaths in group 1. The denominator is the sum of the variances of D_{1j}, according to the hypergeometric law.

2. The Peto-Wilcoxon test.

The Peto-Wilcoxon test uses the following test statistic:

$$W = \frac{\left(\sum_{j=1}^{r} n_j (d_{1j} - e_{1j})\right)^2}{\sum_{j=1}^{r} \dfrac{n_{1j} n_{2j} d_j (n_j - d_j)}{n_j - 1}} \quad \sim \quad \chi_1^2 \text{ (under } H_0). \tag{9.21}$$

This statistic differs from 9.20 on the factor n_j that weighs the differences between observed and expected group 1 deaths.

The Log-Rank test is more appropriate then the Peto-Wilcoxon test when the alternative hypothesis is that the hazard of death for an individual in one group is proportional to the hazard at that time for a similar individual in the other group. The validity of this *proportional hazard assumption* can be elucidated by looking at the survivor functions of both groups. If they clearly do not cross each other then the proportional hazard assumption is quite probably true, and the Log-Rank test should be used. In other cases, the Peto-Wilcoxon test is used instead.

Example 9.7

Q: Consider the fatigue test results for iron and aluminium specimens, subject to low amplitude sinusoidal load (Group 1), given in the Fatigue dataset. Compare the survival times of the iron and aluminium specimens using the Log-Rank and the Peto-Wilcoxon tests.

A: With SPSS or STATISTICA one must fill in a datasheet with columns for the "time", censored and group data. In SPSS one must run the test within the Kaplan-Meier option and select the appropriate test in the Compare Factor window. Note that SPSS calls the Peto-Wilcoxon test as Breslow test.

In R the survdiff function for the log-rank test (default value for rho, rho = 0), is applied as follows:

```
> survdiff(Surv(cycles,break==1) ~ group)
Call:
survdiff(formula = Surv(cycles, cens == 1) ~ group)

          N Observed Expected (O-E)^2/E (O-E)^2/V
group=1 39       23     24.6    0.1046     0.190
group=2 48       32     30.4    0.0847     0.190

 Chisq= 0.2  on 1 degrees of freedom, p= 0.663
```

The Peto-Wilcoxon test is performed by setting `rho = 1`.

SPSS, STATISTICA and R report observed significances of 0.66 and 0.89 for the Log-Rank and Peto-Wilcoxon tests, respectively.

Looking at the survivor functions shown in Figure 9.6, drawn with values computed with STATISTICA, we observe that they practically do not cross. Therefore, the proportional hazard assumption is probably true and the Log-Rank is more appropriate than the Peto-Wilcoxon test. With $p = 0.66$, the null hypothesis of equal hazard functions is not rejected.   □

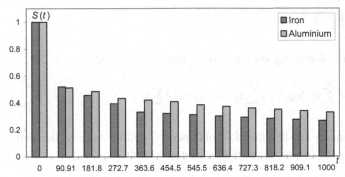

Figure 9.6. Life-table estimates of the survivor functions for the iron and aluminium specimens (Group 1). (Plot obtained with EXCEL using SPSS results.)

9.4 Models for Survival Data

9.4.1 The Exponential Model

The simplest distribution model for survival data is the exponential distribution (see B.2.3). It is an appropriate distribution when the hazard function is constant, $h(t) = \lambda$, i.e., the age of the object has no effect on its probability of surviving (*lack of memory* property). Using 9.2 one can write the hazard function 9.5 as:

$$h(t) = \frac{-dS(t)/dt}{S(t)} = -\frac{d \ln S(t)}{dt}.$$ 9.22

Equivalently:

$$S(t) = \exp\left[-\int_0^t h(u)du \right].$$ 9.23

Thus, when $h(t) = \lambda$, we obtain the exponential distribution:

$$S(t) = e^{-\lambda t} \quad \Rightarrow \quad f(t) = \lambda e^{-\lambda t}.$$ 9.24

The exponential model can be fitted to the data using a maximum likelihood procedure (see Appendix C). Concretely, let the data consist of n survival times, t_1, t_2, ..., t_n, of which r are death times and $n - r$ are censored times. Then, the likelihood function is:

$$L(\lambda) = \prod_{i=1}^{n} (\lambda e^{-\lambda t_i})^{\delta_i} (e^{-\lambda t_i})^{1-\delta_i} \text{ with } \delta_i = \begin{cases} 0 & i\text{th individual is censored} \\ 1 & \text{otherwise} \end{cases} . \quad 9.25$$

Equivalently:

$$L(\lambda) = \prod_{i=1}^{n} \lambda^{\delta_i} e^{-\lambda t_i} , \quad\quad\quad 9.26$$

from where the following log-likelihood formula is derived:

$$\log L(\lambda) = \sum_{i=1}^{n} \delta_i \log \lambda - \lambda \sum_{i=1}^{n} t_i = r \log \lambda - \lambda \sum_{i=1}^{n} t_i . \quad\quad 9.27$$

The maximum log-likelihood is obtained by setting to zero the derivative of 9.27, yielding the following estimate of the parameter λ:

$$\hat{\lambda} = \frac{1}{r} \sum_{i=1}^{n} t_i . \quad\quad\quad 9.28$$

The standard error of this estimate is $\hat{\lambda}/r$. The following statistics are easily derived from 9.24:

$$\hat{t}_{0.5} = \ln 2 / \hat{\lambda} . \quad\quad\quad 9.29a$$

$$\hat{t}_p = \ln(1/(1-p)) / \hat{\lambda} . \quad\quad\quad 9.29b$$

The standard error of these estimates is $\hat{t}_p / \sqrt{r}$.

Example 9.8

Q: Consider the survival data of Example 9.5 (Heart Valve dataset). Determine the exponential estimate of the survivor function and assess the validity of the model. What are the 95% confidence intervals of the parameter λ and of the median time until an event occurs?

A: Using STATISTICA, we obtain the survival and hazard functions estimates shown in Figure 9.7. STATISTICA uses a weighted least square estimate of the model function instead of the log-likelihood procedure. The exponential model fit shown in Figure 9.7 is obtained using weights $n_i h_i$, where n_i is the number of observations at risk in interval i of width h_i. Note that the life-table estimate of the hazard function is suggestive of a constant behaviour. The chi-square goodness of fit test yields an observed significance of 0.59; thus, there is no evidence leading to the rejection of the null, goodness of fit, hypothesis.

STATISTICA computes the estimated parameter as $\hat{\lambda} = 9.8 \times 10^{-5}$ (day^{-1}), with standard error $s = 1 \times 10^{-5}$. Therefore, the 95% confidence interval, under the normality assumption, is $[7.84 \times 10^{-5}, 11.76 \times 10^{-5}]$.

Applying formula 9.29, the median is estimated as $\ln 2 / \hat{\lambda} = 3071$ days $= 8.4$ years. Since there are $r = 106$ events, the standard error of this estimate is 0.8 years. Therefore, the 95% confidence interval of the median event-free time, under the normality assumption, is [6.8, 10] years.   □

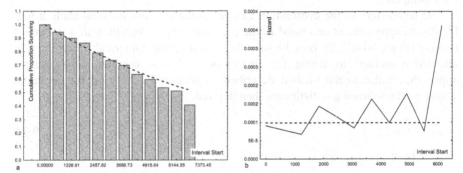

Figure 9.7. Survivor function (a) and hazard function (b) for the Heart Valve dataset with the fitted exponential estimates shown with dotted lines. Plots obtained with STATISTICA

9.4.2 The Weibull Model

The Weibull distribution offers a more general model for describing survival data than the exponential model does. Instead of a constant hazard function, it uses the following parametric form, with positive parameters λ and γ, of the hazard function:

$$h(t) = \lambda \gamma \, t^{\gamma-1}.$$ 9.30

The exponential model corresponds to the particular case $\gamma = 1$. For $\gamma > 1$, the hazard increases monotonically with time, whereas for $\gamma < 1$, the hazard function decreases monotonically. Taking into account 9.23, one obtains:

$$S(t) = e^{-\lambda t^{\gamma}}.$$ 9.31

The probability density function of the survival time is given by the derivative of $F(t) = 1 - S(t)$. Thus:

$$f(t) = \lambda \gamma \, t^{\gamma-1} e^{-\lambda t^{\gamma}}.$$ 9.32

This is the Weibull density function with shape parameter γ and scale parameter $\sqrt[\gamma]{1/\lambda}$ (see B.2.4):

$$f(t) = w_{\gamma,\sqrt[\gamma]{1/\lambda}}(t).$$ 9.33

Figure B.11 illustrates the influence of the shape and scale parameters of the Weibull distribution. Note that in all cases the distribution is positively skewed, i.e., the probability of survival in a given time interval always decreases with increasing time.

The parameters of the distribution can be estimated from the data using a log-likelihood approach, as described in the previous section, resulting in a system of two equations, which can only be solved by an iterative numerical procedure. An alternative method to fitting the distribution uses a weighted least squares approach, similar to the method described in section 7.1.2. From the estimates $\hat{\lambda}$ and $\hat{\gamma}$, the following statistics are then derived:

$$\hat{t}_{0.5} = \left(\ln 2 / \hat{\lambda}\right)^{1/\hat{\gamma}}.$$ 9.34

$$\hat{t}_p = \left(\ln(1/(1-p))/\hat{\lambda}\right)^{1/\hat{\gamma}}.$$

The standard error of these estimates has a complex expression (see e.g. Collet D, 1994 or Kleinbaum DG, Klein M, 2005).

In the assessment of the suitability of a particular distribution for modelling the data, one can resort to the comparison of the survivor function obtained from the data, using the Kaplan-Meier estimate, $\hat{S}(t)$, with the survivor function prescribed by the model, $S(t)$. From 9.31 we have:

$$\ln(-\ln S(t)) = \ln \lambda + \gamma \ln t.$$ 9.35

If $S(t)$ is close to $\hat{S}(t)$, the *log-cumulative hazard plot* of $\ln(-\ln \hat{S}(t))$ against $\ln t$ will be almost a straight line.

An alternative way to assessing the suitability of the model uses the χ^2 goodness of fit test described in section 5.1.3.

Example 9.9

Q: Consider the amount of time until breaking of aluminium specimens submitted to high amplitude sinusoidal loads in fatigue tests, a sample of which is given in the Fatigue dataset. Determine the Weibull estimate of the survivor function and assess the validity of the model. What is the point estimate of the median time until breaking?

A: Figure 9.8 shows the Weibull estimate of the survivor function, determined with STATISTICA (Life tables & Distributions, Number of intervals = 12), using a weighted least square approach similar to the one mentioned in Example 9.8 (Weight 3). Note that the t values are divided, as in

Example 9.3, by 10^4. The observed probability of the chi-square goodness of fit test is very high: $p = 0.96$. The model parameters computed by STATISTICA are:

$$\hat{\lambda}_1 = 0.187; \quad \hat{\gamma} = 0.703.$$

Figure 9.7 also shows the log-cumulative hazard plot obtained with EXCEL and computed from the values of the Kaplan-Meier estimate. From the straight-line fit of this plot, one can compute another estimate of the parameter $\hat{\gamma} = 0.639$. Inspection of this plot and the previous chi-square test result are indicative of a good fit to the Weibull distribution. The point estimate of the median time until breaking is computed with formula 9.34:

$$\hat{t}_{0.5} = \left(\ln 2 / \hat{\lambda}\right)^{1/\hat{\gamma}} = \left(\frac{0.301}{0.1867}\right)^{1.42} = 1.97.$$

Thus, taking into account the 10^4 scale factor used for the t axis, a median number of 1970020 cycles is estimated for the time until breaking of the aluminium specimens.   ☐

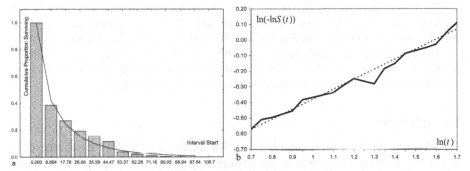

Figure 9.8. Fitting the Weibull model to the time until breaking of aluminium specimens submitted to high amplitude sinusoidal loads in fatigue tests: a) Life-table estimate of the survivor function with Weibull estimate (solid line); b) Log-cumulative hazard plot (solid line) with fitted regression line (dotted line).

9.4.3 The Cox Regression Model

When analysing survival data, one is often interested in elucidating the influence of *explanatory variables* in the survivor and hazard functions. For instance, when analysing the Heart Valve dataset, one is probably interested in knowing the influence of a patient's age on chances of surviving.

Let $h_1(t)$ and $h_2(t)$ be the hazards of death at time t, for two groups: 1 and 2. The *Cox regression model* allows elucidating the influence of the group variable using the *proportional hazards assumption*, i.e., the assumption that the hazards can be expressed as:

$$h_1(t) = \psi h_2(t), \tag{9.36}$$

where the positive constant ψ is known as the *hazard ratio*, mentioned in 9.3.

Let X be an indicator variable such that its value for the ith individual, x_i, is 1 or 0, according to the group membership of the individual. In order to impose a positive value to ψ, we rewrite formula 9.36 as:

$$h_i(t) = e^{\beta x_i} h_0(t). \tag{9.37}$$

Thus $h_2(t) = h_0(t)$ and $\psi = e^\beta$. This model can be generalised for p explanatory variables:

$$h_i(t) = e^{\eta_i} h_0(t), \text{ with } \eta_i = \beta_1 x_{1i} + \beta_2 x_{2i} + \ldots + \beta_p x_{pi}, \tag{9.38}$$

where η_i is known as the *risk score* and $h_0(t)$ is the *baseline hazard function*, i.e., the hazard that one would obtain if all independent explanatory variables were zero.

The Cox regression model is the most general of the regression models for survival data since it does not assume any particular underlying survival distribution. The model is fitted to the data by first estimating the risk score using a log–likelihood approach and finally computing the baseline hazard by an iterative procedure. As a result of the model fitting process, one can obtain parameter estimates and plots for specific values of the explanatory variables.

Example 9.10

Q: Determine the Cox regression solution for the Heart Valve dataset (event-free survival time), using Age as the explanatory variable. Compare the survivor functions and determine the estimated percentages of an event-free 10-year post-operative period for the mean age and for 20 and 60 years-old patients as well.

A: STATISTICA determines the parameter $\beta_{Age} = 0.0214$ for the Cox regression model. The chi-square test under the null hypothesis of "no Age influence" yields an observed $p = 0.004$. Therefore, variable Age is highly significant in the estimation of survival times, i.e., is an explanatory variable.

Figure 9.9a shows the baseline survivor function. Figures 9.9b, c and d, show the survivor function plots for 20, 47.17 (mean age) and 60 years, respectively. As expected, the probability of a given post-operative event-free period decreases with age (survivor curves lower with age). From these plots, we see that the estimated percentages of patients with post-operative event-free 10-year periods are 80%, 65% and 59% for 20, 47.17 (mean age) and 60 year-old patients, respectively.

□

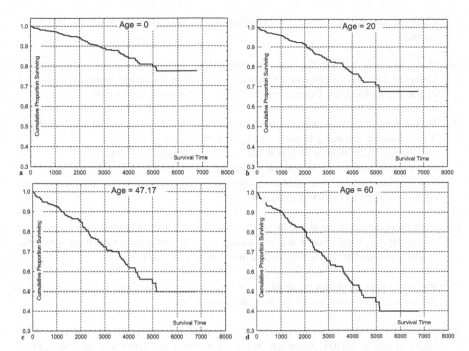

Figure 9.9. Baseline survivor function (a) and survivor functions for different patient ages (b, c and d) submitted to heart valve implant (Heart Valve dataset), obtained by Cox regression in STATISTICA. The survival times are in days. The Age = 47.17 (years) corresponds to the sample mean age.

Exercises

9.1 Determine the probability of having no complaint in the first year for the Car Sale dataset using the life table and Kaplan-Meier estimates of the survivor function.

9.2 Redo Example 9.3 for the iron specimens submitted to high loads using the Kaplan-Meier estimate of the survivor function.

9.3 Redo the previous Exercise 9.2 for the aluminium specimens submitted to low and high loads. Compare the results.

9.4 Consider the Heart Valve dataset. Compute the Kaplan-Meier estimate for the following events: death after 1st operation, death after 1st or 2nd operations, re-operation and endocarditis occurrence. Compute the following statistics:
a) Percentage of patients surviving 5 years.
b) Percentage of patients without endocarditis in the first 5 years.
c) Median survival time with 95% confidence interval.

9.5 Compute the median time until breaking for all specimen types of the Fatigue dataset.

9.6 Redo Example 9.7 for the high amplitude load groups of the Fatigue dataset. Compare the survival times of the iron and aluminium specimens using the Log-Rank or Peto-Wilcoxon tests. Discuss which of these tests is more appropriate.

9.7 Consider the following two groups of patients submitted to heart valve implant (Heart Valve dataset), according to the pre-surgery heart functional class:
   i.   Patients with mild or no symptoms before the operation (PRE C < 3).
   ii.   Patients with severe symptoms before the operation (PRE C ≥ 3).
   Compare the survival time until death of these two groups using the most appropriate of the Log-Rank or Peto-Wilcoxon tests.

9.8 Determine the exponential and Weibull estimates of the survivor function for the Car Sale dataset. Verify that a Weibull model is more appropriate than the exponential model and compute the median time until complaint for that model.

9.9 Redo Example 9.9 for all group specimens of the Fatigue dataset. Determine which groups are better modelled by the Weibull distribution.

9.10 Consider the Weather dataset (Data 1) containing daily measurements of wind speed in m/s at 12H00. Assuming that a wind stroke at 12H00 was used to light an electric lamp by means of an electric dynamo, the time that the lamp would glow is proportional to the wind speed. The wind speed data can thus be interpreted as survival data. Fit a Weibull model to this data using $n = 10$, 20 and 30 time intervals. Compare the corresponding parameter estimates.

9.11 Compare the survivor functions for the wind speed data of the previous Exercise 9.11 for the groups corresponding to the two seasons: winter and summer. Use the most appropriate of the Log-Rank or Peto-Wilcoxon tests.

9.12 Using the Heart Valve dataset, determine the Cox regression solution for the survival time until death of patients undergoing heart valve implant with Age as the explanatory variable. Determine the estimated percentage of a 10-year survival time after operation for 30 years-old patients.

9.13 Using the Cox regression model for the time until breaking of the aluminium specimens of the Fatigue dataset, verify the following results:
   a)   The load amplitude (AMP variable) is an explanatory variable, with chi-square $p = 0$.
   b)   The probability of surviving 2 million cycles for amplitude loads of 80 and 100 MPa is 0.6 and 0.17, respectively (point estimates).

9.14 Using the Cox regression model, show that the load amplitude (AMP variable) cannot be accepted as an explanatory variable for the time until breaking of the iron specimens of the Fatigue dataset. Verify that the survivor functions are approximately the same for different values of AMP.

10 Directional Data

The analysis and interpretation of *directional data* requires specific data representations, descriptions and distributions. Directional data occurs in many areas, namely the Earth Sciences, Meteorology and Medicine. Note that directional data is an "interval type" data: the position of the "zero degrees" is arbitrary. Since usual statistics, such as the arithmetic mean and the standard deviation, do not have this rotational invariance, one must use other statistics. For example, the mean direction between 10° and 350° is not given by the arithmetic mean 180°.

In this chapter, we describe the fundamentals of statistical analysis and the interpretation of directional data, for both the circle and the sphere. SPSS, STATISTICA, MATLAB and R do not provide specific tools for dealing with directional data; therefore, the needed software tools have to be built up from scratch. MATLAB and R offer an adequate environment for this purpose. In the following sections, we present a set of "directional data"-functions – developed in MATLAB and R and included in the CD Tools –, and explain how to apply them to practical problems.

10.1 Representing Directional Data

Directional data is analysed by means of unit length vectors, i.e., by representing the angular observations as points on the unit radius circle or sphere.

For circular data, the angle, ϕ, is usually specified in [−180°, 180°] or in [0°, 360°]. Spherical data is represented in polar form by specifying the *azimuth* (or *declination*) and the *latitude* (or *inclination*). The azimuth, ϕ, is given in [−180°, 180°]. The latitude (also called *elevation angle*), θ, is specified in [−90°, 90°]. Instead of an azimuth and latitude, a *longitude* angle in [0°, 360°] and a *co-latitude* angle in [0°, 180°] are often used.

When dealing with directional data, one often needs, e.g. for representational purposes, to obtain the Cartesian co-ordinates of vectors with specified length and angular directions or, vice-versa, to convert Cartesian co-ordinates to angular, *polar* or *spherical* form. The conversion formulas for azimuths and latitudes are given in Table 10.1 with the angles expressed in radians through multiplication of the values in degrees by $\pi/180$.

The MATLAB and R functions for performing these conversions, with the angles expressed in radians, are given in Commands 10.1.

Example 10.1

Q: Consider the `Joints'` dataset, containing measurements of azimuth and pitch in degrees for several joint surfaces of a granite structure. What are the Cartesian co-ordinates of the unit length vector representing the first measurement?

A: Since the pitch is a descent angle, we use the following MATLAB instructions (see Commands 10.1 for R instructions), where `joints` is the original data matrix (azimuth in the first column, pitch in the second column):

```
» j = joints*pi/180;              % convert to radians
» [x,y,z]=sph2cart(j(1,1),-j(1,2),1)
x =
      0.1162
y =
     -0.1290
z =
     -0.9848
```
                                                                            □

Table 10.1. Conversion formulas from Cartesian to polar or spherical co-ordinates (azimuths and latitudes) and vice-versa.

	Polar to Cartesian	Cartesian to Polar
Circle	$(\phi, \rho) \rightarrow (x, y)$	$(x, y) \rightarrow (\phi, \rho)$
	$x = \rho \cos\phi; \quad y = \rho \sin\phi$	$\phi = \text{atan2}(y,x)$ [a]; $\rho = (x^2 + y^2)^{\frac{1}{2}}$
Sphere	$(\phi, \theta, \rho) \rightarrow (x, y, z)$	$(x, y, z) \rightarrow (\phi, \theta, \rho)$
	$x = \rho \cos\theta \cos\phi; \quad y = \rho \cos\theta \sin\phi;$ $z = \rho \sin\theta$	$\theta = \arctan(z / (x^2 + y^2)^{\frac{1}{2}});$ $\phi = \text{atan2}(y,x); \rho = (x^2 + y^2 + z^2)^{\frac{1}{2}}$

[a] atan2(y,x) denotes the arc tangent of y/x with correction of the angle for $x < 0$ (see formula 10.4).

Commands 10.1. MATLAB and R functions converting from Cartesian to polar or spherical co-ordinates and vice-versa.

MATLAB	```[x,y]=pol2cart(phi,rho)```
	```[phi,rho]=cart2pol(x,y)```
	```[x,y,z]=sph2cart(phi,theta,rho)```
	```[phi,theta,rho]=cart2sph(x,y,z)```

R	```pol2cart(phi,rho)```
	```cart2pol(x,y)```
	```sph2cart(phi,theta,rho)```
	```cart2sph(x,y,z)```

The R functions work in the same way as their MATLAB counterparts. They all return a matrix whose columns have the same information as the returned MATLAB vectors. For instance, the conversion to spherical co-ordinates in Example 10.1 can be obtained with:

```
> m <- sph2cart(phi*pi/180,-pitch*pi/180,1)
```

where `phi` and `pitch` are the columns of the attached `joints` data frame. The columns of matrix m are the vectors x, y and z.   ∎

In the following sections we assume, unless stated otherwise, that circular data is specified in [0°, 360°] and spherical data is specified by the pair (longitude, co-latitude). We will call these specifications the *standard format* for directional data. The MATLAB and R-implemented functions `convazi` and `convlat` (see Commands 10.3) perform the azimuth and latitude conversions to standard format.

Also in all MATLAB and R functions described in the following sections, the directional data is represented by a matrix (often denoted as a), whose first column contains the circular or longitude data, and the second column, when it exists, the co-latitudes and both in degrees.

Circular data is usually plotted in *circular plots* with a marker for each direction plotted over the corresponding point in the unit circle. Spherical data is conveniently represented in *spherical plots*, showing a projection of the unit sphere with markers over the points corresponding to the directions.

For circular data, a popular histogram plot is the *rose diagram*, which shows circular slices whose height is proportional to the frequency of occurrence in a specified angular bin.

Commands 10.2 lists the MATLAB and R functions used for obtaining these plots.

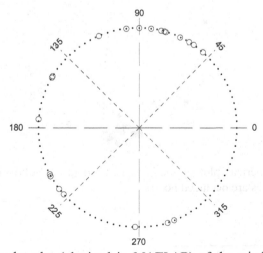

Figure 10.1. Circular plot (obtained in MATLAB) of the wind direction WDB sample included in the `Weather` dataset.

Example 10.2

Q: Plot the March, 1999 wind direction WDB sample, included in the Weather dataset (datasheet Data 3).

A: Figure 10.1 shows the circular plot of this data obtained with polar2d. Visual inspection of the plot suggests a *multimodal* distribution with dispersed data and a mean direction somewhere near 135°.

□

Example 10.3

Q: Plot the Joints' dataset consisting of azimuth and pitch of granite joints of a city street in Porto, Portugal. Assume that the data is stored in the joints matrix whose first column is the azimuth and the second column is the pitch (descent angle)[1].

A: Figure 10.2 shows the spherical plot obtained in MATLAB with:

```
» j=convlat([joints(:,1),-joints(:,2)]);
» polar3d(j);
```

Figure 10.2 suggests a *unimodal* distribution with the directions strongly concentrated around a modal co-latitude near 180°. We then expect the anti-mode (distribution minimum) to be situated near 0°.

□

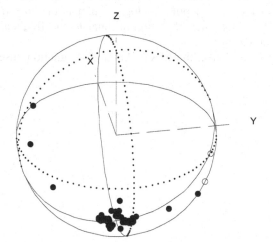

Figure 10.2. Spherical plot of the Joints' dataset. Solid circles are visible points; open circles are occluded points.

---

[1] Note that strictly speaking the joints' data is an example of *axial data*, since there is no difference between the symmetrical directions (ϕ, θ) and ($\phi + \pi, -\theta$). We will treat it, however, as spherical data.

Example 10.4

Q: Represent the rose diagram of the angular measurements H of the VCG dataset.

A: Let vcg denote the data matrix whose first column contains the H measurements. Figure 10.3 shows the rose diagram using the MATLAB rose command:

```
» rose(vcg(:,1)*pi/180,12) % twelve bins
```

Using [t,r]=rose(vcg(:,1)*pi/180,12), one can confirm that 70/120 = 58% of the measurements are in the [−60°, 60°] interval. The same results are obtained with R rose function.   □

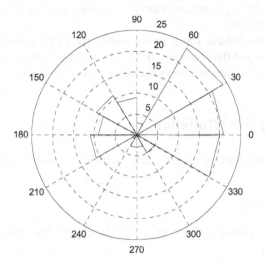

Figure 10.3. Rose diagram (obtained with MATLAB) of the angular H measurements of the VCG dataset.

Commands 10.2. MATLAB and R functions for representing and graphically assessing directional data.

MATLAB	[phi, r] = rose(a,n) *polar2d(a, mark)* ; *polar3d(a)* *unifplot(a)* h=colatplot(a,kl) ; h=longplot(a)
R	*rose(a)* *polar2d(a)*

The MATLAB function rose(a,n) plots the rose diagram of the circular data vector a (radians) with n bins; [phi, r]=rose(a,n) returns the vectors phi and r such that polar(phi, r) is the histogram (no plot is drawn in this case).

The polar2d and polar3d functions are used to obtain circular and spherical plots, respectively. The argument a is, as previously mentioned, either a column vector for circular data or a matrix whose first column contains the longitudes, and the second column the co-latitudes (in degrees).

The unifplot command draws a uniform probability plot of the circular data vector a (see section 10.4). The colatplot and longplot commands are used to assess the von Misesness of a spherical distribution (see section 10.4). The returned value h is 1 if the von Mises hypothesis is rejected at 1% significance level, and 0 otherwise. The parameter kl of colatplot must be 1 for assessing von Misesness with large concentration distributions and 0 for assessing uniformity with low concentration.

The R functions behave much in the same way as their equivalent MATLAB functions. The only differences are: the rose function always uses 12 histogram bins; the polar2d function always uses open circles as marks.   ∎

10.2 Descriptive Statistics

Let us start by considering circular data, with data points represented by a unit length vector:

$$\mathbf{x} = [\cos\theta \quad \sin\theta]'. \tag{10.1}$$

The *mean direction* of n observations can be obtained in Cartesian co-ordinates, in the usual way:

$$\bar{c} = \sum_{i=1}^{n} \cos\theta_i / n; \quad \bar{s} = \sum_{i=1}^{n} \sin\theta_i / n. \tag{10.2}$$

The vector $\bar{\mathbf{r}} = [\bar{c} \quad \bar{s}]'$ is the *mean resultant vector* of the n observations, with *mean resultant length*:

$$\bar{r} = \sqrt{\bar{c}^2 + \bar{s}^2} \in [0, 1], \tag{10.3}$$

and *mean direction* (for $\bar{r} \neq 0$):

$$\bar{\theta} = \begin{cases} \arctan(\bar{s}/\bar{c}), & \text{if} \quad \bar{c} \geq 0; \\ \arctan(\bar{s}/\bar{c}) + \pi \operatorname{sgn}(\bar{s}), & \text{if} \quad \bar{c} < 0. \end{cases} \tag{10.4}$$

Note that the arctangent function (MATLAB and R atan function) takes value in $[-\pi/2, \pi/2]$, whereas $\bar{\theta}$ takes value in $[-\pi, \pi]$, the same as using the MATLAB and R function atan2(y,x) with y representing the vertical component $\bar{s}$ and x the horizontal component $\bar{c}$. Also note that $\bar{r}$ and $\bar{\theta}$ are invariant under rotation.

The mean resultant vector can also be obtained by computing the *resultant* of the n unit length vectors. The resultant, $\mathbf{r} = [\, n\bar{c} \quad n\bar{s} \,]'$, has the same angle, $\bar{\theta}$, and a vector length of $r = n\bar{r} \in [0, n]$. The unit length vector representing the mean direction, called the *mean direction vector*, is $\bar{\mathbf{x}}_0 = [\cos\bar{\theta} \quad \sin\bar{\theta}\,]'$.

The mean resultant length $\bar{r}$, point estimate of the population mean length ρ, can be used as a measure of distribution *concentration*. If the vector directions are uniformly distributed around the unit circle, then there is no preferred direction and the mean resultant length is zero. On the other extreme, if all the vectors are concentrated in the same direction, the mean resultant length is maximum and equal to 1. Based on these observations, the following sample *circular variance* is defined:

$$v = 2(1 - \bar{r}) \in [0, 2].$$
$\qquad$ 10.5

The sample *circular standard deviation* is defined as:

$$s = \sqrt{-2\ln\bar{r}},$$
$\qquad$ 10.6

reducing to approximately $\sqrt{v}$ for small v. The justification for this definition lies in the analysis of the distribution of the wrapped random variable X_w:

$$X \sim n_{\mu,\sigma}(x) \quad \Rightarrow \quad X_w = X(\text{mod } 2\pi) \sim w_{\mu,\rho}(x_w) = \sum_{k=-\infty}^{\infty} n_{\mu,\sigma}(x + 2\pi k). \quad 10.7$$

The wrapped normal density, $w_{\mu,\rho}$, has ρ given by:

$$\rho - \exp(-\sigma^2/2) \quad \Rightarrow \quad \sigma = \sqrt{-2\ln\rho}.$$
$\qquad$ 10.8

For spherical directions, we consider the data points represented by a unit length vector, with the x, y, z co-ordinates computed as in Table 10.1.

The mean resultant vector co-ordinates are then computed in a similar way as in formula 10.2. The definitions of *spherical mean direction*, $(\bar{\theta}, \bar{\phi})$, and *spherical variance* are the direct generalisation to the sphere of the definitions for the circle, using the three-dimensional resultant vector. In particular, the mean direction vector is:

$$\bar{\mathbf{x}}_0 = [\sin\bar{\theta}\cos\bar{\phi} \quad \sin\bar{\theta}\sin\bar{\phi} \quad \cos\bar{\theta}\,]'.$$
$\qquad$ 10.9

Example 10.5

Q: Consider the data matrix j of Example 10.3 (Joints' dataset). Compute the longitude, co-latitude and length of the resultant, as well as the mean resultant length and the standard deviation.

A: We use the function `resultant` (see Commands 10.3) in MATLAB, as follows:

```
» [x,y,z,f,t,r] = resultant(j)
...
f =
    65.4200              % longitude
t =
   178.7780              % co-latitude
r =
    73.1305              % resultant length
» rbar=r/size(j,1)
rbar =
     0.9376              % mean resultant length
» s=sqrt(-2*log(rbar))
s =
     0.3591              % standard deviation in radians
```

Note that the mean co-latitude (178.8°) does indeed confirm the visual observations of Example 10.3. The data is highly concentrated ($\bar{r}$ =0.94, near 1). The standard deviation corresponds to an angle of 20.6°.

□

Commands 10.3. MATLAB and R functions for computing descriptive statistics and performing simple operations with directional data.

MATLAB	`as=convazi(a) ; as=convlat(a)`
	`[x,y,z,f,t,r] = resultant(a)`
	`m = meandir(a,alpha1)`
	`[m,rw,rhow]=pooledmean(a)`
	`v=rotate(a); t=scattermx(a); d=dirdif(a,b)`

R	`convazi(a) ; convlat(a)` `resultant(a) ; dirdif(a,b)`

Functions `convazi` and `convlat` convert azimuth into longitude and latitude into co-latitude, respectively.

Function `resultant` determines the resultant of unit vectors whose angles are the elements of a (in degrees). The Cartesian co-ordinates of the resultant are returned in x, y and z. The polar co-ordinates are returned in f (ϕ), t (θ) and r.

Function `meandir` determines the mean direction of the observations a. The angles are returned in m(1) and m(2). The mean direction length $\bar{r}$ is returned in m(3). The standard deviation in degrees is returned in m(4). The deviation angle corresponding to a confidence level indicated by `alpha1`, assuming a von Mises distribution (see section 10.3), is returned in m(5). The allowed values of `alpha1` (alpha level) are 1, 2 3 and 4 for α = 0.001, 0.01, 0.05 and 0.1, respectively.

Function `pooledmean` computes the pooled mean (see section 10.6.2) of independent samples of circular or spherical observations, a. The last column of a contains the group codes, starting with 1. The mean resultant length and the weighted resultant length are returned through `rw` and `rhow`, respectively.

Function `rotate` returns the spherical data matrix v (standard format), obtained by rotating a so that the mean direction maps onto the North Pole.

Function `scattermx` returns the scatter matrix t of the spherical data a (see section 10.4.4).

Function `dirdif` returns the directional data of the differences of the unit vectors corresponding to a and b (standard format).

The R functions behave in the same way as their equivalent MATLAB functions. For instance, Example 10.5 is solved in R with:

```
j <- convlat(cbind(j[,1],-j[,2]))
> o <- resultant(j)
> o
[1]    0.6487324   1.4182647  -73.1138435   65.4200379
[5]  178.7780083  73.1304754
```

■

10.3 The von Mises Distributions

The importance of the von Mises distributions (see B.2.10) for directional data is similar to the importance of the normal distribution for linear data. As mentioned in B.2.10, several physical phenomena originate von Mises distributions. These enjoy important properties, namely their proximity with the normal distribution as mentioned in properties 3, 4 and 5 of B.2.10. The convolution of von Mises distributions does not produce a von Mises distribution; however, it can be well approximated by a von Mises distribution.

The *generalised $(p-1)$-dimensional von Mises density function*, for a vector of observations **x**, can be written as:

$$m_{\mu,\kappa,p}(\mathbf{x}) = C_p(\kappa)e^{\kappa\,\mu'\mathbf{x}}, \qquad\qquad 10.10$$

where μ is the mean vector, κ is the concentration parameter, and $C_p(\kappa)$ is a normalising factor with the following values:

$C_2(\kappa) = 1/(2\pi I_0(\kappa))^2$, for the circle $(p=2)$;

$C_3(\kappa) = \kappa/(4\pi \sinh(\kappa))$, for the sphere $(p=3)$.

---

[2] I_p denotes the modified Bessel function of the first kind and order p (see B.2.10).

For $p = 2$, one obtains the circular distribution first studied by R. von Mises; for $p = 3$, one obtains the spherical distribution studied by R. Fisher (also called *von Mises-Fisher* or *Langevin* distribution).

Note that for low concentration values, the von Mises distributions approximate the uniform distribution as illustrated in Figure 10.4 for the circle and in Figure 10.5 for the sphere. The sample data used in these figures was generated with the vmises2rnd and vmises3rnd functions, respectively (see Commands 10.4).

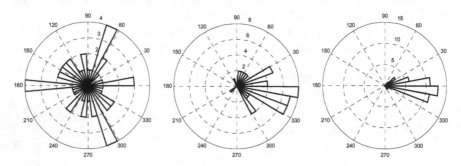

Figure 10.4. Rose diagrams of 50-point samples of circular von Mises distribution around $\mu = 0$, and $\kappa = 0.1, 2, 10$, from left to right, respectively.

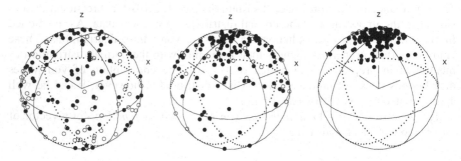

Figure 10.5. Spherical plots of 150-point-samples with von Mises-Fisher distribution around [0 0 1]', and $\kappa = 0.001, 2, 10$, from left to right, respectively.

Given a von Mises distribution $M_{\mu,\kappa,p}$, the maximum likelihood estimation of μ is precisely the mean direction vector. On the other hand, the sample resultant mean length $\bar{r}$ is the maximum likelihood estimation of the population mean resultant length, a function of the concentration parameter, $\rho = A_p(\kappa)$, given by:

$$\rho = A_2(k) = I_1(\kappa) / I_0(\kappa) \text{ , for the circle;}$$
$$\rho = A_3(k) = \coth \kappa - 1/\kappa \text{ , for the sphere.}$$

Thus, the maximum likelihood estimation of the concentration parameter κ is obtained by the inverse function of A_p:

$$\hat{\kappa} = A_p^{-1}(\bar{r}).\qquad\qquad 10.11$$

Values of $\hat{\kappa} = A_p^{-1}(\bar{r})$ for $p = 2, 3$ are given in tables in the literature (see e.g. Mardia KV, Jupp PE, 2000). The function ainv, built in MATLAB, implements 10.11 (see Commands 10.4). The estimate of κ can also be derived from the sample variance, when it is low (large $\bar{r}$):

$$\hat{\kappa} \cong (p-1)/v.\qquad\qquad 10.12$$

As a matter of fact, it can be shown that the inflection points of $m_{\mu,\kappa,2}$ are given by:

$$\frac{1}{\sqrt{\kappa}} \cong \sigma, \text{ for large } \kappa.\qquad\qquad 10.13$$

Therefore, we see that $1/\sqrt{\kappa}$ influences the von Mises distribution in the same way as σ influences the linear normal distribution.

Once the ML estimate of κ has been determined, the circular or spherical region around the mean, corresponding to a $(1-\alpha)$ probability of finding a random direction, can also be computed using tables of the von Mises distribution function. The MATLAB-implemented function vmisesinv gives the respective deviation angle, δ, for several values of α. Function vmises2cdf gives the left tail area of the distribution function of a circular von Mises distribution. These functions use exact-value tables and are listed and explained in Commands 10.4.

Approximation formulas for estimating the concentration parameter, the deviation angles of von Mises distributions and the circular von Mises distribution function can also be found in the literature.

Example 10.6

Q: Assuming that the Joints' dataset (Example 10.3) is well approximated by the von Mises-Fisher distribution, determine the concentration parameter and the region containing 95% of the directions.

A: We use the following sequence of commands:

```
» k=ainv(rbar,3)          %using rbar from Example 10.5
k =
    16.0885
» delta=vmisesinv(k,3,3)      %alphal=3 --> alpha=0.05
delta =
    35.7115
```

Thus, the region containing 95% of the directions is a spherical cap with $\delta = 35.7°$ aperture from the mean (see Figure 10.6).

Note that using formula 10.12, one obtains an estimate of $\hat{\kappa} = 16.0181$. For the linear normal distribution, this corresponds to $\hat{\sigma} = 0.2499$, using formula 10.13. For the equal variance bivariate normal distribution, the 95% percentile corresponds to $2.448\sigma \approx 2.448\,\hat{\sigma} = 0.1617$ radians $= 35.044°$. The approximation to the previous value of δ is quite good.

▯

We will now consider the estimation of a confidence interval for the mean direction $\overline{\mathbf{x}}_0$, using a sample of n observations, $\mathbf{x}_1$, $\mathbf{x}_2$, ..., $\mathbf{x}_n$, from a von Mises distribution. The joint distribution of $\mathbf{x}_1$, $\mathbf{x}_2$, ..., $\mathbf{x}_n$ is:

$$f(\mathbf{x}_1, \mathbf{x}_2, ..., \mathbf{x}_n) = \left(C_p(\kappa)\right)^n \exp(n\kappa\,\overline{r}\,\mathbf{\mu'}\overline{\mathbf{x}}_0) \,. \qquad 10.14$$

From 10.10, it follows that the confidence interval of $\overline{\mathbf{x}}_0$, at α level, is obtained from the von Mises distribution with the concentration parameter $n\kappa\overline{r}$. Function meandir (see Commands 10.3) uses precisely this result.

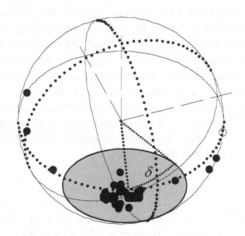

Figure 10.6. Spherical plot of the Joints' dataset with the spherical cap around the mean direction (shaded area) enclosing 95% of the observations ($\delta = 35.7°$).

Example 10.7

Q: Compute the deviation angle of the mean direction of the Joints' dataset for a 95% confidence interval.

A: Using the meandir command we obtain $\delta = 4.1°$, reflecting the high concentration of the data.

▯

Example 10.8

Q: A circular distribution of angles follows the von Mises law with concentration $\kappa=2$. What is the probability of obtaining angles deviating more than 20° from the mean direction?

A: Using 2*vmises2cdf(-20,2) we obtain a probability of 0.6539.

□

Commands 10.4. MATLAB functions for operating with von Mises distributions.

MATLAB	`k=ainv(rbar,p)`
	`delta=vmisesinv(k, p, alpha1)`
	`a=vmises2rnd(n,mu,k)  ;  a=vmises3rnd(n,k)`
	`f=vmises2cdf(a,k)`

Function ainv returns the concentration parameter, k, of a von Mises distribution of order p (2 or 3) and mean resultant length rbar. Function vmisesinv returns the deviation angle delta of a von Mises distribution corresponding to the α level indicated by alpha1. The valid values of alpha1 are 1, 2, 3 and 4 for $\alpha = 0.001$, 0.01, 0.05 and 0.1, respectively.

Functions vmises2rnd and vmises3rnd generate n random points with von Mises distributions with concentration k, for the circle and the sphere, respectively. For the circle, the distribution is around mu; for the sphere around [0 0 1]'. These functions implement algorithms described in (Mardia JP, Jupp PE, 2000) and (Wood, 1994), respectively.

Function vmises2cdf(a,k) returns a vector, f, containing the left tail areas of a circular von Mises distribution, with concentration k, for the vector a angles in [−180°, 180°], using the algorithm described in (Hill GW, 1977).

■

10.4 Assessing the Distribution of Directional Data

10.4.1 Graphical Assessment of Uniformity

An important step in the analysis of directional data is determining whether or not the hypothesis of uniform distribution of the data is significantly supported. As a matter of fact, if the data can be assumed uniformly distributed in the circle or in the sphere, there is no mean direction and the directional concentration is zero.

It is usually convenient to start the assessment of uniformity by graphic inspection. For circular data, one can use a *uniform probability plot*, where the sorted observations $\theta_i/(2\pi)$ are plotted against $i/(n+1)$, $i = 1, 2, ..., n$. If the θ_i come from a uniform distribution, then the points should lie near a unit slope straight line passing through the origin.

Example 10.9

Q: Use the uniform probability plot to assess the uniformity of the wind direction WDB sample of Example 10.2.

A: Figure 10.7 shows the uniform probability plot of the data using command `unifplot` (see Commands 10.2). Visual inspection suggests a sensible departure from uniformity.

□

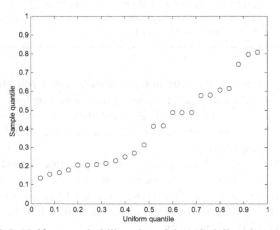

Figure 10.7. Uniform probability plot of the wind direction WDB data.

Let us now turn to the spherical data. In a uniform distribution situation the longitudes are also uniformly distributed in $[0, 2\pi[$, and their uniformity can be graphically assessed with the uniform probability plot. In what concerns the co-latitudes, their distribution is not uniform. As a matter of fact, one can see the uniform distribution as the limit case of the von Mises-Fisher distribution. By property 6 of B.2.10, the co-latitude is independently distributed from the longitude and its density $f_\kappa(\theta)$ will tend to the following density for $\kappa \to 0$:

$$f_\kappa(\theta) \underset{\kappa \to 0}{\to} f(\theta) = \frac{1}{2}\sin\theta \;\Rightarrow\; F(\theta) = \frac{1}{2}(1 - \cos\theta). \qquad 10.15$$

One can graphically assess this distribution by means of a *co-latitude plot* where the sorted observations θ_i are plotted against $\arccos(1-2(i/n))$, $i = 1, 2, \ldots, n$. In case of uniformity, one should obtain a unit slope straight line passing through the origin.

Example 10.10

Q: Consider a spherical data sample as represented in Figure 10.5 with $\kappa = 0.001$. Assess its uniformity.

A: Let a represent the data matrix. We use `unifplot(a)` and `colatplot(a,0)` (see Commands 10.2) to obtain the graphical plots shown in Figure 10.8. We see that both plots strongly suggest a uniform distribution on the sphere.   ☐

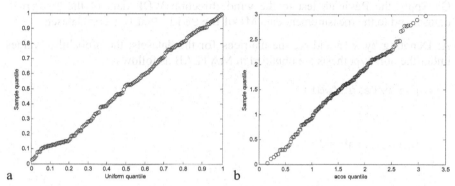

Figure 10.8. Longitude plot (a) and co-latitude plot (b) of the von Mises-Fisher distributed data of Figure 10.5 with $\kappa = 0.001$.

10.4.2 The Rayleigh Test of Uniformity

Let ρ denote the population mean resultant length, i.e., the population concentration, whose sample estimate is $\bar{r}$. The null hypothesis, H_0, for the Rayleigh's test of uniformity is: $\rho = 0$ (zero concentration).

For circular data the Rayleigh test statistic is:

$$z = n\bar{r}^2 = r^2/n.$$ 10.16

Critical values of the sampling distribution of z can be computed using the following approximation (Wilkie D, 1983):

$$P(z \geq k) = \exp\left(\sqrt{1 + 4n + 4(n^2 - nk)} - (1 + 2n)\right).$$ 10.17

For spherical data, the Rayleigh test statistic is:

$$z = 3n\bar{r}^2 = 3r^2/n.$$ 10.18

Using the modified test statistic:

$$z^* = (1 - 1/(2n))z + z^2 /(10 n), \qquad\qquad 10.19$$

it can be proven that the distribution of z^* is asymptotically χ_3^2 with an error decreasing as $1/n$ (Mardia KV, Jupp PE, 2000).

The Rayleigh test is implemented in MATLAB and R function rayleigh (see Commands 10.5)

Example 10.11

Q: Apply the Rayleigh test to the wind direction WDF data of the Weather dataset and to the measurement data M1 of the Soil Pollution dataset.

A: Denoting by wdf and m1 the matrices for the datasets, the probability values under the null hypothesis are obtained in MATLAB as follows:

```
» p=rayleigh(wdf)
p =
    0.1906

» p=rayleigh(m1)
p =
    0
```

Thus, we accept the null hypothesis of uniformity at the 5% level for the WDF data, and reject it for the soil pollution M1 data (see Figure 10.9).

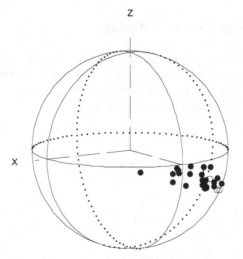

Figure 10.9. Measurement set M1 (negative gradient of Pb-tetraethyl concentration in the soil) of the Soil Pollution dataset.

Commands 10.5. MATLAB and R functions for computing statistical tests of directional data.

MATLAB	`p=rayleigh(a)`
	`[u2,uc]=watson(a,f,alphal)`
	`[u2,uc]=watsonvmises(a,alphal)`
	`[fo,fc,k1,k2]=watswill(a1,a2,alpha)`
	`[w,wc]=unifscores(a,alpha)`
	`[gw,gc]=watsongw(a,alpha)`

R	`rayleigh(a)` `unifscores(a,alpha)`

Function `rayleigh(a)` implements the Rayleigh test of uniformity for the data matrix a (circular or spherical data).

Function `watson` implements the Watson goodness-of-fit test, returning the test statistic u2 and the critical value uc computed for the data vector a (circular data) with theoretical distribution values in f. Vector a must be previously sorted in ascending order (and f accordingly). The valid values of alphal are 1, 2, 3, 4 and 5 for α = 0.1, 0.05, 0.025, 0.01 and 0.005, respectively.

The `watsonvmises` function implements the Watson test assessing von Misesness at alphal level. No previous sorting of the circular data a is necessary.

Function `watswill` implements the Watson-Williams two-sample test for von Mises populations, using samples a1 and a2 (circular or spherical data), at a significance level alpha. The observed test statistic and theoretical value are returned in fo and fc, respectively; k1 and k2 are the estimated concentrations.

Function `unifscores` implements the uniform scores test at alpha level, returning the observed statistic w and the critical value wc. The first column of input matrix a must contain the circular data of all independent groups; the second column must contain the group codes from 1 through the highest code number.

Function `watsongw` implements the Watson test of equality of means for independent spherical data samples. The first two columns of input matrix a contain the longitudes and colatitudes. The last column of a contains group codes, starting with 1. The function returns the observed test statistic gw and the critical value gc at alpha significance value.

The R functions behave in the same way as their equivalent MATLAB functions. For instance, Example 10.11 is solved in R with:

```
> rayleigh(wdf)
[1] 0.1906450

> rayleigh(m1)
[1] 1.242340e-13
```

■

10.4.3 The Watson Goodness of Fit Test

The Watson's U^2 goodness of fit test for circular distributions is based on the computation of the mean square deviation between the empirical and the theoretical distribution.

Consider the n angular values sorted by ascending order: $\theta_1 \leq \theta_2 \leq \ldots \leq \theta_n$. Let $V_i = F(\theta_i)$ represent the value of the theoretical distribution for the angle θ_i, and $\overline{V}$ represent the average of the V_i. The test statistic is:

$$U_n^2 = \sum_{i=1}^{n} V_i^2 - \sum_{i=1}^{n} \frac{(2i-1)V_i}{n} + n\left[\frac{1}{3} - \left(\overline{V} - \frac{1}{2}\right)^2\right].$$

10.20

Critical values of U_n^2 can be found in tables (see e.g. Kanji GK, 1999).

Function watson, implemented in MATLAB (see Commands 10.5), can be used to apply the Watson goodness of fit test to any circular distribution. It is particularly useful for assessing the goodness of fit to the von Mises distribution, using the mean direction and concentration factor estimated from the sample.

Example 10.12

Q: Assess, at the 5% significance level, the von Misesness of the data represented in Figure 10.4 with $\kappa = 2$ and the wind direction data WDB of the Weather dataset.

A: The watson function assumes that the data has been previously sorted. Let us denote the data of Figure 10.4 with $\kappa = 2$ by a. We then use the following sequence of commands:

```
» a = sort(a);
» m = meandir(a);
» k = ainv(m(3),2)
k =
      2.5192

» f = vmises2cdf(a,k)
» [u2,uc] = watson(a,f,2)
u2 =
      0.1484
uc =
      0.1860
```

Therefore, we do not reject the null hypothesis, at the 5% level, that the data follows a von Mises distribution since the observed test statistic u2 is lower than the critical value uc.

Note that the function vmises2cdf assumes a distribution with $\mu = 0$. In general, one should therefore previously refer the data to the estimated mean.

Although data matrix a was generated with $\mu = 0$, its estimated mean is not zero; using the data referred to the estimated mean, we obtain a smaller u2 = 0.1237.

Also note that when using the function vmises2cdf, the input data a must be specified in the [$-180°$, $180°$] interval.

Function watsonvmises (see Commands 10.5) implements all the above operations taking care of all the necessary data recoding for an input data matrix in standard format. Applying watsonvmises to the WDB data, the von Mises hypothesis is not rejected at the 5% level (u2= 0.1042; uc= 0.185). This contradicts the suggestion obtained from visual inspection in Example 10.2 for this low concentrated data ($\bar{r} = 0.358$). □

10.4.4 Assessing the von Misesness of Spherical Distributions

When analysing spherical data it is advisable to first obtain an approximate idea of the distribution shape. This can be done by analysing the eigenvalues of the following *scatter matrix* of the points about the origin:

$$\overline{\mathbf{T}} = \frac{1}{n}\sum_{i=1}^{n}\mathbf{x}_i\mathbf{x}_i{}'.$$ 10.21

Let the eigenvalues be denoted by λ_1, λ_2 and λ_3 and the eigenvectors by $\mathbf{t}_1$, $\mathbf{t}_2$ and $\mathbf{t}_3$, respectively. The shape of the distribution can be inferred from the magnitudes of the eigenvalues as shown in Table 10.2 (for details, see Mardia KV, Jupp PE, 2000). The scatter matrix can be computed with the scattermx function implemented in MATLAB (see Commands 10.3).

Table 10.2. Distribution shapes of spherical distributions according to the eigenvalues and mean resultant length, $\bar{r}$.

Magnitudes	Type of Distribution
$\lambda_1 \approx \lambda_2 \approx \lambda_3$	Uniform
λ_1 large; $\lambda_2 \neq \lambda_3$ small	Unimodal if $\bar{r} \approx 1$, bimodal otherwise
λ_1 large; $\lambda_2 \approx \lambda_3$ small	Unimodal if $\bar{r} \approx 1$, bimodal otherwise with rotational symmetry about $\mathbf{t}_1$
$\lambda_1 \neq \lambda_2$ large; λ_3 small	Girdle concentrated about circle in plane of $\mathbf{t}_1$, $\mathbf{t}_2$
$\lambda_1 \approx \lambda_2$ large; λ_3 small	Girdle with rotational symmetry about $\mathbf{t}_3$

Example 10.13

Q: Analyse the shape of the distribution of the gradient measurement set M1 of the Soil Pollution dataset (see Example 10.11 and Figure 10.9) using the scatter matrix. Assume that the data is stored in m1 in standard format.

A: We first run the following sequence of commands:

```
» m = meandir(m1);
» rbar = m(3)
rbar =
    0.9165

» t = scattermx(m1);
» [v,lambda] = eig(t)

v =
   -0.3564    -0.8902     0.2837
    0.0952    -0.3366    -0.9368
    0.9295    -0.3069     0.2047

lambda =
    0.0047         0          0
         0    0.1379          0
         0         0     0.8574
```

We thus conclude that the distribution is unimodal without rotational symmetry.

□

The von Misesness of a distribution can be graphically assessed, after rotating the data so that the mean direction maps onto [0 0 1]' (using function rotate described in Commands 10.3), by the following plots:

1. *Co-latitude plot*: plots the ordered values of $1 - \cos\theta_i$ against $-\ln(1-(i - 0.5)/n)$. For a von Mises distribution and a not too small κ (say, $\kappa > 2$), the plot should be a straight line through the origin and with slope $1/\kappa$.

2. *Longitude plot*: plots the ordered values of ϕ_i against $(i - 0.5)/n$. For a von Mises distribution, the plot should be a straight line through the origin with unit slope.

The plots are implemented in MATLAB (see Commands 10.2) and denoted colatplot and longplot. These functions, which internally perform a rotation of the data, also return a value indicating whether or not the null hypothesis should be rejected at the 1% significance level, based on test statistics described in (Fisher NI, Best DJ, 1984).

Example 10.14

Q: Using the co-latitude and longitude plots, assess the von Misesness of the gradient measurement set M1 of the Soil Pollution dataset.

A: Figure 10.10 shows the respective plots obtained with MATLAB functions `colatplot` and `longplot`. Both plots suggest an important departure from von Misesness. The `colatplot` and `longplot` results also indicate the rejection of the null hypothesis for the co-latitude (h = 1) and the non-rejection for the longitude (h = 0).   ⧠

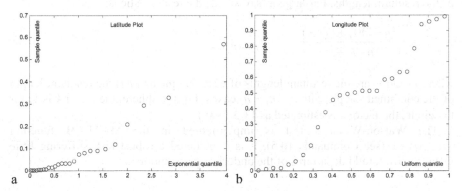

Figure 10.10. Co-latitude plot (a) and longitude plot (b) for the gradient measurement set M1 of the soil pollution dataset.

10.5 Tests on von Mises Distributions

10.5.1 One-Sample Mean Test

The most usual one-sample test is the mean direction test, which uses the same approach followed in the determination of confidence intervals for the mean direction, described in section 10.3.

Example 10.15

Q: Consider the `Joints'` dataset, containing directions of granite joints measured from a city street in Porto, Portugal. The mean direction of the data was studied in Example 10.5; the 95% confidence interval for the mean was studied in Example 10.7. Assume that a geotectonic theory predicts a 90° pitch for the granites in Porto. Does the `Joints'` sample reject this theory at a 95% confidence level?

A: The mean direction of the sample has a co-latitude $\theta = 178.8°$ (see Example 10.5). The 95% confidence interval of the mean direction corresponds to a deviation of 4.1° (see Example 10.7). Therefore, the `Joints'` dataset does not reject the theory at 5% significance level, since the 90° pitch corresponds to a co-latitude of 180° which falls inside the [178.8° − 4.1°, 178.8° + 4.1°] interval.

⧠

10.5.2 Mean Test for Two Independent Samples

The Watson-Williams test assesses whether or not the null hypothesis of equal mean directions of two von Mises populations must be rejected based on the evidence provided by two independent samples with n_1 and n_2 directions. The test assumes equal concentrations of the distributions and is based on the comparison of the resultant lengths. For large κ (say $\kappa > 2$) the test statistic is:

$$F^* = k \frac{(n-2)(r_1 + r_2 - r)}{n - r_1 - r_2} \quad \sim \quad F_{p-1,(p-1)(n-2)},$$ 10.22

where r_1 and r_2 are the resultant lengths of each sample and r is the resultant length of the combined sample with $n = n_1 + n_2$ cases. For the sphere, the factor k is 1; for the circle, the factor k is estimated as $1 + 3/(8\hat{\kappa})$.

The Watson-Williams test is implemented in the MATLAB function watswill (see Commands 10.5). It is considered a robust test, suffering little influence from mild departures of the underlying assumptions.

Example 10.16

Q: Consider the wind direction WD data of the Weather dataset (Data 2 datasheet), which represents the wind directions for several days in all seasons, during the years 1999 and 2000, measured at a location in Porto, Portugal. Compare the mean wind direction of Winter (SEASON = 1) vs. Summer (SEASON = 3) assuming that the WD data in every season follows a von Mises distribution, and that the sample is a valid random sample.

A: Using the watswill function as shown below, we reject the hypothesis of equal mean wind directions during winter and summer, at the 5% significance level. Note that the estimated concentrations have close values.

```
[fo,fc,k1,k2]=watswill(wd(1:25),wd(50:71),0.05)
fo =
    69.7865
fc =
    4.0670
k1 =
    1.4734
k2 =
    1.3581                                                           ☐
```

10.6 Non-Parametric Tests

The von Misessness of directional data distributions is difficult to guarantee in many practical cases[3]. Therefore, non-parametric tests, namely those based on ranking procedures similar to those described in Chapter 5, constitute an important tool when comparing directional data samples.

10.6.1 The Uniform Scores Test for Circular Data

Let us consider q independent samples of circular data, each with n_k cases. The uniform scores test assesses the similarity of the q distributions based on scores of the ordered combined data. For that purpose, let us consider the combined dataset with $n = \sum_{k=1}^{q} n_k$ observations sorted by ascending order. Denoting the ith observation in the kth group by θ_{ik}, we now substitute it by the uniform score:

$$\beta_{ik} = \frac{2\pi\, w_{ik}}{n}, \quad i = 1, \ldots, n_k, \qquad\qquad 10.23$$

where the w_{ik} are linear ranks in $[1, n]$. Thus, the observations are replaced by equally spaced points in the unit circle, preserving their order.

Let r_k represent the resultant length of the kth sample corresponding to the uniform scores. Under the null hypothesis of equal distributions, we expect the β_{ik} to be uniformly distributed in the circle. Using the test statistic:

$$W = 2\sum_{k=1}^{q} \frac{r_k^2}{n_k}, \qquad\qquad 10.24$$

we then reject the null hypothesis for significantly large values of W.

The asymptotic distribution of W, adequate for $n > 20$, is $\chi^2_{2(q-1)}$. For further details see (Mardia KV, Jupp PE, 2000). The uniform scores test is implemented by function unifscores (see Commands 10.5).

Example 10.17

Q: Assess whether the distribution of the wind direction (WD) of the Weather dataset (Data 2 datasheet) can be considered the same for all four seasons.

A: Denoting by wd the matrix whose first column is the wind direction data and whose second column is the season code, we apply the MATLAB unifscores function as shown below and conclude the rejection of equal distributions of the wind direction in all four seasons at the 5% significance level (w > wc).

---

[3] Unfortunately, there is no equivalent of the Central Limit Theorem for directional data.

Similar results are obtained with the R unifscores function.

```
» [w,wc]=unifscores(wd,0.05)
w =
    35.0909
wc =
    12.5916                                              □
```

10.6.2 The Watson Test for Spherical Data

Let us consider q independent samples of spherical data, each with n_i cases. The Watson test assesses the equality of the q mean directions, assuming that the distributions are rotationally symmetric.

The test is based on the estimation of a *pooled mean* of the q samples, using appropriate weights, w_k, summing up to unity. For not too different standard deviations, the weights can be computed as $w_k = n_k/n$ with $n = \sum_{k=1}^{q} n_k$. More complex formulas have to be used in the computation of the pooled mean in the case of very different standard deviations. For details see (Fisher NI, Lewis T, Embleton BJJ (1987). Function pooledmean (see Commands 10.3) implements the computation of the pooled mean of q independent samples of circular or spherical data.

Denoting by $\mathbf{x}_{0k} = [x_{0k}, y_{0k}, z_{0k}]'$ the mean direction of each group, the pooled mean projections are computed as:

$$x_w = \sum_{k=1}^{q} w_k \bar{r}_k x_{0k} \; ; \; y_w = \sum_{k=1}^{q} w_k \bar{r}_k y_{0k} \; ; \; z_w = \sum_{k=1}^{q} w_k \bar{r}_k z_{0k} \; . \qquad 10.25$$

The pooled mean resultant length is:

$$\bar{r}_w = \sqrt{x_w^2 + y_w^2 + z_w^2} \; . \qquad 10.26$$

Under the null hypothesis of equality of means, we would obtain the same value of the pooled mean resultant length simply by weighting the group resultant lengths:

$$\hat{\rho}_w = \sum_{k=1}^{q} w_k \bar{r}_k \; . \qquad 10.27$$

The Watson test rejects the null hypothesis for large values of the following statistic:

$$G_w = 2n(\hat{\rho}_w - \bar{r}_w) \; . \qquad 10.28$$

The asymptotic distribution of G_w is χ^2_{2q-2} (for $n_k \geq 25$). Function watsongw (see Commands 10.5) implements this test.

Example 10.18

Q: Consider the measurements R4, R5 and R6 of the negative gradient of the Soil Pollution dataset, performed in similar conditions. Assess whether the mean gradients above and below 20 m are significantly different at 5% level.

A: We establish two groups of measurements according to the value of variable z (depth) being above or below 20 m. The mean directions of these two groups are:

Group 1:   (156.17°, 117.40°);
Group 2:   (316.99°, 116.25°).

Assuming that the groups are rotationally symmetric and since the sizes are $n_1 = 45$ and $n_2 = 30$, we apply the Watson test at a significance level of 5%, obtaining an observed test statistic of 44.9. Since $\chi^2_{0.95,2} = 5.99$, we reject the null hypothesis of equality of means.   □

10.6.3 Testing Two Paired Samples

The previous two-sample tests assumed that the samples were independent. The two-paired-sample test can be reduced to a one-sample test using the same technique as in Chapter 4 (see section 4.4.3.1), i.e., employing the differences between pair members. If the distributions of the two samples are similar, we expect that the difference sample will be uniformly distributed. The function dirdif implemented in MATLAB (see Commands 10.3) computes the directional data of the difference set in standard format.

Example 10.19

Q: Consider the measurements M2 and M3 of the Soil Pollution dataset. Assess, at the 5% significance level, if one can accept that the two measurement methods yield similar distributions.

A: Let soil denote the data matrix containing all measurements of the Soil Pollution dataset. Measurements M2 and M3 correspond to the column pairs 3-4 and 5-6 of soil, respectively. We use the sequence of R commands shown below and do not reject the hypothesis of similar distributions at the 5% level of significance.

```
> m2<-soil[,3:4]
> m3<-soil[,5:6]
> d<-dirdif(m2,m3)
> p<-rayleigh(d)
> p
[1] 0.1772144
```

                                                                                      □

Exercises

10.1   Compute the mean directions of the wind variable WD (Weather dataset, Data 2) for the four seasons and perform the following analyses:
  a)   Assess the uniformity of the measurements both graphically and with the Rayleigh test. Comment on the relation between the uniform plot shape and the observed value of the test statistic. Which set(s) can be accepted as being uniformly distributed at a 1% level of significance?
  b)   Assess the von Misesness of the measurements.

10.2   Consider the three measurements sets, H, A and I, of the VCG dataset. Using a specific methodology, each of these measurement sets represents circular direction estimates of the maximum electrical heart vector in 97 patients.
  a)   Inspect the circular plots of the three sets.
  b)   Assess the uniformity of the measurements both graphically and with the Rayleigh test. Comment on the relation between the uniform plot shape and the observed value of the test statistic. Which set(s) can be accepted as being uniformly distributed at a 1% level of significance?
  c)   Assess the von Misesness of the measurements.

10.3   Which type of test is adequate for the comparison of any pair of measurement sets studied in the previous Exercise 10.2? Perform the respective pair-wise comparison of the distributions.

10.4   Assuming a von Mises distribution, compute the 95% confidence intervals of the mean directions of the measurement sets studied in the previous Exercise 10.2. Plot the data in order to graphically interpret the results.

10.5   In the von Misesness assessment of the WDB measurement set studied in Example 10.12, an estimate of the concentration parameter κ was used. Show that if instead of this estimate we had used the value employed in the data generation ($\kappa = 2$), we still would not have rejected the null hypothesis.

10.6   Compare the wind directions during March on two streets in Porto, using the Weather dataset (Data 3) and assuming that the datasets are valid random samples.

10.7   Consider the Wave dataset containing angular measurements corresponding to minimal acoustic pressure in ultrasonic radiation fields. Perform the following analyses:
  a)   Determine the mean directions of the TRa and TRb measurement sets.
  b)   Show that both measurement sets support at a 5% significance level the hypothesis of a von Mises distribution.
  c)   Compute the 95% confidence interval of the mean direction estimates.
  d)   Compute the concentration parameter for both measurement sets.
  e)   For the two transducers TRa and TRb, compute the angular sector spanning 95% of the measurements, according to a von Mises distribution.

10.8 Compare the two measurement sets, TRa and TRb, studied in the previous Exercise 10.7, using appropriate parametric and non-parametric tests.

10.9 The Pleiades data of the Stars' dataset contains measurements of the longitude and co-latitude of the stars constituting the Pleiades' constellation as well as their photo-visual magnitude. Perform the following analyses:
   a)   Determine whether the Pleiades' data can be modelled by a von Mises distribution.
   b)   Compute the mean direction of the Pleiades' data with the 95% confidence interval.
   c)   Compare the mean direction of the Pleiades' stars with photo-visual magnitude above 12 with the mean direction of the remaining stars.

10.10 The Praesepe data of the Stars' dataset contains measurements of the longitude and co-latitude of the stars constituting the Praesepe constellation obtained by two researchers (Gould and Hall).
   a)   Determine whether the Praesepe data can be modelled by a von Mises distribution.
   b)   Determine the mean direction of the Praesepe data with the 95% confidence interval.
   c)   Compare the mean directions of the Prasepe data obtained by the two researchers.

Appendix A - Short Survey on Probability Theory

In Appendix A we present a short survey on Probability Theory, emphasising the most important results in this area in order to afford a better understanding of the statistical methods described in the book. We skip proofs of Theorems, which can be found in abundant references on the subject.

A.1 Basic Notions

A.1.1 Events and Frequencies

Probability is a measure of uncertainty attached to the outcome of a *random experiment*, the word "experiment" having a broad meaning, since it can, for instance, be a thought experiment or the comprehension of a set of given data whose generation could be difficult to guess. The main requirement is being able to view the outcomes of the experiment as being composed of *single events*, such as A, B, ... The measure of certainty must, however, satisfy some conditions, presented in section A.1.2.

In the frequency approach to fixing the uncertainty measure, one uses the *absolute frequencies* of occurrence, n_A, n_B, ..., of the single events in n independent outcomes of the experiment. We then measure, for instance, the uncertainty of A in n outcomes using the *relative frequency* (or *frequency* for short):

$$f_A = \frac{n_A}{n}.$$ A. 1

In a long run of outcomes, i.e., with $n \to \infty$, the relative frequency is expected to stabilise, "converging" to the uncertainty measure known as *probability*. This will be a real number in [0, 1], with the value 0 corresponding to an event that never occurs (the *impossible* event) and the value 1 corresponding to an event that always occurs (the *sure* event). Other ways of obtaining probability measures in [0, 1], besides this classical "event frequency" approach have also been proposed.

We will now proceed to describe the mathematical formalism for operating with probabilities. Let $\mathcal{E}$ denote the set constituted by the single events E_i of a random experiment, known as the *sample space*:

$$\mathcal{E} = \{E_1, E_2, ...\}.$$ A. 2

Subsets of $\mathcal{E}$ correspond to *events* of the random experiment, with singleton subsets corresponding to single events. The empty subset, ϕ, denotes the *impossible event*. The usual operations of union ($\cup$), intersection ($\cap$) and complement ($^{-}$) can be applied to subsets of $\mathcal{E}$.

Consider a collection of events, $\mathcal{A}$, defined on $\mathcal{E}$, such that:

i.   If $A_i \in \mathcal{A}$ then $\overline{A}_i = \mathcal{E} - A_i \in \mathcal{A}$.

ii. Given the finite or denumerably infinite sequence $A_1, A_2, \ldots$, such that $A_i \in \mathcal{A}, \forall i$, then $\bigcup_i A_i \in \mathcal{A}$.

Note that $\mathcal{E} \in \mathcal{A}$ since $\mathcal{E} = A \cup \overline{A}$. In addition, using the well-known De Morgan's law ($\overline{A_i \cup A_j} = \overline{A}_i \cap \overline{A}_j$), it is verified that $\bigcap A_i \in \mathcal{A}$ as well as $\phi \in \mathcal{A}$. The collection $\mathcal{A}$ with the operations of union, intersection and complement constitutes what is known as a *Borel algebra*.

A.1.2 Probability Axioms

To every event $A \in \mathcal{A}$, of a Borel algebra, we assign a real number $P(A)$, satisfying the following *Kolmogorov's axioms of probability*:

1. $0 \le P(A) \le 1$.
2. Given the finite or denumerably infinite sequence $A_1, A_2, \ldots$, such that *any two events are mutually exclusive*, $A_i \cap A_j = \phi, \forall i, j$, then

$$P\left(\bigcup_i A_i\right) = \sum_i P(A_i).$$

3. $P(\mathcal{E}) = 1$.

The triplet $(\mathcal{E}, \mathcal{A}, P)$ is called a *probability space*.

Let us now enumerate some important consequences of the axioms:

i. $P(\overline{A}) = 1 - P(A); \quad P(\phi) = 1 - P(\mathcal{E}) = 0$.

ii. $A \subset B \Rightarrow P(A) \le P(B)$.

iii. $A \cap B \ne \phi \Rightarrow P(A \cup B) = P(A) + P(B) - P(A \cap B)$.

iv. $P\left(\bigcup_{i=1}^{n} P(A_i)\right) \le \sum_{i=1}^{n} P(A_i)$.

If the set $\mathcal{E} = \{E_1, E_2, \ldots, E_k\}$ of all possible outcomes is finite, and if all outcomes are equally likely, $P(E_i) = p$, then the triplet $(\mathcal{E}, \mathcal{A}, P)$ constitutes a *classical probability space*. We then have:

$$1 = P(\mathcal{E}) = P\left(\bigcup_{i=1}^{k} E_i\right) = \sum_{i=1}^{k} P(E_i) = kp \quad \Rightarrow \quad p = \frac{1}{k}. \qquad \text{A. 3}$$

Furthermore, if A is the union of m elementary events, one has:

$$P(A) = \frac{m}{k}, \qquad \text{A. 4}$$

corresponding to the classical approach of defining probability, also known as *Laplace rule*: *ratio of the number of favourable events over the number of possible events, considered equiprobable.*

One often needs to use the main operations of *combinatorial analysis* in order to compute the number of favourable events and of possible events.

Example A. 1

Q: Two dice are thrown. What is the probability that the sum of their faces is four?

A: When throwing two dice there are 6×6 equiprobable events. From these, only the events (1,3), (3,1), (2,2) are favourable. Therefore:

$$p(A) = \frac{3}{36} = 0.083 .$$

Thus, in the frequency interpretation of probability we expect to obtain four as sum of the faces roughly 8% of the times in a long run of two-dice tossing.

□

Example A. 2

Q: Two cards are drawn from a deck of 52 cards. Compute the probability of obtaining two aces, when drawing with and without replacement.

A: When drawing a card from a deck, there are 4 possibilities of obtaining an ace out of the 52 cards. Therefore, with replacement, the number of possible events is 52×52 and the number of favourable events is 4×4. Thus:

$$P(A) = \frac{4 \times 4}{52 \times 52} = 0.0059 .$$

When drawing without replacement we are left, in the second drawing, with 51 possibilities, only 3 of which are favourable. Thus:

$$P(A) = \frac{4 \times 3}{52 \times 51} = 0.0045 .$$ □

Example A. 3

Q: N letters are put randomly into N envelopes. What is the probability that the right letters get into the envelopes?

A: There are N distinct ways to put one of the letters (the first) in the right envelope. The next (second) letter has now a choice of $N-1$ free envelopes, and so on. We have, therefore, a total number of *factorial of N*, $N! = N(N-1)(N-2)...1$ *permutations* of possibilities for the N letters. Thus:

$$P(A) = 1/N! .$$ □

A.2 Conditional Probability and Independence

A.2.1 Conditional Probability and Intersection Rule

If in n outcomes of an experiment, the event B has occurred n_B times and among them the event A has occurred n_{AB} times, we have:

$$f_B = \frac{n_B}{n}; \quad f_{A \cap B} = \frac{n_{AB}}{n}.$$ A. 5

We define the *conditional frequency* of occurring A given that B has occurred as:

$$f_{A|B} = \frac{n_{AB}}{n_B} = \frac{f_{A \cap B}}{f_B}.$$ A. 6

Likewise, we define the *conditional probability* of A given that B has occurred – denoted $P(A \mid B)$ –, with $P(B) > 0$, as the ratio:

$$P(A \mid B) = \frac{P(A \cap B)}{P(B)}.$$ A. 7

We have, similarly, for the conditional probability of B given A:

$$P(B \mid A) = \frac{P(A \cap B)}{P(A)}.$$ A. 8

From the definition of conditional probability, the following rule of *compound probability* results:

$$P(A \cap B) = P(A)P(B \mid A) = P(B)P(A \mid B),$$ A. 9

which generalizes to the following *rule of event intersection*:

$$P(A_1 \cap A_2 \cap ... \cap A_n) =$$
$$P(A_1)P(A_2 \mid A_1)P(A_3 \mid A_1 \cap A_2)...P(A_n \mid A_1 \cap A_2 \cap ... \cap A_{n-1}).$$ A. 10

A.2.2 Independent Events

If the occurrence of B has no effect on the occurrence of A, both events are said to be *independent*, and we then have, for non-null probabilities of A and B:

$$P(A \mid B) = P(A) \quad \text{and} \quad P(B \mid A) = P(B).$$ A. 11

Therefore, using the intersection rule A.9, we define two events as being *independent* when the following multiplication rule holds:

$$P(A \cap B) = P(A)P(B).$$ A. 12

Given a set of n events, they are *jointly or mutually independent* if the multiplication rule holds for:

- Pairs: $P(A_i \cap A_j) = P(A_i)P(A_j)$, $1 \le i, j \le n$;
- Triplets: $P(A_i \cap A_j \cap A_k) = P(A_i)P(A_j)P(A_k)$, $1 \le i, j, k \le n$;
  and so on,
- until n: $P(A_1 \cap A_2 \cap ... \cap A_n) = P(A_1)P(A_2)...P(A_n)$.

If the independence is only verified for pairs of events, they are said to be *pairwise independent*.

Example A. 4

Q: What is the probability of winning the football lottery composed of 13 matches with three equiprobable outcomes: "win", "loose", or "even"?

A: The outcomes of the 13 matches are jointly independent, therefore:

$$P(A) = \underbrace{\frac{1}{3} \cdot \frac{1}{3} \cdots \frac{1}{3}}_{13 \text{ times}} = \frac{1}{3^{13}}.$$

□

Example A. 5

Q: An airplane has a probability of $1/3$ to hit a railway with a bomb. What is the probability that the railway is destroyed when 3 bombs are dropped?

A: The probability of not hitting the railway with one bomb is $2/3$. Assuming that the events of not hitting the railway are independent, we have:

$$P(A) = 1 - \left(\frac{2}{3}\right)^3 = 0.7.$$

□

Example A. 6

Q: What is the probability of obtaining 2 sixes when throwing a dice 6 times?

A: For any sequence with 2 sixes out of 6 throws the probability of its occurrence is:

$$P(A) = \left(\frac{1}{6}\right)^2 \left(\frac{5}{6}\right)^4.$$

In order to compute how many such sequences exist we just notice that this is equivalent to choosing two positions of the sequence, out of six possible positions. This is given by

$$\binom{6}{2} = \frac{6!}{2!\,4!} = \frac{6 \times 5}{2} = 15 \text{; therefore, } P(6,2) = 15P(A) = 0.2.$$

□

A.3 Compound Experiments

Let $\mathcal{E}_1$ and $\mathcal{E}_2$ be two sample spaces. We then form the space of the Cartesian product $\mathcal{E}_1 \times \mathcal{E}_2$, corresponding to the *compound experiment* whose elementary events are the pairs of elementary events of $\mathcal{E}_1$ and $\mathcal{E}_2$.

We now have the triplet $(\mathcal{E}_1 \times \mathcal{E}_2, \mathcal{A}, P)$ with:

$$\begin{cases} P(A_i, B_j) = P(A_i)P(B_j), & \text{if } A_i \in \mathcal{E}_1, B_j \in \mathcal{E}_2 \text{ are independent;} \\ P(A_i, B_j) = P(A_i)P(B_j \mid A_i), & \text{otherwise.} \end{cases}$$

This is generalized in a straightforward way to a compound experiment corresponding to the Cartesian product of n sample spaces.

Example A. 7

Q: An experiment consists in drawing two cards from a deck, with replacement, and noting down if the cards are: ace, figure (king, queen, jack) or number (2 to 10). Represent the sample space of the experiment composed of two "drawing one card" experiments, with the respective probabilities.

A: Since the card drawing is made with replacement, the two card drawings are jointly independent and we have the representation of the compound experiment shown in Figure A.1.   □

Notice that the sums along the rows and along the columns, the so-called *marginal probabilities*, yield the same value: the probabilities of the single experiment of drawing one card. We have:

$$P(A_i) = \sum_{j=1}^{k} P(A_i)P(B_j \mid A_i) = \sum_{j=1}^{k} P(A_i)P(B_j); \qquad \text{A. 13}$$

$$\sum_{i=1}^{k}\sum_{j=1}^{k} P(A_i)P(B_j) = 1.$$

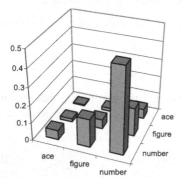

	ace	figure	number	
ace	0.006	0.018	0.053	0.077
figure	0.018	0.053	0.160	0.231
number	0.053	0.160	0.479	0.692
	0.077	0.231	0.692	1.000

Figure A.1. Sample space and probabilities corresponding to the compound card drawing experiment.

The first rule, A.13, is known as the *total probability rule*, which applies whenever one has a *partition* of the sample space into a finite or denumerably infinite sequence of events, C_1, C_2, ..., with non-null probability, mutually disjoint and with $P(\bigcup C_i) = 1$.

A.4 Bayes' Theorem

Let C_1, C_2, ... be a partition, to which we can apply the total probability rule as previously mentioned in A.13. From this rule, the following *Bayes' Theorem* can then be stated:

$$P(C_k \mid A) = \frac{P(C_k)P(A \mid C_k)}{\sum_j P(C_j)P(A \mid C_j)} \quad k = 1, 2, \ldots \quad . \qquad \text{A. 14}$$

Notice that $\sum_k P(C_k \mid A) = 1$.

In classification problems the probabilities $P(C_k)$ are called the "*a priori*" probabilities, *priors* or *prevalences*, and the $P(C_k \mid A)$ the "*a posteriori*" or *posterior* probabilities.

Often the C_k are the "causes" and A is the "effect". The Bayes' Theorem allows us then to infer the probability of the causes, as in the following example.

Example A. 8

Q: The probabilities of producing a defective item with three machines M_1, M_2, M_3 are 0.1, 0.08 and 0.09, respectively. At any instant, only one of the machines is being operated, in the following percentage of the daily work, respectively: 30%, 30%, 40%. An item is randomly chosen and found to be defective. Which machine most probably produced it?

A: Denoting the defective item by A, the total probability breaks down into:

$P(M_1)P(A \mid M_1) = 0.3 \times 0.1 \,;$
$P(M_2)P(A \mid M_2) = 0.3 \times 0.08 \,;$
$P(M_3)P(A \mid M_3) = 0.4 \times 0.09 \,.$

Therefore, the total probability is 0.09 and using Bayes' Theorem we obtain: $P(M_1 \mid A) = 0.33$; $P(M_2 \mid A) = 0.27$; $P(M_3 \mid A) = 0.4$. The machine that most probably produced the defective item is M_3. Notice that $\sum_k P(M_k) = 1$ and $\sum_k P(M_k \mid A) = 1$. □

Example A. 9

Q: An urn contains 4 balls that can either be white or black. Four extractions are made with replacement from the urn and found to be all white. What can be said about the composition of the urn if: a) all compositions are equally probable; b) the compositions are in accordance to the extraction with replacement of 4 balls from another urn, with an equal number of white and black balls?

A: There are five possible compositions, C_i, for the urn: zero white balls (C_0) , 1 white ball (C_1), ..., 4 white balls (C_4). Let us first solve situation "a", equally probable compositions. Denoting by $P_k = P(C_k)$ the probability of each composition, we have: $P_0 = P_1 = ... = P_4 = 1/5$. The probability of the event A, consisting in the extraction of 4 white balls, for each composition, is:

$$P(A \mid C_0) = 0, \quad P(A \mid C_1) = \left(\frac{1}{4}\right)^4, \dots, P(A \mid C_4) = \left(\frac{4}{4}\right)^4 = 1.$$

Applying Bayes Theorem, the probability that the urn contains 4 white balls is:

$$P(C_4 \mid A) = \frac{P(C_4)P(A \mid C_4)}{\sum_j P(C_j)P(A \mid C_j)} = \frac{4^4}{1^4 + 2^4 + 3^4 + 4^4} = 0.723.$$

This is the largest "a posteriori" probability one can obtain. Therefore, for situation "a", the most probable composition is C_4.

In situation "b" the "a priori" probabilities of the composition are in accordance to the binomial probabilities of extracting 4 balls from the second urn. Since this urn has an equal number of white and black balls, the prevalences are therefore proportional to the binomial coefficients $\binom{4}{k}$. For instance, the probability of C_4 is:

$$P(C_4 \mid A) = \frac{P(C_4)P(A \mid C_4)}{\sum_j P(C_j)P(A \mid C_j)} = \frac{4^4}{4.1^4 + 6.2^4 + 4.3^4 + 1.4^4} = 0.376.$$

This is, however, smaller than the probability for C_3: $P(C_3 \mid A) = 0.476$. Therefore, C_3 is the most probable composition, illustrating the drastic effect of the prevalences.   □

A.5 Random Variables and Distributions

A.5.1 Definition of Random Variable

A real function $X \equiv X(E_i)$, defined on the sample space $\mathcal{E} = \{E_i\}$ of a random experiment, is a *random variable* for the probability space ($\mathcal{E}$, $\mathcal{A}$, P), if for every real number z, the subset:

$$\{X \le z\} = \{E_i; \quad X(E_i) \le z\},$$ A. 15

is a member of the collection of events $\mathcal{A}$. Particularly, when $z \to \infty$, one obtains $\mathcal{E}$ and with $z \to -\infty$, one obtains ϕ.

From the definition, one determines the event corresponding to an interval $]a, b]$ as:

$$\{a < X \le b\} = \{E_i; \ X(E_i) \le b\} - \{E_i; \ X(E_i) \le a\}. \qquad \text{A. 16}$$

Example A. 10

Consider the random experiment of throwing two dice, with sample space $\mathcal{E} = \{(a, b); 1 \le a, b \le 6\} = \{(1,1), (1,2), \ldots, (6,6)\}$ and the collection of events $\mathcal{A}$ that is a Borel algebra defined on $\{ \{(1,1)\}, \{(1,2), (2,1)\}, \{(1,3), (2,2), (3,1)\}, \{(1,4), (2,3), (3,2), (4,1)\}, \{(1,5), (2,4), (3,3), (4,2), (5,1)\}, \ldots, \{(6,6)\} \}$. The following variables $X(\mathcal{E})$ can be defined:

$X(a, b) = a+b$. This is a random variable for the probability space $(\mathcal{E}, \mathcal{A}, P)$. For instance, $\{X \le 4.5\} = \{(1,1), (1,2), (2,1), (1,3), (2,2), (3,1)\} \in \mathcal{A}$.

$X(a, b) = ab$. This is not a random variable for the probability space $(\mathcal{E}, \mathcal{A}, P)$. For instance, $\{X \le 3.5\} = \{(1,1), (1,2), (2,1), (1,3), (3,1)\} \notin \mathcal{A}$. □

A.5.2 Distribution and Density Functions

The *probability distribution function* (PDF) of a random variable X is defined as:

$$F_X(x) = P(X \le x). \qquad \text{A. 17}$$

We usually simplify the notation, whenever no confusion can arise from its use, by writing $F(x)$ instead of $F_X(x)$.

Figure A.2 shows the distribution function of the random variable $X(a, b) = a + b$ of Example A.10.

Until now we have only considered examples involving sample spaces with a finite number of elementary events, the so-called discrete sample spaces to which *discrete random variables* are associated. These can also represent a denumerably infinite number of elementary events.

For discrete random variables, with probabilities p_j assigned to the singleton events of $\mathcal{A}$, the following holds:

$$F(x) = \sum_{x_j \le x} p_j. \qquad \text{A. 18}$$

For instance, in Example A.10, we have $F(4.5) = p_1 + p_2 + p_3 = 0.17$ with $p_1 = P(\{(1,1)\})$, $p_2 = P(\{(1,2), (2,1)\})$ and $p_3 = P(\{(1,3), (2,2), (3,1)\})$. The p_j sequence is called a *probability distribution*.

When dealing with non-denumerable infinite sample spaces, one needs to resort to *continuous random variables*, characterized by a continuous *distribution function* $F_X(x)$, differentiable everywhere (except perhaps at a finite number of points).

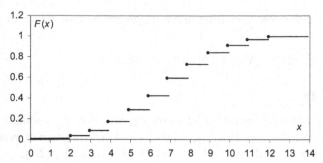

Figure A.2. Distribution function of the random variable associated to the sum of the faces in the two-dice throwing experiment. The solid circles represent point inclusion.

The function $f_X(x) = dF_X(x)/dx$ (or simply $f(x)$) is called the *probability density function* (pdf) of the continuous random variable X. The properties of the density function are as follows:

i. $f(x) \geq 0$ (where defined) ;

ii. $\int_{-\infty}^{\infty} f(t)dt = 1$;

iii. $F(x) = \int_{-\infty}^{x} f(t)dt$.

The event corresponding to $]\,a, b]$ has the following probability:

$$P(a < X \leq b) = P(X \leq b) - P(X \leq a) = F(b) - F(a) = \int_a^b f(t)dt. \qquad \text{A. 19}$$

This is the same as $P(a \leq X \leq b)$ in the absence of a discontinuity at a. For an infinitesimal interval we have:

$$P(a < X \leq a + \Delta a) = F(a + \Delta a) - F(a) = f(a)\Delta a \;\Rightarrow$$

$$f(a) = \frac{F(a + \Delta a) - F(a)}{\Delta a} = \frac{P([a, a + \Delta a])}{\Delta a}, \qquad \text{A. 20}$$

which justifies the name density function, since it represents the "mass" probability corresponding to the interval Δa, measured at a, per "unit length" of the random variable (see Figure A.3a).

The solution $X = x_\alpha$ of the equation:

$$F_X(x) = \alpha, \qquad \text{A. 21}$$

is called the *α-quantile* of the random variable X. For $\alpha = 0.1$ and 0.01, the quantiles are called *deciles* and *percentiles*. Especially important are also the *quartiles* ($\alpha = 0.25$) and the *median* ($\alpha = 0.5$) as shown in Figure A.3b. Quantiles are useful location measures; for instance, the *inter-quartile range*, $x_{0.75} - x_{0.25}$, is often used to locate the central tendency of a distribution.

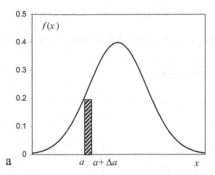 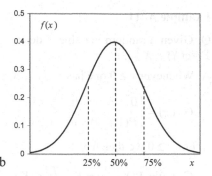

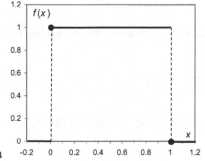 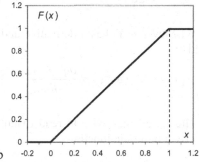

Figure A.3. a) A pdf example; the shaded area in $[a, a+\Delta a]$ is an infinitesimal probability mass. b) Interesting points of a pdf: *lower quartile* (25% of the total area); *median* (50% of the total area); *upper quartile* (75% of the total area).

Figure A.4. Uniform random variable: a) Density function (the circles indicate point inclusion); b) Distribution function.

Figure A.4 shows the uniform density and distribution functions defined in [0, 1]. Note that $P(a < X \leq a+w) = w$ for every a such that $[a, a+w] \subset [0, 1]$, which justifies the name *uniform* distribution.

A.5.3 Transformation of a Random Variable

Let X be a random variable defined in the probability space $(\mathcal{E}, \mathcal{A}, P)$, whose distribution function is:

$$F_X(x) = P(X \leq x).$$

Consider the variable $Y = g(X)$ such that every interval $-\infty < Y \leq y$ maps into an event S_y of the collection $\mathcal{A}$. Then Y is a random variable whose distribution function is:

$$G_Y(y) = P(Y \leq y) = P(g(X) \leq y) = P(x \in S_y).$$ A. 22

Example A. 11

Q: Given a random variable X determine the distribution and density functions of $Y = g(X) = X^2$.

A: Whenever $y \geq 0$ one has $-\sqrt{y} \leq X \leq \sqrt{y}$. Therefore:

$$G_Y(y) = \begin{cases} 0 & \text{if } y < 0 \\ P(Y \leq y) & \text{if } y \geq 0 \end{cases}.$$

For $y \geq 0$ we then have:

$$G_Y(y) = P(Y \leq y) = P(-\sqrt{y} \leq X \leq \sqrt{y}) = F_X(\sqrt{y}) - F_X(-\sqrt{y}).$$

If $F_X(x)$ is continuous and differentiable, we obtain for $y > 0$:

$$g_Y(y) = \frac{1}{2\sqrt{y}} \left[f_X(\sqrt{y}) + f_X(-\sqrt{y}) \right]. \qquad\qquad \square$$

Whenever $g(X)$ has a derivative and is strictly monotonic, the following result holds:

$$g_Y(y) = f_X(g^{-1}(y)) \left| \frac{dg^{-1}(y)}{dy} \right|$$

The reader may wish to redo Example A.11 by first considering the following strictly monotonic function:

$$g(X) = \begin{cases} 0 & \text{if } X < 0 \\ X^2 & \text{if } X \geq 0 \end{cases}$$

A.6 Expectation, Variance and Moments

A.6.1 Definitions and Properties

Let X be a random variable and $g(X)$ a new random variable resulting from transforming X with the function g. The *expectation* of $g(X)$, denoted $E[g(X)]$, is defined as:

$$E[g(X)] = \sum_i g(x_i) P(X = x_i), \quad \text{if } X \text{ is discrete (and the sum exists);} \qquad \text{A.23a}$$

$$E[g(X)] = \int_{-\infty}^{\infty} g(x) f(x) dx, \quad \text{if } X \text{ is continuous (and the integral exists).} \qquad \text{A.23b}$$

Example A. 12

Q: A gambler throws a dice and wins 1€ if the face is odd, loses 2.5€ if the face is 2 or 4, and wins 3€ if the face is 6. What is the gambler's expectation?

A: We have:

$$g(x) = \begin{cases} 1 & \text{if} \quad X = 1, 3, 5; \\ -2.5 & \text{if} \quad X = 2, 4; \\ 3 & \text{if} \quad X = 6. \end{cases}$$

Therefore: $E[g(X)] = 3\dfrac{1}{6} - 2\dfrac{2.5}{6} + \dfrac{3}{6} = \dfrac{1}{6}$.

The word "expectation" is somewhat misleading since the gambler will only expect to get close to winning 1/6 € in a long run of throws. ☐

The following cases are worth noting:

1. $g(X) = X$: *Expected value*, *mean* or *average* of X.

$\mu = E[X] = \sum_i x_i P(X = x_i)$, if X is discrete (and the sum exists);   A.24a

$\mu = E[X] = \int_{-\infty}^{\infty} x f(x) dx$, if X is continuous (and the integral exists).   A.24b

The mean of a distribution is the probabilistic mass center (center of gravity) of the distribution.

Example A. 13

Q: Consider the Cauchy distribution, with: $f_X(x) = \dfrac{a}{\pi} \dfrac{1}{a^2 + x^2}$, $x \in \Re$. What is its mean?

A: We have:

$E[X] = \dfrac{a}{\pi} \int_{-\infty}^{\infty} \dfrac{x}{a^2 + x^2} dx$. But $\int \dfrac{x}{a^2 + x^2} dx = \dfrac{1}{2} \ln(a^2 + x^2)$, therefore the integral diverges and the mean does not exist. ☐

Properties of the mean (for arbitrary real constants a, b):

  i. $E[aX + b] = aE[X] + b$ (linearity);

  ii. $E[X + Y] = E[X] + E[Y]$ (additivity);

  iii. $E[XY] = E[X]E[Y]$ if X and Y are independent.

The mean reflects the "central tendency" of a distribution. For a data set with n values x_i occurring with frequencies f_i, the mean is estimated as (see A.24a):

$$\hat{\mu} \equiv \bar{x} = \sum_{i=1}^{n} x_i f_i \ .$$

A. 25

This is the so-called *sample mean*.

Example A. 14

Q: Show that the random variable $X - \mu$ has zero mean.

A. Applying the linear property to $\mathrm{E}[X - \mu]$ we have:

$$\mathrm{E}[X - \mu] = \mathrm{E}[X] - \mu = \mu - \mu = 0 \ .$$

□

2. $g(X) = X^k$: *Moments of order k of X.*

$$\mathrm{E}[X^k] = \sum_i x_i^k P(X = x_i) \ , \quad \text{if } X \text{ is discrete (and the sum exists);} \qquad \text{A.26a}$$

$$\mathrm{E}[X^k] = \int_{-\infty}^{\infty} (x - \mu)^k f(x) dx \ , \text{if } X \text{ is continuous (and the integral exists).A.26b}$$

Especially important, as explained below, is the moment of order two: $\mathrm{E}[X^2]$.

3. $g(X) = (X - \mu)^k$: *Central moments of order k of X.*

$$m_k = \mathrm{E}[(X - \mu)^k] = \sum_i (x_i - \mu)^k P(X = x_i) \ ,$$

if X is discrete (and the sum exists);   A.27a

$$m_k = \mathrm{E}[(X - \mu)^k] = \int_{-\infty}^{\infty} (x - \mu)^k f(x) dx$$

if X is continuous (and the integral exists).   A.27b

Of particular importance is the central moment of order two, m_2 (we often use $\mathrm{V}[X]$ instead), and known as *variance*. Its square root is the *standard deviation*: $\sigma_X = \{\mathrm{V}[X]\}^{\frac{1}{2}}$.
Properties of the variance:

i. $\mathrm{V}[X] \geq 0$;

ii. $\mathrm{V}[X] = 0$ iff X is a constant;

iii. $\mathrm{V}[aX + b] = a^2 \mathrm{V}[X]$;

iv. $\mathrm{V}[X + Y] = \mathrm{V}[X] + \mathrm{V}[Y]$ if X and Y are independent.

The variance reflects the "data spread" of a distribution. For a data set with n values x_i occurring with frequencies f_i, and estimated mean $\bar{x}$, the variance can be estimated (see A.27a) as:

$$v \equiv \hat{V}[X] = \sum_{i=1}^{n} (x_i - \bar{x})^2 f_i .$$ A. 28

This is the so-called *sample variance*. The square root of v, $s = \sqrt{v}$, is the *sample standard deviation*. In Appendix C we present a better estimate of v.

The variance can be computed using the second order moment, observing that:

$$V[X] = E[(X - \mu)^2] = E[X^2] - 2\mu E[X] + \mu^2 = E[X^2] - \mu^2 .$$ A. 29

4. Gauss' approximation formulae:

i. $E[g(X)] \approx g(E[X])$;

ii. $V[g(X)] \approx V[X] \left[\dfrac{dg}{dx} \bigg|_{E[X]} \right]^2 .$

A.6.2 Moment-Generating Function

The *moment-generating function* of a random variable X, is defined as the expectation of e^{tX} (when it exists), i.e.:

$$\psi_X (t) = E[e^{tX}] .$$ A. 30

The importance of this function stems from the fact that one can derive all moments from it, using the result:

$$E[X^k] = \dfrac{d^n \psi_X (t)}{dt^n} \bigg|_{t=0} .$$ A. 31

A distribution function is uniquely determined by its moments as well as by its moment-generating function.

Example A. 15

Q: Consider a random variable with the Poisson probability function $P(X = k) = e^{-\lambda} \lambda^k / k!$, $k \geq 0$. Determine its mean and variance using the moment-generating function approach.

A: The moment-generating function is:

$$\psi_X (t) = E[e^{tX}] = \sum_{k=0}^{\infty} e^{tk} e^{-\lambda} \lambda^k / k! = e^{-\lambda} \sum_{k=0}^{\infty} (\lambda e^t)^k / k!.$$

Since the power series expansion of the exponential is $e^x = \sum_{k=0}^{\infty} x^k / k!$ one can write:

$$\psi_X (t) = e^{-\lambda} e^{\lambda e^t} = e^{\lambda(e^t - 1)} .$$

Hence: $\mu = \dfrac{d\psi_X(t)}{dt}\bigg|_{t=0} = \lambda e^t e^{\lambda(e^t-1)}\bigg|_{t=0} = \lambda$;

$E[X^2] = \dfrac{d^2\psi_X(t)}{dt^2}\bigg|_{t=0} = (\lambda e^t + 1)\lambda e^t e^{\lambda(e^t-1)}\bigg|_{t=0} = \lambda^2 + \lambda \implies V[X] = \lambda.$ □

A.6.3 Chebyshev Theorem

The Chebyshev Theorem states that for any random variable X and any real constant k, the following holds:

$$P(|X - \mu| > k\sigma) \le \frac{1}{k^2}.$$ A. 32

Since it is applicable to any probability distribution, it is instructive to see the proof (for continuous variables) of this surprising and very useful Theorem; from the definition of variance, and denoting by S the domain where $(X - \mu)^2 > a$, we have:

$$\mu_2 = E[(X-\mu)^2] = \int_{-\infty}^{\infty}(x-\mu)^2 f(x)dx \ge$$
$$\int_S (x-\mu)^2 f(x)dx \ge a\int_S f(x)dx = aP((X-\mu)^2 > a).$$

Taking $a = k^2\sigma^2$, we get:

$$P((X - \mu)^2 > k^2\sigma^2) \le \frac{1}{k^2} ,$$

from where the above result is obtained.

Example A. 16

Q: A machine produces resistances of nominal value 100 Ω (ohm) with a standard deviation of 1 Ω. What is an upper bound for the probability of finding a resistance deviating more than 3 Ω from the mean?

A: The 3 Ω tolerance corresponds to three standard deviations; therefore, the upper bound is $1/9 = 0.11$.   □

A.7 The Binomial and Normal Distributions

A.7.1 The Binomial Distribution

One often needs to compute the probability that a certain event occurs k times in a sequence of n events. Let the probability of the interesting event (the *success*) be p.

The probability of the complement (the *failure*) is, therefore, $q = 1 - p$. The random variable associated to the occurrence of k successes in n trials, X_n, has the *binomial probability distribution* (see Example A.6):

$$P(X_n = k) = \binom{n}{k} p^k q^{n-k}, \qquad 0 \le k \le n.$$ A. 33

By studying the $P(X_n = k+1) / P(X_n = k)$ ratios one can prove that the largest probability value occurs at the integer value close to $np - q$ or np. Figure A.5 shows the binomial probability function for two different values of n.

For the binomial distribution, one has:

Mean: $\mu = np$;   Variance: $\sigma^2 = npq$.

Given the fast growth of the factorial function, it is often convenient to compute the binomial probabilities using the *Stirling formula*:

$$n! = n^n e^{-n} \sqrt{2\pi n}(1 + \varepsilon_n).$$ A. 34

The quantity ε_n tends to zero with large n, with $n\varepsilon_n$ tending to $1/_{12}$. The convergence is quite fast: for $n = 20$ the error of the approximation is already below 0.5%.

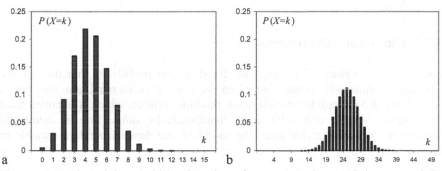

Figure A.5. Binomial probability functions for $p = 0.3$: a) $n = 15$ ($np - q = 3.8$); b) $n = 50$.

A.7.2 The Laws of Large Numbers

The following important result, known as *Weak Law of Large Numbers*, or Bernoulli Theorem, can be proved using the binomial distribution:

$$P\left(\left|\frac{k}{n} - p\right| \ge \varepsilon\right) \le \frac{pq}{\varepsilon^2 n} \quad \text{or, equivalently,} \quad P\left(\left|\frac{k}{n} - p\right| < \varepsilon\right) \ge 1 - \frac{pq}{\varepsilon^2 n}.$$ A. 35

Therefore, in order to obtain a certainty $1 - \alpha$ (*confidence level*) that a relative frequency deviates from the probability of an event less than ε (*tolerance* or *error*), one would need a sequence of n trials, with:

$$n \geq \frac{pq}{\varepsilon^2 \alpha}.$$ A. 36

Note that $\lim_{n \to \infty} P\left(\left|\frac{k}{n} - p\right| \geq \varepsilon\right) = 0$.

A stronger result is provided by the *Strong Law of Large Numbers*, which states the convergence of k/n to p with probability one.

These results clarify the assumption made in section A.1 of the convergence of the relative frequency of an event to its probability, in a long sequence of trials.

Example A. 17

Q: What is the tolerance of the percentage, p, of favourable votes on a certain market product, based on a sample enquiry of 2500 persons, with a confidence level of at least 95%?

A: As we do not know the exact value of p, we assume the worst-case situation for A.36, occurring at $p = q = \frac{1}{2}$. We then have:

$$\varepsilon = \sqrt{\frac{pq}{n\alpha}} = 0.045.$$ ☐

A.7.3 The Normal Distribution

For increasing values of n and with fixed p, the probability function of the binomial distribution becomes flatter and the position of its maximum also grows (see Figure A.5). Consider the following random variable, which is obtained from the random variable with a binomial distribution by *subtracting its mean and dividing by its standard deviation* (the so-called *standardised* random variable or *z-score*):

$$Z = \frac{X_n - np}{\sqrt{npq}}.$$ A. 37

It can be proved that for large n and not too small p and q (say, with np and nq greater than 5), the standardised discrete variable is well approximated by a continuous random variable having density function $f(z)$, with the following asymptotic result:

$$P(Z) \underset{n \to \infty}{\to} f(z) = \frac{1}{\sqrt{2\pi}} e^{-z^2/2}.$$ A. 38

This result, known as *De Moivre's Theorem*, can be proved using the above Stirling formula A.34. The density function $f(z)$ is called the *standard normal* (or

Gaussian) density and is represented in Figure A.7 together with the distribution function, also known as *error function*. Notice that, taking into account the properties of the mean and variance, this new random variable has zero mean and unit variance.

The approximation between normal and binomial distributions is quite good even for not too large values of n. Figure A.6 shows the situation with $n = 50$, $p = 0.5$. The maximum deviation between binomial and normal approximation occurs at the middle of the distribution and is 0.056. For $n = 1000$, the deviation is 0.013. In practice, when np or nq are larger than 25, it is reasonably safe to use the normal approximation of the binomial distribution.

Note that:

$$Z = \frac{X_n - np}{\sqrt{npq}} \sim N_{0,1} \quad \Rightarrow \quad \hat{P} = \frac{X_n}{n} \sim N_{p,\sqrt{pq/n}},$$ A. 39

where $N_{\mu,\sigma}$ is the Gaussian distribution with mean μ and standard deviation σ, and the following density function:

$$f(x) = \frac{1}{\sqrt{2\pi}\sigma} e^{-(x-\mu)^2/2\sigma^2}.$$ A. 40

Both binomial and normal distribution values are listed in tables (see Appendix D) and can also be obtained from software tools (such as EXCEL, SPSS, STATISTICA, MATLAB and R).

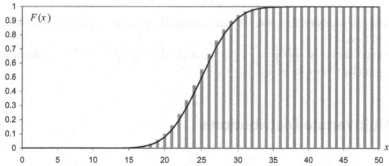

Figure A.6. Normal approximation (solid line) of the binomial distribution (grey bars) for $n = 50$, $p = 0.5$.

Example A. 18

Q: Compute the tolerance of the previous Example A.17 using the normal approximation.

A: Like before, we consider the worst-case situation with $p = q = \frac{1}{2}$. Since $\sigma_{\hat{p}} = \sqrt{1/4n} = 0.01$, and the 95% confidence level corresponds to the interval

$[-1.96\sigma, 1.96\sigma]$ (see normal distribution tables), we then have: $\varepsilon = 1.96\sigma = 0.0196$ (smaller than the previous "model-free" estimate).   □

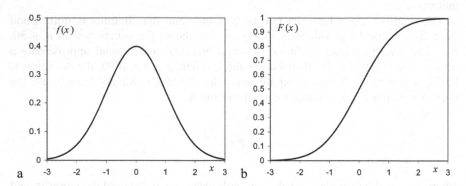

Figure A.7. The standard normal density (a) and distribution (b) functions.

Example A. 19

Q: Let X be a standard normal variable. Determine the density of $Y = X^2$ and its expectation.

A: Using the previous result of Example A.11:

$$g(y) = \frac{1}{2\sqrt{y}}\left[f(\sqrt{y}) + f(-\sqrt{y})\right] = \frac{1}{\sqrt{2\pi y}}\,e^{-y/2} \qquad y > 0.$$

This is the density function of the so-called *chi-square distribution* with one degree of freedom.

The expectation is: $E[Y] = \int_0^\infty yg(y)dy = \left(1/\sqrt{2\pi}\right)\int_0^\infty \sqrt{y}\,e^{-y/2}dy$. Substituting y by x^2, it can be shown to be 1.   □

A.8 Multivariate Distributions

A.8.1 Definitions

A sequence of random variables X_1, X_2,..., X_d, can be viewed as a vector $\mathbf{x} = [X_1, X_2, \dots X_d]$ with d components. The *multivariate* (or *joint*) distribution function is defined as:

$$F(x_1, x_2, \dots x_d) = P(X_1 \le x_1, X_2 \le x_2, \dots, X_d \le x_d). \qquad \text{A. 41}$$

The following results are worth mentioning:

1.   If for a fixed j, $1 \leq j \leq d$, $X_j \to \infty$, then $F(x_1, x_2, \ldots x_d)$ converges to a function of $d - 1$ variables which is the distribution function $F(x_1, \ldots, x_{j-1}, x_{j+1}, \ldots, x_d)$, the so-called jth *marginal distribution*.

2.   If the d-fold partial derivative:

$$f(x_1, x_2, \ldots, x_d) = \frac{\partial^{(d)} F(x_1, x_2, \ldots, x_d)}{\partial x_1 \partial x_2 \ldots \partial x_d},$$
         A. 42

exists, then it is called the *density function* of **x**. We then have:

$$P\big((X_1, X_2, \ldots, X_d) \in S\big) = \int \int \ldots \int_S f(x_1, x_2, \ldots, x_d) dx_1 dx_2 \ldots dx_d .$$ A. 43

Example A. 20

For the Example A.7, we defined the bivariate random vector $\mathbf{x} = \{X_1, X_2\}$, where each X_j performs the mapping: $X_j(\text{ace})=0$; $X_j(\text{figure})=1$; $X_j(\text{number})=2$. The joint distribution is shown in Figure A.8, computed from the probability function (see Figure A.1).   ☐

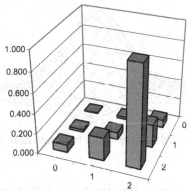

Figure A.8. Joint distribution of the bivariate random experiment of drawing two cards, with replacement from a deck, and categorising them as ace (0), figure (1) and number (2).

Example A. 21

Q: Consider the bivariate density function:

$$f(x_1, x_2) = \begin{cases} 2 & \text{if} \quad 0 \leq x_1 \leq x_2 \leq 1; \\ 0 & \text{otherwise.} \end{cases}$$

Compute the marginal distributions and densities as well as the probability corresponding to $x_1, x_2 \leq \tfrac{1}{2}$.

A: Note first that the domain where the density is non-null corresponds to a triangle of area ½. Therefore, the total volume under the density function is 1 as it should be. The marginal distributions and densities are computed as follows:

$$F_1(x_1) = \int_{-\infty}^{x_1} \int_{-\infty}^{\infty} f(u,v)dudv = \int_0^{x_1} \left(\int_u^1 2dv \right) du = 2x_1 - x_1^2$$

$$\Rightarrow \quad f_1(x_1) = \frac{dF_1(x_1)}{dx_1} = 2 - 2x_1$$

$$F_2(x_2) = \int_{-\infty}^{\infty} \int_{-\infty}^{x_2} f(u,v)dudv = \int_0^{x_2} \left(\int_0^v 2du \right) dv = x_2^2 \Rightarrow f_2(x_2) = \frac{dF_2(x_2)}{dx_2} = 2x_2.$$

The probability is computed as:

$$P(X_1 \le \tfrac{1}{2}, X_2 \le \tfrac{1}{2}) = \int_{-\infty}^{\frac{1}{2}} \int_{-\infty}^{v} 2dudv = \int_0^{\frac{1}{2}} 2vdv = \tfrac{1}{4}.$$

The same result could be more simply obtained by noticing that the domain has an area of 1/8. □

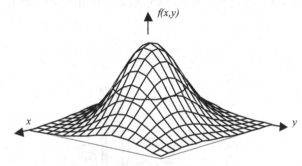

Figure A.9. Bell-shaped surface of the bivariate normal density function.

The bivariate normal density function has a bell-shaped surface as shown in Figure A.9. The equidensity curves in this surface are circles or ellipses (an example of which is also shown in Figure A.9). The probability of the event $(x_1 \le X < x_2,\ y_1 \le Y < y_2)$ is computed as the volume under the surface in the mentioned interval of values for the random variables X and Y.

The equidensity surfaces of a trivariate normal density function are spheres or ellipsoids, and in general, the equidensity hypersurfaces of a d-variate normal density function are hyperspheres or hyperellipsoids in the d-dimensional space, $\Re^d$.

A.8.2 Moments

The moments of multivariate random variables are a generalisation of the previous definition for single variables. In particular, for bivariate distributions, we have the central moments:

$$m_{kj} = E[(X - \mu_x)^k (Y - \mu_y)^j].$$ A. 44

The following central moments are worth noting:

$m_{20} = \sigma_X^2$: *variance of X*; $m_{02} = \sigma_Y^2$: *variance of Y*;

$m_{11} \equiv \sigma_{XY} = \sigma_{YX}$: *covariance of X and Y, with* $m_{11} = E[XY] - \mu_X \mu_Y$.

For multivariate d-dimensional distributions we have a symmetric positive definite *covariance matrix*:

$$\Sigma = \begin{bmatrix} \sigma_1^2 & \sigma_{12} & \cdots & \sigma_{1d} \\ \sigma_{21} & \sigma_2^2 & \cdots & \sigma_{2d} \\ \cdots & \cdots & \cdots & \cdots \\ \sigma_{d1} & \sigma_{d2} & \cdots & \sigma_d^2 \end{bmatrix}.$$ A. 45

The *correlation coefficient*, which is a measure of linear association between X and Y, is defined by the relation:

$$\rho \equiv \rho_{XY} = \frac{\sigma_{XY}}{\sigma_X . \sigma_Y}.$$ A. 46

Properties of the correlation coefficient:

i. $-1 \leq \rho \leq 1$;

ii. $\rho_{XY} = \rho_{YX}$;

iii. $\rho = \pm 1$ iff $(Y - \mu_Y)/\sigma_Y = \pm (X - \mu_X)/\sigma_X$;

iv. $\rho_{aX+b,cY+d} = \rho_{XY}$, $ac > 0$; $\rho_{aX+b,cY+d} = -\rho_{XY}$, $ac < 0$.

If $m_{11} = 0$, the random variables are said to be *uncorrelated*. Since $E[XY] = E[X]E[Y]$ if the variables are independent, then they are also uncorrelated. The converse statement is not generally true. However, it is true in the case of normal distributions, where uncorrelated variables are also independent.

The definitions of covariance and correlation coefficient have a straightforward generalisation for the d-variate case.

A.8.3 Conditional Densities and Independence

Assume that the bivariate random vector $[X, Y]$ has a density function $f(x, y)$. Then, the conditional distribution of X given Y is defined, whenever $f(y) \neq 0$, as:

$$F(x \mid y) = P(X \le x \mid Y = y) = \lim_{\Delta y \to 0} P(X \le x \mid y < Y \le y + \Delta y).$$ A. 47

From the definition it can be proved that the following holds true:

$$f(x, y) = f(x \mid y) f(y).$$ A. 48

In the case of discrete Y, $F(x \mid y)$ can be computed directly. It can also be proved the Bayes' Theorem version for this case:

$$P(y_i \mid x) = \frac{P(y_i) f(x \mid y_i)}{\sum_k P(y_k) f(x \mid y_k)}.$$ A. 49

Note the mixing of discrete prevalences with values of conditional density functions.

A set of random variables $X_1, X_2, \ldots, X_d$ are independent if the following applies:

$$F(x_1, x_2, \ldots x_d) = F(x_1) F(x_2) \ldots F(x_d);$$ A.50a

$$f(x_1, x_2, \ldots x_d) = f(x_1) f(x_2) \ldots f(x_d).$$ A.50b

For two independent variables, we then have:

$$f(x, y) = f(x) f(y); \text{ therefore, } f(x \mid y) = f(x); \quad f(y \mid x) = f(y).$$ A. 51

Also: $\mathrm{E}[XY] = \mathrm{E}[X] \mathrm{E}[Y].$ A. 52

It is readily seen that the random variables in correspondence with the bivariate density of Example A.21 are not independent since $f(x_1, x_2) \ne f(x_1) f(x_2)$.

Consider two independent random variables, X_1, X_2, with Gaussian densities and parameters (μ_1, σ_1), (μ_2, σ_2) respectively. The joint density is the product of the marginal Gaussian densities:

$$f(x_1, x_2) = \frac{1}{2\pi\sigma_1\sigma_2} e^{-\left[\frac{(x_1 - \mu_1)^2}{2\sigma_1^2} + \frac{(x_2 - \mu_2)^2}{2\sigma_2^2}\right]}.$$ A. 53

In this case it can be proved that $\rho_{12} = 0$, i.e., for Gaussian distributions, independent variables are also uncorrelated. In this case, the equidensity curves in the (X_1, X_2) plane are ellipsis aligned with the axes.

If the distributions are not independent (and are, therefore, correlated) one has:

$$f(x_1, x_2) = \frac{1}{2\pi\sigma_1\sigma_2\sqrt{1-\rho^2}} e^{-\frac{1}{2(1-\rho^2)}\left[\frac{(x_1 - \mu_1)^2}{2\sigma_1^2} + \frac{(x_2 - \mu_2)^2}{2\sigma_2^2} - \frac{2\rho X_1 X_2}{\sigma_1\sigma_2}\right]}.$$ A. 54

For the d-variate case, this generalises to:

$$f(x_1,\ldots x_d) \equiv N_d(\mathbf{\mu}, \mathbf{\Sigma}) = \frac{1}{(2\pi)^{d/2}\sqrt{\det(\mathbf{\Sigma})}} \exp\left(-\frac{1}{2}(\mathbf{x}-\mathbf{\mu})'\mathbf{\Sigma}^{-1}(\mathbf{x}-\mathbf{\mu})\right), \quad \text{A. 55}$$

where $\mathbf{\Sigma}$ is the symmetric matrix of the covariances with determinant $\det(\mathbf{\Sigma})$ and $\mathbf{x} - \mathbf{\mu}$ is the difference vector between the d-variate random vector and the mean vector. The equidensity surfaces for the correlated normal variables are ellipsoids, whose axes are the eigenvectors of $\mathbf{\Sigma}$.

A.8.4 Sums of Random Variables

Let X and Y be two independent random variables. Then, the distribution of their *sum* corresponds to:

$$P(X+Y=s) = \sum_{x_i+y_j=s} P(X=x_i)P(Y=y_j), \quad \text{if they are discrete;} \qquad \text{A.56a}$$

$$f_{X+Y}(z) = \int_{-\infty}^{\infty} f_X(u)f_Y(z-u)du, \quad \text{if they are continuous.} \qquad \text{A.56b}$$

The roles of $f_X(u)$ and $f_Y(u)$ can be interchanged. The operation performed on the probability or density functions is called a *convolution* operation. By analysing the integral A.56b, it is readily seen that the *convolution* operation can be interpreted as multiplying one of the densities by the reflection of the other as it slides along the domain variable u.

Figure A.10 illustrates the effect of the convolution operation when adding discrete random variables for both symmetrical and asymmetrical probability functions. Notice how successive convolutions will tend to produce a bell-shaped probability function, displacing towards the right, even when the initial probability function is asymmetrical.

Consider the arithmetic mean, $\overline{X}$, of n i.i.d. random variables with mean μ and standard deviation σ.

$$\overline{X} = \sum_{i=1}^{n} X_i / n \qquad \text{A. 57}$$

As can be expected the probability or density function of $\overline{X}$ will tend to a bell-shaped curve for large n. Also, taking into account the properties of the mean and the variance, mentioned in A.6.1, it is clearly seen that the following holds:

$$E[\overline{X}] = \mu; \quad V[\overline{X}] = \sigma^2/n. \qquad \text{A.58a}$$

Therefore, the distribution of $\overline{X}$ will have the same mean as the distribution of X and a standard deviation (spread or dispersion) that decreases with $\sqrt{n}$. Note that for any variables the additive property of the means is always verified but for the variance this is not true:

$$V\left[\sum_i c_i X_i\right] = \sum_i c_i^2 V[X_i] + 2\sum_{i<j} c_i c_j \sigma_{X_i X_j} \qquad \text{A.58b}$$

A.8.5 Central Limit Theorem

We have previously seen how multiple addition of the same random variable tends to produce a bell-shaped probability or density function. The *Central Limit Theorem* (also known as Levy-Lindeberg Theorem) states that the sum of n independent random variables, all with the same mean, μ, and the same standard deviation $\sigma \neq 0$ has a density that is asymptotically Gaussian, with mean $n\mu$ and $\sigma_n = \sigma\sqrt{n}$. Equivalently, the random variable:

$$X_n = \frac{X_1 + ... + X_n - n\mu}{\sigma\sqrt{n}} = \sum_{i=1}^{n} \frac{X_i - \mu}{\sigma\sqrt{n}},$$
A. 59

is such that $\lim_{n\to\infty} F_{X_n}(x) = \frac{1}{\sqrt{2\pi}} \int_{-\infty}^{\infty} e^{-x^2/2} dx$.

In particular the $X_1, ..., X_n$ may be n independent copies of X.

Let us now consider the sequence of n mutually independent variables $X_1, ..., X_n$ with means μ_k and variances σ_k^2. Then, the sum $S = X_1 + ... + X_n$ has mean and variance given by $\mu = \mu_1 + ... + \mu_n$ and $\sigma^2 = \sigma_1^2 + ... + \sigma_n^2$, respectively.

We say that the sequence obeys the Central Limit Theorem if for every fixed $\alpha < \beta$, the following holds:

$$P\left(\alpha < \frac{S - \mu}{\sigma} < \beta\right) \underset{n\to\infty}{\to} N_{0,1}(\beta) - N_{0,1}(\alpha).$$
A. 60

As it turns out, a surprisingly large number of distributions satisfy the Central Limit Theorem. As a matter of fact, a necessary and sufficient condition for this result to hold is that the X_k are mutually independent and uniformly bounded, i.e., $|X_k| < A$ (see Galambos, 1984, for details). In practice, many phenomena can be considered the result of the addition of many independent causes, yielding then, by the Central Limit Theorem, a distribution that closely approximates the normal distribution, even for a moderate number of additive causes (say above 5).

Example A. 22

Consider the following probability functions defined for the domain {1, 2, 3, 4, 5, 6, 7} (zero outside):

$P_X = \{0.183, 0.270, 0.292, 0.146, 0.073, 0.029, 0.007\}$;
$P_Y = \{0.2, 0.2, 0.2, 0.2, 0.2, 0, 0\}$;
$P_Z = \{0.007, 0.029, 0.073, 0.146, 0.292, 0.270, 0.183\}$.

Figure A.11 shows the resulting probability function of the sum $X + Y + Z$. The resemblance with the normal density function is manifest.   □

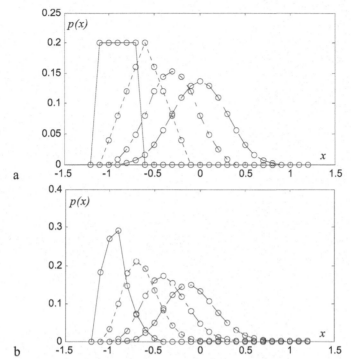

a

b

Figure A.10. Probability function of the sum of $k = 1,.., 4$ i.i.d. discrete random variables: a) Equiprobable random variable (symmetrical); b) Asymmetrical random variable. The solid line shows the univariate probability function; all other curves correspond to probability functions with a coarser dotted line for growing k. The circles represent the probability values.

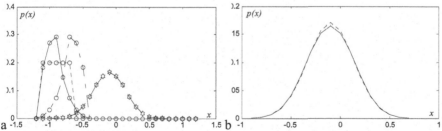

a

b

Figure A.11. a) Probability function (curve with stars) resulting from the addition of three random variables with distinct distributions; b) Comparison with the normal density function (dotted line) having the same mean and standard deviation (the peaked aspect is solely due to the low resolution used).

Appendix B - Distributions

B.1 Discrete Distributions

B.1.1 Bernoulli Distribution

<u>Description</u>: Success or failure in one trial. The probability of dichotomised events was studied by Jacob Bernoulli (1645-1705), hence the name. A dichotomous trial is also called a Bernoulli trial.

<u>Sample space</u>: $\{0, 1\}$, with $0 \equiv$ failure (no success) and $1 \equiv$ success.

<u>Probability function</u>:

$p(x) \equiv P(X = x) = p^x (1 - p)^{1-x}$, or putting it more simply,

$$p(x) = \begin{cases} 1 - p = q, & x = 0 \\ p, & x = 1 \end{cases}.$$ B.1

<u>Mean</u>: $\mu = p$.

<u>Variance</u>: $\sigma^2 = pq,$

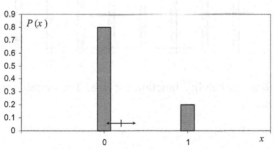

Figure B.1. Bernoulli probability function for $p = 0.2$. The double arrow corresponds to the $\mu \pm \sigma$ interval .

Example B. 1

Q: A train waits 5 minutes at the platform of a railway station, where it arrives at regular half-hour intervals. Someone unaware of the train timetable arrives randomly at the railway station. What is the probability that he will catch the train?

A: The probability of a success in the single "train-catching" trial is the percentage of time that the train waits at the platform in the inter-arrival period, i.e., $p = 5/30 = 0.17$.   □

B.1.2 Uniform Distribution

<u>Description</u>: Probability of occurring one out of n equiprobable events.

<u>Sample space</u>: $\{1, 2, ..., n\}$.

<u>Probability function</u>:

$$u(k) \equiv P(X = k) = \frac{1}{n}, \quad 1 \le k \le n \ . \qquad \qquad \text{B. 2}$$

<u>Distribution function</u>:

$$U(k) = \sum_{i=1}^{k} u(i) \ . \qquad \qquad \text{B. 3}$$

<u>Mean</u>: $\mu = (n+1)/2$.

<u>Variance</u>: $\sigma^2 = [(n+1)(2n+1)]/6$.

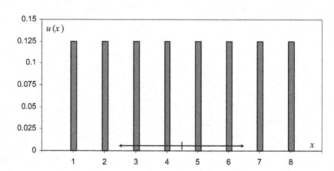

Figure B.2. Uniform probability function for $n=8$. The double arrow corresponds to the $\mu \pm \sigma$ interval.

Example B. 2

Q: A card is randomly drawn out of a deck of 52 cards until it matches a previous choice. What is the probability function of the random variable representing the number of drawn cards until a match is obtained?

A: The probability of a match at the first drawn card is 1/52. For the second drawn card it is $(51/52)(1/51)=1/52$. In general, for a match at the kth trial, we have:

$$p(k) = \underbrace{\frac{51}{52}\frac{50}{51}\cdots\frac{52-(k-1)}{52-(k-2)}}_{\text{wrong card in the first }k-1\text{ trials}}\frac{1}{52} = \frac{1}{52}.$$

Therefore the random variable follows a uniform law with $n = 52$.   ☐

B.1.3 Geometric Distribution

<u>Description</u>: Probability of an event occurring for the first time at the kth trial, in a sequence of independent Bernoulli trials, when it has a probability p of occurrence in one trial.

<u>Sample space</u>: $\{1, 2, 3, ...\}$.

<u>Probability function</u>:

$$g_p(k) \equiv P(X = k) = (1-p)^{k-1}p, \quad x \in \{1, 2, 3, ...\} \text{ (0, otherwise)}. \qquad \text{B. 4}$$

<u>Distribution function</u>:

$$G_p(k) = \sum_{i=1}^{k} g_p(i). \qquad \text{B. 5}$$

<u>Mean</u>: $1/p$.

<u>Variance</u>: $(1-p)/p^2$.

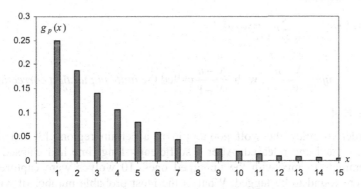

Figure B.3. Geometric probability function for $p = 0.25$. The mean occurs at $x = 4$.

Example B. 3

Q: What is the probability that one has to wait at least 6 trials before obtaining a certain face when tossing a dice?

A: The probability of obtaining a certain face is 1/6 and the occurrence of that face at the kth Bernoulli trial obeys the geometric distribution, therefore: $P(X \geq 6) = 1 - G_{1/6}(5) = 1 - 0.6 = 0.4$. □

B.1.4 Hypergeometric Distribution

Description: Probability of obtaining k items, of one out of two categories, in a sample of n items extracted without replacement from a population of N items that has $D = pN$ items of that category (and $(1-p)N = qN$ items from the other category). In quality control, the category of interest is usually one of the defective items.

Sample space: $\{\max(0, n - N + D), \ldots, \min(n,D)\}$.

Probability function:

$$h_{N,D,n}(k) \equiv P(X = k) = \frac{\binom{D}{k}\binom{N-D}{n-k}}{\binom{N}{n}} = \frac{\binom{Np}{k}\binom{Nq}{n-k}}{\binom{N}{n}},$$

$$\text{B. 6}$$

$$k \in \{\max(0, n-N+D), \ldots, \min(n,D)\}.$$

From the $\binom{N}{n}$ possible samples of size n, extracted from the population of N items, their composition consists of k items from the interesting category and $n - k$ items from the complement category. There are $\binom{D}{k}\binom{N-D}{n-k}$ possibilities of such compositions; therefore, one obtains the previous formula.

Distribution function:

$$H_{N,D,n}(k) = \sum_{i=\max(0,n-N+D)}^{k} h_{N,D,n}(i).$$

$$\text{B. 7}$$

Mean: np.

Variance: $npq\left(\dfrac{N-n}{N-1}\right)$, with $\dfrac{N-n}{N-1}$ called the *finite population correction*.

Example B. 4

Q: In order to study the wolf population in a certain region, 13 wolves were captured, tagged and released. After a sufficiently long time had elapsed for the tagged animals to mix with the untagged ones, 10 wolves were captured, 2 of which were found to be tagged. What is the most probable number of wolves in that region?

A: Let N be the size of the population of wolves of which $D = 13$ are tagged. The number of tagged wolves in the second capture sample is distributed according to the hypergeometric law. By studying the $h_{N,D,n} / h_{(N-1),D,n}$ ratio, it is found that the value of N that maximizes $h_{N,D,n}$ is:

$$N = D\frac{n}{k} = 13\frac{10}{2} = 65.$$
 □

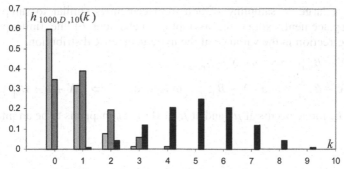

Figure B. 4. Hypergeometric probability function for $N = 1000$ and $n = 10$, for: $D = 50$ ($p = 0.05$) (light grey); $D = 100$ ($p = 0.1$) (dark grey); $D = 500$ ($p = 0.5$) (black).

B.1.5 Binomial Distribution

<u>Description</u>: Probability of k successes in n independent and constant probability Bernoulli trials.

<u>Sample space</u>: $\{0, 1, \ldots, n\}$.

<u>Probability function</u>:

$$b_{n,p}(k) \equiv P(X = k) = \binom{n}{k} p^k (1-p)^{n-k} = \binom{n}{k} p^k q^{n-k} ,$$ B. 8

with $k \in \{0, 1, \ldots, n\}$.

<u>Distribution function</u>: $B_{n,p}(k) = \sum_{i=0}^{k} b_{n,p}(i)$. B. 9

A binomial random variable can be considered as a sum of n Bernoulli random variables, and when sampling from a finite population, arises only when the sampling is done with replacement. The name comes from the fact that B.8 is the kth term of the binomial expansion of $(p + q)^n$.

For a sequence of k successes in n trials – since they are independent and the success in any trial has a constant probability p –, we have:

$$P(k \text{ successes in } n \text{ trials}) = p^k q^{n-k} .$$

Since there are $\binom{n}{k}$ such sequences, the formula above is obtained.

<u>Mean</u>: $\mu = np$.

<u>Variance</u>: $\sigma^2 = npq$

<u>Properties</u>:

1. $\lim_{N \to \infty} h_{N,D,n}(k) = b_{n,p}(k)$.

For large N, sampling without replacement is similar to sampling with replacement. Notice the asymptotic behaviour of the finite population correction in the variance of the hypergeometric distribution.

2. $X \sim B_{n,p} \Rightarrow n - X \sim B_{n,1-p}$.

3. $X \sim B_{m_1,p}$ and $Y \sim B_{n_2,p}$ independent $\Rightarrow X + Y \sim B_{m_1+n_2, p}$.

4. The mode occurs at μ (and at $\mu - 1$ if $(n+1)p$ happens to be an integer).

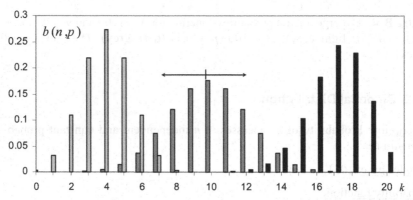

Figure B.5. Binomial probability functions: $B_{8,\,0.5}$ (light grey); $B_{20,\,0.5}$ (dark grey); $B_{20,\,0.85}$ (black). The double arrow indicates the $\mu \pm \sigma$ interval for $B_{20,\,0.5}$.

Example B. 5

Q: The cardiology service of a Hospital screens patients for myocardial infarction. In the population with heart complaints arriving at the service, the probability of having that disease is 0.2. What is the probability that at least 5 out of 10 patients do not have myocardial infarction?

A: Let us denote by p the probability of not having myocardial infarction, i.e., $p = 0.8$. The probability we want to compute is then:

$$P = \sum_{k=5}^{10} b_{10,\,0.8}(k) = 1 - B_{10,\,0.8}(4) = 0.9936. \qquad \square$$

B.1.6 Multinomial Distribution

<u>Description</u>: Generalisation of the binomial law when there are more than two categories of events in n independent trials with constant probability, p_i (for $i = 1$, 2, ..., k categories), throughout the trials.

<u>Sample space</u>: $\{0, 1, ..., n\}^k$.

Probability function:

$$m_{n,p_1,\dots,p_k}(n_1,\dots,n_k) \equiv P(X_1 = n_1,\dots,X_k = n_k) = \frac{n!}{n_1!\dots n_k!}p_1^{m_1}\dots p_k^{n_k}$$

with $\sum_{i=1}^{k} p_i = 1$; $n_i \in \{0, 1, \dots, n\}$, $\sum_{i=1}^{k} n_i = n$.   B.10

Distribution function:

$$M_{n,p_1,\dots,p_k}(n_1,\dots,n_k) = \sum_{i_1=0}^{n_1}\dots\sum_{i_k=0}^{n_k} m_{n,p_1,\dots,p_k}(i_1,\dots,i_k),\ \sum_{i=1}^{k} n_i = n.$$ B. 11

Mean: $\mu_i = np_i$

Variance: $\sigma_i^2 = np_i q_i$

Properties:

1. $X \sim m_{n,p_1,p_2} \quad \Rightarrow \quad X \sim b_{n,p_1}$.

2. $\rho(X_i, X_j) = -\sqrt{\dfrac{p_i p_j}{(1-p_i)(1-p_j)}}$.

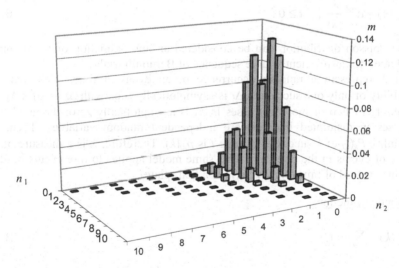

Figure B.6. Multinomial probability function for the card-hand problem of Example B.6. The mode is $m(0, 2, 8) = 0.1265$.

Example B. 6

Q: Consider a hand of 10 cards drawn randomly from a 52 cards deck. Compute the probability that the hand has at least one ace or two figures (king, dame or valet).

A: The probabilities of one single random draw of an ace (X_1), a figure (X_2) and a number (X_3) are $p_1 = 4/52$, $p_2 = 12/52$ and $p_3 = 36/52$, respectively. In order to compute the probability of getting at least one ace or two figures, in the 10-card hand, we use the multinomial probability function $m(n_1, n_2, n_3) \equiv m_{10, p_1, p_2, p_3}(n_1, n_2, n_3)$, shown in Figure B.6, as follows:

$$P(X_1 \geq 1 \cup X_2 \geq 2) = 1 - P(X_1 < 1 \cap X_2 < 2) = 1 - m(0,0,10) - m(0,1,9) =$$
$$1 - 0.025 - 0.084 = 0.89.$$ □

B.1.7 Poisson Distribution

Description: Probability of obtaining k events when the probability of an event is very small and occurs at an average rate of λ events per unit time or space (probability of rare events). The name comes from Siméon Poisson (1781-1840), who first studied this distribution.

Sample space: 0, 1, 2, …, ∞[.

Probability function:

$$p_\lambda(k) = e^{-\lambda} \frac{\lambda^x}{k!}, \quad k \geq 0.$$ B. 12

The Poisson distribution can be considered an approximation of the binomial distribution for a sufficiently large sequence of Bernoulli trials.

Let X represent a random occurrence of an event in time, such that: the probability of only one success in Δt is asymptotically (i.e., with $\Delta t \to 0$) $\lambda \Delta t$; the probability of two or more successes in Δt is asymptotically zero; the number of successes in disjointed intervals are independent random variables. Then, the probability $P_k(t)$ of k successes in time t is $p_{\lambda t}(k)$. Therefore, λ is a measure of the density of events in the interval t. This same model applies to rare events in other domains, instead of time (e.g. bacteria in a Petri plate).

Distribution function:

$$P_\lambda(k) = \sum_{i=0}^{k} p_\lambda(i).$$ B. 13

Mean: λ.

Variance: λ.

Properties:

1. For small probability of the success event, assuming $\mu = np$ is constant, the binomial distribution converges to the Poisson distribution, i.e.,
    $$b_{n,p} \xrightarrow[n \to \infty; \ np<5]{} p_\lambda, \quad \lambda = np.$$

2. $b_{n,p}(k)/b_{n,p}(k-1) \quad \underset{n\to\infty;\ np<5}{\longrightarrow} \quad \dfrac{\lambda}{k}$.

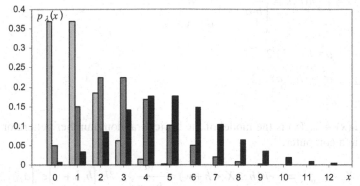

Figure B.7. Probability function of the Poisson distribution for $\lambda = 1$ (light grey), $\lambda = 3$ (dark grey) and $\lambda = 5$ (black). Note the asymmetry of the distributions.

Example B. 7

Q: A radioactive substance emits alpha particles at a constant rate of one particle every 2 seconds, in the conditions stated above for applying the Poisson distribution model. What is the probability of detecting at most 1 particle in a 10-second interval?

A: Assuming the second as the time unit, we have $\lambda = 0.5$. Therefore $\lambda t = 5$ and we have:

$$P(X \le 1) = p_5(0) + p_5(1) = e^{-5}(1 + \frac{5}{1!}) = 0.04.$$ ∎

B.2 Continuous Distributions

B.2.1 Uniform Distribution

Description: Equiprobable equal-length sub-intervals of an interval. Approximation of discrete uniform distributions, an example of which is the random number generator routine of a computer.

Sample space: $\mathfrak{R}$.

Density function: $u_{a,b}(x) = \begin{cases} \dfrac{1}{b-a}, & a \le x < b \\ 0, & \text{otherwise} \end{cases}$.   B. 14

Distribution function:

$$U_{a,b}(x) = \int_{-\infty}^{x} u(t)dt = \begin{cases} 0 & \text{if } x < a; \\ \dfrac{x-a}{b-a} & \text{if } a \le x < b; \\ 1 & \text{if } x \ge b. \end{cases}$$ B. 15

Mean: $\mu = (a + b)/2.$

Variance: $\sigma^2 = (b - a)^2/12.$

Properties:

1. $u(x) \equiv u_{0,1}(x)$ is the model of the typical random number generator routine in a computer.

2. $X \sim u_{a,b} \Rightarrow P(h \le X < h + w) = \dfrac{w}{b-a}, \quad \forall h, [h, h+w] \subset [a,b].$

Example B. 8

Q: In a cathode ray tube, the electronic beam sweeps a 10 cm line at constant high speed, from left to right. The return time of the beam from right to left is negligible. At random instants the beam is stopped by means of a switch. What is the most probable 2σ-interval to find the beam?

A: Since for every equal length interval in the swept line there is an equal probability to find the beam, we use the uniform distribution and compute the most probable interval within one standard deviation as $\mu \pm \sigma = 5 \pm 2.9$ cm (see formulas above).   □

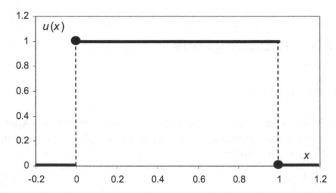

Figure B.8. The uniform distribution in [0, 1[, model of the usual random number generator routine in computers. The solid circle means point inclusion.

B.2.2 Normal Distribution

Description: The normal distribution is an approximation of the binomial distribution for large n and not too small p and q and was first studied by Abraham de Moivre (1667-1754) and Pierre Simon de Laplace (1749-1855). It is also an approximation of large sums of random variables acting independently (Central Limit Theorem). Measurement errors often fall into this category. Sequences of measurements whose deviations from the mean satisfy this distribution law, the so-called *normal sequences*, were studied by Karl F. Gauss (1777-1855).

Sample space: $\Re$.

Density function:

$$n_{\mu,\sigma}(x) = \frac{1}{\sqrt{2\pi}\sigma} e^{-\frac{(x-\mu)^2}{2\sigma^2}} .$$ B. 16

Distribution function:

$$N_{\mu,\sigma}(x) = \int_{-\infty}^{x} n_{\mu,\sigma}(t)\, dt .$$ B. 17

$N_{0,1}$ (zero mean, unit variance) is called the *standard normal distribution*. Note that one can always compute $N_{\mu,\sigma}(x)$ by referring it to the standard distribution:

$$N_{\mu,\sigma}\left(\frac{x-\mu}{\sigma}\right) = N_{0,1}(x) .$$

Mean: μ .

Variance: σ^2.

Properties:

1. $X \sim B_{n,p} \Rightarrow X \underset{n\to\infty}{\sim} N_{np,\sqrt{npq}}$; $\dfrac{X-np}{\sqrt{npq}} \underset{n\to\infty}{\sim} N_{0,1}$.

2. $X \sim B_{n,p} \Rightarrow f = \dfrac{X}{n} \underset{n\to\infty}{\sim} N_{p,\sqrt{pq/n}}$.

3. $X_1, X_2, \ldots, X_n \sim n_{\mu,\sigma}$ independent $\Rightarrow X = \sum_{i=1}^{n} X_i \sim n_{\mu,\sigma^2/n}$

4. $N_{0,1}(-x) = 1 - N_{0,1}(x)$.

5. $N_{0,1}(x_\alpha) = \alpha \Rightarrow N_{0,1}(x_{\alpha/2}) - N_{0,1}(-x_{\alpha/2}) = P(-x_{\alpha/2} < X \le x_{\alpha/2}) = 1-\alpha$

6. The points $\mu \pm \sigma$ are inflexion points (points where the derivative changes of sign) of $n_{\mu,\sigma}$.

7. $n_{0,1}(x)/x - n_{0,1}(x)/x^3 < 1 - N_{0,1}(x) < n_{0,1}(x)/x$, for every $x > 0$.

8. $1 - N_{0,1}(x) \underset{x\to\infty}{\approx} n_{0,1}(x)/x$.

9. $N_{0,1}(x) = \dfrac{x(4.4-x)}{10} + \dfrac{1}{2} + \varepsilon$ with $|\varepsilon| \le 0.005$.

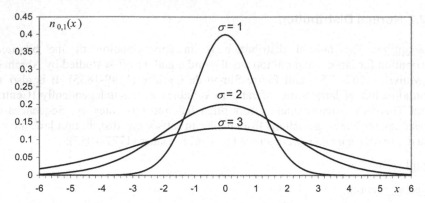

Figure B.9. Normal density function with zero mean for three different values of σ.

Values of interest for $P(X > x_\alpha) = \alpha$ with $X \sim n_{0,1}$:

α	0.0005	0.001	0.005	0.01	0.025	0.05	0.10
x_α	3.29	3.09	2.58	2.33	1.96	1.64	1.28

Example B. 9

Q: The length of screws produced by a machine follows a normal law with average value of 1 inch and standard deviation of 0.05 inch. In a large stock of screws, what is the percentage of screws one may expect to exceed 1.15 inches?

A: Referring to the standard normal distribution we determine:

$$P\left(X > \frac{1.15-1}{0.05}\right) = P(X > 3) \cong 0.1\%. \qquad \square$$

B.2.3 Exponential Distribution

<u>Description</u>: Distribution of decay phenomena, where the rate of decay is constant, such as in radioactivity phenomena.

<u>Sample space:</u> $\Re^+$.

<u>Density function:</u>

$$\varepsilon_\lambda(x) = \lambda e^{-\lambda x} \ , \ x \geq 0 \quad (0, \text{ otherwise}). \qquad \text{B. 18}$$

λ is the so-called *spread factor*.

Distribution function:

$$E_\lambda(x) = \int_0^x \varepsilon_\lambda(t)\,dt = 1 - e^{-\lambda x} \quad \text{(if } x \ge 0; 0, \text{ otherwise)} \qquad \text{B. 19}$$

Mean: $\mu = 1/\lambda.$

Variance: $\sigma^2 = 1/\lambda^2.$

Properties:

1. Let X be a random variable with a Poisson distribution, describing the event of a success in time. The random variable associated to the event that the interval between consecutive successes is $\le t$ follows an exponential distribution.

2. Let the probability of an event lasting more than $t + s$ seconds be represented by $P(t + s) = P(t)P(s)$, i.e., the probability of lasting s seconds after t does not depend on t. Then $P(t + s)$ follows an exponential distribution.

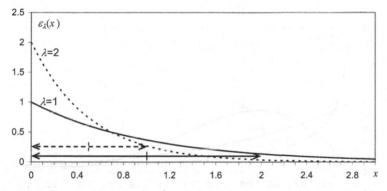

Figure B.10. Exponential density function for two values of λ. The double arrows indicate the $\mu \pm \sigma$ intervals.

Example B. 10

Q: The lifetime of a micro-organism in a certain culture follows an exponential law with an average lifetime of 1 hour. In a culture of such micro-organisms, how long must one wait until finding more than 80% of the micro-organisms dead?

A: Let X represent the lifetime of the micro-organisms, modelled by the exponential law with $\lambda = 1$. We have:

$$P(X \le t) = 0.8 \quad \Rightarrow \quad \int_0^t e^{-x}\,dx = 0.8 \quad \Rightarrow \quad t = 1.6 \text{ hours.} \qquad \square$$

B.2.4 Weibull Distribution

<u>Description</u>: The Weibull distribution describes the failure rate of equipment and the wearing-out of materials. Named after W. Weibull, a Swedish physicist specialised in applied mechanics.

<u>Sample space</u>: $\Re^+$.

<u>Density function</u>:

$$w_{\alpha,\beta}(x) = \frac{\alpha}{\beta}(x/\beta)^{\alpha-1}e^{-(x/\beta)^\alpha}, \quad \alpha,\beta > 0 \ (0, \text{otherwise}), \qquad \text{B. 20}$$

where α and β are known as the *shape* and *scale* parameters, respectively.

<u>Distribution function</u>:

$$W_{\alpha,\beta}(x) = \int_0^x w_{\alpha,\beta}(t)dt = 1 - e^{-(x/\beta)^\alpha}. \qquad \text{B. 21}$$

<u>Mean</u>: $\mu = \beta\Gamma((1+\alpha)/\alpha)$

<u>Variance</u>: $\sigma^2 = \beta^2\{\Gamma((2+\alpha)/\alpha) - [\Gamma((1+\alpha)/\alpha)]^2\}$

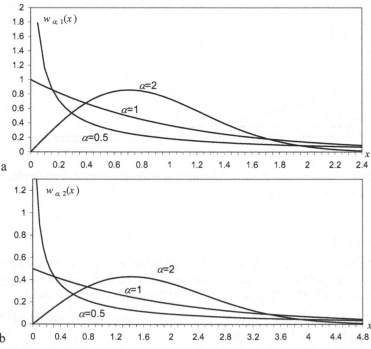

Figure B.11. Weibull density functions for fixed $\beta=1$ (a), $\beta=2$ (b), and three values of α. Note the scaling and shape effects. For $\alpha=1$ the exponential density function is obtained.

Properties:

1. $w_{1,1/\lambda}(x) \equiv \varepsilon_{\lambda}(x)$.

2. $w_{2,1/\lambda}(x)$ is the so-called Rayleigh distribution.

3. $X \sim \varepsilon_{\lambda} \implies \sqrt[\alpha]{X} \sim w_{\alpha, \sqrt[\alpha]{1/\lambda}}$.

4. $X \sim w_{\alpha, \sqrt[\alpha]{1/\lambda}} \implies X^{\alpha} \sim \varepsilon_{\lambda}$.

Example B. 11

Q: Consider that the time in years that an implanted prosthesis survives without needing replacement follows a Weibull distribution with parameters $\alpha = 2$, $\beta = 10$. What is the expected percentage of patients needing a replacement of the prosthesis after 6 years?

A: $P = W_{0.5,1}(6) = 30.2\%$.   ☐

B.2.5 Gamma Distribution

Description: The Gamma distribution is a sort of generalisation of the exponential distribution, since the sum of independent random variables, each with the exponential distribution, follows the Gamma distribution. Several continuous distributions can be regarded as a generalisation of the Gamma distribution.

Sample space: $\Re^{+}$.

Density function:

$$\gamma_{a,p}(x) = \frac{1}{a^{p}\Gamma(p)} e^{-x/a} x^{p-1}, \quad a, p > 0 \quad (0, \text{otherwise}),$$ B. 22

with $\Gamma(p)$, the *gamma function*, defined as $\Gamma(p) = \int_{0}^{\infty} e^{-x} x^{p-1} dx$, constituting a generalization of the notion of factorial, since $\Gamma(1)=1$ and $\Gamma(p) = (p-1)\Gamma(p-1)$. Thus, for integer p, one has: $\Gamma(p) = (p-1)!$

Distribution function:

$$\Gamma_{a,p}(x) = \int_{0}^{x} \gamma_{a,p}(t)dt .$$ B. 23

Mean: $\mu = a\,p$.

Variance: $\sigma^{2} = a^{2}p$.

Properties:

1. $\gamma_{a,1}(x) \equiv \varepsilon_{1/a}(x)$.

2. Let $X_1, X_2, ..., X_n$ be a set of n independent random variables, each with exponential distribution and spread factor λ. Then, $X = X_1 + X_2 + ... + X_n \sim \gamma_{1/\lambda, n}$.

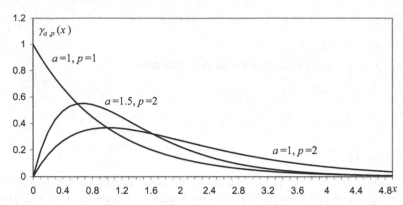

Figure B.12. Gamma density functions for three different pairs of a, p. Notice that p works as a *shape parameter* and a as a *scale parameter*.

Example B. 12

Q: The lifetime in years of a certain model of cars, before a major motor repair is needed, follows a gamma distribution with $a = 0.2$, $p = 3.5$. In the first 6 years, what is the probability that a given car needs a major motor repair?

A: $\Gamma_{0.2,3.5}(6) = 0.066$.   ◻

B.2.6 Beta Distribution

<u>Description</u>: The Beta distribution is a continuous generalization of the binomial distribution.

<u>Sample space</u>: [0, 1].

<u>Density function</u>:

$$\beta_{p,q}(x) = \frac{1}{B(p,q)} x^{p-1}(1-x)^{q-1} , \quad x\in[0, 1] \quad (0, \text{otherwise}), \qquad \text{B. 24}$$

with $B(p,q) = \dfrac{\Gamma(p)\Gamma(q)}{\Gamma(p+q)}, \quad p,q > 0$, the so-called *beta function*.

<u>Distribution function</u>:

$$B_{p,q}(x) = \int_{-\infty}^{x} \beta_{p,q}(t)dt \qquad \text{B. 25}$$

<u>Mean:</u> $\mu = p/(p+q)$. The sum $c = p + q$ is called *concentration parameter*.

<u>Variance:</u> $\sigma^2 = pq /\left[(p+q)^2 (p+q+1) \right] = \mu(1-\mu)/(c+1)$.

Properties:

1. $\beta_{1,1}(x) \equiv u(x)$.
2. $X \sim B_{n,p}(k) \Rightarrow P(X \geq a) = B_{a,n-a+1}(p)$.
3. $X \sim \beta_{p,q} \Rightarrow 1/X \sim \beta_{q,p}$.
4. $X \sim \beta_{p,q} \Rightarrow \sqrt{c+1} \dfrac{(X-\mu)}{\sqrt{\mu(1-\mu)}} \underset{\text{large } c}{\sim} n_{0,1}$.

Example B. 13

Q: Assume that in a certain country the wine consumption in daily litres per capita (for the above 18 year old population) follows a beta distribution with $p = 1.3$, $q = 0.6$. Find the median value of this distribution.

A: The median value of the wine consumption is computed as:

$$P_{1.3,\,0.6}(X \leq 0.5) = 0.26 \text{ litres.} \qquad \square$$

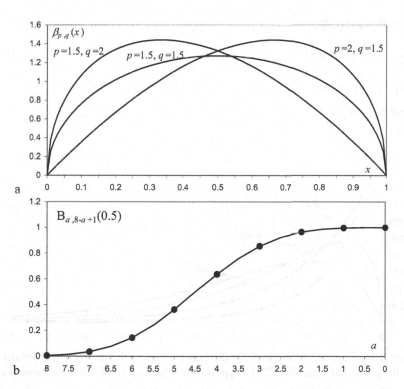

Figure B.13. a) Beta density functions for different values of p, q; b) $P(X \geq a)$ assuming the binomial distribution with $n = 8$ and $p = 0.5$ (circles) and the beta distribution according to property 2 (solid line).

B.2.7 Chi-Square Distribution

<u>Description</u>: The sum of squares of independent random variables, with standard normal distribution, follows the chi-square (χ^2) distribution. The number n of added terms is the so-called number of degrees of freedom[1], $df = n$ (number of terms that can vary independently, achieving the same sum).

<u>Sample space</u>: $\Re^+$.

<u>Density function</u>:

$$\chi^2_{df}(x) = \frac{1}{2^{df/2}\Gamma(df/2)} x^{(df/2)-1} e^{-x/2}, x \geq 0 \quad (0, \text{otherwise}), \qquad \text{B. 26}$$

with df degrees of freedom.

All density functions are skew, but the larger df is, the more symmetric the distribution.

<u>Distribution function</u>:

$$X^2_{df}(x) = \int_0^x \chi^2_{df}(t)dt. \qquad \text{B. 27}$$

<u>Mean</u>: $\mu = df$.

<u>Variance</u>: $\sigma^2 = 2\,df$.

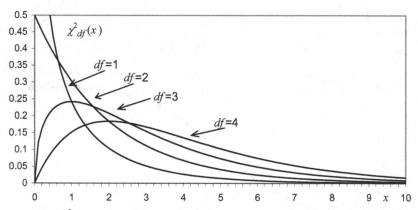

Figure B.14. χ^2 density functions for different values of the degrees of freedom, df. Note the particular cases for $df = 1$ (hyperbolic behaviour) and $df = 2$ (exponential).

---

[1] Also denoted υ in the literature.

Properties:

1. $\chi^2_{df}(x) = \gamma_{df/2,2}(x)$; in particular, $df = 2$ yields the exponential distribution with $\lambda = \frac{1}{2}$.

2. $X = \sum_{i=1}^{n} X_i^2$, X_i independent $\sim n_{0,1}$ $\Rightarrow$ $X \sim \chi^2_n$

3. $X = \sum_{i=1}^{n} (X_i - \bar{x})^2$, X_i independent $\sim n_{0,1}$ $\Rightarrow$ $X \sim \chi^2_{n-1}$

4. $X = \frac{1}{\sigma^2} \sum_{i=1}^{n} (X_i - \mu)^2$, X_i independent $\sim n_{\mu,\sigma}$ $\Rightarrow$ $X \sim \chi^2_n$

5. $X = \frac{1}{\sigma^2} \sum_{i=1}^{n} (X_i - \bar{x})^2$, X_i independent $\sim n_{\mu,\sigma}$ $\Rightarrow$ $X \sim \chi^2_{n-1}$

6. $X \sim \chi^2_{df_1}$, $Y \sim \chi^2_{df_2}$ $\Rightarrow$ $X + Y \sim \chi^2_{df_1 + df_2}$ (convolution of two χ^2 results in a χ^2).

Example B. 14

Q: The electric current passing through a $10\ \Omega$ resistance shows random fluctuations around its nominal value that can be well modelled by $n_{0,\sigma}$ with $\sigma = 0.1$ Ampere. What is the probability that the heat power generated in the resistance deviates more than 0.1 Watt from its nominal value?

A: The heat power is $p = 10\,i^2$, where i is the current passing through the $10\ \Omega$ resistance. Therefore:

$P(p > 0.1) = P(10\,i^2 > 0.1) = P(100\,i^2 > 1)$.

But: $i \sim n_{0,0.1}$ $\Rightarrow$ $\frac{1}{\sigma^2} i^2 = 100 i^2 \sim \chi^2_1$.

Hence: $P(p > 0.1) = P(\chi^2_1 > 1) = 0.317$.   ▯

B.2.8 Student's *t* Distribution

Description: The Student's *t* distribution is the distribution followed by the ratio of the mean deviations over the sample standard deviation. It was derived by the English brewery chemist W.S. Gosset (pen-name "Student") at the beginning of the 20th century.

Sample space: $\Re$.

Density function:

$$t_{df}(x) = \frac{\Gamma((df+1)/2)}{\sqrt{df\pi}\ \Gamma(df/2)} \left(1 + \frac{x^2}{df}\right)^{-(df+1)/2}, \quad \text{with } df \text{ degrees of freedom.} \quad \text{B. 28}$$

Distribution function:

$$T_{df}(x) = \int_{-\infty}^x t_{df}(t)dt.$$ B. 29

Mean: $\mu = 0$.

Variance: $\sigma^2 = df/(df-2)$ for $df > 2$.

Properties:

1. $t_{df} \xrightarrow[df\to\infty]{} n_{0,1}$.

2. $X \sim n_{0,1}$, $Y \sim \chi^2_{df}$, X and Y independent $\Rightarrow \dfrac{X}{\sqrt{Y/df}} \sim t_{df}$.

3. $X = \dfrac{\overline{X} - \mu}{s/\sqrt{n}}$ with $\overline{X} = \dfrac{\sum_{i=1}^n X_i}{n}$, $s^2 = \dfrac{\sum_{i=1}^n (X_i - \overline{X})^2}{n-1}$,

X_i independent $\sim n_{\mu,\sigma} \Rightarrow X \sim t_{n-1}$.

Example B. 15

Q: A sugar factory introduced a new packaging machine for the production of 1Kg sugar packs. In order to assess the quality of the machine, 15 random packs were picked up and carefully weighted. The mean and standard deviation were computed and found to be $m = 1.1$ Kg and $s = 0.08$ Kg. What is the probability that the true mean value is at least 1Kg?

A: $P(\mu \ge 1) = P(m - \mu \le 0.1) = P\left(\dfrac{m-\mu}{0.08\sqrt{15}} \le 0.323\right) = P(t_{14} \le 0.323) = 0.62$.

◻

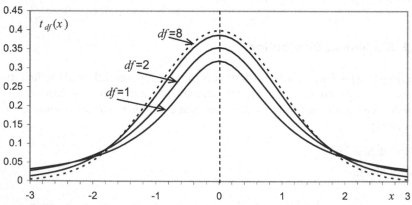

Figure B.15. Student's t density functions for several values of the degrees of freedom, df. The dotted line corresponds to the standard normal density function.

B.2.9 F Distribution

Description: The F distribution was introduced by Ronald A. Fisher (1890-1962), in order to study the ratio of variances, and is named after him. The ratio of two independent Gamma-distributed random variables, each divided by its mean, also follows the F distribution.

Sample space: $\Re^+$.

Density function:

$$f_{df_1,df_2}(x) = \frac{\Gamma\left(\dfrac{df_1 + df_2}{2}\right)\left(\dfrac{df_1}{df_2}\right)^{df_1/2}}{\Gamma(df_1/2)\Gamma(df_2/2)} \frac{x^{(df_1-2)/2}}{\left(1 + \dfrac{df_1}{df_2}x^2\right)^{(df_1+df_2)/2}}, \, x \geq 0,$$

with df_1, df_2 degrees of freedom.   B. 30

Distribution function:

$$F_{df_1,df_2}(x) = \int_0^x f_{df_1,df_2}(t)dt .$$ B. 31

Mean: $\mu = \dfrac{df_2}{df_2 - 2}$, $df_2 > 2.$

Variance: $\sigma^2 = \dfrac{2df_2^2(df_1 + df_2 - 2)}{df_1(df_2 - 2)^2(df_2 - 4)}$, for $df_2 > 4.$

Properties:

1. $X_1 \sim \gamma_{a_1,p_1}, X_2 \sim \gamma_{a_2,p_2} \quad \Rightarrow \quad \dfrac{X_1/(a_1 p_1)}{X_2/(a_2 p_2)} \sim f_{2a_1,2a_2}$.

2. $X \sim \beta_{a,b} \quad \Rightarrow \quad \dfrac{X/a}{(1-X)/b} = \dfrac{X/\mu}{(1-X)/(1-\mu)} \sim f_{2a,2b}$.

3. $X \sim f_{a,b} \quad \Rightarrow \quad 1/X \sim f_{b,a}$, as can be derived from the properties of the beta distribution.

4. $X \sim \chi^2_{m_1}$, $Y \sim \chi^2_{m_2}$, X, Y independent $\Rightarrow \dfrac{X/n_1}{Y/n_2} \sim f_{m_1,m_2}$.

5. Let $X_1,..., X_n$ and $Y_1,..., Y_m$ be $n + m$ independent random variables such that $X_i \sim n_{\mu_1,\sigma_1}$ and $Y_i \sim n_{\mu_2,\sigma_2}$.

    Then $\left(\sum_{i=1}^n (X_i - \mu_1)^2/(n\sigma_1^2)\right) / \left(\sum_{i=1}^m (Y_i - \mu_2)^2/(m\sigma_2^2)\right) \sim f_{n,m}$.

6. Let $X_1,..., X_n$ and $Y_1,..., Y_m$ be $n + m$ independent random variables such that $X_i \sim n_{\mu_1,\sigma_1}$ and $Y_i \sim n_{\mu_2,\sigma_2}$.

    Then $\left(\sum_{i=1}^n (X_i - \bar{x})^2/((n-1)\sigma_1^2)\right) / \left(\sum_{i=1}^m (Y_i - \bar{y})^2/((m-1)\sigma_2^2)\right) \sim f_{n-1,m-1}$,

    where $\bar{x}$ and $\bar{y}$ are sample means.

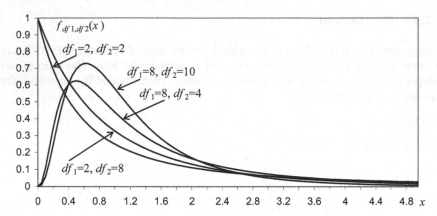

Figure B. 16. F density functions for several values of the degrees of freedom, df_1, df_2.

Example B. 16

Q: A factory is using two machines, M1 and M2, for the production of rods with a nominal diameter of 1 cm. In order to assess the variability of the rods' diameters produced by the two machines, a random sample of each machine was collected as follows: $n_1 = 15$ rods produced by M1 and $n_2 = 21$ rods produced by M2. The diameters were measured and the standard deviations computed and found to be: $s_1 = 0.012$ cm, $s_2 = 0.01$ cm. What is the probability that the variability of the rod diameters produced by M2 is smaller than the one referring to M1?

A: Denote by σ_1, σ_2 the standard deviations corresponding to M1 and M2, respectively. We want to compute:

$$P(\sigma_2 < \sigma_1) = P\left(\frac{\sigma_2}{\sigma_1} < 1\right).$$

According to property 6, we have:

$$P\left(\frac{\sigma_2^2}{\sigma_1^2} < 1\right) = P\left(\frac{s_1^2 / \sigma_1^2}{s_2^2 / \sigma_2^2} < \frac{s_1^2}{s_2^2}\right) = P\left(\frac{s_1^2 / \sigma_1^2}{s_2^2 / \sigma_2^2} < 1.44\right) = F_{14,20}(1.44) = 0.78. \quad \square$$

B.2.10 Von Mises Distributions

Description: The *von Mises* distributions are the normal distribution equivalent for circular and spherical data. The von Mises distribution for circular data, also known as *circular normal* distribution, was introduced by R. von Mises (1918) in order to study the deviations of measured atomic weights from integer values. It also describes many other physical phenomena, e.g. diffusion processes on the circle, as well as in the plane, for particles with Brownian motion hitting the unit circle. The *von Mises-Fisher* distribution is a generalization for a $(p-1)$-dimensional unit-radius hypersphere S^{p-1} ($\mathbf{x'x} = 1$), embedded in $\Re^p$.

Sample space: $S^{p-1} \subset \Re^p$.

Density function:

$$m_{\mu,\kappa,p}(\mathbf{x}) = \left(\frac{\kappa}{2}\right)^{p/2-1} \frac{1}{\Gamma(p/2)I_{p/2-1}(\kappa)} e^{\kappa \mu'\mathbf{x}},$$ B. 32

where μ is the unit length mean vector (also called *polar vector*), $\kappa \geq 0$ is the *concentration* parameter and I_ν is the modified Bessel function of the first kind and order ν.

For $p = 2$ one obtains the original von Mises circular distribution:

$$m_{\mu,\kappa}(\theta) = \frac{1}{2\pi I_0(\kappa)} e^{\kappa \cos(\theta-\mu)},$$ B. 33

where I_0 denotes the modified Bessel function of the first kind and order 0, which can be computed by the following power series expansion:

$$I_0(\kappa) = \sum_{r=0}^{\infty} \frac{1}{(r!)^2} \left(\frac{\kappa}{2}\right)^{2r}.$$ B. 34

For $p = 3$ one obtains the spherical *Fisher* distribution:

$$m_{\mu,\kappa,3}(\mathbf{x}) = \frac{\kappa}{2\sinh\kappa} e^{\kappa \mu'\mathbf{x}}.$$ B. 35

Mean: μ.

Circular Variance:

$$v = 1 - I_1(\kappa)/I_0(\kappa) = \frac{\kappa}{2}\left\{1 - \frac{\kappa^2}{9} + \frac{\kappa^4}{48} - \frac{11\kappa^6}{3072} + ...\right\}.$$

Spherical Variance:

$$v = 1 - \coth\kappa - 1/\kappa.$$

Properties:

1. $m_{\mu,\kappa}(\theta + 2\pi) = m_{\mu,\kappa}(\theta)$.

2. $M_{\mu,\kappa}(\theta+2\pi) - M_{\mu,\kappa}(\theta) = 1$, where $M_{\mu,\kappa}$ is the circular von Mises distribution function.

3. $M_{\mu,\kappa}, \kappa \to \infty \sim N_{0,1}$ (approx.).

4. $M_{\mu,\kappa} \cong WN_{\mu,A(\kappa)}$ with $A(\kappa) = I_1(\kappa)/I_0(k)$, and $WN_{\mu,\sigma}$ the wrapped normal distribution (wrapping around 2π).

5. Let $\mathbf{x} = r(\cos\theta, \sin\theta)'$ have a bivariate normal distribution with mean $\mu = (\cos\mu, \sin\mu)'$ and equal variance $1/\kappa$. Then, the conditional distribution of θ given $r = 1$ is $M_{\mu,\kappa}$.

6. Let the unit length vector $\mathbf{x}$ be expressed in polar co-ordinates in $\mathfrak{R}^3$, i.e., $\mathbf{x} = (\cos\theta, \sin\theta\cos\phi, \sin\theta\sin\phi)'$, with θ the co-latitude and ϕ the azimuth. Then, θ and ϕ are independently distributed, with:

$$f_\kappa(\theta) = \frac{\kappa}{2\sinh\kappa} e^{\kappa\cos\theta} \sin\theta, \quad \theta \in [0, \pi];$$

$h(\phi) = 1/(2\pi)$, $\phi \in [0, 2\pi[$, is the uniform distribution.

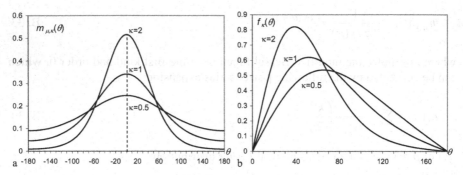

Example B. 17. a) Density function of the circular von Mises distribution for $\mu = 0$ and several values of κ; b) Density function of the co-latitude of the spherical von Mises distribution for several values of κ.

Appendix C - Point Estimation

In Appendix C, we present a few introductory concepts on point estimation and on results regarding the estimation of the mean and the variance.

C.1 Definitions

Let $F_X(x)$ be a distribution function of a random variable X, dependent on a certain parameter θ. We assume that there is available a random sample $\mathbf{x} = [x_1, x_2, \ldots, x_n]'$ and build a function $t_n(\mathbf{x})$ that gives us an estimate of the parameter θ, a *point estimate* of θ. Note that, by definition, in a random sample from a population with a density function $f_X(x)$, the random variables associated with the values $x_1, \ldots, x_n$, are i.i.d., i.e., the random sample has a joint density given by:

$$f_{X_1, X_2, \ldots, X_n}(x_1, x_2, \ldots, x_n) = f_X(x_1) f_X(x_2) \ldots f_X(x_n).$$

The estimate $t_n(\mathbf{x})$ is considered a value of a random variable, called a *point estimator* or *statistic*, $T_n \equiv t_n(X_n)$, where X_n denotes the n-dimensional random variable corresponding to the sampling process.

The following properties are desirable for a point estimator:

– *Unbiased ness*. A point estimator is said to be unbiased if its expectation is θ.

$$\mathrm{E}[\,T_n\,] \equiv \mathrm{E}[t_n(X_n)] = \theta.$$

– *Consistency*. A point estimator is said to be consistent if the following holds:

$$\forall \varepsilon > 0, \quad P\left(\left|T_n - \theta\right| > \varepsilon\right) \underset{n \to \infty}{\longrightarrow} 0.$$

As illustrated in Figure C.1, a biased point estimator yields a mean value different from the true value of the distribution parameter. Figure C.1 also illustrates the notion of consistency.

When comparing two unbiased and consistent point estimators $T_{n,1}$ and $T_{n,2}$, it is reasonable to prefer the one that has a smaller variance, say $T_{n,1}$:

$$\mathrm{V}[T_{n,1}] \le \mathrm{V}[T_{n,2}].$$

The estimator $T_{n,1}$ is then said to be more *efficient* than $T_{n,2}$.

There are several methods to construct point estimator functions. A popular one is the *maximum likelihood* (ML) method, which is applied by first constructing for sample **x**, the following *likelihood function*:

$$L(\mathbf{x} \mid \theta) = f(x_1 \mid \theta) f(x_2 \mid \theta) \dots f(x_3 \mid \theta) = \prod_{i=1}^{n} f(x_i \mid \theta),$$

where $f(x_i|\theta)$ is the density function (probability function in the discrete case) evaluated at x_i, given the value θ of the parameter to be estimated.

Next, the value that maximizes $L(\theta)$ (within the domain of values of θ) is obtained. The ML method will be applied in the next section. Its popularity derives from the fact that it will often yield a consistent point estimator, which, when biased, is easy to adjust by using a simple corrective factor.

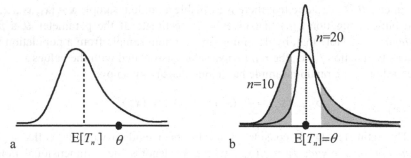

Figure C.1. a) Density function of a biased point estimator (expected mean is different from the true parameter value); b) Density functions of an unbiased and consistent estimator for two different values of *n*: the probability of a $\pm\ \varepsilon$ deviation from the true parameter value – shaded area – tends to zero with growing *n*.

Example C. 1

Q: A coin is tossed *n* times until head turns up for the first time. What is the maximum likelihood estimate of the probability of head turning up?

A: Let us denote by p and $q = 1 - p$ the probability of turning up head or tail, respectively. Denoting $X_1, \dots, X_n$ the random variables associated to the coin tossing sequence, the likelihood is given by:

$$L(p) = P\!\left(X_1 = \text{tail} \mid p\right)\!.P\!\left(X_2 = \text{tail} \mid p\right)\!\dots P\!\left(X_n = \text{head} \mid p\right) = q^{n-1} p$$

The maximum likelihood estimate is therefore given by:

$$\frac{dL(p)}{dp} = q^{n-1} - (n-1)q^{n-2}p = 0 \quad \Rightarrow \quad \hat{p} = 1/n$$

This estimate is biased and inconsistent. We see that the ML method does not always provide good estimates. □

Example C. 2

Q: Let us now modify the previous example assuming that in n tosses of the coin heads turned up k times. What is the maximum likelihood estimate of the probability of heads turning up?

A: Using the same notation as before we now have:

$$L(p) = q^{n-k} p^k$$

Hence:

$$\frac{dL(p)}{dp} = q^{n-k-1} p^{k-1} \left[-(n-k)p + kq \right] = 0 \quad \Rightarrow \quad \hat{p} = k/n \ \text{(for } p \neq 0, 1)$$

This is the well-known unbiased and consistent estimate. □

C.2 Estimation of Mean and Variance

Let X be a normal random variable with mean μ and variance v:

$$f(x) = \frac{1}{\sqrt{2\pi v}} e^{-\frac{(x-\mu)^2}{2v}} \ .$$

Assume that we were given a sample of size n from X and were asked to derive the ML point estimators of μ and variance v. We would then have:

$$L(\mathbf{x} | \theta) = \prod_{i=1}^{n} f(x_i | \theta) = (2\pi v)^{-n/2} e^{-\frac{1}{2} \sum_{i=1}^{n} (x_i - \mu)^2 / v} \ .$$

Instead of maximizing $L(\mathbf{x}|\theta)$ we may, equivalently, maximize its logarithm:

$$\ln L(\mathbf{x} | \theta) = -(n/2) \ln(2\pi v) - \frac{1}{2} \sum_{i=1}^{n} (x_i - \mu)^2 / v \ .$$

Therefore, we obtain:

$$\frac{\partial \ln L(\mathbf{x} \mid \theta)}{\partial \mu} = 0 \quad \Rightarrow \quad m \equiv \bar{x} = \sum_{i=1}^{n} x_i / n$$

$$\frac{\partial \ln L(\mathbf{x} \mid \theta)}{\partial v} = 0 \quad \Rightarrow \quad s^2 = \sum_{i=1}^{n} (x_i - m)^2 / n$$

Let us now comment on these results. The point estimate of the mean, given by the arithmetic mean, $\bar{x}$, is unbiased and consistent. This is a general result, valid not only for normal random variables but for any random variables as well. As a matter of fact, from the properties of the arithmetic mean (see Appendix A) we know that it is unbiased (A.58a) and consistent, given the inequality of Chebyshev and the expression of the variance (A.58b). As a consequence, the unbiased and consistent point estimator of a proportion is readily seen to be:

$$p = \frac{k}{n},$$

where k is the number of times the "success" event has occurred in the n i.i.d. Bernoulli trials. This results from the fact that the summation of x_i for the Bernoulli trials is precisely k. The reader can also try to obtain this same estimator by applying the ML method to a binomial random experiment.

Let us now consider the point estimate of the variance. We have:

$$\mathrm{E}[\sum (x_i - m)^2] = \mathrm{E}[\sum (x_i - \mu)^2] - n\mathrm{E}[(m - \mu)^2]$$

$$= n\mathrm{V}[X] - n\mathrm{V}[\bar{X}] = n\sigma^2 - n\frac{\sigma^2}{n} = (n-1)\sigma^2$$

Therefore, the unbiased estimator of the variance is:

$$s^2 = \frac{1}{n-1} \sum_{i=1}^{n} (x_i - \bar{x})^2 .$$

This corresponds to multiplying the previous ML estimator by the corrective factor $n/(n-1)$ (only noticeable for small n). The point estimator of the variance can also be proven to be consistent.

Appendix D - Tables

D.1 Binomial Distribution

The following table lists the values of $B_{n,p}(k)$ (see B.1.2).

						p					
n	k	0.05	0.10	0.15	0.20	0.25	0.30	0.35	0.40	0.45	0.50
1	0	0.9500	0.9000	0.8500	0.8000	0.7500	0.7000	0.6500	0.6000	0.5500	0.5000
	1	1.0000	1.0000	1.0000	1.0000	1.0000	1.0000	1.0000	1.0000	1.0000	1.0000
2	0	0.9025	0.8100	0.7225	0.6400	0.5625	0.4900	0.4225	0.3600	0.3025	0.2500
	1	0.9975	0.9900	0.9775	0.9600	0.9375	0.9100	0.8775	0.8400	0.7975	0.7500
	2	1.0000	1.0000	1.0000	1.0000	1.0000	1.0000	1.0000	1.0000	1.0000	1.0000
3	0	0.8574	0.7290	0.6141	0.5120	0.4219	0.3430	0.2746	0.2160	0.1664	0.1250
	1	0.9928	0.9720	0.9393	0.8960	0.8438	0.7840	0.7183	0.6480	0.5748	0.5000
	2	0.9999	0.9990	0.9966	0.9920	0.9844	0.9730	0.9571	0.9360	0.9089	0.8750
	3	1.0000	1.0000	1.0000	1.0000	1.0000	1.0000	1.0000	1.0000	1.0000	1.0000
4	0	0.8145	0.6561	0.5220	0.4096	0.3164	0.2401	0.1785	0.1296	0.0915	0.0625
	1	0.9860	0.9477	0.8905	0.8192	0.7383	0.6517	0.5630	0.4752	0.3910	0.3125
	2	0.9995	0.9963	0.9880	0.9728	0.9492	0.9163	0.8735	0.8208	0.7585	0.6875
	3	1.0000	0.9999	0.9995	0.9984	0.9961	0.9919	0.9850	0.9744	0.9590	0.9375
	4	1.0000	1.0000	1.0000	1.0000	1.0000	1.0000	1.0000	1.0000	1.0000	1.0000
5	0	0.7738	0.5905	0.4437	0.3277	0.2373	0.1681	0.1160	0.0778	0.0503	0.0313
	1	0.9774	0.9185	0.8352	0.7373	0.6328	0.5282	0.4284	0.3370	0.2562	0.1875
	2	0.9988	0.9914	0.9734	0.9421	0.8965	0.8369	0.7648	0.6826	0.5931	0.5000
	3	1.0000	0.9995	0.9978	0.9933	0.9844	0.9692	0.9460	0.9130	0.8688	0.8125
	4	1.0000	1.0000	0.9999	0.9997	0.9990	0.9976	0.9947	0.9898	0.9815	0.9688
	5	1.0000	1.0000	1.0000	1.0000	1.0000	1.0000	1.0000	1.0000	1.0000	1.0000
6	0	0.7351	0.5314	0.3771	0.2621	0.1780	0.1176	0.0754	0.0467	0.0277	0.0156
	1	0.9672	0.8857	0.7765	0.6554	0.5339	0.4202	0.3191	0.2333	0.1636	0.1094
	2	0.9978	0.9842	0.9527	0.9011	0.8306	0.7443	0.6471	0.5443	0.4415	0.3438
	3	0.9999	0.9987	0.9941	0.9830	0.9624	0.9295	0.8826	0.8208	0.7447	0.6563
	4	1.0000	0.9999	0.9996	0.9984	0.9954	0.9891	0.9777	0.9590	0.9308	0.8906
	5	1.0000	1.0000	1.0000	0.9999	0.9998	0.9993	0.9982	0.9959	0.9917	0.9844
	6	1.0000	1.0000	1.0000	1.0000	1.0000	1.0000	1.0000	1.0000	1.0000	1.0000

n	k	0.05	0.10	0.15	0.20	p 0.25	0.30	0.35	0.40	0.45	0.50
7	0	0.6983	0.4783	0.3206	0.2097	0.1335	0.0824	0.0490	0.0280	0.0152	0.0078
	1	0.9556	0.8503	0.7166	0.5767	0.4449	0.3294	0.2338	0.1586	0.1024	0.0625
	2	0.9962	0.9743	0.9262	0.8520	0.7564	0.6471	0.5323	0.4199	0.3164	0.2266
	3	0.9998	0.9973	0.9879	0.9667	0.9294	0.8740	0.8002	0.7102	0.6083	0.5000
	4	1.0000	0.9998	0.9988	0.9953	0.9871	0.9712	0.9444	0.9037	0.8471	0.7734
	5	1.0000	1.0000	0.9999	0.9996	0.9987	0.9962	0.9910	0.9812	0.9643	0.9375
	6	1.0000	1.0000	1.0000	1.0000	0.9999	0.9998	0.9994	0.9984	0.9963	0.9922
	7	1.0000	1.0000	1.0000	1.0000	1.0000	1.0000	1.0000	1.0000	1.0000	1.0000
8	0	0.6634	0.4305	0.2725	0.1678	0.1001	0.0576	0.0319	0.0168	0.0084	0.0039
	1	0.9428	0.8131	0.6572	0.5033	0.3671	0.2553	0.1691	0.1064	0.0632	0.0352
	2	0.9942	0.9619	0.8948	0.7969	0.6785	0.5518	0.4278	0.3154	0.2201	0.1445
	3	0.9996	0.9950	0.9786	0.9437	0.8862	0.8059	0.7064	0.5941	0.4770	0.3633
	4	1.0000	0.9996	0.9971	0.9896	0.9727	0.9420	0.8939	0.8263	0.7396	0.6367
	5	1.0000	1.0000	0.9998	0.9988	0.9958	0.9887	0.9747	0.9502	0.9115	0.8555
	6	1.0000	1.0000	1.0000	0.9999	0.9996	0.9987	0.9964	0.9915	0.9819	0.9648
	7	1.0000	1.0000	1.0000	1.0000	1.0000	0.9999	0.9998	0.9993	0.9983	0.9961
	8	1.0000	1.0000	1.0000	1.0000	1.0000	1.0000	1.0000	1.0000	1.0000	1.0000
9	0	0.6302	0.3874	0.2316	0.1342	0.0751	0.0404	0.0207	0.0101	0.0046	0.0020
	1	0.9288	0.7748	0.5995	0.4362	0.3003	0.1960	0.1211	0.0705	0.0385	0.0195
	2	0.9916	0.9470	0.8591	0.7382	0.6007	0.4628	0.3373	0.2318	0.1495	0.0898
	3	0.9994	0.9917	0.9661	0.9144	0.8343	0.7297	0.6089	0.4826	0.3614	0.2539
	4	1.0000	0.9991	0.9944	0.9804	0.9511	0.9012	0.8283	0.7334	0.6214	0.5000
	5	1.0000	0.9999	0.9994	0.9969	0.9900	0.9747	0.9464	0.9006	0.8342	0.7461
	6	1.0000	1.0000	1.0000	0.9997	0.9987	0.9957	0.9888	0.9750	0.9502	0.9102
	7	1.0000	1.0000	1.0000	1.0000	0.9999	0.9996	0.9986	0.9962	0.9909	0.9805
	8	1.0000	1.0000	1.0000	1.0000	1.0000	1.0000	0.9999	0.9997	0.9992	0.9980
	9	1.0000	1.0000	1.0000	1.0000	1.0000	1.0000	1.0000	1.0000	1.0000	1.0000
10	0	0.5987	0.3487	0.1969	0.1074	0.0563	0.0282	0.0135	0.0060	0.0025	0.0010
	1	0.9139	0.7361	0.5443	0.3758	0.2440	0.1493	0.0860	0.0464	0.0233	0.0107
	2	0.9885	0.9298	0.8202	0.6778	0.5256	0.3828	0.2616	0.1673	0.0996	0.0547
	3	0.9990	0.9872	0.9500	0.8791	0.7759	0.6496	0.5138	0.3823	0.2660	0.1719
	4	0.9999	0.9984	0.9901	0.9672	0.9219	0.8497	0.7515	0.6331	0.5044	0.3770
	5	1.0000	0.9999	0.9986	0.9936	0.9803	0.9527	0.9051	0.8338	0.7384	0.6230
	6	1.0000	1.0000	0.9999	0.9991	0.9965	0.9894	0.9740	0.9452	0.8980	0.8281
	7	1.0000	1.0000	1.0000	0.9999	0.9996	0.9984	0.9952	0.9877	0.9726	0.9453
	8	1.0000	1.0000	1.0000	1.0000	1.0000	0.9999	0.9995	0.9983	0.9955	0.9893
	9	1.0000	1.0000	1.0000	1.0000	1.0000	1.0000	1.0000	0.9999	0.9997	0.9990
	10	1.0000	1.0000	1.0000	1.0000	1.0000	1.0000	1.0000	1.0000	1.0000	1.0000

n	k	0.05	0.10	0.15	0.20	p 0.25	0.30	0.35	0.40	0.45	0.50
11	0	0.5688	0.3138	0.1673	0.0859	0.0422	0.0198	0.0088	0.0036	0.0014	0.0005
	1	0.8981	0.6974	0.4922	0.3221	0.1971	0.1130	0.0606	0.0302	0.0139	0.0059
	2	0.9848	0.9104	0.7788	0.6174	0.4552	0.3127	0.2001	0.1189	0.0652	0.0327
	3	0.9984	0.9815	0.9306	0.8389	0.7133	0.5696	0.4256	0.2963	0.1911	0.1133
	4	0.9999	0.9972	0.9841	0.9496	0.8854	0.7897	0.6683	0.5328	0.3971	0.2744
	5	1.0000	0.9997	0.9973	0.9883	0.9657	0.9218	0.8513	0.7535	0.6331	0.5000
	6	1.0000	1.0000	0.9997	0.9980	0.9924	0.9784	0.9499	0.9006	0.8262	0.7256
	7	1.0000	1.0000	1.0000	0.9998	0.9988	0.9957	0.9878	0.9707	0.9390	0.8867
	8	1.0000	1.0000	1.0000	1.0000	0.9999	0.9994	0.9980	0.9941	0.9852	0.9673
	9	1.0000	1.0000	1.0000	1.0000	1.0000	1.0000	0.9998	0.9993	0.9978	0.9941
	10	1.0000	1.0000	1.0000	1.0000	1.0000	1.0000	1.0000	1.0000	0.9998	0.9995
	11	1.0000	1.0000	1.0000	1.0000	1.0000	1.0000	1.0000	1.0000	1.0000	1.0000
12	0	0.5404	0.2824	0.1422	0.0687	0.0317	0.0138	0.0057	0.0022	0.0008	0.0002
	1	0.8816	0.6590	0.4435	0.2749	0.1584	0.0850	0.0424	0.0196	0.0083	0.0032
	2	0.9804	0.8891	0.7358	0.5583	0.3907	0.2528	0.1513	0.0834	0.0421	0.0193
	3	0.9978	0.9744	0.9078	0.7946	0.6488	0.4925	0.3467	0.2253	0.1345	0.0730
	4	0.9998	0.9957	0.9761	0.9274	0.8424	0.7237	0.5833	0.4382	0.3044	0.1938
	5	1.0000	0.9995	0.9954	0.9806	0.9456	0.8822	0.7873	0.6652	0.5269	0.3872
	6	1.0000	0.9999	0.9993	0.9961	0.9857	0.9614	0.9154	0.8418	0.7393	0.6128
	7	1.0000	1.0000	0.9999	0.9994	0.9972	0.9905	0.9745	0.9427	0.8883	0.8062
	8	1.0000	1.0000	1.0000	0.9999	0.9996	0.9983	0.9944	0.9847	0.9644	0.9270
	9	1.0000	1.0000	1.0000	1.0000	1.0000	0.9998	0.9992	0.9972	0.9921	0.9807
	10	1.0000	1.0000	1.0000	1.0000	1.0000	1.0000	0.9999	0.9997	0.9989	0.9968
	11	1.0000	1.0000	1.0000	1.0000	1.0000	1.0000	1.0000	1.0000	0.9999	0.9998
	12	1.0000	1.0000	1.0000	1.0000	1.0000	1.0000	1.0000	1.0000	1.0000	1.0000
13	0	0.5133	0.2542	0.1209	0.0550	0.0238	0.0097	0.0037	0.0013	0.0004	0.0001
	1	0.8646	0.6213	0.3983	0.2336	0.1267	0.0637	0.0296	0.0126	0.0049	0.0017
	2	0.9755	0.8661	0.6920	0.5017	0.3326	0.2025	0.1132	0.0579	0.0269	0.0112
	3	0.9969	0.9658	0.8820	0.7473	0.5843	0.4206	0.2783	0.1686	0.0929	0.0461
	4	0.9997	0.9935	0.9658	0.9009	0.7940	0.6543	0.5005	0.3530	0.2279	0.1334
	5	1.0000	0.9991	0.9925	0.9700	0.9198	0.8346	0.7159	0.5744	0.4268	0.2905
	6	1.0000	0.9999	0.9987	0.9930	0.9757	0.9376	0.8705	0.7712	0.6437	0.5000
	7	1.0000	1.0000	0.9998	0.9988	0.9944	0.9818	0.9538	0.9023	0.8212	0.7095
	8	1.0000	1.0000	1.0000	0.9998	0.9990	0.9960	0.9874	0.9679	0.9302	0.8666
	9	1.0000	1.0000	1.0000	1.0000	0.9999	0.9993	0.9975	0.9922	0.9797	0.9539
	10	1.0000	1.0000	1.0000	1.0000	1.0000	0.9999	0.9997	0.9987	0.9959	0.9888
	11	1.0000	1.0000	1.0000	1.0000	1.0000	1.0000	1.0000	0.9999	0.9995	0.9983
	12	1.0000	1.0000	1.0000	1.0000	1.0000	1.0000	1.0000	1.0000	1.0000	0.9999
	13	1.0000	1.0000	1.0000	1.0000	1.0000	1.0000	1.0000	1.0000	1.0000	1.0000

						p					
n	k	0.05	0.10	0.15	0.20	0.25	0.30	0.35	0.40	0.45	0.50
14	0	0.4877	0.2288	0.1028	0.0440	0.0178	0.0068	0.0024	0.0008	0.0002	0.0001
	1	0.8470	0.5846	0.3567	0.1979	0.1010	0.0475	0.0205	0.0081	0.0029	0.0009
	2	0.9699	0.8416	0.6479	0.4481	0.2811	0.1608	0.0839	0.0398	0.0170	0.0065
	3	0.9958	0.9559	0.8535	0.6982	0.5213	0.3552	0.2205	0.1243	0.0632	0.0287
	4	0.9996	0.9908	0.9533	0.8702	0.7415	0.5842	0.4227	0.2793	0.1672	0.0898
	5	1.0000	0.9985	0.9885	0.9561	0.8883	0.7805	0.6405	0.4859	0.3373	0.2120
	6	1.0000	0.9998	0.9978	0.9884	0.9617	0.9067	0.8164	0.6925	0.5461	0.3953
	7	1.0000	1.0000	0.9997	0.9976	0.9897	0.9685	0.9247	0.8499	0.7414	0.6047
	8	1.0000	1.0000	1.0000	0.9996	0.9978	0.9917	0.9757	0.9417	0.8811	0.7880
	9	1.0000	1.0000	1.0000	1.0000	0.9997	0.9983	0.9940	0.9825	0.9574	0.9102
	10	1.0000	1.0000	1.0000	1.0000	1.0000	0.9998	0.9989	0.9961	0.9886	0.9713
	11	1.0000	1.0000	1.0000	1.0000	1.0000	1.0000	0.9999	0.9994	0.9978	0.9935
	12	1.0000	1.0000	1.0000	1.0000	1.0000	1.0000	1.0000	0.9999	0.9997	0.9991
	13	1.0000	1.0000	1.0000	1.0000	1.0000	1.0000	1.0000	1.0000	1.0000	0.9999
	14	1.0000	1.0000	1.0000	1.0000	1.0000	1.0000	1.0000	1.0000	1.0000	1.0000
15	0	0.4633	0.2059	0.0874	0.0352	0.0134	0.0047	0.0016	0.0005	0.0001	0.0000
	1	0.8290	0.5490	0.3186	0.1671	0.0802	0.0353	0.0142	0.0052	0.0017	0.0005
	2	0.9638	0.8159	0.6042	0.3980	0.2361	0.1268	0.0617	0.0271	0.0107	0.0037
	3	0.9945	0.9444	0.8227	0.6482	0.4613	0.2969	0.1727	0.0905	0.0424	0.0176
	4	0.9994	0.9873	0.9383	0.8358	0.6865	0.5155	0.3519	0.2173	0.1204	0.0592
	5	0.9999	0.9978	0.9832	0.9389	0.8516	0.7216	0.5643	0.4032	0.2608	0.1509
	6	1.0000	0.9997	0.9964	0.9819	0.9434	0.8689	0.7548	0.6098	0.4522	0.3036
	7	1.0000	1.0000	0.9994	0.9958	0.9827	0.9500	0.8868	0.7869	0.6535	0.5000
	8	1.0000	1.0000	0.9999	0.9992	0.9958	0.9848	0.9578	0.9050	0.8182	0.6964
	9	1.0000	1.0000	1.0000	0.9999	0.9992	0.9963	0.9876	0.9662	0.9231	0.8491
	10	1.0000	1.0000	1.0000	1.0000	0.9999	0.9993	0.9972	0.9907	0.9745	0.9408
	11	1.0000	1.0000	1.0000	1.0000	1.0000	0.9999	0.9995	0.9981	0.9937	0.9824
	12	1.0000	1.0000	1.0000	1.0000	1.0000	1.0000	0.9999	0.9997	0.9989	0.9963
	13	1.0000	1.0000	1.0000	1.0000	1.0000	1.0000	1.0000	1.0000	0.9999	0.9995
	14	1.0000	1.0000	1.0000	1.0000	1.0000	1.0000	1.0000	1.0000	1.0000	1.0000
	15	1.0000	1.0000	1.0000	1.0000	1.0000	1.0000	1.0000	1.0000	1.0000	1.0000
16	0	0.4401	0.1853	0.0743	0.0281	0.0100	0.0033	0.0010	0.0003	0.0001	0.0000
	1	0.8108	0.5147	0.2839	0.1407	0.0635	0.0261	0.0098	0.0033	0.0010	0.0003
	2	0.9571	0.7892	0.5614	0.3518	0.1971	0.0994	0.0451	0.0183	0.0066	0.0021
	3	0.9930	0.9316	0.7899	0.5981	0.4050	0.2459	0.1339	0.0651	0.0281	0.0106
	4	0.9991	0.9830	0.9209	0.7982	0.6302	0.4499	0.2892	0.1666	0.0853	0.0384
	5	0.9999	0.9967	0.9765	0.9183	0.8103	0.6598	0.4900	0.3288	0.1976	0.1051
	6	1.0000	0.9995	0.9944	0.9733	0.9204	0.8247	0.6881	0.5272	0.3660	0.2272
	7	1.0000	0.9999	0.9989	0.9930	0.9729	0.9256	0.8406	0.7161	0.5629	0.4018
	8	1.0000	1.0000	0.9998	0.9985	0.9925	0.9743	0.9329	0.8577	0.7441	0.5982

n	k	0.05	0.10	0.15	0.20	p 0.25	0.30	0.35	0.40	0.45	0.50
16	9	1.0000	1.0000	1.0000	0.9998	0.9984	0.9929	0.9771	0.9417	0.8759	0.7728
	10	1.0000	1.0000	1.0000	1.0000	0.9997	0.9984	0.9938	0.9809	0.9514	0.8949
	11	1.0000	1.0000	1.0000	1.0000	1.0000	0.9997	0.9987	0.9951	0.9851	0.9616
	12	1.0000	1.0000	1.0000	1.0000	1.0000	1.0000	0.9998	0.9991	0.9965	0.9894
	13	1.0000	1.0000	1.0000	1.0000	1.0000	1.0000	1.0000	0.9999	0.9994	0.9979
	14	1.0000	1.0000	1.0000	1.0000	1.0000	1.0000	1.0000	1.0000	0.9999	0.9997
	15	1.0000	1.0000	1.0000	1.0000	1.0000	1.0000	1.0000	1.0000	1.0000	1.0000
	16	1.0000	1.0000	1.0000	1.0000	1.0000	1.0000	1.0000	1.0000	1.0000	1.0000
17	0	0.4181	0.1668	0.0631	0.0225	0.0075	0.0023	0.0007	0.0002	0.0000	0.0000
	1	0.7922	0.4818	0.2525	0.1182	0.0501	0.0193	0.0067	0.0021	0.0006	0.0001
	2	0.9497	0.7618	0.5198	0.3096	0.1637	0.0774	0.0327	0.0123	0.0041	0.0012
	3	0.9912	0.9174	0.7556	0.5489	0.3530	0.2019	0.1028	0.0464	0.0184	0.0064
	4	0.9988	0.9779	0.9013	0.7582	0.5739	0.3887	0.2348	0.1260	0.0596	0.0245
	5	0.9999	0.9953	0.9681	0.8943	0.7653	0.5968	0.4197	0.2639	0.1471	0.0717
	6	1.0000	0.9992	0.9917	0.9623	0.8929	0.7752	0.6188	0.4478	0.2902	0.1662
	7	1.0000	0.9999	0.9983	0.9891	0.9598	0.8954	0.7872	0.6405	0.4743	0.3145
	8	1.0000	1.0000	0.9997	0.9974	0.9876	0.9597	0.9006	0.8011	0.6626	0.5000
	9	1.0000	1.0000	1.0000	0.9995	0.9969	0.9873	0.9617	0.9081	0.8166	0.6855
	10	1.0000	1.0000	1.0000	0.9999	0.9994	0.9968	0.9880	0.9652	0.9174	0.8338
	11	1.0000	1.0000	1.0000	1.0000	0.9999	0.9993	0.9970	0.9894	0.9699	0.9283
	12	1.0000	1.0000	1.0000	1.0000	1.0000	0.9999	0.9994	0.9975	0.9914	0.9755
	13	1.0000	1.0000	1.0000	1.0000	1.0000	1.0000	0.9999	0.9995	0.9981	0.9936
	14	1.0000	1.0000	1.0000	1.0000	1.0000	1.0000	1.0000	0.9999	0.9997	0.9988
	15	1.0000	1.0000	1.0000	1.0000	1.0000	1.0000	1.0000	1.0000	1.0000	0.9999
	16	1.0000	1.0000	1.0000	1.0000	1.0000	1.0000	1.0000	1.0000	1.0000	1.0000
	17	1.0000	1.0000	1.0000	1.0000	1.0000	1.0000	1.0000	1.0000	1.0000	1.0000
18	0	0.3972	0.1501	0.0536	0.0180	0.0056	0.0016	0.0004	0.0001	0.0000	0.0000
	1	0.7735	0.4503	0.2241	0.0991	0.0395	0.0142	0.0046	0.0013	0.0003	0.0001
	2	0.9419	0.7338	0.4797	0.2713	0.1353	0.0600	0.0236	0.0082	0.0025	0.0007
	3	0.9891	0.9018	0.7202	0.5010	0.3057	0.1646	0.0783	0.0328	0.0120	0.0038
	4	0.9985	0.9718	0.8794	0.7164	0.5187	0.3327	0.1886	0.0942	0.0411	0.0154
	5	0.9998	0.9936	0.9581	0.8671	0.7175	0.5344	0.3550	0.2088	0.1077	0.0481
	6	1.0000	0.9988	0.9882	0.9487	0.8610	0.7217	0.5491	0.3743	0.2258	0.1189
	7	1.0000	0.9998	0.9973	0.9837	0.9431	0.8593	0.7283	0.5634	0.3915	0.2403
	8	1.0000	1.0000	0.9995	0.9957	0.9807	0.9404	0.8609	0.7368	0.5778	0.4073
	9	1.0000	1.0000	0.9999	0.9991	0.9946	0.9790	0.9403	0.8653	0.7473	0.5927
	10	1.0000	1.0000	1.0000	0.9998	0.9988	0.9939	0.9788	0.9424	0.8720	0.7597
	11	1.0000	1.0000	1.0000	1.0000	0.9998	0.9986	0.9938	0.9797	0.9463	0.8811
	12	1.0000	1.0000	1.0000	1.0000	1.0000	0.9997	0.9986	0.9942	0.9817	0.9519
	13	1.0000	1.0000	1.0000	1.0000	1.0000	1.0000	0.9997	0.9987	0.9951	0.9846

						p					
n	k	0.05	0.10	0.15	0.20	0.25	0.30	0.35	0.40	0.45	0.50
18	14	1.0000	1.0000	1.0000	1.0000	1.0000	1.0000	1.0000	0.9998	0.9990	0.9962
	15	1.0000	1.0000	1.0000	1.0000	1.0000	1.0000	1.0000	1.0000	0.9999	0.9993
	16	1.0000	1.0000	1.0000	1.0000	1.0000	1.0000	1.0000	1.0000	1.0000	0.9999
	17	1.0000	1.0000	1.0000	1.0000	1.0000	1.0000	1.0000	1.0000	1.0000	1.0000
	18	1.0000	1.0000	1.0000	1.0000	1.0000	1.0000	1.0000	1.0000	1.0000	1.0000
19	0	0.3774	0.1351	0.0456	0.0144	0.0042	0.0011	0.0003	0.0001	0.0000	0.0000
	1	0.7547	0.4203	0.1985	0.0829	0.0310	0.0104	0.0031	0.0008	0.0002	0.0000
	2	0.9335	0.7054	0.4413	0.2369	0.1113	0.0462	0.0170	0.0055	0.0015	0.0004
	3	0.9868	0.8850	0.6841	0.4551	0.2631	0.1332	0.0591	0.0230	0.0077	0.0022
	4	0.9980	0.9648	0.8556	0.6733	0.4654	0.2822	0.1500	0.0696	0.0280	0.0096
	5	0.9998	0.9914	0.9463	0.8369	0.6678	0.4739	0.2968	0.1629	0.0777	0.0318
	6	1.0000	0.9983	0.9837	0.9324	0.8251	0.6655	0.4812	0.3081	0.1727	0.0835
	7	1.0000	0.9997	0.9959	0.9767	0.9225	0.8180	0.6656	0.4878	0.3169	0.1796
	8	1.0000	1.0000	0.9992	0.9933	0.9713	0.9161	0.8145	0.6675	0.4940	0.3238
	9	1.0000	1.0000	0.9999	0.9984	0.9911	0.9674	0.9125	0.8139	0.6710	0.5000
	10	1.0000	1.0000	1.0000	0.9997	0.9977	0.9895	0.9653	0.9115	0.8159	0.6762
	11	1.0000	1.0000	1.0000	1.0000	0.9995	0.9972	0.9886	0.9648	0.9129	0.8204
	12	1.0000	1.0000	1.0000	1.0000	0.9999	0.9994	0.9969	0.9884	0.9658	0.9165
	13	1.0000	1.0000	1.0000	1.0000	1.0000	0.9999	0.9993	0.9969	0.9891	0.9682
	14	1.0000	1.0000	1.0000	1.0000	1.0000	1.0000	0.9999	0.9994	0.9972	0.9904
	15	1.0000	1.0000	1.0000	1.0000	1.0000	1.0000	1.0000	0.9999	0.9995	0.9978
	16	1.0000	1.0000	1.0000	1.0000	1.0000	1.0000	1.0000	1.0000	0.9999	0.9996
	17	1.0000	1.0000	1.0000	1.0000	1.0000	1.0000	1.0000	1.0000	1.0000	1.0000
	18	1.0000	1.0000	1.0000	1.0000	1.0000	1.0000	1.0000	1.0000	1.0000	1.0000
	19	1.0000	1.0000	1.0000	1.0000	1.0000	1.0000	1.0000	1.0000	1.0000	1.0000
20	0	0.3585	0.1216	0.0388	0.0115	0.0032	0.0008	0.0002	0.0000	0.0000	0.0000
	1	0.7358	0.3917	0.1756	0.0692	0.0243	0.0076	0.0021	0.0005	0.0001	0.0000
	2	0.9245	0.6769	0.4049	0.2061	0.0913	0.0355	0.0121	0.0036	0.0009	0.0002
	3	0.9841	0.8670	0.6477	0.4114	0.2252	0.1071	0.0444	0.0160	0.0049	0.0013
	4	0.9974	0.9568	0.8298	0.6296	0.4148	0.2375	0.1182	0.0510	0.0189	0.0059
	5	0.9997	0.9887	0.9327	0.8042	0.6172	0.4164	0.2454	0.1256	0.0553	0.0207
	6	1.0000	0.9976	0.9781	0.9133	0.7858	0.6080	0.4166	0.2500	0.1299	0.0577
	7	1.0000	0.9996	0.9941	0.9679	0.8982	0.7723	0.6010	0.4159	0.2520	0.1316
	8	1.0000	0.9999	0.9987	0.9900	0.9591	0.8867	0.7624	0.5956	0.4143	0.2517
	9	1.0000	1.0000	0.9998	0.9974	0.9861	0.9520	0.8782	0.7553	0.5914	0.4119
	10	1.0000	1.0000	1.0000	0.9994	0.9961	0.9829	0.9468	0.8725	0.7507	0.5881
	11	1.0000	1.0000	1.0000	0.9999	0.9991	0.9949	0.9804	0.9435	0.8692	0.7483
	12	1.0000	1.0000	1.0000	1.0000	0.9998	0.9987	0.9940	0.9790	0.9420	0.8684
	13	1.0000	1.0000	1.0000	1.0000	1.0000	0.9997	0.9985	0.9935	0.9786	0.9423
	14	1.0000	1.0000	1.0000	1.0000	1.0000	1.0000	0.9997	0.9984	0.9936	0.9793

D.2 Normal Distribution

The following table lists the values of $N_{0,1}(x)$ (see B.1.2).

x	0	0.01	0.02	0.03	0.04	0.05	0.06	0.07	0.08	0.09
0	0.5	0.50399	0.50798	0.51197	0.51595	0.51994	0.52392	0.5279	0.53188	0.53586
0.1	0.53983	0.5438	0.54776	0.55172	0.55567	0.55962	0.56356	0.56749	0.57142	0.57535
0.2	0.57926	0.58317	0.58706	0.59095	0.59483	0.59871	0.60257	0.60642	0.61026	0.61409
0.3	0.61791	0.62172	0.62552	0.6293	0.63307	0.63683	0.64058	0.64431	0.64803	0.65173
0.4	0.65542	0.6591	0.66276	0.6664	0.67003	0.67364	0.67724	0.68082	0.68439	0.68793
0.5	0.69146	0.69497	0.69847	0.70194	0.7054	0.70884	0.71226	0.71566	0.71904	0.7224
0.6	0.72575	0.72907	0.73237	0.73565	0.73891	0.74215	0.74537	0.74857	0.75175	0.7549
0.7	0.75804	0.76115	0.76424	0.7673	0.77035	0.77337	0.77637	0.77935	0.7823	0.78524
0.8	0.78814	0.79103	0.79389	0.79673	0.79955	0.80234	0.80511	0.80785	0.81057	0.81327
0.9	0.81594	0.81859	0.82121	0.82381	0.82639	0.82894	0.83147	0.83398	0.83646	0.83891
1	0.84134	0.84375	0.84614	0.84849	0.85083	0.85314	0.85543	0.85769	0.85993	0.86214
1.1	0.86433	0.8665	0.86864	0.87076	0.87286	0.87493	0.87698	0.879	0.881	0.88298
1.2	0.88493	0.88686	0.88877	0.89065	0.89251	0.89435	0.89617	0.89796	0.89973	0.90147
1.3	0.9032	0.9049	0.90658	0.90824	0.90988	0.91149	0.91308	0.91466	0.91621	0.91774
1.4	0.91924	0.92073	0.9222	0.92364	0.92507	0.92647	0.92785	0.92922	0.93056	0.93189
1.5	0.93319	0.93448	0.93574	0.93699	0.93822	0.93943	0.94062	0.94179	0.94295	0.94408
1.6	0.9452	0.9463	0.94738	0.94845	0.9495	0.95053	0.95154	0.95254	0.95352	0.95449
1.7	0.95543	0.95637	0.95728	0.95818	0.95907	0.95994	0.9608	0.96164	0.96246	0.96327
1.8	0.96407	0.96485	0.96562	0.96638	0.96712	0.96784	0.96856	0.96926	0.96995	0.97062
1.9	0.97128	0.97193	0.97257	0.9732	0.97381	0.97441	0.975	0.97558	0.97615	0.9767
2	0.97725	0.97778	0.97831	0.97882	0.97932	0.97982	0.9803	0.98077	0.98124	0.98169
2.1	0.98214	0.98257	0.983	0.98341	0.98382	0.98422	0.98461	0.985	0.98537	0.98574
2.2	0.9861	0.98645	0.98679	0.98713	0.98745	0.98778	0.98809	0.9884	0.9887	0.98899
2.3	0.98928	0.98956	0.98983	0.9901	0.99036	0.99061	0.99086	0.99111	0.99134	0.99158
2.4	0.9918	0.99202	0.99224	0.99245	0.99266	0.99286	0.99305	0.99324	0.99343	0.99361
2.5	0.99379	0.99396	0.99413	0.9943	0.99446	0.99461	0.99477	0.99492	0.99506	0.9952
2.6	0.99534	0.99547	0.9956	0.99573	0.99585	0.99598	0.99609	0.99621	0.99632	0.99643
2.7	0.99653	0.99664	0.99674	0.99683	0.99693	0.99702	0.99711	0.9972	0.99728	0.99736
2.8	0.99744	0.99752	0.9976	0.99767	0.99774	0.99781	0.99788	0.99795	0.99801	0.99807
2.9	0.99813	0.99819	0.99825	0.99831	0.99836	0.99841	0.99846	0.99851	0.99856	0.99861

D.3 Student's t Distribution

The following table lists the values of (see B.1.2): $P(t_{df} \leq x) = 1 - \int_{-\infty}^{x} t_{df}(t)dt$.

						df							
x	1	3	5	7	9	11	13	15	17	19	21	23	25
0	0.5	0.5	0.5	0.5	0.5	0.5	0.5	0.5	0.5	0.5	0.5	0.5	0.5
0.1	0.532	0.537	0.538	0.538	0.539	0.539	0.539	0.539	0.539	0.539	0.539	0.539	0.539
0.2	0.563	0.573	0.575	0.576	0.577	0.577	0.578	0.578	0.578	0.578	0.578	0.578	0.578
0.3	0.593	0.608	0.612	0.614	0.615	0.615	0.616	0.616	0.616	0.616	0.616	0.617	0.617
0.4	0.621	0.642	0.647	0.649	0.651	0.652	0.652	0.653	0.653	0.653	0.653	0.654	0.654
0.5	0.648	0.674	0.681	0.684	0.685	0.687	0.687	0.688	0.688	0.689	0.689	0.689	0.689
0.6	0.672	0.705	0.713	0.716	0.718	0.72	0.721	0.721	0.722	0.722	0.723	0.723	0.723
0.7	0.694	0.733	0.742	0.747	0.749	0.751	0.752	0.753	0.753	0.754	0.754	0.755	0.755
0.8	0.715	0.759	0.77	0.775	0.778	0.78	0.781	0.782	0.783	0.783	0.784	0.784	0.784
0.9	0.733	0.783	0.795	0.801	0.804	0.806	0.808	0.809	0.81	0.81	0.811	0.811	0.812
1	0.75	0.804	0.818	0.825	0.828	0.831	0.832	0.833	0.834	0.835	0.836	0.836	0.837
1.1	0.765	0.824	0.839	0.846	0.85	0.853	0.854	0.856	0.857	0.857	0.858	0.859	0.859
1.2	0.779	0.842	0.858	0.865	0.87	0.872	0.874	0.876	0.877	0.878	0.878	0.879	0.879
1.3	0.791	0.858	0.875	0.883	0.887	0.89	0.892	0.893	0.895	0.895	0.896	0.897	0.897
1.4	0.803	0.872	0.89	0.898	0.902	0.905	0.908	0.909	0.91	0.911	0.912	0.913	0.913
1.5	0.813	0.885	0.903	0.911	0.916	0.919	0.921	0.923	0.924	0.925	0.926	0.926	0.927
1.6	0.822	0.896	0.915	0.923	0.928	0.931	0.933	0.935	0.936	0.937	0.938	0.938	0.939
1.7	0.831	0.906	0.925	0.934	0.938	0.941	0.944	0.945	0.946	0.947	0.948	0.949	0.949
1.8	0.839	0.915	0.934	0.943	0.947	0.95	0.952	0.954	0.955	0.956	0.957	0.958	0.958
1.9	0.846	0.923	0.942	0.95	0.955	0.958	0.96	0.962	0.963	0.964	0.964	0.965	0.965
2	0.852	0.93	0.949	0.957	0.962	0.965	0.967	0.968	0.969	0.97	0.971	0.971	0.972
2.1	0.859	0.937	0.955	0.963	0.967	0.97	0.972	0.973	0.975	0.975	0.976	0.977	0.977
2.2	0.864	0.942	0.96	0.968	0.972	0.975	0.977	0.978	0.979	0.98	0.98	0.981	0.981
2.3	0.869	0.948	0.965	0.973	0.977	0.979	0.981	0.982	0.983	0.984	0.984	0.985	0.985
2.4	0.874	0.952	0.969	0.976	0.98	0.982	0.984	0.985	0.986	0.987	0.987	0.988	0.988
2.5	0.879	0.956	0.973	0.98	0.983	0.985	0.987	0.988	0.989	0.989	0.99	0.99	0.99
2.6	0.883	0.96	0.976	0.982	0.986	0.988	0.989	0.99	0.991	0.991	0.992	0.992	0.992
2.7	0.887	0.963	0.979	0.985	0.988	0.99	0.991	0.992	0.992	0.993	0.993	0.994	0.994
2.8	0.891	0.966	0.981	0.987	0.99	0.991	0.992	0.993	0.994	0.994	0.995	0.995	0.995
2.9	0.894	0.969	0.983	0.989	0.991	0.993	0.994	0.995	0.995	0.995	0.996	0.996	0.996
3	0.898	0.971	0.985	0.99	0.993	0.994	0.995	0.996	0.996	0.996	0.997	0.997	0.997
3.1	0.901	0.973	0.987	0.991	0.994	0.995	0.996	0.996	0.997	0.997	0.997	0.997	0.998
3.2	0.904	0.975	0.988	0.992	0.995	0.996	0.997	0.997	0.997	0.998	0.998	0.998	0.998
3.3	0.906	0.977	0.989	0.993	0.995	0.996	0.997	0.998	0.998	0.998	0.998	0.998	0.999
3.4	0.909	0.979	0.99	0.994	0.996	0.997	0.998	0.998	0.998	0.998	0.999	0.999	0.999
3.5	0.911	0.98	0.991	0.995	0.997	0.998	0.998	0.998	0.999	0.999	0.999	0.999	0.999

D.4 Chi-Square Distribution

Table of the one-sided chi-square probability: $P(\chi^2_{df} > x) = 1 - \int_0^x \chi^2_{df}(t)dt$.

						df							
x	1	3	5	7	9	11	13	15	17	19	21	23	25
1	0.317	0.801	0.963	0.995	0.999	1.000	1.000	1.000	1.000	1.000	1.000	1.000	1.000
2	0.157	0.572	0.849	0.960	0.991	0.998	1.000	1.000	1.000	1.000	1.000	1.000	1.000
3	0.083	0.392	0.700	0.885	0.964	0.991	0.998	1.000	1.000	1.000	1.000	1.000	1.000
4	0.046	0.261	0.549	0.780	0.911	0.970	0.991	0.998	0.999	1.000	1.000	1.000	1.000
5	0.025	0.172	0.416	0.660	0.834	0.931	0.975	0.992	0.998	0.999	1.000	1.000	1.000
6	0.014	0.112	0.306	0.540	0.740	0.873	0.946	0.980	0.993	0.998	0.999	1.000	1.000
7	0.008	0.072	0.221	0.429	0.637	0.799	0.902	0.958	0.984	0.994	0.998	0.999	1.000
8	0.005	0.046	0.156	0.333	0.534	0.713	0.844	0.924	0.967	0.987	0.995	0.998	0.999
9	0.003	0.029	0.109	0.253	0.437	0.622	0.773	0.878	0.940	0.973	0.989	0.996	0.999
10	0.002	0.019	0.075	0.189	0.350	0.530	0.694	0.820	0.904	0.953	0.979	0.991	0.997
11	0.001	0.012	0.051	0.139	0.276	0.443	0.611	0.753	0.857	0.924	0.963	0.983	0.993
12	0.001	0.007	0.035	0.101	0.213	0.364	0.528	0.679	0.800	0.886	0.940	0.970	0.987
13	0.000	0.005	0.023	0.072	0.163	0.293	0.448	0.602	0.736	0.839	0.909	0.952	0.977
14	0.000	0.003	0.016	0.051	0.122	0.233	0.374	0.526	0.667	0.784	0.870	0.927	0.962
15	0.000	0.002	0.010	0.036	0.091	0.182	0.307	0.451	0.595	0.723	0.823	0.895	0.941
16	0.000	0.001	0.007	0.025	0.067	0.141	0.249	0.382	0.524	0.657	0.770	0.855	0.915
17	0.000	0.001	0.004	0.017	0.049	0.108	0.199	0.319	0.454	0.590	0.711	0.809	0.882
18	0.000	0.000	0.003	0.012	0.035	0.082	0.158	0.263	0.389	0.522	0.649	0.757	0.842
19	0.000	0.000	0.002	0.008	0.025	0.061	0.123	0.214	0.329	0.457	0.585	0.701	0.797
20	0.000	0.000	0.001	0.006	0.018	0.045	0.095	0.172	0.274	0.395	0.521	0.642	0.747
21	0.000	0.000	0.001	0.004	0.013	0.033	0.073	0.137	0.226	0.337	0.459	0.581	0.693
22	0.000	0.000	0.001	0.003	0.009	0.024	0.055	0.108	0.185	0.284	0.400	0.520	0.636
23	0.000	0.000	0.000	0.002	0.006	0.018	0.042	0.084	0.149	0.237	0.344	0.461	0.578
24	0.000	0.000	0.000	0.001	0.004	0.013	0.031	0.065	0.119	0.196	0.293	0.404	0.519
25	0.000	0.000	0.000	0.001	0.003	0.009	0.023	0.050	0.095	0.161	0.247	0.350	0.462
26	0.000	0.000	0.000	0.001	0.002	0.006	0.017	0.038	0.074	0.130	0.206	0.301	0.408
27	0.000	0.000	0.000	0.000	0.005	0.005	0.012	0.029	0.058	0.105	0.171	0.256	0.356
28	0.000	0.000	0.000	0.000	0.001	0.003	0.009	0.022	0.045	0.083	0.140	0.216	0.308
29	0.000	0.000	0.000	0.000	0.001	0.002	0.007	0.016	0.035	0.066	0.114	0.180	0.264
30	0.000	0.000	0.000	0.000	0.000	0.002	0.005	0.012	0.026	0.052	0.092	0.149	0.224
31	0.000	0.000	0.000	0.000	0.000	0.001	0.003	0.009	0.020	0.040	0.074	0.123	0.189
32	0.000	0.000	0.000	0.000	0.000	0.001	0.002	0.006	0.015	0.031	0.059	0.100	0.158
33	0.000	0.000	0.000	0.000	0.000	0.001	0.002	0.005	0.011	0.024	0.046	0.081	0.131
34	0.000	0.000	0.000	0.000	0.000	0.000	0.001	0.003	0.008	0.018	0.036	0.065	0.108
35	0.000	0.000	0.000	0.000	0.000	0.000	0.001	0.002	0.006	0.014	0.028	0.052	0.088
36	0.000	0.000	0.000	0.000	0.000	0.000	0.001	0.002	0.005	0.011	0.022	0.041	0.072
37	0.000	0.000	0.000	0.000	0.000	0.000	0.000	0.001	0.003	0.008	0.017	0.033	0.058

D.5 Critical Values for the F Distribution

For $\alpha = 0.99$:

df_2	1	2	3	4	6	df_1 8	10	15	20	30	40	50
1	4052	4999	5404	5624	5859	5981	6056	6157	6209	6260	6286	6302
2	98.50	99.00	99.16	99.25	99.33	99.38	99.40	99.43	99.45	99.47	99.48	99.48
3	34.12	30.82	29.46	28.71	27.91	27.49	27.23	26.87	26.69	26.50	26.41	26.35
4	21.20	18.00	16.69	15.98	15.21	14.80	14.55	14.20	14.02	13.84	13.75	13.69
5	16.26	13.27	12.06	11.39	10.67	10.29	10.05	9.72	9.55	9.38	9.29	9.24
6	13.75	10.92	9.78	9.15	8.47	8.10	7.87	7.56	7.40	7.23	7.14	7.09
7	12.25	9.55	8.45	7.85	7.19	6.84	6.62	6.31	6.16	5.99	5.91	5.86
8	11.26	8.65	7.59	7.01	6.37	6.03	5.81	5.52	5.36	5.20	5.12	5.07
9	10.56	8.02	6.99	6.42	5.80	5.47	5.26	4.96	4.81	4.65	4.57	4.52
10	10.04	7.56	6.55	5.99	5.39	5.06	4.85	4.56	4.41	4.25	4.17	4.12

For $\alpha = 0.95$:

df_2	1	2	3	4	6	df_1 8	10	15	20	30	40	50
1	161.45	199.50	215.71	224.58	233.99	238.88	241.88	245.95	248.02	250.10	251.14	251.77
2	18.51	19.00	19.16	19.25	19.33	19.37	19.40	19.43	19.45	19.46	19.47	19.48
3	10.13	9.55	9.28	9.12	8.94	8.85	8.79	8.70	8.66	8.62	8.59	8.58
4	7.71	6.94	6.59	6.39	6.16	6.04	5.96	5.86	5.80	5.75	5.72	5.70
5	6.61	5.79	5.41	5.19	4.95	4.82	4.74	4.62	4.56	4.50	4.46	4.44
6	5.99	5.14	4.76	4.53	4.28	4.15	4.06	3.94	3.87	3.81	3.77	3.75
7	5.59	4.74	4.35	4.12	3.87	3.73	3.64	3.51	3.44	3.38	3.34	3.32
8	5.32	4.46	4.07	3.84	3.58	3.44	3.35	3.22	3.15	3.08	3.04	3.02
9	5.12	4.26	3.86	3.63	3.37	3.23	3.14	3.01	2.94	2.86	2.83	2.80
10	4.96	4.10	3.71	3.48	3.22	3.07	2.98	2.85	2.77	2.70	2.66	2.64
11	4.84	3.98	3.59	3.36	3.09	2.95	2.85	2.72	2.65	2.57	2.53	2.51
12	4.75	3.89	3.49	3.26	3.00	2.85	2.75	2.62	2.54	2.47	2.43	2.40
13	4.67	3.81	3.41	3.18	2.92	2.77	2.67	2.53	2.46	2.38	2.34	2.31
14	4.60	3.74	3.34	3.11	2.85	2.70	2.60	2.46	2.39	2.31	2.27	2.24
15	4.54	3.68	3.29	3.06	2.79	2.64	2.54	2.40	2.33	2.25	2.20	2.18
16	4.49	3.63	3.24	3.01	2.74	2.59	2.49	2.35	2.28	2.19	2.15	2.12
17	4.45	3.59	3.20	2.96	2.70	2.55	2.45	2.31	2.23	2.15	2.10	2.08
18	4.41	3.55	3.16	2.93	2.66	2.51	2.41	2.27	2.19	2.11	2.06	2.04
19	4.38	3.52	3.13	2.90	2.63	2.48	2.38	2.23	2.16	2.07	2.03	2.00
20	4.35	3.49	3.10	2.87	2.60	2.45	2.35	2.20	2.12	2.04	1.99	1.97
30	4.17	3.32	2.92	2.69	2.42	2.27	2.16	2.01	1.93	1.84	1.79	1.76
40	4.08	3.23	2.84	2.61	2.34	2.18	2.08	1.92	1.84	1.74	1.69	1.66
60	4.00	3.15	2.76	2.53	2.25	2.10	1.99	1.84	1.75	1.65	1.59	1.56

Appendix E - Datasets

Datasets included in the book CD are presented in the form of Microsoft EXCEL files with a description worksheet.

E.1 Breast Tissue

The Breast Tissue.xls file contains 106 electrical impedance measurements performed on samples of freshly excised breast tissue. Six classes of tissue were studied:

CAR: Carcinoma (21 cases)   FAD: Fibro-adenoma (15 cases)
MAS: Mastopathy (18 cases)   GLA: Glandular (16 cases)
CON: Connective (14 cases)   ADI: Adipose (22 cases)

Impedance measurements were taken at seven frequencies and plotted in the real-imaginary plane, constituting the impedance spectrum from which the following features were computed:

I0:   Impedance at zero frequency (Ohm)
PA500:   Phase angle at 500 KHz
HFS: High-frequency slope of the phase angle
DA: Impedance distance between spectral ends
AREA: Area under the spectrum
A/DA: Area normalised by DA
MAX IP: Maximum amplitude of the spectrum
DR: Distance between I0 and the real part of the maximum frequency
            point
P: Length of the spectral curve

Source: J Jossinet, INSERM U.281, Lyon, France.

E.2 Car Sale

The Car Sale.xls file contains data on 22 cars that was collected between 12 September, 2000 and 31 March, 2002.

The variables are:

Sale1:   Date that a car was sold.
Complaint: Date that a complaint about any type of malfunctioning was presented for the first time.
Sale2:   Last date that a car accessory was purchased (unrelated to the complaint).
Lost: Lost contact during the study? True = Yes; False = No.
End: End date of the study.
Time: Number of days until event (Sale2, Complaint or End).

Source: New and Used Car Stand in Porto, Portugal.

E.3 Cells

The Cells.xls file has the following two datasheets:

1. CFU Datasheet

The data consists of counts of "colony forming units", CFUs, in mice infected with a mycobacterium. Bacterial load is studied at different time points in three target organs: the spleen, the liver and the lungs.

After the mice are dissected, the target organs are homogenised and plated for bacterial counts (CFUs).

There are two groups for each time point:

1   Anti-inflammatory protein deficient group (knock-out group, KO).
2   Normal control group (C).

The two groups (1 and 2) dissected at different times are independent.

2. SPLEEN Datasheet

The data consists of stained cell counts from infected mice spleen, using two biochemical markers: CD4 and CD8.

Cell counting is performed with a flow cytometry system. The two groups (K and C) dissected at different times are independent.

Source: S Lousada, IBMC (Instituto de Biologia Molecular e Celular), Porto, Portugal.

E.4 Clays

The Clays.xls file contains the analysis results of 94 clay samples from probes collected in an area with two geological formations (in the region of Anadia, Portugal). The following variables characterise the dataset:

Age: Geological age: 1 - pliocenic (good quality clay); 2 - pliocenic
                    (bad quality clay); 3 - holocenic.
Level: Probe level (m).
Grading: LG (%) - low grading: < 2 microns;
                    MG (%) - medium grading: ≥ 2 , < 62 microns;
                    HG (%) - high grading: ≥ 62 microns.
Minerals: Ilite, pyrophyllite, caolinite, lepidolite, quartz, goethite, K-
                    feldspar, Na-feldspar, hematite (%).
BS: Bending strength (Kg/cm2).
Contraction: v/s (%) - volume contraction, 1st phase;
                    s/c (%) - volume contraction, 2nd phase;
                    tot (%) - volume contraction, total.
Chemical analysis results: SiO_2, Al_2O_3, Fe_2O_3, FeO, CaO, MgO, Na_2O, K_2O,
            TiO_2 (%).

Source: C Carvalho, IGM - Instituto Geológico-Mineiro, Porto, Portugal.

E.5 Cork Stoppers

The Cork Stoppers.xls file contains measurements of cork stopper defects.
These were automatically obtained by an image processing system on 150 cork
stoppers belonging to three classes.

The first column of the Cork Stoppers.xls datasheet contains the class
labels assigned by human experts:

1:    Super Quality   (n_1 = 50 cork stoppers)
2:    Normal Quality   (n_2 = 50 cork stoppers)
3:    Poor Quality   (n_3 = 50 cork stoppers)

The following columns contain the measurements:

N: Total number of defects.
PRT: Total perimeter of the defects (in pixels).
ART: Total area of the defects (in pixels).
PRM: Average perimeter of the defects (in pixels) = PRT/N-
ARM: Average area of the defects (in pixels) = ART/N.
NG: Number of big defects (area bigger than an adequate threshold).
PRTG: Total perimeter of big defects (in pixels).
ARTG: Total area of big defects (in pixels).
RAAR: Area ratio of the defects = ARTG/ART.
RAN: Big defects ratio = NG/N.

Source: A Campilho, Dep. Engenharia Electrotécnica e de Computadores,
Faculdade de Engenharia, Universidade do Porto, Porto, Portugal.

E.6 CTG

The `CTG.xls` file contains measurements and classification results of cardiotocographic (CTG) examinations of 2126 foetuses. The examinations were performed at São João Hospital, Porto, Portugal. Cardiotocography is a popular diagnostic method in Obstetrics, consisting of the analysis and interpretation of the following signals: foetal heart rate; uterine contractions; foetal movements.

The measurements included in the `CTG.xls` file correspond only to foetal heart rate (FHR) features (e.g., basal value, accelerative/decelerative events), computed by an automatic system on FHR signals. The classification corresponds to a diagnostic category assigned by expert obstetricians independently of the CTG.

The following cardiotocographic features are available in the `CTG.xls` file:

LBE	Baseline value (medical expert)	LB	Baseline value (system)
AC	No. of accelerations	FM	No. of foetal movements
UC	No. of uterine contractions	DL	No. of light decelerations
DS	No. of severe decelerations	DP	No. of prolonged decelerations
DR	No. of repetitive decelerations	MIN	Low freq. of the histogram
MAX	High freq. of the histogram	MEAN	Histogram mean
NZER	Number of histogram zeros	MODE	Histogram mode
NMAX	Number of histogram peaks	VAR	Histogram variance
MEDIAN	Histogram median	WIDTH	Histogram width
TEND	Histogram tendency: -1= left assym.; 0 = symm.; 1 = right assym.		
ASTV	Percentage of time with abnormal short term (beat-to-beat) variability		
MSTV	Mean value of short term variability		
ALTV	Percentage of time with abnormal long term (one minute) variability		
MLTV	Mean value of long term variability		

Features AC, FM, UC, DL, DS, DP and DR should be converted to per unit time values (e.g. per minute) using the duration time of the analysed signal segment computed from start and end times given in columns B and E (in seconds).

The data is classified in ten classes:

A: Calm sleep
B: Rapid-eye-movement sleep
C: Calm vigilance
D: Active vigilance
SH: Shift pattern (A or SUSP with shifts)
AD: Accelerative/decelerative pattern (stress situation)
DE: Decelerative pattern (vagal stimulation)
LD: Largely decelerative pattern
FS: Flat-sinusoidal pattern (pathological state)
SUSP: Suspect pattern

A column containing the codes of Normal (1), Suspect (2) and Pathologic (3) classification is also included.

Source: J Bernardes, Faculdade de Medicina, Universidade do Porto, Porto, Portugal.

E.7 Culture

The Culture.xls file contains percentages of the "culture budget" assigned to different cultural activities in 165 Portuguese boroughs in 1995.
The boroughs constitute a sample of 3 regions:

Region: 1 - Alentejo province;
           2 - Center provinces;
           3 - Northern provinces.

The cultural activities are:

Cine: Cinema and photography
Halls: Halls for cultural activities
Sport: Games and sport activities
Music: Musical activities
Literat: Literature
Heritage: Cultural heritage (promotion, maintenance, etc.)
Theatre: Performing Arts
Fine Arts: Fine Arts (promotion, support, etc.)

Source: INE - Instituto Nacional de Estatística, Portugal.

E.8 Fatigue

The Fatigue.xls file contains results of fatigue tests performed on aluminium and iron specimens for the car industry. The specimens were subject to a sinusoidal load (20 Hz) until breaking or until a maximum of 10^7 (ten million) cycles was reached. There are two datasheets, one for the aluminium specimens and the other for the iron specimens.
The variables are:

Ref: Specimen reference.
Amp: Amplitude of the sinusoidal load in MPa.
NC: Number of cycles.
DFT: Defect type.
Break: Yes/No according to specimen having broken or not.
AmpG: Amplitude group: 1 - Low; 2 - High.

Source: Laboratório de Ensaios Tecnológicos, Dep. Engenharia Mecânica, Faculdade de Engenharia, Universidade do Porto, Porto, Portugal.

E.9 FHR

The FHR.xls file contains measurements and classifications performed on 51 foetal heart rate (FHR) signals with 20-minute duration, and collected from pregnant women at intra-partum stage.

All the signals were analysed by an automatic system (SP≡SisPorto system) and three human experts (E1≡Expert 1, E2≡Expert 2 and E3≡Expert 3).

The analysis results correspond to the following variables:

Baseline: The baseline value represents the stable value of the foetal heart rate (in beats per minute). The variables are SPB, E1B, E2B, E3B.

Class: The classification columns (variables SPC, E1C, E2C, E3C) have the following values:

N (=0) - Normal; S (=1) - Suspect; P (=2) - Pathologic.

Source: J Bernardes, Faculdade de Medicina, Universidade do Porto, Porto, Portugal.

E.10 FHR-Apgar

The FHR-Apgar.xls file contains 227 measurements of foetal heart rate (FHR) tracings recorded just previous to birth, and the respective Apgar index, evaluated by obstetricians according to a standard clinical procedure one minute and five minutes after birth. All data was collected in Portuguese hospitals following a strict protocol. The Apgar index is a ranking index in the [0, 10] interval assessing the wellbeing of the newborn babies. Low values (below 5) are considered bad prognosis. Normal newborns have an Apgar above 6.

The following measurements are available in the FHR-Apgar.xls file:

Apgar1:   Apgar measured at 1 minute after birth.
Apgar5:   Apgar measured at 5 minutes after birth.
Duration: Duration in minutes of the FHR tracing.
Baseline: Basal value of the FHR in beat/min.
Acelnum: Number of FHR accelerations.
Acelrate: Number of FHR accelerations per minute.
ASTV: Percentage of time with abnormal short term variability.
MSTV: Average duration of abnormal short term variability.
ALTV: Percentage of time with abnormal long term variability.
MLTV: Average duration of abnormal long term variability.

Source: D Ayres de Campos, Faculdade de Medicina, Universidade do Porto, Porto, Portugal.

E.11 Firms

The Firms.xls file contains values of the following economic indicators relative to 838 Portuguese firms during the year 1995:

Branch: 1 = Services; 2 = Commerce; 3 = Industry; 4 = Construction.
GI: Gross Income (millions of Portuguese Escudos).
CAP: Invested Capital (millions of Portuguese Escudos).
CA: Capital + Assets.
NI: Net Income (millions of Portuguese Escudos) = GI – (wages + taxes).
NW: Number of workers.
P: Apparent Productivity = GI/NW.
GIR: Gross Income Revenue = NI/GI.
CAPR: Capital Revenue = NI/CAP.
A/C: Assets share = (CA–CAP)/CAP %.
DEPR: Depreciations + provisions.

Source: Jornal de Notícias - Suplemento, Nov. 1995, Porto, Portugal.

E.12 Flow Rate

The Flow Rate.xls file contains daily measurements of river flow (m^3/s), during December 1985 and January 1986. Measurements were performed at two river sites in the North of Portugal: AC - Alto Cávado Dam; T - Toco Dam.

Source: EDP - Electricidade de Portugal, Portugal.

E.13 Foetal Weight

The Foetal Weight.xls file contains echographic measurements obtained from 414 newborn babies shortly before delivery at four Portuguese hospitals. Obstetricians use such measurements in order to predict foetal weight and related delivery risk.
The following measurements, all obtained under a strict protocol, are available:

MW	Mother's weight	MH	Mother's height
GA	Gestation age in weeks	DBMB	Days between meas. and birth
BPD	Biparietal diameter	CP	Cephalic perimeter
AP	Abdominal perimeter	FL	Femur length
FTW	Foetal weight at birth	FTL	Foetal length at birth
CPB	Cephalic perimeter at birth		

Source: A Matos, Hospital de São João, Porto, Portugal.

E.14 Forest Fires

The Forest Fires.xls file contains data on the number of fires and area of burnt forest in continental Portugal during the period 1943-1978. The variables are:

Year: 1943 -1978.
Nr: Number of forest fires.
Area: Area of burnt forest in ha.

Source: INE - Instituto Nacional de Estatística, Portugal.

E.15 Freshmen

The Freshmen.xls file summarises the results of an enquiry carried out at the Faculty of Engineering, Porto University, involving 132 freshmen. The enquiry intention was to evaluate the freshmen attitude towards the "freshmen initiation rites".
    The variables are:

SEX: 1 = Male; 2 = Female.
AGE: Freshman age in years.
CS: Civil status: 1 = single; 2 = married.
COURSE: 1 = civil engineering; 2 = electrical and computer engineering; 3 = informatics; 4 = mechanical engineering; 5 = material engineering; 6 = mine engineering; 7 = industrial management engineering; 8 = chemical engineering.
DISPL: Displacement from the local of origin: 1 = Yes; 2 = No.
ORIGIN: 1 = Porto; 2 = North; 3 = South; 4 = Center; 5 = Islands; 6 = Foreign.
WS: Work status: 1 = Only studies; 2 = Part-time work; 3 = Full-time work.
OPTION: Preference rank when choosing the course: 1...4.
LIKE: Attitude towards the course: 1 = Like; 2 = Dislike; 3 = No comment.
EXAM 1-5:   Scores in the first 5 course examinations, measured in [0, 20].
EXAMAVG: Average of the examination scores.
INIT: Whether or not the freshman was initiated: 1 = Yes; 2 = No.

Questions:
Q1:   Initiation makes it easier to integrate in the academic life.
Q2:   Initiation is associated to a political ideology.
Q3:   Initiation quality depends on who organises it.
Q4:   I liked to be initiated.
Q5:   Initiation is humiliating.
Q6:   I felt compelled to participate in the Initiation.

Q7:   I participated in the Initiation on my own will.
Q8:   Those that do not participate in the Initiation feel excluded.

All the answers were scored as: 1 = Fully disagree; 2 = Disagree; 3 = No comment; 4 = Agree; 5 = Fully agree. The missing value is coded 9.
The file contains extra variables in order to facilitate the data usage. These are:

Positive average: 1, if the average is at least 10; 0, otherwise.
Q1P, ..., Q8P: The same as Q1, ..., Q8 if the average is positive; 0, otherwise.

Source: H Rebelo, Serviço de Apoio Psicológico, Faculdade de Engenharia, Universidade do Porto, Porto, Portugal.

E.16 Heart Valve

The Heart Valve.xls file contains data from a follow-up study of 526 patients submitted to a heart valve implant at São João Hospital, Porto, Portugal.
The variables are:

VALVE: Valve type.
SIZE: Size of the prosthesis.
AGE: Patient age at time of surgery.
EXCIRC: Extra body circulation in minutes.
CLAMP: Time of aorta clamp.
PRE_C: Pre-surgery functional class, according to NYHA (New York Heart Association): 0 = No symptoms; 1, 2 = Mild symptoms; 3, 4 = Severe symptoms).
POST_C: Post-surgery functional class, according to NYHA.
ACT_C: Functional class at last consultation, according to NYHA.
DATE_OP: Date of the operation.
DDOP: Death during operation (TRUE, FALSE).
DATE_DOP: Date of death due to operation complications.
DCAR: Death by cardiac causes in the follow-up (TRUE, FALSE).
DCARTYPE: Type of death for DCAR = TRUE: 1 - Sudden death; 2 – Cardiac failure; 3 -Death in the re-operation.
NDISF: Normo-disfunctional valve (morbility factor): 1 = No; 2 = Yes.
VALVESUB: Subject to valve substitution in the follow-up (TRUE, FALSE).
LOST: Lost in the follow-up (not possible to contact).
DATE_EC: Date of endocarditis (morbility factor).
DATE_ECO: Date of last echocardiogram (usually the date used for follow-up when there is no morbility factor) or date of last consultation.
DATE_LC: Date of the last consultation (usually date of follow-up when no morbility is present).
DATE_FU: Date of death in the follow-up.
REOP: Re-operation? (TRUE, FALSE).

DATE_REOP: Re-operation date.

The Survival Data worksheet contains the data needed for the "time-until-event" study and includes the following variables computed from the previous ones:

EC: TRUE, if endocarditis has occurred; FALSE, otherwise.
EVENT: True, if an event (re-operation, death, endocarditis) has occurred
DATE_STOP: Final date for the study, computed either from the events (EVENT=TRUE) or as the maximum of the other dates (last consultation, etc.) (EVENT=FALSE).

Source: Centro de Cirurgia Torácica, Hospital de São João, Porto, Portugal.

E.17 Infarct

The Infarct.xls file contains the following measurements performed on 64 patients with myocardial infarction:

EF: Ejection Fraction = (dyastolic volume - systolic volume)/dyastolic volume, evaluated on echocardiographic images.
CK: Maximum value of creatinokynase enzyme (measuring the degree of muscular necrosis of the heart).
IAD: Integral of the amplitude of the QRS spatial vector during abnormal depolarization, measured on the electrocardiogram. The QRS spatial vector is the electrical vector during the stimulation of the heart left ventricle.
GRD: Ventricular gradient = integral of the amplitude of the QRST spatial vector. The QRST spatial vector is the electrical vector during the stimulation of the heart left ventricle, followed by its relaxation back down to the restful state.
SCR: Score (0 to 5) of the necrosis severeness, based on the vectocardiogram.

Source: C Abreu-Lima, Faculdade de Medicina, Universidade do Porto, Porto, Portugal.

E.18 Joints

The Joints.xls file contains 78 measurements of joint surfaces in the granite structure of a Porto street. The variables are:

Phi: Azimuth (°) of the joint.
Theta: Pitch (°) of the joint.

x, y, z: Cartesian co-ordinates corresponding to (Phi, Theta).

Source: C Marques de Sá, Dep. Geologia, Faculdade de Ciências, Universidade do Porto, Porto, Portugal.

E.19 Metal Firms

The Metal Firms.xls file contains benchmarking study results concerning the Portuguese metallurgical industry. The sample is composed of eight firms considered representative of the industrial branch. The data includes scores, percentages and other enquiry results in the following topics:

Leadership;   Process management;
Policy and Strategy;   Client satisfaction;
Social impact;
People management - organizational structure;
People management - policies;
People management - evaluation and development of competence;
Assets management - financial;
Results (objectives, rentability, productivity, investment, growth).

Source: L Ribeiro, Dep. Engenharia Metalúrgica e de Materiais, Faculdade de Engenharia, Universidade do Porto, Porto, Portugal.

E.20 Meteo

The Meteo.xls file contains data of weather variables reported by 25 meteorological stations in the continental territory of Portugal. The variables are:

Pmax: Maximum precipitation (mm) in 1980.
RainDays: Number of rainy days.
T80:   Maximum temperature (ºC) in the year 1980.
T81:   Maximum temperature (ºC) in the year 1981.
T82:   Maximum temperature (ºC) in the year 1982.

Source: INE - Instituto Nacional de Estatística, Portugal.

E.21 Moulds

The Moulds.xls file contains paired measurements performed on 100 moulds of bottle bottoms using three methods:

RC: Ring calibre;

CG: Conic gauge;
EG: End gauges.

Source: J Rademaker, COVEFA, Leerdam, The Netherlands.

E.22 Neonatal

The Neonatal.xls file contains neonatal mortality rates in a sample of 29 Portuguese localities (1980 data). The variables are:

MORT-H: Neonatal mortality rate at home (in 1/1000)
MORT-I: Neonatal mortality rate at Health Centre (in 1/1000)

Source: INE - Instituto Nacional de Estatística, Portugal.

E.23 Programming

The Programming.xls file contains data collected for a pedagogical study concerning the teaching of "Programming in Pascal" to first year Electrical Engineering students. As part of the study, 271 students were enquired during the years 1986-88. The results of the enquiry are summarised in the following variables:

SCORE: Final score in the examinations ([0, 20]).
F: Freshman? 0 = No, 1= Yes.
O: Was Electrical Engineering your first option? 0 = no, 1 = yes.
PROG: Did you learn programming at the secondary school? 0 = no; 1 = scarcely; 2 = a lot.
AB: Did you learn Boole's Algebra in secondary school? 0 = no; 1 = scarcely; 2 = a lot.
BA: Did you learn binary arithmetic in secondary school? 0 = no; 1 = scarcely; 2 = a lot.
H: Did you learn digital systems in secondary school? 0 = no; 1 = scarcely; 2 = a lot.
K: Knowledge factor: 1 if (Prog + AB + BA + H) ≥ 5; 0 otherwise.
LANG: If you have learned programming in the secondary school, which language did you use? 0 = Pascal; 1 = Basic; 2 = other.

Source: J Marques de Sá, Dep. Engenharia Electrotécnica e de Computadores, Faculdade de Engenharia, Universidade do Porto, Porto, Portugal.

E.24 Rocks

The Rocks.xls file contains a table of 134 Portuguese rocks with names, classes, code numbers, values of oxide composition in percentages (SiO_2, ..., TiO_2) and measurements obtained from physical-mechanical tests:

RMCS: Compression breaking load, DIN 52105/E226 standard (kg/cm2).
RCSG: Compression breaking load after freezing/thawing tests, DIN 52105/E226 standard (kg/cm2).
RMFX: Bending strength, DIN 52112 standard (kg/cm2).
MVAP: Volumetric weight, DIN 52102 standard (Kg/m3).
AAPN: Water absorption at NP conditions, DIN 52103 standard (%).
PAOA: Apparent porosity, LNEC E-216-1968 standard (%).
CDLT: Thermal linear expansion coefficient (x 10^{-6}/°C).
RDES: Abrasion test, NP-309 (mm).
RCHQ: Impact test: minimum fall height (cm).

Source: IGM - Instituto Geológico-Mineiro, Porto, Portugal, collected by J Góis, Dep. Engenharia de Minas, Faculdade de Engenharia, Universidade do Porto, Porto, Portugal.

E.25 Signal & Noise

The Signal+Noise worksheet of the Signal & Noise.xls file contains 100 equally spaced values of a noise signal generated with a chi-square distribution, to which were added impulses with arrival times following a Poisson distribution. The amplitudes of the impulses were also generated with a chi-square distribution. The resulting signal with added noise is shown in the Signal+Noise variable.

A threshold value (variable THRESHOLD) can be specified in order to detect the signal impulses. Changing the value of the threshold will change the number of true (Correct Detections variable) and false impulse detections.

The computed sensibility and specificity are shown at the bottom of the Signal+Noise datasheet.

The Data worksheet of the Signal & Noise.xls file contains the data used for ROC curve studies, with column A containing 8 times the signal + noise sequence and column B the true detections for 8 different thresholds (0.8, 1, 2, 3, 4, 5, 6, 7).

Source: J Marques de Sá, Dep. Engenharia Electrotécnica e de Computadores, Faculdade de Engenharia, Universidade do Porto, Porto, Portugal.

E.26 Soil Pollution

The Soil Pollution.xls file contains thirty measurements of Pb-tetraethyl concentrations in ppm (parts per million) collected at different points in the soil of a petrol processing plant in Portugal. The variables are:

x, y, z: Space coordinates in metres (geo-references of Matosinhos Town Hall); z is a depth measurement.

c: Pb-tetraethyl concentration in ppm.

xm, ym: x, y referred to the central (mean) point.

The following computed variables were added to the datasheet:

phi, theta: Longitude and co-latitude of the negative of the local gradient at each point, estimated by 6 methods (M1, M2, M3, R4, R5, R6): M1, M2 and M3 use the resultant of 1, 2 and 3 fastest descent vectors; R4, R5, R6: use linear interpolation of the concentration in 4, 5, and 6 nearest points. A zero value codes a missing value.

Source: A Fiúza, Dep. Engenharia de Minas, Faculdade de Engenharia, Universidade do Porto, Porto, Portugal. The phi and theta angles were computed by J Marques de Sá, Dep. Engenharia Electrotécnica e de Computadores, Faculdade de Engenharia, Universidade do Porto, Porto, Portugal.

E.27 Stars

The Stars.xls file contains measurements of star positions. The stars are from two constellations, Pleiades and Praesepe. To each constellation corresponds a datasheet:

Pleiades (positions of the Pleiades' stars in 1969). Variables:

Hertz   Hertzsprung catalog number
PTV   Photo-visual magnitude
RAh   Right Ascension (h)
RAm   Right Ascension (min)
RAs   Right Ascension (s)
DEd   Declination (deg)
DEm   Declination (arcmin)
DEs   Declination (arcsec)
PHI   Longitude (computed from RAh, RAm and RAs)
THETA   Latitude (computed from DEd, DEm and DEs)

PTV is a dimensionless measure given by $-2.5\log(\text{light energy}) + \text{constant}$. The higher PTV is the lower is the star shine.

Source: Warren Jr WH, US Naval Observatory Pleiades Catalog, 1969.

Praesepe (positions of Praesepe stars measured by Gould BA and Hall A). Variables:

Gah	Right Ascension (h) measured by Gould
Gam	Right Ascension (min) measured by Gould
Gas	Right Ascension (s) measured by Gould
Hah	Right Ascension (h) measured by Hall
Ham	Right Ascension (min) measured by Hall
Has	Right Ascension (s) measured by Hall
Gdh	Declination (deg) measured by Gould
Gdm	Declination (min) measured by Gould
Gds	Declination (s) measured by Gould
Hdh	Declination (deg) measured by Hall
Hdm	Declination (min) measured by Hall
Hds	Declination (s) measured by Hall
Gphi	Longitude according to Gould
Gtheta	Latitude according to Gould
Hphi	Longitude according to Hall
Htheta	Latitude according to Hall

Source: Chase EHS, The Astronomical Journal, 1889.

E.28 Stock Exchange

The Stock Exchange.xls file contains data from daily share values of Portuguese enterprises listed on the Lisbon Stock Exchange Bourse, together with important economic indicators, during the period of June 1, 1999 through August 31, 2000. The variables are:

Lisbor6M:	Bank of Portugal Interest Rate for 6 months.
Euribor6M:	European Interest Rate for 6 months.
BVL30:	Lisbon Stock Exchange index ("Bolsa de Valores de Lisboa").
BCP:	Banco Comercial Português.
BESC:	Banco Espírito Santo.
BRISA:	Road construction firm.
CIMPOR:	Cement firm.
EDP:	Electricity of Portugal Utilities Company.
SONAE:	Large trade firm.
PTEL:	Portuguese telephones.
CHF:	Swiss franc (exchange rate in Euros).
JPY:	Japanese yen (exchange rate in Euros).
USD:	US dollar (exchange rate in Euros).

Source: Portuguese bank newsletter bulletins.

E.29 VCG

The VCG.xls file contains measurements of the mean QRS vector performed in a set of 120 vectocardiograms (VCG).

QRS designates a sequence of electrocardiographic waves occurring during ventricular activation. As the electrical heart vector evolves in time, it describes a curve in a horizontal plane. The mean vector, during the QRS period, is commonly used for the diagnosis of right ventricular hypertrophy.

The mean vector was measured in 120 patients by the following three methods:

H: Half area: the vector that bisects the QRS loop into two equal areas.
A: Amplitude: the vector computed with the maximum amplitudes in two orthogonal directions (x, y).
I: Integral: The vector computed with the signal areas along (x, y).

Source: C Abreu-Lima, Faculdade de Medicina, Universidade do Porto, Porto, Portugal.

E.30 Wave

The Wave.xls file contains eleven angular measurements corresponding to the direction of minimum acoustic pressure in an ultrasonic radiation field, using two types of transducers: TRa and TRb.

Source: D Freitas, Dep. Engenharia Electrotécnica e de Computadores, Faculdade de Engenharia, Universidade do Porto, Porto, Portugal.

E.31 Weather

The Weather.xls file contains measurements of several meteorological variables made in Porto at 12H00 and grouped in the following datasheets:

Data 1:
Weather data refers to the period of January 1, 1999 through August 23, 2000. All measurements were made at 12H00, at "Rua dos Bragas" (Bragas Street), Porto, Portugal. The variables are:

T: Temperature (°C);
H: Humidity (%);
WS: Wind speed (m/s),
WD: Wind direction (anticlockwise, relative to North);
NS: Projection of WD in the North-South direction;
EW: Projection of WD in the East-West direction.

Data 2:
Wind direction measured at "Rua dos Bragas", Porto, Portugal, over several days in the period January 1, 1999 through August 23, 2000 (12H00). The variables are:

WD: Wind direction (anticlockwise, relative to North);
SEASON: 0 = Winter; 1 = Spring; 2 = Summer; 3 = Autumn.

Data 3:
Wind direction measured during March, 1999 at 12H00 in two locations in Porto, Portugal:

WDB: "Bragas" Street, Porto;   WDF: "Formosa" Street, Porto.

Data 4:
Time of occurrence of the maximum daily temperature at "Rua dos Bragas", Porto, for the following months: January, February and July, 2000. The variables are:

Tmax: Maximum temperature (°C).
Time: Time of occurrence of maximum temperature.
TimeNr: Number codifying the time in [0, 1], with 0 = 0:00:00 (12:00:00 AM) and 1 = 23:59:59 (11:59:59 P.M).

Source: "Estação Meteorológica da FEUP" and "Direcção Regional do Ambiente", Porto, Portugal. Compiled by J Góis, Dep. Engenharia de Minas, Faculdade de Engenharia, Universidade do Porto, Porto, Portugal.

E.32 Wines

The Wines.xls file contains the results of chemical analyses performed on 67 Portuguese wines. The WINE column is a label, with the VB code for the white wines (30 cases) and the VT code for the red wines (37 cases). The data sheet gives the concentrations (mg/l) of:

ASP: Aspartame;	GLU: Glutamate;	ASN: Asparagine;
SER: Serine;	GLN: Glutamine;	HIS: Histidine;
GLY: Glycine;	THR: Threonine;	CIT: Citruline;
ARG: Arginine;	ALA: Alanine;	GABA: γ-aminobutyric acid;
TYR: Tyrosine;	ETA: Ethanolamine;	VAL: Valine;
MET: Methionine;	HISTA: Histamine;	TRP: Tryptophan;
METIL: Methylamine;	PHE: Phenylalanine;	ILE: Isoleucine;
LEU: Leucine;	ORN: Ornithine;	LYS: Lysine;
ETIL: Ethylamine;	TIRA: Thyramine;	PUT: Putrescine;
ISO: Isoamilamine;	PRO: Proline;	
TRY+FEN: Tryptamine+β-phenylethylamine		

Source: P Herbert, Dep. Engenharia Química, Faculdade de Engenharia, Universidade do Porto, Porto, Portugal.

Appendix F - Tools

F.1 MATLAB Functions

The functions below, implemented in MATLAB, are available in files with the same function name and suffix ".m". Usually these files should be copied to the MATLAB work directory. All function codes have an explanatory header.

Function (used as)	Described In
k = ainv(rbar,p)	Commands 10.4
[c,df,sig]=chi2test(x)	Commands 5.4
[p,l,u]=ciprop(n0,n1,alpha)	Commands 3.4
[l,u]=civar(v,n,alpha)	Commands 3.5
[r,l,u]=civar2(v1,n1,v2,n2,alpha)	Commands 3.6
c=classmatrix(x,y)	Commands 6.1
h=colatplot(a,kl)	Commands 10.2
as=convazi(a)	Commands 10.3
as=convlat(a)	Commands 10.3
[r,t,tcrit]=corrtest(x,y,alpha)	Commands 4.2
d=dirdif(a,b)	Commands 10.3
g=gammacoef(t)	Commands 2.10
[ko,z,zc]=kappa(x,alpha)	Commands 2.11
h=longplot(a)	Commands 10.2
m = meandir(a,alphal)	Commands 10.3
c=pccorr(x)	Commands 8.1
polar2d(a,mark)	Commands 10.2
polar3d(a)	Commands 10.2
[m,rw,rhow]=pooledmean(a)	Commands 10.2
p=rayleigh(a)	Commands 10.5
[x,y,z,f,t,r] = resultant(a)	Commands 10.3
v=rotate(a)	Commands 10.3
[n1,n2,r,x1,x2]=runs(x,alpha)	Commands 5.1

t=scattermx(a)	Commands 10.3
unifplot(a)	Commands 10.2
[w,wc]=unifscores(a,alpha)	Commands 10.5
f=velcorr(x,icov)	Commands 8.1
f=vmises2cdf(a,k)	Commands 10.4
a=vmises2rnd(n,mu,k)	Commands 10.4
a=vmises3rnd(n,k)	Commands 10.4
delta=vmisesinv(k, p, alpha1)	Commands 10.4
[u2,uc]=watson(a,f,alpha1)	Commands 10.5
[gw,gc]=watsongw(a, alpha)	Commands 10.5
[u2,uc]=watsonvmises(a,alpha1)	Commands 10.5
[fo,fc,k1,k2]=watswill(a1,a2,alpha)	Commands 10.5

F.2 R Functions

The functions below, implemented in R, are available in text files with the same function name and suffix ".txt". An expedite way to use these functions is to copy the respective text and paste it into the R console. All function codes have an explanatory header.

Function (used as)	Described In
o<-cart2pol(x,y); o≡[phi,rho]	Commands 10.1
o<-cart2sph(x,y,z); o≡[phi,theta,rho]	Commands 10.1
o<-cimean(x,alpha=0.05); o≡[l, u]	Commands 3.1
o<-ciprop(n0,n1,alpha); o≡[p,l,u]	Commands 3.4
o<-civar(v,n,alpha=0.05); o≡[l, u]	Commands 3.5
o<-civar2(v1,n1,v2,n2,alpha); o≡[r,l,u]	Commands 3.6
o<-classify(sample,train,group)	Commands 6.1
cm<-classmatrix(x,y)	Commands 6.1
as<-convazi(a)	Commands 10.3
as<-convlat(a)	Commands 10.3
d<-dirdif(a,b)	Commands 10.3
g<-gammacoef(t)	Commands 2.10
o<-kappa(x,alpha); o≡[ko,z,zc]	Commands 2.11
k<-kurtosis(x)	Commands 2.8

r<-pccorr(x)	Commands 8.1
o<-pol2cart(phi,rho); o≡[x, y, z]	Commands 10.1
polar2d(a)	Commands 10.2
p<-rayleigh(a) (resultant(a) must be loaded)	Commands 10.5
o<-resultant(a); o≡[x,y,z,f,t,r]	Commands 10.3
rose(a)	Commands 10.2
o<-runs(x,alpha); o≡[n1,n2,r,x1,x2]	Commands 5.1
s<-skewness(x)	Commands 2.8
o<-sph2cart(phi,theta,rho); o≡[x,y,z]	Commands 10.1
o<-unifscores(a,alpha); o≡[w,wc] (resultant(a) must be loaded)	Commands 10.5
f<-velcorr(x,icov)	Commands 8.1

F.3 Tools EXCEL File

The Tools.xls file has the following data sheets:

Nr of Bins
> Computes the number of histogram bins using the criteria of Sturges, Larson and Scott (see section 2.2.2, for details).

Confidence Intervals
> Computes confidence intervals for a proportion and a variance (see sections 3.3 and 3.4, for details).

Correlation Test
> Computes the 5% critical value for the correlation test (see section 4.4.1, for details).

Broken Stick
> Computes the expected length percentage of the kth largest segment of a stick, with total length one, randomly broken into d segments (see section 8.2, for details).

The Macros of the Tools.xls EXCEL file must be enabled in order to work adequately (use security level Medium in the Macro Security button of the EXCEL Options menu).

F.4 SCSize Program

The SCSize program displays a picture box containing graphics of the following variables, for a two-class linear classifier with specified Battacharrya distance

(Mahalanobis distance of the means) and for several values of the dimensionality ratio, n/d:

Bayes error;
Expected design set error (resubstitution method);
Expected test set error (holdout method).

Both classes are assumed to be represented by the same number of patterns per class, n.

The user only has to specify the dimension d and the square of the Battacharrya distance (computable by several statistical software products).

For any chosen value of n/d, the program also displays the standard deviations of the error estimates when the mouse is clicked over a selected point of the picture box.

The expected design and test set errors are computed using the formulas presented in the work of Foley (Foley, 1972). The formula for the expected test set error is an approximation formula, which can produce slightly erroneous values, below the Bayes error, for certain n/d ratios.

The program is installed in the Windows standard way.

References

Chapters 1 and 2

Anderson TW, Finn JD (1996), The New Statistical Analysis of Data. Springer-Verlag New York, Inc.
Beltrami E (1999), What is Random? Chance and Order in Mathematics and Life. Springer-Verlag New York, Inc.
Bendat JS, Piersol AG (1986), Random Data Analysis and Measurement Procedures. Wiley, Interscience.
Biran A, Breiner M (1995), MATLAB for Engineers, Addison-Wesley Pub. Co. Inc.
Blom G (1989), Probability and Statistics, Theory and Applications. Springer-Verlag New York Inc.
Buja A, Tukey PA (1991), Computing and Graphics in Statistics. Springer-Verlag.
Chatfield C (1981), Statistics for Technology. Chapman & Hall Inc.
Cleveland WS (1984), Graphical Methods for Data Presentation: Full Scale Breaks, Dot Charts, and Multibased Logging. The American Statistician, 38:270-280.
Cleveland WS (1984), Graphs in Scientific Publications. The American Statistician, 38, 270-280.
Cox DR, Snell EJ (1981), Applied Statistics. Chapman & Hall Inc.
Dalgaard P (2002), Introductory Statistics with R. Springer-Verlag.
Dixon WJ, Massey Jr. FJ (1969), Introduction to Statistical Analysis. McGraw Hill Pub. Co.
Foster JJ (1993), Starting SPSS/PC+ and SPSS for Windows. Sigma Press.
Gilbert N (1976), Statistics. W. B. Saunders Co.
Green SB, Salkind NJ, Akey TM (1997), Using SPSS for Windows. Analyzing and Understanding Data. Prentice-Hall, Inc.
Hoel PG (1976), Elementary Statistics. John Wiley & Sons Inc., Int. Ed.
Iversen GR (1997), Statistics, The Conceptual Approach. Springer-Verlag.
Jaffe AJ, Spirer HF (1987), Misused Statistics, Straight Talk for Twisted Numbers. Marcel Dekker, Inc.
Johnson RA, Bhattacharyya GK (1987), Statistics, Principles & Methods. John Wiley & Sons, Inc.
Johnson RA, Wichern DW (1992) Applied Multivariate Statistical Analysis. Prentice-Hall International, Inc.
Larson HJ (1975), Statistics: An Introduction. John Wiley & Sons, Inc.
Martinez WL, Martinez AR (2002), Computational Statistics Handbook with MATLAB®. Chapman & Hall/CRC.
Meyer SL (1975), Data Analysis for Scientists and Engineers. John Wiley & Sons, Inc.
Milton JS, McTeer PM, Corbet JJ (2000), Introduction to Statistics. McGraw Hill Coll. Div.
Montgomery DC (1984), Design and Analysis of Experiments. John Wiley & Sons, Inc.
Mood AM, Graybill FA, Boes DC (1974), Introduction to the Theory of Statistics. McGraw-Hill Pub. Co.

Nie NH, Hull CH, Jenkins JG, Steinbrenner K, Bent DH (1970), Statistical Package for the Social Sciences. McGraw Hill Pub. Co.

Salsburg D (2001), The Lady Tasting Tea: How Statistics Revolutionized Science in the Twentieth Century. W H Freeman & Co.

Sanders DH (1990), Statistics. A Fresh Approach. McGraw-Hill Pub. Co.

Scott DW (1979), On Optimal and Data-Based Histograms. Biometrika, 66:605-610.

Sellers GR (1977), Elementary Statistics. W. B. Saunders Co.

Spiegel MR, Schiller J, Srinivasan RA (2000), Schaum's Outline of Theory and Problems of Probability and Statistics. McGraw-Hill Pub. Co.

Sturges HA (1926), The Choice of a Class Interval. J. Am. Statist. Assoc., 21:65-66.

Venables WN, Smith DM and the R Development Core Team (2005), An Introduction to R. http://www.r-project.org/

Waller RA (1979), Statistics. An Introduction to Numerical Reasoning. Holden-Day Inc.

Wang C (1993), Sense and Nonsense of Statistical Inference, Controversy, Misuse and Subtlety. Marcel Dekker, Inc.

Chapters 3, 4 and 5

Andersen EB (1997), Introduction to the Statistical Analysis of Categorical Data. Springer-Verlag.

Anderson TW, Finn JD (1996), The New Statistical Analysis of Data. Springer-Verlag New York, Inc.

Barlow RE, Proschan F (1975), Statistical Theory of Reliability and Life Testing. Holt, Rinehart & Winston, Inc.

Beltrami E (1999), What is Random? Chance and Order in Mathematics and Life. Springer-Verlag New York, Inc.

Bishop YM, Fienberg SE, Holland PW (1975), Discrete Multivariate Analysis, Theory and Practice. The MIT Press.

Blom G (1989), Probability and Statistics, Theory and Applications. Springer-Verlag New York Inc.

Box GEP, Hunter JS, Hunter WG (1978), Statistics for Experimenters: An Introduction to Design, Data Analysis and Model Building. John Wiley & Sons, Inc.

Chow SL (1996), Statistical Significance, Rationale, Validity and Utility. Sage Publications Ltd.

Cohen J (1983), Statistical Power Analysis for the Behavioral Sciences (2nd ed.). Lawrence Erlbaum Associates, Publishers.

Conover WJ (1980), Practical Nonparametric Statistics. John Wiley & Sons, Inc.

D'Agostino RB, Stephens MA (1986), Goodness-of-Fit Techniques. Marcel Dekker Inc.

Dixon WJ, Massey Jr. FJ (1969), Introduction to Statistical Analysis. McGraw-Hill Pub. Co.

Dodge Y (1985), Analysis of Experiments with Missing Data. John Wiley & Sons, Inc.

Dudewicz EJ, Mishra SN (1988), Modern Mathematical Statistics. John Wiley & Sons, Inc.

Efron B (1979), Bootstrap Methods: Another Look at the Jackknife. Ann. Statist., 7, pp. 1-26.

Efron B (1982), The Jackknife, the Bootstrap and Other Resampling Plans. Society for Industrial and Applied Mathematics, SIAM CBMS-38.

Efron B, Tibshirani RJ (1993), An Introduction to the Bootstrap. Chapman & Hall/CRC.

Everitt BS (1977), The Analysis of Contingency Tables. Chapman & Hall, Inc.

Gardner MJ, Altman DG (1989), Statistics with Confidence – Confidence Intervals and Statistical Guidelines. British Medical Journal.

Gibbons JD (1985), Nonparametrical Statistic Inference. Marcel Dekker, Inc.

Hesterberg T, Monaghan S, Moore DS, Clipson A, Epstein R (2003), Bootstrap Methods and Permutation Tests. Companion Chapter 18 to the Practice of Business Statistics. W. H. Freeman and Co.

Hettmansperger TP (1984), Statistical Inference Based on Ranks. John Wiley & Sons, Inc.

Hoel PG (1976), Elementary Statistics. John Wiley & Sons, Inc., Int. Ed.

Hollander M, Wolfe DA (1973), Nonparametric Statistical Methods. John Wiley & Sons, Inc.

Iversen GR (1997), Statistics. The Conceptual Approach. Springer-Verlag.

James LR, Mulaik SA, Brett JM (1982), Causal Analysis. Assumptions, Models and Data. Sage Publications Ltd.

Kachigan SK (1986), Statistical Analysis. Radius Press.

Kanji GK (1999), 100 Statistical Tests. Sage Publications Ltd.

Kenny DA (1979), Correlation and Causality. John Wiley & Sons, Inc.

Lavalle IH (1970), An Introduction to Probability, Decision and Inference. Holt, Rinehart & Winston, Inc.

Lindman HR (1974), Analysis of Variance in Complex Experimental Designs. W.H. Freeman & Co.

Mason RL, Gunst RF, Hess JL (1989), Statistical Design and Analysis of Experiments with Applications to Engineering and Science. John Wiley & Sons, Inc.

Milton JS, McTeer PM, Corbet JJ (2000), Introduction to Statistics. McGraw Hill College Div.

Montgomery DC (1984), Design and Analysis of Experiments. John Wiley & Sons, Inc.

Montgomery DC (1991), Introduction to Statistical Quality Control. John Wiley & Sons, Inc.

Mood AM, Graybill FA, Boes DC (1974), Introduction to the Theory of Statistics. McGraw-Hill Pub. Co.

Murphy KR, Myors B (1998), Statistical Power Analysis. Lawrence Erlbaum Associates, Publishers.

Randles RH, Wolfe DA (1979), Introduction to the Theory of Nonparametric Statistics. Wiley.

Sachs L (1982), Applied Statistics. Springer-Verlag New York, Inc.

Sanders DH (1990), Statistics, A Fresh Approach. McGraw-Hill Pub. Co.

Sellers GR (1977), Elementary Statistics. W. B. Saunders Co.

Shapiro SS, Wilk SS, Chen SW (1968), A comparative study of various tests for normality. J Am Stat Ass, 63:1343-1372.

Siegel S, Castellan Jr. NJ (1998), Nonparametric Statistics for the Behavioral Sciences. McGraw Hill Book Co.

Spanos A (1999), Probability Theory and Statistical Inference – Econometric Modeling with Observational Data. Cambridge University Press.

Spiegel MR, Schiller J, Srinivasan RA (2000), Schaum's Outline of Theory and Problems of Probability and Statistics. McGraw-Hill Pub. Co.

Sprent P (1993), Applied Non-Parametric Statistical Methods. CRC Press.

Waller RA (1979), Statistics, An Introduction to Numerical Reasoning. Holden-Day, Inc.

Wang C (1993), Sense and Nonsense of Statistical Inference, Controversy, Misuse and Subtlety. Marcel Dekker, Inc.

Wilcox RR (2001), Fundamentals of Modern Statistical Methods. Springer-Verlag.

Chapter 6

Argentiero P, Chin R, Baudet P (1982), An Automated Approach to the Design of Decision Tree Classifiers. IEEE Tr. Patt. An. Mach. Intel., 4:51-57.

Bell DA (1978), Decision Trees, Tables and Lattices. In: Batchelor BG (ed) Case Recognition, Ideas in Practice. Plenum Press, New York, pp. 119-141.

Breiman L, Friedman JH, Olshen RA, Stone CJ (1993), Classification and Regression Trees. Chapman & Hall / CRC.

Centor RM (1991), Signal Detectability: The Use of ROC Curves and Their Analyses. Medical Decision Making, 11:102-106.

Chang CY (1973), Dynamic Programming as Applied to Feature Subset Selection in a Pattern Recognition System. IEEE Tr. Syst. Man and Cybern., 3:166-171.

Cooley WW, Lohnes PR (1971), Multivariate Data Analysis. Wiley.

Devijver PA (1982), Statistical Pattern Recognition. In: Fu KS (ed) Applications of Case Recognition, CRC Press Inc., pp. 15-35.

Duda RO, Hart PE (1973), Pattern Classification and Scene Analysis. J. Wiley & Sons, Inc.

Dudewicz EJ, Mishra SN (1988), Modern Mathematical Statistics. John Wiley & Sons, Inc.

Foley DH (1972), Considerations of Sample and Feature Size. IEEE Tr. Info. Theory, 18:618-626.

Fu KS (1982), Introduction. In: Fu KS (ed) Applications of Pattern Recognition. CRC Press Inc., pp. 2-13.

Fukunaga K (1969), Calculation of Bayes' Recognition Error for Two Multivariate Gaussian Distributions. IEEE Tr. Comp., 18:220-229.

Fukunaga K (1990), Introduction to Statistical Pattern Recognition. Academic Press.

Fukunaga K, Hayes RR (1989a), Effects of Sample Size in Classifier Design. IEEE Tr. Patt. Anal. Mach. Intel., 11:873-885.

Fukunaga K, Hayes RR (1989b), Estimation of Classifier Performance. IEEE Tr. Patt. Anal. Mach. Intel., 11:1087-1101.

Jain AK, Chandrasekaran B (1982), Dimensionality and Sample Size Considerations in Pattern Recognition. In: Krishnaiah PR, Kanal LN (eds) Handbook of Statistics, 2, North Holland Pub. Co., pp. 835-855.

Jain AK, Duin RPW, Mao J (2000), Statistical Pattern Recognition: A Review. IEEE Tr. Patt. Anal. Mach. Intel., 1:4-37.

Kittler J (1978), Feature Set Search Algorithms. In (Chen CH ed): Pattern Recognition and Signal Processing, Noordhoff Pub. Co.

Klecka WR (1980), Discriminant Analysis. Sage Publications Ltd.

Loh, WY, Shih YS (1997), Split Selection Methods for Classification Trees. Statistica Sinica, vol. 7, 815-840.

Lusted L (1978), General Problems in Medical Decision Making with Comments on ROC Analysis. Seminars in Nuclear Medicine, 8:299-306.

Marques de Sá JP (2001), Patten Recognition, Concepts, Methods and Applications. Springer-Verlag.

Metz CE (1978), Basic Principles of ROC Analysis. Seminars in Nuclear Medicine, 8:283-298.

Metz CE, Goodenough DJ, Rossmann K (1973), Evaluation of Receiver Operating Characteristic Curve Data in Terms of Information Theory, with Applications in Radiography. Radiology, 109:297-304.

Mucciardi AN, Gose EE (1971), A Comparison of Seven Techniques for Choosing Subsets of Pattern Recognition Properties. IEEE Tr. Comp., 20:1023-1031.

Raudys S, Pikelis V (1980), On dimensionality, sample size, classification error and complexity of classification algorithm in case recognition. IEEE Tr. Patt. Anal. Mach. Intel., 2:242-252.

Sharma S (1996), Applied Multivariate Techniques. John Wiley & Sons, Inc.

Swain PH (1977), The Decision Tree Classifier: Design and Potential. IEEE Tr. Geosci. Elect., 15:142-147.

Swets JA (1973), The Relative Operating Characteristic in Psychology. Science, 182:990-1000.

Tabachnick BG, Fidell LS (1989), Using Multivariate Statistics. Harper & Row Pub., Inc.

Toussaint GT (1974), Bibliography on Estimation of Misclassification. IEEE Tr. Info. Theory, 20:472-479.

Chapter 7

Aldrich JH, Nelson FD (1984), Linear Probability, Logit, and Probit models. Sage Publications Ltd.

Anderson JM (1982), Logistic Discrimination. In: Krishnaiah PR, Kanal LN (eds) Handbook of Statistics vol. 2, North Holland Pub. Co., 169-191.

Bates DM, Watts DG (1988), Nonlinear Regression Analysis and its Applications. John Wiley & Sons, Inc.

Bronson R (1991), Matrix Methods. An Introduction. Academic Press, Inc.

Box GE, Hunter JS, Hunter WG (1978), Statistics for Experimenters: An Introduction to Design, Data Analysis and Model Building. John Wiley & Sons.

Cooley WW, Lohnes PR (1971), Multivariate Data Analysis. Wiley.

Darlington RB (1990), Regression and Linear Models. McGraw-Hill Pub. Co..

Dixon WJ, Massey FJ (1983), Introduction to Statistical Analysis. McGraw Hill Pub. Co.

Draper NR, Smith H (1966), Applied Regression Analysis. John Wiley & Sons, Inc.

Dudewicz EJ, Mishra SN (1988), Modern Mathematical Statistics. John Wiley & Sons, Inc.

Kleinbaum DG, Kupper LL, Muller KE (1988), Applied Regression Analysis and Other Multivariate Methods (2nd Edition). PWS-KENT Pub. Co.

Mason RL, Gunst RF, Hess JL (1989), Statistical Design and Analysis of Experiments with Applications to Engineering and Science. John Wiley & Sons, Inc.

Mendenhall W, Sincich T (1996), A Second Course in Business Statistics – Regression Analysis. Prentice Hall, Inc.

Seber GA, Wild CJ (1989) Nonlinear Regression. John Wiley & Sons, Inc.

Tabachnick BG, Fidell LS (1989) Using Multivariate Statistics. Harper & Row Pub., Inc.

Chapter 8

Cooley WW, Lohnes PR (1971), Multivariate Data Analysis. Wiley.

Fukunaga K (1990), Introduction to Statistical Pattern Recognition. Academic Press, Inc.

Jambu M (1991), Exploratory and Multivariate Data Analysis. Academic Press, Inc.

Jackson JE (1991), A User's Guide to Principal Components. John Wiley & Sons, Inc.

Johnson M (1991), Exploratory and Multivariate Data Analysis. Academic Press, Inc.

Johnson RA, Wichern DW (1992), Applied Multivariate Statistical Analysis. Prentice-Hall International, Inc.

Jolliffe IT (2002), Principal Component Analysis (2nd ed.).Springer Verlag.

Loehlin JC (1987), Latent Variable Models: An Introduction to Latent, Path, and Structural Analysis. Erlbaum Associates, Publishers.

Manly BF (1994), Multivariate Statistical Methods. A Primer. Chapman & Hall, Inc.

Morisson DF (1990), Multivariate Statistical Methods. McGraw-Hill Pub. Co.

Sharma S (1996), Applied Multivariate Techniques. John Wiley & Sons, Inc.

Velicer WF, Jackson DN (1990), Component Analysis vs. Factor Analysis: Some Issues in Selecting an Appropriate Procedure. Multivariate Behavioral Research, 25, 1-28.

Chapter 9

Chatfield C (1981), Statistics for Technology (2nd Edition). Chapman & Hall, Inc.

Collet D (1994), Modelling Survival Data in Medical Research. Chapman & Hall, Inc.

Cox DR, Oakes D (1984), Analysis of Survival Data. Chapman & Hall, Inc.

Dawson-Saunders B, Trapp RG (1994), Basic & Clinical Biostatistics. Appleton & Lange.

Dudewicz EJ, Mishra SN (1988), Modern Mathematical Statistics. John Wiley & Sons, Inc.

Elandt-Johnson RC, Johnson NL (1980), Survival Models and Data Analysis. John Wiley & Sons, Inc.

Feigl P, Zelen M (1965), Estimation of Exponential Survival Probabilities with Concomitant Information. Biometrics, 21, 826- 838.

Gehan EA, Siddiqui MM (1973), Simple Regression Methods for Survival Time Studies. Journal Am. Stat. Ass., 68, 848-856.

Gross AJ, Clark VA (1975), Survival Distributions: Reliability Applications in the Medical Sciences. John Wiley & Sons, Inc.

Hahn GJ, Shapiro SS (1967), Statistical Models in Engineering. John Wiley & Sons, Inc.

Kleinbaum DG, Klein M (2005), Survival Analysis, A Self-Learning Text (2nd ed.). Springer Verlag

Miller R (1981), Survival Data. John Wiley & Sons, Inc.

Rosner B (1995), Fundamentals of Biostatistics. Duxbury Press, Int. Thomson Pub. Co.

Chapter 10

Fisher NI, Best DJ (1984), Goodness-of-Fit Tests for Fisher's Distribution on the Sphere. Austral. J. Statist., 26:142-150.

Fisher NI, Lewis T, Embleton BJJ (1987), Statistical Analysis of Spherical Data. Cambridge University Press.

Greenwood JA, Durand D (1955), The Distribution of Length and Components of the Sum of n Random Unit Vectors. Ann. Math. Statist., 26:233-246.

Gumbel EJ, Greenwood JA, Durand D (1953), The Circular Normal Distribution: Theory and Tables. J. Amer. Statist. Assoc., 48:131:152.

Hill GW (1976), New Approximations to the von Mises Distribution. Biometrika, 63:676-678.

Hill GW (1977), Algorithm 518. Incomplete Bessel Function I_0: The Von Mises Distribution. ACM Tr. Math. Software, 3:270-284.

Jammalamadaka SR (1984), Nonparametric Methods in Directional Data Analysis. In: Krishnaiah PR, Sen PK (eds), Handbook of Statistics, vol. 4, Elsevier Science B.V., 755-770.

Kanji GK (1999), 100 Statistical Tests. Sage Publications Ltd.

Mardia KV, Jupp PE (2000), Directional Statistics. John Wiley and Sons, Inc.

Schou G (1978,) Estimation of the Concentration Parameter in von Mises Distributions. Biometrika, 65:369-377.

Upton GJG (1973), Single-Sample Test for the von Mises Distribution. Biometrika, 60:87-99.

Upton GJG (1986), Approximate Confidence Intervals for the Mean Direction of a von Mises Distribution. Biometrika, 73:525-527.

Watson GS, Williams EJ (1956), On the Construction of Significance Tests on the Circle and the Sphere. Biometrika, 48:344-352.

Wilkie D (1983), Rayleigh Test for Randomness of Circular Data. Appl. Statist., 7:311-312.

Wood, ATA (1994), Simulation of the von Mises Fisher Distribution. Comm. Statist. Simul., 23:157-164.

Zar JH (1996), Biostatistical Analysis. Prentice Hall, Inc.

Appendices A, B and C

Aldrich, JH, Nelson FD (1984), Linear probability, logit, and probit models. Sage Publications Ltd.

Blom G (1989), Probability and Statistics, Theory and Applications. Springer-Verlag New York Inc.

Borel E, Deltheil R, Huron R (1964), Probabilités. Erreurs. Collection Armand Colin.

Brunk HD (1975), An Introduction to Mathematical Statistics. Xerox College Pub.

Burington RS, May DC (1970), Handbook of Probability and Statistics with Tables. McGraw-Hill Pub. Co.

Chatfield C (1981), Statistics for Technology. Chapman & Hall Inc.

Dudewicz EJ, Mishra SN (1988), Modern Mathematical Statistics. John Wiley & Sons, Inc.

Dwass M (1970), Probability. Theory and Applications. W. A. Benjamin, Inc.

Feller W (1968), An Introduction to Probability Theory and its Applications. John Wiley & Sons, Inc.

Galambos J (1984), Introduction to Probability Theory. Marcel Dekker, Inc.

Johnson NL, Kotz S (1970), Discrete Distributions. John Wiley & Sons, Inc.

Johnson NL, Kotz S (1970), Continuous Univariate Distributions (vols 1, 2). John Wiley & Sons, Inc.

Lavalle IH (1970), An Introduction to Probability, Decision and Inference. Holt, Rinehart & Winston, Inc.

Mardia KV, Jupp PE (1999), Directional Statistics. John Wiley & Sons, Inc.

Papoulis A (1965), Probability, Random Variables and Stochastic Processes, McGraw-Hill Pub. Co.

Rényi A (1970), Probability Theory. North Holland Pub. Co.

Ross SM (1979), Introduction to Probability Models. Academic Press, Inc.

Spanos A (1999), Probability Theory and Statistical Inference – Econometric Modeling with Observational Data. Cambridge University Press.

Wilcox RR (2001), Fundamentals of Modern Statistical Methods. Springer-Verlag.

Index

A

accuracy, 82
actuarial table, 355
adjusted prevalences, 239
alternative hypothesis, 111
ANOVA, 142
  one-way, 146
  tests, 285
  two-way, 156
AS analysis, 120
average risk, 238, 240

B

backward search, 254, 304
bar graph, 43
baseline hazard, 372
Bayes classifier, 235
Bayes' Theorem, 409, 426
Bernoulli trial, 92, 431
beta coefficient, 275
beta function, 446
Bhattacharyya distance, 242, 254
bias, 82, 223, 455
binomial distribution, 419
bins, 47
bootstrap
  confidence interval, 101
  distribution, 100
  sample, 100
  standard error, 100
broken stick model, 337

C

CART method, 264
cases, 5, 29
category, 133, 143, 156
causality, 129
censored cases, 355
Central Limit Theorem, 428
central moments, 416

chaos, 3
Chebyshev Theorem, 418
circular plot, 377
circular variance, 381, 453
classification
  matrix, 230
  risk, 237
coefficient of determination, 276
co-latitude, 375
  plot, 388, 394
Commands
  2.1 (freq. tables), 41
  2.2 (bar graphs), 43
  2.3 (histograms), 51
  2.4 (cross tables), 54
  2.5 (scatter plots), 54
  2.6 (box plot), 57
  2.7 (measures of location), 58
  2.8 (spread and shape), 62
  2.9 (association), 69
  2.10 (assoc. of ordinal var.), 72
  2.11 (assoc. of nominal var.), 73
  3.1 (conf. int. of mean), 89
  3.2 (case selection), 90
  3.3 (quantiles), 92
  3.4 (conf. int. prop.), 95
  3.5 (conf. int. variance), 97
  3.6 (conf. int. var. ratio), 99
  3.7 (bootstrap), 106
  4.1 (single mean t test), 124
  4.2 (correlation test), 128
  4.3 (independent samples t test), 137
  4.4 (paired samples t test), 141
  4.5 (one-way ANOVA), 149
  4.6 (two-way ANOVA), 165
  5.1 (runs test), 174
  5.2 (case weighing), 177
  5.3 (binomial test), 178
  5.4 (chi-square test), 183
  5.5 (goodness of fit), 185
  5.6 (distribution plots), 186
  5.7 (contingency table tests), 192
  5.8 (two indep. samples tests), 201

5.9 (two paired samples tests), 205
5.10 (Kruskal-Wallis test), 212
5.11 (Friedmann test), 215
6.1 (discriminant analysis), 233
6.2 (ROC curve), 252
6.3 (tree classifiers), 268
7.1 (simple linear regression), 277
7.2 (ANOVA test), 286
7.3 (polynomial, non-linear regr.), 301
7.4 (stepwise regression), 305
7.5 (regression diagnostics), 307
7.6 (ridge regression), 322
7.7 (logit, probit regression), 327
8.1 (pc and factor analysis), 335
9.1 (survival analysis), 358
10.1 (direct. data conversion), 376
10.2 (directional data plots), 379
10.3 (direct. data descriptives), 382
10.4 (von Mises distributions), 387
10.5 (directional data tests), 391
common factors, 347
communality, 347
compound experiment, 408
concentration parameter, 381, 446, 453
concordant pair, 71
conditional distribution, 425
conditional probability, 406
confidence
    interval, 83
    level, 13, 83, 113, 420
    limits, 83
    risk, 83
consistency, 455
contingency table, 52, 189
continuity correction, 175
continuous random variable, 411
contrasts, 151, 162
control chart, 88
control group, 133
convolution, 427
Cook's distance, 307
correlation, 66, 425
    coefficient, 66
    matrix, 67
    Pearson, 127
    rank, 69
    Spearman, 69, 198
covariance, 330, 425
    matrix, 228, 425
Cox regression, 371
critical
    region, 114
    value, 125

cross table, 52, 54
cross-validation, 257, 258
cumulative distribution, 184

D

data
    deterministic, 1
    discrete, 40
    grouped, 56
    missing, 31, 40
    random, 2
    rank, 10
    sorting, 35
    spreadsheet, 29
    transposing, 37
dataset
    Breast Tissue, 152, 260, 469
    Car Sale, 354, 469
    Cells, 470
    Clays, 213, 324, 470
    Cork Stoppers, 48, 60, 63, 67, 70, 87,
        88, 96, 146, 181, 214, 226, 254,
        274, 332, 341, 471
    CTG, 98, 472
    Culture, 473
    Fatigue, 358, 366, 473
    FHR, 76, 209, 217, 474
    FHR-Apgar, 161, 252, 474
    Firms, 475
    Flow Rate, 475
    Foetal Weight, 291, 304, 315, 475
    Forest Fires, 173, 476
    Freshmen, 52, 74, 94, 177, 181, 191,
        194, 214, 476
    Heart Valve, 361, 368, 477
    Infarct, 478
    Joints, 376, 378, 385, 395, 478
    Metal Firms, 207, 216, 479
    Meteo, 29, 40, 123, 126, 127, 479
    Moulds, 479
    Neonatal, 480
    Programming, 196, 204, 247, 480
        247, 480
    Rocks, 339, 345, 481
    Signal & Noise, 249, 481
    Soil Pollution, 394, 399, 482
    Stars, 482
    Stock Exchange, 302, 483
    VCG, 379, 484
    Wave, 484
    Weather, 378, 390, 396, 484
    Wines, 135, 204, 485

De Moivre's Theorem, 420
decile, 60
decision
   function, 223
   region, 223
   rule, 223, 261
   threshold, 112
   tree, 259
declination, 375
degrees of freedom, 63, 96, 448, 451
deleted residuals, 307
density function, 13, 412
dependent samples, 133
dimensional reduction, 330, 337
dimensionality ratio, 243
discordant pair, 71
discrete random variable, 411
distribution
   Bernoulli, 431
   Beta, 446
   binomial, 12, 93, 419, 435
   chi-square, 96, 180, 448
   circular normal, 452
   exponential, 353, 367, 442
   F, 97, 129, 146, 451
   function, 11, 13, 411
   Gamma, 445
   Gauss, 13, 420, 441
   geometric, 433
   hypergeometric, 365, 434
   multimodal, 60
   multinomial, 179, 436
   normal, 13, 420, 441
   Poisson, 438
   Rayleigh, 445
   Student's t, 86, 118, 122, 449
   uniform, 413, 432, 439
   von Mises, 383, 452
   von Mises-Fisher, 383, 453
   Weibull, 353, 369, 444
dynamic search, 254

E

effects, 132, 142
   additive, 157
   interaction, 159
eigenvalue, 331, 393
eigenvector, 331, 393
elevation, 375
empirical distribution, 183
equality of variance, 143

ergodic process, 8
error, 272
   bias, 256
   experimental, 143, 157, 159
   function, 421
   mean square, 144
   probability, 242
   proportional reduction of, 75
   root mean square, 63
   standard deviation, 244
   sum of squares, 143, 275
   test set, 243
   training set, 230, 243
   type I, 113
   type II, 115
   variance, 256
expectation, 414
explanatory variable, 371
exponential regression, 301
exposed group, 364
extra sums of squares, 296

F

factor (R), 150
factor loadings, 339, 347
factorial experiment, 158
factors, 132, 142, 156
failure rate, 353
feature selection, 253
Fisher coefficients, 230
fixed factors, 142
forward search, 253, 304
frequency, 7
   absolute, 11, 40, 59, 403
   relative, 11, 40, 403
   table, 48
full model, 287, 299

G

gamma function, 445
gamma statistic, 198
Gauss' approximation formulae, 417
Gaussian distribution, 420
generalised variance, 332
Gini index, 263
Goodman and Kruskal lambda, 199
goodness of fit, 179, 183, 187
grand total, 160
Greenwood's formula, 362
group variable, 132
Guttman-Kaiser criterion, 337

H

hazard function, 353
hazard ratio, 372
hierarchical classifier, 259
histogram, 48, 51
holdout method, 257
Hotteling's T^2, 333
hyperellisoid, 228
hyperplane, 224, 226

I

inclination, 375
independent events, 406
independent samples, 132
index of association, 199
intercept, 272
inter-quartile range, 57, 60, 62, 412
interval estimate, 14, 81
interval estimation
    one-sided, 83
    two-sided, 83

J

joint distribution, 422

K

Kaiser criterion, 337
Kaplan-Meier estimate, 359
kappa statistic, 200
Kolmogorov axioms, 404
Kruskal-Wallis test, 212
kurtosis, 65

L

lack of fit sum of squares, 287
Laplace rule, 405
large sample, 87
Larson's formula, 49
latent variable, 348
Law of Large Numbers, 419
least square error, 273
leave-one-out method, 257
life table, 355
likelihood, 235
likelihood function, 456
linear
    classifier, 232
    discriminant, 224
    regression, 272

log-cumulative hazard, 370
logit model, 322
log-likelihood, 324
longitude plot, 394
loss matrix, 238
lower control limit, 88
LSE, 273

M

Mahalanobis distance, 228
manifest variables, 348
Mantel-Haenszel procedure, 365
marginal distribution, 423
matched samples, 133
maximum likelihood estimate, 456
mean, 13, 58, 415
    direction, 380
    ensemble, 8
    estimate, 85
    global, 158
    population, 7
    response, 273
    resultant, 380
    sample, 7
    temporal, 8
    trimmed, 59
median, 57, 59, 60, 412
merit criterion, 253
minimum risk, 238
ML estimate, 456
mode, 60
modified Levene test, 309
moment generating function, 417
moments, 416, 425
MSE, 275
MSR, 285
multicollinearity, 300, 307
multiple
    correlation, 254, 293
    R square, 276
    regression, 289
multivariate distribution, 422

N

new observations, 283
node impurity, 263
node splitting, 265
non-linear regression, 301
normal
    distribution, 420
    equations, 273

probability plot, 184
regression, 279
sequences, 441
null hypothesis, 111

O

observed significance, 114, 124
orthogonal experiment, 157
orthonormal matrix, 331
outliers, 306

P

paired differences, 139
paired samples, 132
parameter estimation, 81
partial correlations, 297
partial F test, 299
partition, 409
partition method, 257
pc scores, 331
pdf, 412
PDF, 411
Pearson correlation, 276
percentile, 60, 122
phi coefficient, 199
plot
    3D plot, 54, 55
    box plot, 57
    box-and-whiskers plot, 57
    categorized, 56
    scatter plot, 54, 55
point estimate, 14, 81, 82, 455
point estimator, 82, 455
polar vector, 453
polynomial regression, 300
pooled
    covariance, 241
    mean, 398
    variance, 131
posterior probability, 239, 409
post-hoc comparison, 150, 151
power, 115
    curve, 116
    one-way ANOVA, 154
    two-way ANOVA, 164
power-efficiency, 171
predicted values, 273
predictor, 271
predictor correlations, 291
prevalence, 234, 409

principal component, 330
principal factor, 348
prior probability, 409
probability, 404
    density, 12
    function, 11
    space, 404
    distribution, 411
probit model, 322
product-limit estimate, 359
proportion estimate, 92
proportion reduction of error, 199
proportional hazard, 366, 371
prototype, 225
pure error sum of squares, 287

Q

quadratic classifier, 232, 241
quality control, 333
quantile, 60, 412
quartile, 60, 412

R

random
    data, 2
    error, 82
    number, 4
    process, 2
    sample, 7, 81
    variable, 5, 8, 410
    experiment, 403
range, 62
rank correlation, 69
reduced model, 287, 299
regression, 271
regression sum of squares, 285
reliability function, 353
repeated measurements, 158
repeated samples, 282
replicates, 286
residuals, 273
response, 205
resubstitution method, 257
risk, 238
ROC
    curve, 246, 250
    threshold, 251
rose diagram, 377
RS analysis, 119

S

sample, 5
  mean, 416
  size, 14
  space, 403
  standard deviation, 417
  variance, 417
sample mean
  global, 143
sampled population, 81
samples
  independent, 132
  paired, 132
sampling distribution, 14, 83, 114
  correlation, 127
  gamma, 198
  kappa statistic, 200
  Mann-Whitney W, 203
  mean, 86, 122
  phi coefficient, 199
  proportion, 175
  range of means, 151
  Spearman's correlation, 198
  two independent samples, 134
  two paired samples, 139
  variance, 96, 126
  variance ratio, 97, 129
scale parameter, 444
scatter matrix, 393
Scott's formula, 49
scree test, 337
semistudentised residuals, 306
sensibility, 247
sequential search, 253
shape parameter, 444
sigmoidal functions, 323
significance level, 13, 111, 114
significant digits, 61
skewness, 64
  negative, 64
  positive, 64
slope, 272
small sample, 87
Spearman's correlation, 69, 198
specificity, 247
spherical mean direction, 381
spherical plot, 377
spherical variance, 381, 453
split criterion, 263
SSE, 275
SSPE, 287

SSR, 285
SST, 276
standard
  deviation, 13, 57, 63, 416
  error, 86, 123, 275
  normal distribution, 441
  residuals, 306
standardised
  effect, 117, 154
  model, 275, 291
  random variable, 420
statistic, 5, 7, 82, 455
  descriptive, 29, 58
  gamma, 71
  kappa, 75
  lambda, 75
statistical inference, 81
Statistical Quality Control, 88
Stirling formula, 419
studentised statistic, 122, 280
Sturges' formula, 49
sum of squares
  between-class, 143
  between-group, 143
  columns, 157
  error, 143
  mean between-group, 144
  mean classification, 144
  model, 158
  residual, 157
  rows, 157
  subtotal, 158
  total, 143
  within-class, 143
  within-group, 143
survival data, 353
survivor function, 353
systematic error, 82

T

target population, 81
test
  binomial, 174
  Cochran Q, 217
  correlation, 127
  equality of variance, 129
  error, 115
  Friedman, 215
  Kolmogorov-Smirnov one-sample,
    183

Kolmogorov-Smirnov two-sample, 201
lack of fit, 286
Levene, 130
Lilliefors, 187
log-rank, 365
Mann-Whitney, 202
McNemar, 205
non-parametric, 171
one-sided, 119
one-tail, 119
one-way ANOVA, 143
operational characteristic, 116
parametric, 111
Peto-Wilcoxon, 366
power, 115
proportion, 174
rank-sum, 202
Rayleigh, 389
robust, 130
runs, 172
Scheffé, 150, 152
set, 230
Shapiro-Wilk, 187
sign, 207
single variance, 125
t, 115, 122, 131, 135, 139, 146, 175
two means (indep. samples), 134
two means (paired samples), 139
two-sided, 119
two-tail, 119
uniform scores, 397
variance ratio, 129
Watson, 398
Watson U^2, 392
Watson-Williams, 396
Wilcoxon, 209
χ^2 2×2 contingency table, 191
χ^2 goodness of fit, 180
χ^2 of independence, 195
χ^2 r×c contingency table, 194
test of hypotheses, 81, 111
single mean, 121
tolerance, 14, 84, 420
level, 254
Tools, 127
total probability, 235, 409
total sum of squares, 276
training set, 223
tree
branch, 261
classifier, 259
pruning, 264

U

unbiased estimates, 273
unexposed group, 364
uniform probability plot, 387
univariate split, 263
upper control limit, 88

V

variable
continuous, 9, 12
creating, 34
dependent, 271
discrete, 9, 10
grouping, 56, 132
hidden, 329
independent, 271
interval-type, 10
nominal, 9
ordinal, 9
random, 29
ratio-type, 10
variance, 62, 416
analysis, 142, 145
between-group, 144
estimate, 95
inflation factors, 307
of the means, 145
pooled, 131, 144
ratio, 97
total, 143
within-group, 144
varimax procedure, 349
Velicer partial correlation, 337
VIF, 307

W

warning line, 88
weights, 223
Wilks' lambda, 253
workbook, 42
wrapped normal, 381

Y

Yates' correction, 191

Z

z score, 113, 420